Lecture Notes in Computer Science 14855

The series Lecture Notes in Computer Science (LNCS), including its subseries Lecture Notes in Artificial Intelligence (LNAI) and Lecture Notes in Bioinformatics (LNBI), has established itself as a medium for the publication of new developments in computer science and information technology research, teaching, and education.

LNCS enjoys close cooperation with the computer science R & D community, the series counts many renowned academics among its volume editors and paper authors, and collaborates with prestigious societies. Its mission is to serve this international community by providing an invaluable service, mainly focused on the publication of conference and workshop proceedings and postproceedings. LNCS commenced publication in 1973.

Makoto Onizuka · Jae-Gil Lee · Yongxin Tong ·
Chuan Xiao · Yoshiharu Ishikawa ·
Sihem Amer-Yahia · H. V. Jagadish · Kejing Lu
Editors

Database Systems for Advanced Applications

29th International Conference, DASFAA 2024
Gifu, Japan, July 2–5, 2024
Proceedings, Part VI

Springer

Editors
Makoto Onizuka
Osaka University
Suita, Osaka, Japan

Jae-Gil Lee
KAIST
Daejeon, Korea (Republic of)

Yongxin Tong
Beihang University
Beijing, China

Chuan Xiao
Osaka University
Osaka, Japan

Yoshiharu Ishikawa
Nagoya University
Nagoya, Japan

Sihem Amer-Yahia
University of Grenoble Alpes
Saint-Martin d'Hères, France

H. V. Jagadish
University of Michigan
Ann Arbor, MI, USA

Kejing Lu
Nagoya University
Nagoya, Japan

ISSN 0302-9743 ISSN 1611-3349 (electronic)
Lecture Notes in Computer Science
ISBN 978-981-97-5571-4 ISBN 978-981-97-5572-1 (eBook)
https://doi.org/10.1007/978-981-97-5572-1

This Springer imprint is published by the registered company Springer Nature Singapore Pte Ltd.
The registered company address is: 152 Beach Road, #21-01/04 Gateway East, Singapore 189721, Singapore

Preface

It is our great pleasure to present the proceedings of the 29th International Conference on Database Systems for Advanced Applications (DASFAA 2024), held during July 2–5, 2024, in Gifu, Japan. DASFAA is an annual international database conference which showcases state-of-the-art R&D activities in database systems and advanced applications. It provides a premier international forum for technical presentations and discussions among database researchers, developers, and users from both academia and industry.

This year we received a record high number of 971 research paper submissions. We conducted a double-blind review following the tradition of DASFAA, and constructed a large committee consisting of 43 Senior Program Committee (SPC) members and 327 Program Committee (PC) members. Each valid submission was reviewed mostly by three PC members and meta-reviewed by one SPC member, who also led the discussion with the PC members. We, the PC co-chairs, considered the recommendations from the SPC members and investigated each submission as well as its reviews to make the final decisions. As a result, 147 full papers (acceptance ratio of 15.1%) and 85 short papers (acceptance ratio of 8.8%) were accepted. The review process was supported by the Microsoft CMT system. During the three main conference days, these full papers were presented in 25 research sessions and short papers were presented in three poster sessions. The dominant keywords for the accepted papers included graph, recommendation, contrastive, model, knowledge, efficient, data, framework, network, prediction, detection, neural, and federated. In addition, we included 14 industry papers (three of them are short papers), 18 demo papers, and 6 tutorials in the program. Finally, to shed light on the direction in which the database field is headed, the conference program included three invited keynote presentations by Tova Milo (Tel Aviv University, Israel), Paolo Papotti (EURECOM, France), and Jeffrey Xu Yu (Chinese University of Hong Kong, China).

Six tutorials were selected by the tutorial co-chairs: Compressing Large Language Models by U Kang (Seoul National University), Mining Multi-layer Graphs: Models, Algorithms and Applications by Zhaonian Zou (Harbin Institute of Technology), Dandan Liu (Harbin Institute of Technology), and Run-An Wang (Harbin Institute of Technology), Redefining Generative Modeling in Data-Limited Environments by Divya Saxena (Hong Kong Polytechnic University), Jiannong Cao (Hong Kong Polytechnic University), and Tarun Kulshrestha (Hong Kong Polytechnic University), Data Fabric: Technologies and Applications by Kamalakar Karlapalem (IIIT Hyderabad), P. Radha Krishna (National Institute of Technology, Warangal), and Satyanarayana R. Valluri (Databricks Inc.), Secure and Private Federated Learning with Differential Privacy by Tsubasa Takahashi (LY Corp.), and Real-Time Forecasting of Big Time-Series Data: Foundations and Challenges by Yasushi Sakurai (Osaka University) and Yasuko Matsubara (Osaka University).

Four workshops were selected by the workshop co-chairs to be held in conjunction with DASFAA 2024, which were the 10th International Workshop on Big Data Management and Service (BDMS 2024), the 9th International Workshop on Big Data Quality Management (BDQM 2024), the 8th International Workshop on Graph Data Management and Analysis (GDMA 2024), and the Workshop on Emerging Results in Data Science and Engineering (ERDSE 2024). The workshop papers are included in a separate volume of the proceedings also published by Springer in its Lecture Notes in Computer Science series.

We are grateful to the general chairs, Yoshiharu Ishikawa (Nagoya University, Japan), Sihem Amer-Yahia (University of Grenoble Alpes, France), and H.V. Jagadish (University of Michigan, USA), all SPC members, PC members, and external reviewers who contributed their time and expertise to the DASFAA 2024 paper-reviewing process. We would like to thank all the members of the Organizing Committee, and the many volunteers, for their great support.

June 2024

Makoto Onizuka
Jae-Gil Lee
Yongxin Tong
Chuan Xiao

Organization

Steering Committee Members

Chair

Lei Chen	Hong Kong University of Science and Technology (Guangzhou), China

Vice Chair

Stéphane Bressan	National University of Singapore, Singapore

Treasurer

Yasushi Sakurai	Osaka University, Japan

Secretary

Kyuseok Shim	Seoul National University, South Korea

Members

Aamir Cheema	Monash University, Australia
Arnab Bhattachaya	IIT Kanpur, India
Atsuyuki Morishima	University of Tsukaba, Japan
De-Nian Yang	Academia Sinica, Taiwan
Eenjun Hwang	Korea University, South Korea
Guoliang Li	Tsinghua University, China
Sang-Won Lee	SKKU, South Korea
Sourav Bhowmick	Nanyang Technological University, Singapore
Yang-Sae Moon	Kangwon National University, South Korea
Zhanhuai Li	Northwestern Polytechnical University, China
Zhiyong Peng	Wuhan University, China
Zi Huang	University of Queensland, Australia

Organizing Committee

General Chairs

Yoshiharu Ishikawa	Nagoya University, Japan
H. V. Jagadish	University of Michigan, USA
Sihem Amer-Yahia	University of Grenoble Alpes, France

Program Committee Chairs

Makoto Onizuka	Osaka University, Japan
Jae-Gil Lee	KAIST, South Korea
Yongxin Tong	Beihang University, China

Vice Program Committee Chair

Chuan Xiao	Osaka University, Japan

Industry Program Chairs

Jun Miyazaki	Tokyo Institute of Technology, Japan
Mukesh Mohania	IIIT Delhi, India

Panel Chairs

Toshiyuki Amagasa	University of Tsukuba, Japan
Ladjel Bellatreche	ISAE-ENSMA, France
Sourav S. Bhowmick	Nanyang Technological University, Singapore

Workshop Chairs

Atsuyuki Morishima	University of Tsukuba, Japan
Guoliang Li	Tsinghua University, China

Tutorial Chairs

Masato Oguchi	Ochanomizu University, Japan
Uday Kiran Rage	University of Aizu, Japan
Wenjie Zhang	University of New South Wales, Australia

Demo Chairs

Kazuo Goda	University of Tokyo, Japan
Lin Li	Wuhan University of Technology, China

Publication Chair

Kejing Lu	Nagoya University, Japan

DASFAA Steering Committee Liaison

Lei Chen	Hong Kong University of Science and Technology (Guangzhou), China

Local Arrangement Chair

Yu Suzuki	Gifu University, Japan

Local Arrangement Committee

Takahiro Komamizu	Nagoya University, Japan
Daisuke Yamamoto	Nagoya Institute of Technology, Japan
Yusuke Yamamoto	Nagoya City University, Japan

Registration Chair

Yousuke Watanabe	Nagoya University, Japan

Web Chair

Kento Sugiura	Nagoya University, Japan

Secretary

Eri Oikawa	Nagoya University, Japan

Program Committee

Research Track

Chairs

Makoto Onizuka	Osaka University, Japan
Jae-Gil Lee	KAIST, South Korea
Yongxin Tong	Beihang University, China

Senior Program Committee (SPC) Members

Baihua Zheng	Singapore Management University, Singapore
Bin Cui	Peking University, China
Byung Suk Lee	University of Vermont, USA
Chee-Yong Chan	National University of Singapore, Singapore
Hady Lauw	Singapore Management University, Singapore
Hayato Yamana	Waseda University, Japan
Hiroyuki Toda	Yokohama City University, Japan
Hong Gao	Harbin Institute of Technology, China
Hongzhi Yin	University of Queensland, Australia
Jiaheng Lu	University of Helsinki, Finland
Jianliang Xu	Hong Kong Baptist University, China
Jianyong Wang	Tsinghua University, China
Jun Miyazaki	Tokyo Institute of Technology, Japan
K. Selçuk Candan	Arizona State University, USA
Kyoung-Sook Kim	National Institute of Advanced Industrial Science and Technology, South Korea
Kyuseok Shim	Seoul National University, South Korea
Lei Li	Hong Kong University of Science and Technology (Guangzhou), China
Lina Yao	University of New South Wales, Australia
Ling Liu	Georgia Institute of Technology, USA
Masato Oguchi	Ochanomizu University, Japan
Meihui Zhang	Beijing Institute of Technology, China
Mizuho Iwaihara	Waseda University, Japan
Nikos Bikakis	Athena Research Center, Greece
Norio Katayama	National Institute of Informatics, Japan
Qiang Zhu	University of Michigan - Dearborn, USA
Reynold Cheng	University of Hong Kong, China
Sang-Wook Kim	Hanyang University, South Korea

Satoshi Oyama	Nagoya City University, Japan
Shuai Ma	Beihang University, China
Steven E. Whang	KAIST, South Korea
Vana Kalogeraki	Athens University of Economics and Business, Greece
Wang-Chien Lee	Pennsylvania State University, USA
Wei-Shinn Ku	Auburn University, USA
Wenjie Zhang	University of New South Wales, Australia
Xiang Zhao	National University of Defense Technology, China
Xiaochun Yang	Northeastern University, China
Xiaofang Zhou	Hong Kong University of Science and Technology, China
Xin Wang	Tianjin University, China
Ye Yuan	Beijing Institute of Technology, China
Yon Dohn Chung	Korea University, South Korea
Yunjun Gao	Zhejiang University, China
Zhiyong Peng	Wuhan University, China
Zimu Zhou	City University of Hong Kong, China

Program Committee Members

Akiyoshi Matono	AIST, Japan
Alexander Zhou	Hong Kong University of Science and Technology, China
Alkis Simitsis	Athena Research Center, Greece
An Liu	Soochow University, China
Anne Laurent	University of Montpellier, France
Barbara Catania	University of Genoa, Italy
Beibei Li	Chongqing University, China
Binbin Zhou	Hangzhou City University, China
Bo Xu	Dalian University of Technology, China
Bo Xu	Donghua University, China
Bohan Li	Nanjing University of Aeronautics and Astronautics, China
Cai Xu	Xidian University, China
Changdong Wang	SYSU, China
Chao Huang	University of Notre Dame, USA
Chao Zhang	Tsinghua University, China
Chaokun Wang	Tsinghua University, China
Chen Chen	University of Wollongong, Australia
Chen Liang	Apple, USA

Chen Ma	City University of Hong Kong, China
Chen Shen	Megagon Labs, USA
Chen Xu	East China Normal University, China
Chengcheng Mai	Nanjing Normal University, China
Chengliang Chai	Beijing Institute of Technology, China
Chenxu Wang	Xi'an Jiaotong University, China
Chenyang Wang	Aalto University, Finland
Cheqing Jin	East China Normal University, China
Chih-Ya Shen	National Tsing Hua University, Taiwan
Chingfang Hsu	Central China Normal University, China
Christos Doulkeridis	University of Pireaus, Greece
Chuan Ma	Chongqing Univerisity, China
Chuan Qin	BOSS Zhipin, China
Chuanyu Zong	Shenyang Aerospace University, China
Daichi Amagata	Osaka University, Japan
Dan He	University of Queensland, Australia
Dan Li	Sun Yat-sen University, China
Debarati B. Chakraborty	University of Hull, UK
Denghao Ma	Meituan, China
De-Nian Yang	Academia Sinica, Taiwan
Derong Shen	Northeastern University, China
Dhaval Patel	IBM Research, USA
Di Yao	Institute of Computing Technology, Chinese Academy of Sciences, China
Dieter Pfoser	George Mason University, USA
Dingyuan Shi	Beihang University, China
Divya Saxena	Hong Kong Polytechnic University, China
Dong Wen	University of New South Wales, Australia
Dongjin Yu	Hangzhou Dianzi University, China
Dongxiang Zhang	Zhejiang University, China
Dongyuan Li	Tokyo Institute of Technology, Japan
Faming Li	Northeastern University, China
Fang Zhou	East China Normal University, China
Fei Guo	Central South University, China
Ge Yu	Northeastern University, China
Goce Trajcevski	Iowa State University, USA
Gong Cheng	Nanjing University, China
Guandong Xu	Education University of Hong Kong, China
Guanghui Zhu	Nanjing University, China
Guangyou Zhou	Central China Normal University, China
Guixian Zhang	China University of Mining and Technology, China

Hailong Liu	Northwestern Polytechnical University, China
Haitian Yang	Institute of Information Engineering, Chinese Academy of Sciences, China
Hanbing Zhang	Fudan University, China
Hanchen Wang	University of Technology Sydney, Australia
Hang Yu	Shanghai University, China
Hao Xin	Hong Kong University of Science and Technology, China
Haobing Liu	Ocean University of China, China
Haoyang Li	Hong Kong University of Science and Technology, China
Hengyu Liu	University of Technology Sydney, Australia
Hiroaki Ohshima	University of Hyogo, Japan
Hiroaki Shiokawa	University of Tsukuba, Japan
Hongzhi Wang	Harbin Institute of Technology, China
Huaijie Zhu	Sun Yat-sen University, China
Hui Li	Xidian University, China
Huiqi Hu	East China Normal University, China
Ioannis Konstantinou	University of Thessaly, Greece
Jagat Sesh Challa	BITS Pilani, India
Jen-Wei Huang	National Cheng Kung University, Taiwan
Jia Xu	Guangxi University, China
Jiali Mao	East China Normal University, China
Jianbin Qin	Shenzhen University, China
Jianqiu Xu	Nanjing University of Aeronautics and Astronautics, China
Jianxin Li	Beihang University, China
Jianxiong Guo	Beijing Normal University, China
Jianye Yang	Guangzhou University, China
Jiawei Jiang	Wuhan University, China
Jiawei Sheng	Institute of Information Engineering, Chinese Academy of Sciences, China
Jiaxin Ding	Shanghai Jiao Tong University, China
Jidong Yuan	Beijing Jiaotong University, China
Jie Shao	University of Electronic Science and Technology of China, China
Jilian Zhang	Jinan University, China
Jilin Hu	Aalborg University, Denmark
Jin Wang	Megagon Labs, USA
Jingya Zhou	Soochow University, China
Jithin Vachery	National University of Singapore, Singapore
Jongik Kim	Chungnam National University, South Korea

Ju Fan	Renmin University of China, China
Jun Gao	Peking University, China
Junhu Wang	Griffith University, Australia
Junhua Fang	Soochow University, China
Junhua Zhang	University of New South Wales, Australia
Junjie Yao	East China Normal University, China
Junwei Zhang	Hangzhou Institute of Medicine, Chinese Academy of Sciences, China
Junya Arai	Nippon Telegraph and Telephone Corporation, Japan
Kai Wang	Shanghai Jiao Tong University, China
Kai Zhang	Fudan University, China
Kai Zhang	University of Science and Technology of China, China
Kai Zheng	University of Electronic Science and Technology of China, China
Kaixi Hu	Wuhan University of Technology, China
Kaixin Huang	Shanghai Jiao Tong University, China
Kangzheng Liu	Huazhong University of Science and Technology, China
Kazutoshi Umemoto	University of Tokyo, Japan
Kesheng Wu	LBNL, USA
Krishna Reddy P.	IIIT, Hyderabad, India
Kristian Torp	Aalborg University, Denmark
Kun Yue	Yunnan University, China
Ladjel Bellatreche	ISAE-ENSMA, France
Le Wang	Guangzhou University, China
Lei Duan	Sichuan University, China
Leong Hou U.	University of Macau, China
Li Jiajia	Shenyang Aerospace University, China
Liang Hong	Wuhan University, China
Lianyong Qi	China University of Petroleum (East China), China
Libin Zheng	Sun Yat-sen University, China
Lijun Chang	University of Sydney, Australia
Lin Li	Wuhan University of Technology, China
Ling Huang	South China Agricultural University, China
Linglin Yang	Peking University, China
Lingwei Chen	Wright State University, USA
Linlin Zong	Dalian University of Technology, China
Lizhen Cui	Shandong University, China
Long Yuan	Nanjing University of Science and Technology, China

Longkun Guo	Fuzhou University, China
Lu Chen	Swinburne University of Technology, Australia
Lu Chen	Zhejiang University, China
Manish Kesarwani	IBM Research Lab, USA
Markus Schneider	University of Florida, USA
Marwan Hassani	TU Eindhoven, Netherlands
Meng Liu	National University of Defense Technology, China
Meng Wang	Southeast University, China
Meng Xiao	Chinese Academy of Sciences, China
Mengxuan Zhang	Australian National University, Australia
Michele Melchiori	Università degli Studi di Brescia, Italy
Ming Zhong	Wuhan University, China
Minghao Zhao	East China Normal University, China
Minghe Yu	Northeastern University, China
Mingyuan Fan	East China Normal University, China
Mingyue Cheng	University of Science and Technology of China, China
Mingyue Wang	Harbin Institute of Technology, Shenzhen, China
Mo Li	Liaoning University, China
Muhammad Haris	L3S, Germany
Nannan Wu	Tianjin University, China
Ning Liu	Shandong University, China
Ning Wang	Ocean University of China, China
Ning Wang	Beijing Jiaotong University, China
Ning Yang	Sichuan University, China
Ningning Cui	Chang'an University, China
Ozan Kahramanogullari	Free University of Bozen-Bolzano, Italy
Panagiotis Bouros	Johannes Gutenberg University Mainz, Germany
Peiquan Jin	University of Science and Technology of China, China
Peng Cai	East China Normal University, China
Peng Cheng	East China Normal University, China
Peng Liu	Guangxi Normal University, China
Peng Peng	Hunan University, China
Peng Wang	Southeast University, China
Pengpeng Zhao	Soochow University, China
Ping Lu	Beihang University, China
Qi Liu	University of Science and Technology of China, China
Qi Song	University of Science and Technology of China, China

Qiang Qu	SIAT, China
Qiang Yang	King Abdullah University of Science and Technology, Saudi Arabia
Qiang Yin	Shanghai Jiao Tong University, China
Qianzhen Zhang	National University of Defense Technology, China
Qing Liao	Harbin Institute of Technology (Shenzhen), China
Qing Liu	Data61, CSIRO, Australia
Qingqing Ye	Hong Kong Polytechnic University, China
Qingyun Sun	Beihang University, China
Qingzhi Ma	Soochow University, China
Qiyao Peng	Tianjin University, China
Qiyu Liu	Southwest University, China
Quanqing Xu	Beijing University of Posts and Telecommunications, China
Renjie Sun	East China Normal University, China
Rong Zhu	Alibaba Group, China
Rui Zhou	Swinburne University of Technology, Australia
Rui Zhu	Shenyang Aerospace University, China
Rui Zhu	Indiana University, USA
Ruihong Qiu	University of Queensland, Australia
Ruiyuan Li	Chongqing University, China
Sai Wu	Zhejiang University, China
Sanghyun Park	Yonsei University, South Korea
Sanjay Kumar Madria	Missouri University of Science & Technology, USA
Savong Bou	University of Tsukuba, Japan
Sebastian Link	University of Auckland, New Zealand
Shanshan Feng	Centre for Frontier AI Research, A*STAR, Singapore
Shaoxu Song	Tsinghua University, China
Sheng Wang	Wuhan University, China
Shengan Zheng	Shanghai Jiao Tong University, China
Shengzhong Zhang	Fudan University, China
Shijie Zhang	Tencent, China
Shiyu Yang	Guangzhou University, China
Shoko Wakamiya	Nara Institute of Science and Technology, Japan
Shuai Xu	Nanjing University of Aeronautics and Astronautics, China
Shuhao Zhang	Nanyang Technological University, Singapore
Shuiqiao Yang	UNSW, Australia
Shunmei Meng	Nanjing University of Science and Technology, China

Shuyuan Li	Beihang University, China
Sibo Wang	Chinese University of Hong Kong, China
Silvestro Roberto Poccia	University of Turin, Italy
Siyuan Chen	Guangzhou University, China
Songlei Jian	National University of Defense Technology, China
Songlei Wang	Harbin Institute of Technology, Shenzhen, China
Tangpeng Dan	Renmin University of China, China
Tao Qiu	Shenyang Aerospace University, China
Teng Ma	Tsinghua University, China
Thanh Tam Nguyen	Griffith University, Australia
Theodoros Chondrogiannis	University of Konstanz, Germany
Tianzi Zang	Nanjing University of Aeronautics and Astronautics, China
Tieke He	Nanjing University, China
Tieyun Qian	Wuhan University, China
Tiezheng Nie	Northeastern University, China
Tomoki Yoshihisa	Shiga University, Japan
Tsubasa Takahashi	LY Corporation, Japan
Tsz Nam (Edison) Chan	Shenzhen University, China
Wei Chen	Huazhong University of Science and Technology, China
Wei Hu	Nanjing University, China
Wei Ju	Peking University, China
Wei Li	Harbin Engineering University, China
Wei Liu	Sun Yat-sen University, China
Wei Liu	Huazhong University of Science and Technology, China
Wei Lu	Renmin University of China, China
Wei Shen	Nankai University, China
Wei Song	Wuhan University, China
Wei Wang	Hong Kong University of Science and Technology (Guangzhou), China
Wei Emma Zhang	University of Adelaide, Australia
Weiguo Zheng	Fudan University, China
Weijian Ma	Fudan University, China
Weiren Yu	University of Warwick, UK
Weiwei Sun	Fudan University, China
Weixin Zeng	National University of Defense Technology, China
Weixiong Rao	Tongji University, China
Wen Hua	Hong Kong Polytechnic University, China

Wenjie Li	Tsinghua University, China
Wenqi Sun	Renmin University of China, China
Wentao Li	Hong Kong University of Science and Technology (Guangzhou), China
Wolf-Tilo Balke	TU Braunschweig, Germany
Xi Guo	University of Science and Technology Beijing, China
Xiang Ao	Institute of Computing Technology, Chinese Academy of Sciences, China
Xiangmin Zhou	RMIT University, Australia
Xiangping Zheng	Harbin Engineering University, China
Xiangyu Song	Swinburne University of Technology, Australia
Xiao Luo	University of California, Los Angeles, USA
Xiao Pan	Shijiazhuang Tiedao University, China
Xiaochun Yang	Northeastern University, China
Xiaofan Wang	Civil Aviation Management Institute of China; Institute of Software, Chinese Academy of Sciences, China
Xiaofeng Gao	Shanghai Jiao Tong University, China
Xiaoling Wang	East China Normal University, China
Xiaoou Ding	Harbin Institute of Technology, China
Xiaoxi He	University of Macau, China
Xiaoyang Wang	University of New South Wales, Australia
Xin Cao	University of New South Wales, Australia
Xin Huang	Hong Kong Baptist University, China
Xin Wang	Southwest Petroleum University, China
Xinghua Zhang	Institute of Information Engineering, Chinese Academy of Sciences, China
Xiongxiao Xu	Illinois Institute of Technology, USA
Xiuhua Li	Chongqing University, China
Xiulong Liu	Tianjin University, China
Xu Chen	Renmin University of China, China
Xuyun Zhang	Macquarie University, Australia
Yajun Yang	Tianjin University, China
Yan Zhang	Peking University, China
Yanfeng Zhang	Northeastern University, China
Yang Cao	Tokyo Institute of Technology, Japan
Yang Chen	Fudan University, China
Yangfu Zhu	Beijing University of Posts and Telecommunications, China
Yang-Sae Moon	Kangwon National University, South Korea
Yangtao Wang	Guangzhou University, China
Yanhao Wang	East China Normal University, China

Yanlong Wen	Nankai University, China
Yanmei Hu	Chengdu University of Technology, China
Yanmin Zhu	Shanghai Jiao Tong University, China
Yantong Lai	Institute of Information Engineering, Chinese Academy of Sciences, China
Yexuan Shi	Beihang University, China
Yijun Su	JD.com, China
Yingxia Shao	Beijing University of Posts and Telecommunications, China
Yiqun Zhang	Guangdong University of Technology, China
Yishu Wang	Northeastern University, China
Yixiang Fang	Chinese University of Hong Kong, Shenzhen, China
Yong Zhang	Tsinghua University, China
Yongchao Liu	Ant Group, China
Yonghong Yu	Nanjing University of Posts and Telecommunications, China
Yongpan Sheng	Southwest University, China
Yoshiyuki Shoji	Shizuoka University, Japan
Yu Gu	Northeastern University, China
Yu Liu	Huazhong University of Science and Technology, China
Yu Sun	Nankai University, China
Yu Yang	Hong Kong Polytechnic University, China
Yuanbo Xu	Jilin University, China
YuAng Chen	Chinese University of Hong Kong, China
Yuankai Fan	Fudan University, China
Yuanyuan Zhu	Wuhan University, China
Yue Cui	Hong Kong University of Science and Technology, China
Yue Kou	Northeastern University, China
Yujue Wang	Beihang University, China
Yuliang Ma	Northeastern University, China
Yunpeng Chai	Renmin University of China, China
Yunpeng Weng	Tencent, China
Yurong Cheng	Beijing Institute of Technology, China
Yuxiang Zeng	Hong Kong University of Science and Technology, China
Yuya Sasaki	Osaka University, Japan
Yuyang Dong	NEC, Japan
Yuyu Luo	Hong Kong University of Science and Technology (Guangzhou), China
Zeke Wang	Zhejiang University, China

Zhao Zhang	East China Normal University, China
Zhaojing Luo	Beijing Institute of Technology, China
Zhaonian Zou	Harbin Institute of Technology, China
Zhaoquan Gu	Harbin Institute of Technology (Shenzhen), China
Zheng Liu	Nanjing University of Posts and Telecommunications, China
Zhengyi Yang	University of New South Wales, Australia
Zhenya Huang	University of Science and Technology of China, China
Zhenying He	Fudan University, China
Zhigang Wang	Ocean University of China, China
Zhihui Wang	Fudan University, China
Zhiwei Zhang	Beijing Institute of Technology, China
Zhixu Li	Fudan University, China
Zhiying Deng	Huazhong University of Science and Technology, China
Zhongnan Zhang	Xiamen University, China
Zi Chen	Nanjing University of Aeronautics and Astronautics, China
Ziang Yang	Institute of Information Engineering, Chinese Academy of Sciences; University of Chinese Academy of Sciences, China
Zirui Zhuang	Beijing University of Posts and Telecommunications, China

Industry Track

Chairs

Jun Miyazaki	Tokyo Institute of Technology, Japan
Mukesh Mohania	IIIT Delhi, India

Program Committee Members

Changbo Qu	Simon Fraser University, Canada
Denis Ponomaryov	Ershov Institute of Informatics Systems, Russia
Hieu Hanh Le	Ochanomizu University, Japan
Hiroyuki Toda	Yokohama City University, Japan
Hongzhi Wang	Harbin Institute of Technology, China
Jinglin Peng	Simon Fraser University, Canada
Jinwei Zhu	Huawei, China

Kenta Oku — Ryukoku University, Japan
Kshitij Gulati — University of Rochester, USA
Qiong Chang — Tokyo Institute of Technology, Japan
Rajeev Gupta — Microsoft, USA
Shinsuke Nakajima — Kyoto Sangyo University, Japan
Takeshi Yamamuro — NTT, Japan
Weiyuan Wu — Simon Fraser University, Canada
Xiaoou Ding — Harbin Institute of Technology, China
Yinan Mei — Huawei, China
Yun Liu — National Institute of Advanced Industrial Science and Technology, Japan

Demo Track

Chairs

Kazuo Goda — University of Tokyo, Japan
Lin Li — Wuhan University of Technology, China

Program Committee Members

Chenhao Ma — Chinese University of Hong Kong, Shenzhen, China
Dario Garigliotti — University of Bergen, Norway
Ergute Bao — National University of Singapore, Singapore
Jiayuan He — RMIT University, Australia
Kaiping Zheng — National University of Singapore, Singapore
Kajal Kansal — National University of Singapore, Singapore
Peipei Wang — Qilu University of Technology, China
Qing Xie — Wuhan University of Technology, China
Tong Chen — University of Queensland, Australia
Yuya Sasaki — Osaka University, Japan
Yuyu Luo — Hong Kong University of Science and Technology (Guangzhou), China
Zhanhao Zhao — Renmin University of China, China
Zheng Wang — Huawei Singapore Research Center, Singapore
Zilong Jiang — Guizhou University of Finance and Economics, China

Contents – Part VI

Graph and Network

Hardware Acceleration

Graph and Network

Cascading Graph Convolution Contrastive Learning Networks for Multi-behavior Recommendation

Nan Liu[1], Shunmei Meng[1,2](✉), Yu Jiang[1], Qianmu Li[1], Xiaolong Xu[3], Lianyong Qi[4], and Xuyun Zhang[5]

[1] Nanjing University of Science and Technology, Nanjing, China
{liunan,mengshunmei,jiangyu,qianmu}@njust.edu.cn
[2] State Key Laboratory for Novel Software Technology, Nanjing University, Nanjing, China
[3] School of Computer and Software, Nanjing University of Information Science and Technology, Nanjing, China
xlxu@nuist.edu.cn
[4] College of Computer Science and Technology, China University of Petroleum (East China), Qingdao, China
[5] Macquarie University, Sydney, Australia
xuyun.zhang@mq.edu.au

Abstract. In most real-world recommendation scenarios, there are multiple types of behaviors that tend to follow a specific chain of behaviors (e.g., view->-cart->collect->purchase). To forecast users' possible preferences for the target behavior (e.g., purchase), existing multi-behavior recommendations use auxili-ary behaviors (e.g., view, collect, and cart), and they often emphasize the difference between behaviors. However, different sorts of behavior partially reflect the same user preferences, and similarities between them are largely ignored. At the same time, the subsequent behavior in the chain usually reflects stronger user preferences than the previous behavior, and most multi-behavior models cannot capture the interdependence in the chain of behaviors. To tackle these issues, we propose a Cascading Graph Convolution Contrastive Learning Networks (CGCCN) for Multi-Behavior Recommendation. Specifically, We use LightGCN to learn user and item embeddings, and then we combine multi-task learning with contrastive learning to explicitly exploit behavioral dependence in embeddings learning and capture differences between embeddings. We conduct comprehensive experiments on two real-world datasets to validate the efficiency of our model. The results further demonstrate the rationality and effectiveness of the designed CGCCN, where the maximum improvement can reach to 42.98%.

Keywords: Multi-task learning · Contrastive learning · Multi-behavior recommendation

M. Onizuka et al. (Eds.): DASFAA 2024, LNCS 14855, pp. 3–18, 2024.
https://doi.org/10.1007/978-981-97-5572-1_1

1 Introduction

Personalized recommendation system aims to infer the user's preferences from the user's interaction with the item, and recommend the items matching these preferences for the user. At present, the majority of recommendation models are built based on a single behavior, that is, a type of behavior between users and items (such as the most common purchase behavior). The CF model [1,2] is the most representative among them, which has gradually developed from matrix factorization methods [2,3] to a more mature graph neural network models [4]. The majority CF models only take one kind of behavior into account, but in real-life applications, there are frequently several user behavior, and instead of purchase behavior, users often take other sorts of behavior (e.g., click, collect, and cart) for comparison and screening to help them make a final decision. It is common to refer to the purchase behavior as the target behavior, and other behaviors as auxiliary behaviors. For example, in e-commerce platforms, click behavior makes it easier for users to learn about items, collect behavior can better indicate the popularity of items, and cart behavior allows users to make a final decision. Recently, more and more studies have found that one-behavior recommendation models will face significant data sparsity issues in practical applications, and auxiliary behaviors also reflect user preferences to a certain extent, which is quite useful in predicting target behaviors.

In order to make full use of the role of auxiliary behavior on target behavior, recent years have seen the emergence of certain multi-behavior recommendation models. Although these studies have improved to some extent, there are still the following shortcomings. First of all, most studies mainly learn user preferences for different types of behaviors separately and then integrate multi-behaviors, but they do not consider that user preferences for various behaviors may not always correspond to actual user preferences, and cannot be simply integrated. At the same time, many types of similarities between user behaviors, that is, they ought to represent similar user preferences, and these behaviors often play a decisive role in the target behavior. Therefore, we need to carefully explore the real user preferences caused by various user behaviors, which cannot be simply integrated. At the same time, we need to explore the similarities between auxiliary behaviors and target behaviors, and integrate these similarities into target behaviors. This will improve the quality of embedding and achieve more accurate prediction effects.

Second, most methods do not take advantage of user preference information contained in the sequence of interactions between different behaviors, such as click -> collect -> cart -> buy. In fact, the different interaction behaviors of users and items can indirectly reflect the influence of users by the popularity of the items, such as the higher the popularity of items, the more likely the users are to collect the items. The order of different interaction behaviors often reveals the user's different preferences, such as clicking behavior indicates that users are interested in the items in a certain aspect (e.g., appearance or cover). The purchase behavior indicates that the items can meet the user's preferences well. To the best of our knowledge, there are only a few approaches in existing research to model the dependence between behaviors using multi-behaviors. A recent example is CRGCN [7], which assigns a residual block to each type of behavior to learn behavior characteristics based on the output of the prior block and the interaction data of the behavior, using the relationship between behavior sequences to learn user preference representation. However, no research has explored the similarities and dependencies between various behaviors, instead, they have only considered specific aspects, ignoring their synergistic effects. Therefore, in the field of multi-behavior recommendation, it is preferable to develop a multi-faceted and high-precision recommendation model.

In this paper, we propose a novel cascading graph convolution contrastive learning networks for multi-behavior recommendation model (CGCCN). Specifically, we first construct a cascade sequence according to the user purchasing process, starting with the view behavior and ending with the buy behavior, every type of behavior corresponds to a GCN block that is used to learn the behavior characteristics of the interaction information between the behavior and the output from the prior block. We use the cascade structure to connect multiple GCN blocks and make clear use of the behavior dependence in the chain to output the embedded characteristics under each behavior subgraph. Then multi-task learning is utilized to perform the prediction task of the current behavior on the output of each GCN block, and jointly optimize the learning of auxiliary behavior and target behavior, thereby continuously approaching and learning user preferences. Secondly, considering the similarities between multiple behaviors, we propose a contrastive learning task that compares the embedded features of each auxiliary behavior with the embedded features of the target behavior, which greatly improves the prediction accuracy. Finally, the two tasks are combined for joint optimization. Experimental findings demonstrate that CGCCN can outperform state-of-the-art models in terms of accuracy and efficiency.

To sum up, we make the following contributions in this paper:

- We emphasize the importance of considering both the dependencies within the behavior chain and the similarities between various behaviors in multi-behavior recommendation. The integration of these factors can significantly enhance prediction accuracy.
- We propose a novel multi-behavior recommendation model called CGCCN, which combines contrastive learning with multi-task learning. In particular, the single-behavior connection of each auxiliary behavior and the single-behavior connection of the target behavior are conduct contrastive learning, and the prediction task is performed for each behavior. Finally combine the two tasks for optimization.
- We conduct comprehensive experiments on two real-world datasets to validate the efficiency of our CGCCN model. CGCCN outperforms state-of-the-art multi-behavior recommendation methods, which is a notable advancement, with average relative improvements of 26.09% and 9.99% for Recall@10 and NDCG@10 over the two datasets, respectively.

2 Related Work

Multi-behavior recommendation uses various types of user-item interactions to improve the prediction ability of target behavior, which plays a significant role in alleviating data sparsity and improving recommendation performance. Early methods mostly extended single-behavior recommendation approach to multi-behavior. For example, He *et al.* [4] directly modelled all types of behaviors without considering the differences between behaviors, and apply a single behavior recommendation model one by one. To distinguish the impact of different types of behavior on user preferences for the importance of behavior modeling. Xia *et al.* [9] assigned different learnable weights to different user behaviors to introduce the role of different behaviors. Loni *et al.* [10] proposed to assign different preference weights to various types of behaviors, enhanced the standard BPR [11] when extracting negative samples from different behaviors, and used a new negative sample sampling strategy. Qiu *et al.* [12] created an adaptive sampler in BPR in order to equalize the correlation between behaviors. Chen *et al.* [13] designed a Cold-Start method based Multi-scenario Network. Li *et al.* [14] proposed a intra-behavior and inter-behavior contrastive learning framework to improve MBR.

Meanwhile, more and more research use graph neural network (GNN) in the field of multi-behavior recommendation. Some of these studies provide embedding representations for each type of behavior, participating in graph convolution operations in the form of node embeddings. For example, Cheng *et al.* [8] proposed a cascaded graph convolutional network recommendation model, whereby the subsequent behavior embedding learning through feature transformation uses the embedding learnt from the first behavior as the input feature. High-quality user and item embeddings are also crucial to recommendation performance. He *et al.* [4] took the weighted sum of the embeddings of all layers as the final embedding by learning user and item embeddings in each type of behavior. Currently, creating a single user-item graph based on all behaviors and then applying GCN operations to it to find user-item embeddings is the more common approach for using the GCN model [15]. Chen *et al.* [16] learned user-item embeddings via GCN and then utilize aggregation to predict each behavior individually. Jin *et al.* [6] learned the behavioral contributions on the user-item graph, modeled the behavioral semantics on the item-item collaborative interaction graph, and obtained the final prediction score through behavioral contributions and behavioral semantics. Xia *et al.* [17] modeled different behaviors with graph-based meta-learning to capture behavioral heterogeneity and diversity in graphs. However, existing research still cannot take into account the dependence and similarities between different behaviors. As a result, auxiliary behavioral data cannot be fully utilized and fails to accurately reflect user preferences. On the contrary, our model integrates contrastive learning with multi-task learning. It considers that different behaviors may lead to the same user preference and also considers the dependencies between chain of behavior. Furthermore, it jointly optimizes the auxiliary behavior and the target behavior.

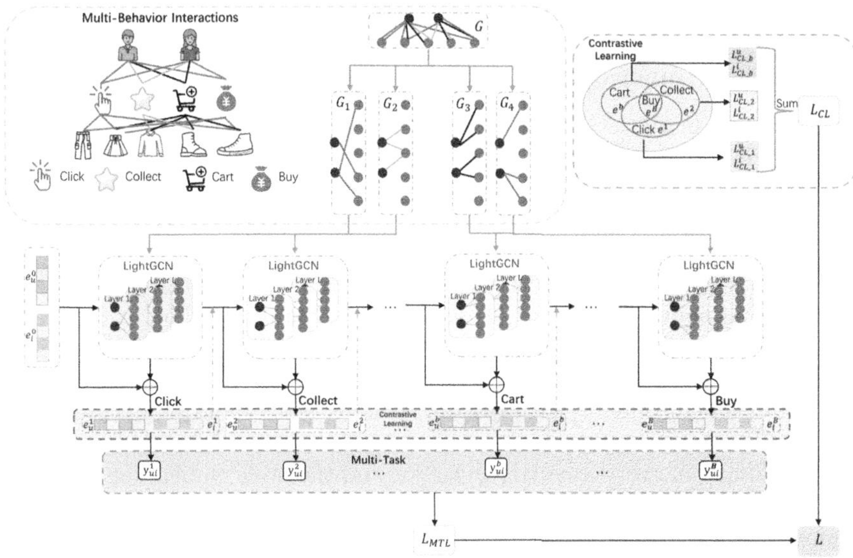

Fig. 1. Overview of our CGCCN model. (We take an example that the first behavior is click, the second behavior is collect, the bth behavior is cart, and the Bth behavior is the target behavior)

3 Problem Definition

Suppose a recommender system that has a user set $\mathcal{U} = \{u_1, ..., u_n\}$ and an item set $\mathcal{I} = \{i_1, ..., i_m\}$. We describe graph $G = (V, E)$, where nodes V is made up of item nodes $i \in \mathcal{I}$ and user nodes $u \in \mathcal{U}$. The edges in E correspond to the user-item interaction in G. The nodes for item and user can be extracted to create a subgraph $G_b = (V, E_b)$ by using the edges under the $b^{th}(1 \leq b \leq B)$ behavior, which are represented as E_b. The set of interaction matrices is denoted by the notation $H = [H^1, ..., H^b, ..., H^B]$, where B^{th} behavior is the target behavior and H^b is the interaction matrix for the bth behavior. Each entry in the matrices of interactions has a value of either 1 or 0, defined as:

$$h_{ui}^{b} = \begin{cases} 1, & \text{if u has interacted with } i \text{under behavior;} \\ 0, & \text{otherwise.} \end{cases} \tag{1}$$

The following formulation represents the goal of our task:

Input: The interaction matrix list H, user set $\mathcal{U}$, item set $\mathcal{I}$.

Output: A prediction model forecasts the possibility that a user u will interact with an item i under the bth behavior(e.g., buy).

4 Proposed Model

4.1 Overall Framework

The three parts of our CGCCN model are depicted in Fig. 1 as its general structure: 1) Node representation learning, based on the user's purchasing process, we construct a cascading sequence that goes from the user's view behavior to the purchase behavior. Every type of behavior corresponds to a GCN block, it not only learns users and items from each type of behavior embedding, but also records the cascading effect of user preferences between different behaviors, obtaining node embeddings under each behavioral subgraph. 2) Multi-task learning, using multi-task learning to perform the prediction task of the current behavior on the output of each GCN block, and jointly learning embeddings from different behaviors. 3) Contrastive learning, since multiple behaviors may lead to the same user preference, we conduct contrastive learning between the node embedding of every auxiliary behavior and the node embedding of the target behavior, which greatly improves the prediction accuracy. Finally, the two learning tasks are combined and optimized together.

4.2 Node Representation Learning

First, we need to initialize the embeddings of user u and item i respectively, represented by e_u^0 and e_i^0 respectively. In the LightGCN of the first behavior, the initialized embeddings serve as the users' and items' input feature.

$$e_u^0 = E \cdot W_u, \quad e_i^0 = F \cdot W_i \tag{2}$$

where $E \in R^{M\times d}$ and $F \in R^{N\times d}$ to respectively denote the users and items embedding matrix, W_u and W_i respectively denote the one-hot embedding matrix for users and items.

Then we build a cascade sequence based on the user's purchasing process, which is composed of LightGCN chain. Every behavior type corresponds to a GCN block, which can not only obtain user preferences from single behaviors, but also capture user preferences cascading relationships between behaviors. The core idea of the cascade GCN block is to initialize the user and item embeddings with basic features, and then improve the embedding quality by combining single-behavior modeling with multi-behavior cascading relationship modeling, and continuously control and learn user preferences. Next, as shown below, the multi-layer message propagation formula for GCNs.

$$e_u^{(b,l+1)} = \delta e_u^{(b,l)} + (1-\delta) \sum_{i\in N_u} \frac{1}{\sqrt{|N_u|}\sqrt{|N_i|}} e_i^{(b,l)} \tag{3}$$

$$e_i^{(b,l+1)} = \delta e_i^{(b,l)} + (1-\delta) \sum_{u\in N_i} \frac{1}{\sqrt{|N_u|}\sqrt{|N_i|}} e_u^{(b,l)} \tag{4}$$

where $e_u^{(b,l)}$ and $e_i^{(b,l)}$ denotes the users and items outputs from the lth layer, $e_u^{(b,l+1)}$ and $e_i^{(b,l+1)}$ denotes the output of the last LightGCN block. When $l = 0$,

$e_u^{(b,0)}$ and $e_i^{(b,0)}$ represent the initialized user and item embeddings. N_u and N_i are the set of neighbors of u and i, respectively, and $\sqrt{|N_u|}\sqrt{|N_i|}$ represents the normalization coefficient. We introduce a hyper-parameter δ to control the amount information about the node itself that is maintained during neighbor aggregation. In order to facilitate the representation of further calculations, we denote the outputs of equation (6) and equation (7) as $\hat{e}_u^b$ and $\hat{e}_i^b$ respectively.

The embedding results obtained above are then normalized so that the features learned by the GCN block have the same range as the input features of each block:

$$\tilde{e}_u^b = \frac{\hat{e}_u^b}{||\hat{e}_u^b||_2}, \tilde{e}_i^b = \frac{\hat{e}_i^b}{||\hat{e}_i^b||_2} \tag{5}$$

We find that each type of behavior has a different impact on user preferences, so simply obtaining the final feature embedding through the cascade effect cannot represent users and items well. We make full use of the role of each behavior and record the representation of each behavior subgraph in order to participate in subsequent tasks, and the output of each GCN block not only contains the user and item embeddings in the current behavior, but also captures the dependence between behavior chains. The following is an example of user embedding:

$$\begin{aligned} e_u^1 &= e_u^0 + \tilde{e}_u^1, \\ &\cdots \\ e_u^b &= e_u^{b-1} + \tilde{e}_u^b, \\ &\cdots \\ e_u^B &= e_u^{B-1} + \tilde{e}_u^B, \end{aligned} \tag{6}$$

where e_u^b represents the normalized behavior feature learned from the bth behavior, e_u^0 is the initial user embedding feature, and e_u^B represents the user feature of the target behavior (i.e., purchase).

Based on the above design, we can obtain the behavioral characteristics of each GCN block. Even if the existing behavior has no interaction data, the embedding features obtained from the previous behavior can be well saved and transferred to the next behavior. In addition, the cascade effect facilitates the transmission and integration of information across different blocks, which helps alleviate the cold start problem and data sparsity.

4.3 Multi-task Learning

Multi-task learning (MTL) is a migration learning method in which multiple tasks assist each other in learning through shared representations. The main task benefits from the training signals of related tasks, which provide domain-specific information, thereby improving generalization effectiveness. For CGCCN, we utilize the output of each GCN block as the task of current behavior prediction. Due to the cascade effect within the behavior chain, not only can the current task GCN block can be trained, but also the GCN blocks of previous tasks can be trained.

We design a loss function to regulate the learning of behavioral features for each behavior. The CGCCN model obtains user embedding feature sets $\{e_u^1, ..., e_u^b, ..., e_u^B\}$ and item embedding feature sets $\{e_i^1, ..., e_i^b, ..., e_i^B\}$ from all blocks, which are learned separately. The prediction score is defined as the inner product of users and items embeddings:

$$y_{ui} = e_u^T e_i \tag{7}$$

In order to optimize the multi-task learning module, we enhance the embedding similarity of linked users and items under the target behavior by using pairwise Bayesian Personalized Ranking (BPR) loss [13]. BPR loss function is as follows:

$$\mathrm{L}_{MTL_b} = \sum_{(u,i,j)\in Q} -ln\sigma(y_{ui} - y_{uj}) \tag{8}$$

where $Q = \{(u,i,j)|(u,i) \in O^+, (u,j) \in O^-\}$ denotes the positive and negative sample pairs. O^+ and O^- denotes observed and unobserved samples in the target behavior, respectively. $\sigma(\cdot)$ denotes the sigmoid function.

Based on MTL, we regard each behavior as a task, and the final loss function of multi-task learning is the sum of all different behavior loss:

$$L_{MTL} = \sum_{b=1}^{B} \mathrm{L}_{MTL_b} \tag{9}$$

4.4 Contrastive Learning

We find that multiple types of behaviors will lead to similar user preferences. In order to capture the similarities between multiple behaviors, we use the contrastive learning method. Previous studies conducted contrastive learning on any two behaviors, which greatly increased the complexity. We only compare auxiliary behaviors with target behaviors, reducing comparison pairs and reducing complexity. The embedding quality under the target behavior can be improved by capturing the similarities between multiple behaviors.

When extracting positive sample pairs, the consistency of the same node in different behavior subgraphs is first considered. But under the target behavior, users or items with similar associated information can be viewed as positive samples. Therefore, we use point-wise mutual information (PMI) [18] to calculate the similarity correlation between any two users or items, taking the user similarity correlation as an example:

$$\mathrm{PMI}(u, u') = \log \frac{P(u, u')}{P(u)P(u')} \tag{10}$$

$$P(u) = \frac{|I(u)|}{|I|} \tag{11}$$

$$P(u, u') = \frac{|I(u) \cap I(u')|}{|I|} \tag{12}$$

where $I(u)$ represents the items associated with users u, $I(u) \cap I(u')$ represents the items associated with users u and u'.

If the calculated similarity correlation between any two users exceeds the threshold ε, the two users are considered a positive sample pair. The selection and calculation method of the positive sample pairs for items is consistent with that for users. Negative samples are randomly selected.

Then we utilize the contrastive loss InfoNCE [19] to maximize the coherence of positive pairs and minimize the consistency of negative pairs. When the Bth target behavior is contrastive learning with the bth auxiliary behavior ($b = [1, ..., B-1]$), the loss function is as follows:

$$L_{CL_b}^u = \sum_{u \in \mathcal{U}} -log \frac{\sum_{u^+ \epsilon \mathcal{U}} \exp\left\{\frac{(r_{uB})^T r_{u^+b}}{\mu}\right\}}{\sum_{u^- \epsilon \mathcal{U}} \exp\left\{\frac{(r_{uB})^T r_{u^-b}}{\mu}\right\}} \tag{13}$$

$$L_{CL_b}^i = \sum_{i \in \mathcal{I}} -log \frac{\sum_{i^+ \epsilon \mathcal{I}} \exp\left\{\frac{(r_{iB})^T r_{i^+b}}{\mu}\right\}}{\sum_{i^- \epsilon \mathcal{I}} \exp\left\{\frac{(r_{iB})^T r_{i^-b}}{\mu}\right\}} \tag{14}$$

where (r_{uB}, r_{u^+b}) and (r_{uB}, r_{u^-b}) represents positive and negative sample pairs, respectively. μ represents hyperparameters.

The total loss function under the contrastive learning task is produced by summing the loss functions under all users and items:

$$L_{CL} = \sum_{b=1}^{B-1} (L_{CL_b}^u + L_{CL_b}^i) \tag{15}$$

4.5 Joint Optimization

Combining the above two tasks to get the final recommendation model, we use the following method to optimize the recommendation model:

$$L = L_{MTL} + \lambda L_{CL} + \varphi \|\vartheta\|^2 \tag{16}$$

where ϑ represents all trainable parameters, φ represents the hyperparameters used to control L_2 regularization to prevent overfitting, and λ represents the hyperparameters used to control the share of the contrastive learning task in the total.

5 Experiment

In this section, we conduct experiments on two publicly available datasets to evaluate the performance of CGCCN model by comparing with various state-of-the-art recommendation techniques. Finally, we conduct a series of parametric, multi-behavioral impact and ablation studies to analyze our framework.

5.1 Experiment Settings

Dataset. On two publicly accessible datasets, we test our model using the same preprocessing and parameter settings as baseline models. The statistical statistics for the two datasets used in our studies are summarized in Table 1.

- Tmall. The data comes from one of China's top e-commerce platforms, including four behavioral interaction types, namely click, collect, cart, purchase. The collected data is from 2017/06/01 to 2017/06/30.
- Jdata. The data from China, one of the largest comprehensive online retail platform also includes four types of behavior, the data collected from 2018/02/01 to 2018/04/15.

Table 1. Statistics of evaluation datasets

Dataset	#User	#Item	#Interaction	#Target Interaction	#Interactive Behavior Type
Tmall	41,738	11,953	2.3×10^6	255,586	{Click,Collect,Cart,Buy}
Jdata	93,334	24,624	2.1×10^6	333,383	{Click,Collect,Cart,Buy}

Evaluation Protocols. To assess model performance, we utilize two commonly used metrics, namely hit ratio (HR@N) and normalized discounted cumulative gain (NDCG@N). The same experimental setup is performed according to the leave-oneout strategy for evaluation [6,20], and one evaluation is reserved for train set and test set split. We construct the test dataset by incorporating the most recent interactive item. The validation set consists of the second-to-last interactive item for each user, which is used for hyperparameter tuning. The remaining data is considered for training.

Baselines. We compare the effectiveness of CGCCN against some state-of-the-art methods, which can be classified into two groups: One-behavior methods include: MFBPR [11] and LightGCN [4,5], Multi-behavior methods include: RGCN [21], GNMR [22], NMTR [23], S-MBRec [24] and CRGCN [7].

Hyper-parameter Settings. Our CGCCN model is implemented in Pytorch. We opti-mized the learning rate in the range of {*1e-2, 5e-3, 1e-4*} using grid search. With a fixed embedding dim of 64 and a training batch size of 1024. The task weight parameter λ is selected from {*0.1, 0.2, 0.5, 0.8, 1.0*}, and the regularization coefficient is searched in ranges of {*1e-2, 1e-3, 3e-4*}. We adjust the behavior parameters separately, considering values in the range of {*0.05, 0.1, 0.2, 0.3 0.5, 0.6, 0.8, 1.0*}.

5.2 Overall Performance

In this section, we conduct numerous experiments and record the results. Tables 2 and 3 display the approach performance on two datasets in relation to HR@N and NDCG@N. In the experiment, we take N = 10, 20 and 80, respectively. We have the following findings:

Table 2. Overall performance comparison between our method and baselines on Tmall dataset

	Dataset	Tmall					
	Method	Recall@10	NDCG@10	Recall@20	NDCG@20	Recall@80	NDCG@80
One	MFBPR	0.0230	0.0124	0.0316	0.0144	0.0541	0.0183
	LightGCN	0.0393	0.0209	0.0538	0.0243	0.0984	0.0322
Multi	RGCN	0.0316	0.0157	0.0489	0.0198	0.1067	0.0300
	GNMR	0.0393	0.0193	0.0619	0.0247	0.1410	0.0388
	NMTR	0.0517	0.0250	0.0847	0.0330	0.1963	0.0531
	MBGCN	0.0549	0.0285	0.0799	0.0345	0.1629	0.0493
	S-MBRec	0.0694	0.0362	0.1009	0.0438	0.1901	0.0601
	CRGCN	0.0840	0.0442	0.1238	0.0540	0.2491	0.0766
	CGCCN	**0.1129**	**0.0632**	**0.1520**	**0.0728**	**0.2546**	**0.0914**
	impv-e%	34.40%	42.98%	22.78%	34.81%	2.21%	19.32%

- Our CGCCN model achieves the best performance, outperforming all baselines. Specifically, in terms of Recall@10 and NDCG@10 metrics, CGCCN outperforms the best baselines on Tmall and Jdata datasets respectively 34.40%, 42.98%, 11.96% and 14.00%. This significant improvement proves the accuracy and effectiveness of CGCCN model recommendation.
- Compared with existing studies, our model not only considers the dependence between behaviors but also considers the similarity between auxiliary behaviors and target behaviors. We combine multi-task learning with contrastive learning, it is not limited by the sparsity problem.
- Multi-behavior models typically outperform single-behavior models in terms of accuracy, which fully shows that the addition of auxiliary behaviors greatly enriches the semantics, and the use of interaction data of multiple types of behaviors plays a positive role in model prediction.
- The results of the model with cascading relationships are often better than those with only fusing behaviors, which indicates that it is very necessary to consider the correlation between behaviors.

5.3 Ablation Study

In order to more fully verify the integrity of the function of the CGCCN model, we conducted the following ablation experiments.

First, we evaluate the effectiveness of the sub-modules of the CGCCN framework, considering two variants of the model: (1) **w/o. CL:** the contrastive learning component has been removed. (2) **w/o. MTL:** without considering the cascade relationship between behavior, only the contrastive learning is performed after simply fusing the data of each behavior. Figure 2 summarizes the performance of CGCCN and its variants. We find that removing any module in the model will degrade the model performance, which fully proves the effectiveness and rationality of combining multi-task learning with contrastive learning.

The combination of models significantly improves the performance of the overall model. Compared with w/o. CL and w/o. MTL, it is evident that the experiment with multi-task learning yields better results.

Table 3. Overall performance comparison between our method and baselines on Jdata dataset

	Dataset	Jdata					
	Method	Recall@10	NDCG@10	Recall@20	NDCG@20	Recall@80	NDCG@80
One	MFBPR	0.1850	0.1238	0.2192	0.1325	0.2890	0.1456
	LightGCN	0.2252	0.1436	0.2825	0.1582	0.4108	0.1822
Multi	RGCN	0.2406	0.1444	0.3418	0.1588	0.5548	0.2008
	GNMR	0.3068	0.1581	0.3694	0.1944	0.5106	0.2114
	NMTR	0.3142	0.1717	0.4086	0.1966	0.5843	0.2304
	MBGCN	0.2803	0.1572	0.3603	0.1790	0.5741	0.2098
	S-MBRec	0.4125	0.2779	0.4957	0.2989	0.6584	0.3295
	CRGCN	0.5001	0.2914	0.6190	0.3225	0.8359	0.3652
	CGCCN	**0.5599**	**0.3322**	**0.6761**	**0.3628**	**0.8608**	**0.3993**
	impv-e%	11.96%	14.00%	9.22%	12.49%	2.98%	9.34%

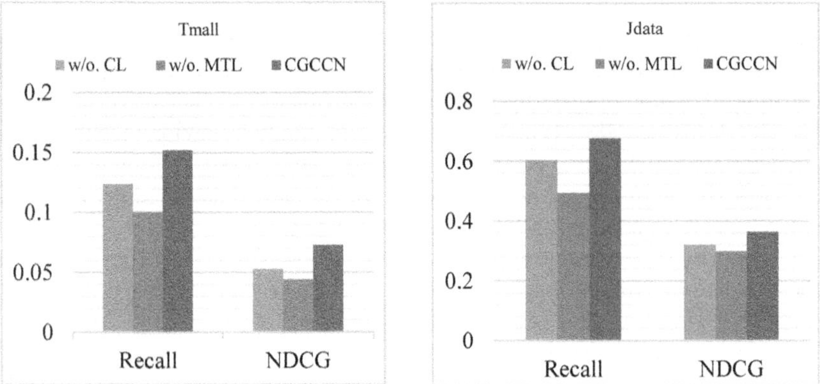

Fig. 2. Performances of different CGCCN variants

Secondly, we explore the impact of the comparison between the target behavior and different auxiliary behaviors on the prediction results. We adopt three different comparison methods: (1) **w/o. click:** represents the lack of comparison between click behavior and target behavior (2) **w/o. collect:** represents the lack of comparison between collect behavior and target behavior (3) **w/o. cart:** represents the lack of comparison between cart behavior and target behavior. The experimental results are shown in Fig. 3, we find that the absence of click behavior compared to the target behavior often yields better prediction results than the other two variants. Conversely, when comparing the absence of collect behavior to the target behavior, the prediction results often fall below those of the other two variants. This indicates that collect behavior shares more similarities with the user's true preferences, exerting a greater influence on user target behavior.

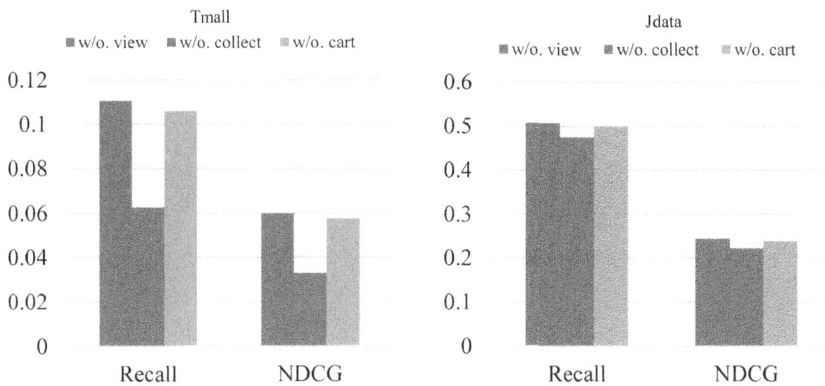

Fig. 3. Performances of different comparison methods

5.4 Hyper-parameter Study

Behavior Parameters. In the InfoNCE loss function of contrastive learning, we set the behavior parameters separately and adjust them in {*0.05, 0.1, 0.2, 0.3 0.5, 0.6, 0.8, 1.0*}, as shown in Fig. 4. According to the experimental results, we selected the optimal parameter settings. In the click and collect behavior, the selected user parameters are greater than the item parameters, and the experimental results are often better. In the cart behavior, the selected item parameters are greater than the user parameters, and the experimental results are better. This shows that in click and collect behavior, the user's subjectivity has a greater impact on the prediction result, while in the cart behavior, the items added to cart has a greater impact on the prediction result.

Hyperparameters. We adjust λ within the range of {*0.1, 0.2, 0.5, 0.8, 1.0*} to compare the results. As shown in Fig. 5, with λ increasing from *0.1* to *0.2*, there is a noticeable improvement in the experimental results, indicating that the balance between the two tasks is increasingly reasonable. When $\lambda \geq 0.2$, the experimental results show a downward trend, which indicated that the proportion of contrast learning tasks is significant, leading to a reduction in the impact of multi-task learning on prediction tasks. With the gradual increase of λ, the experimental results show a more pronounced decrease. In summary, the combination of the two tasks can effectively improve the prediction effect. However, the proportion of each task should be controlled to a certain extent. Specifically, multi-task learning still accounts for a significant portion.

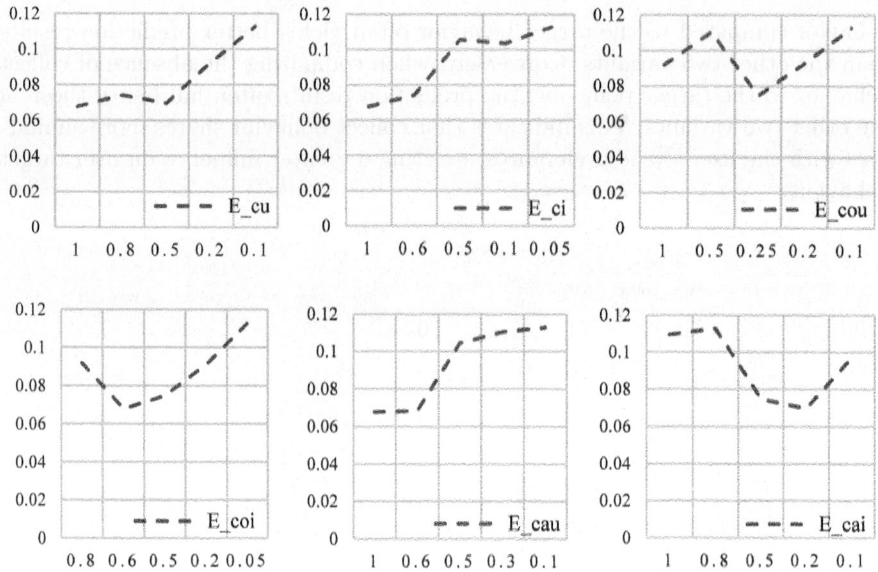

Fig. 4. Performances of different behavior parameters

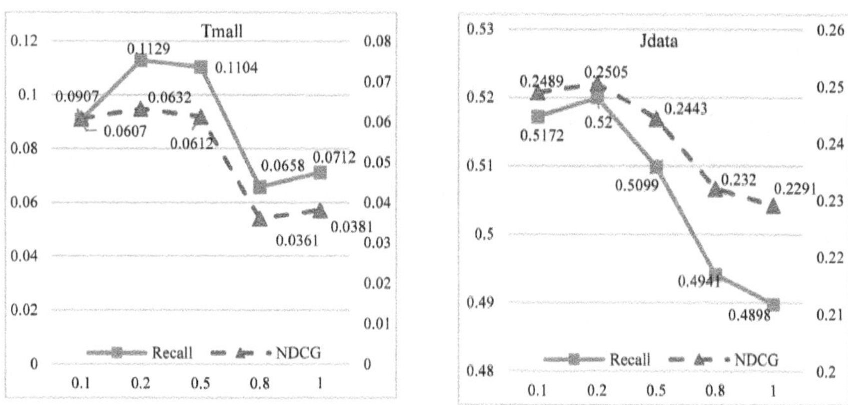

Fig. 5. Impact of λ under two datasets on our model

6 Conclusion

In this paper, we propose a CGCCN framework for multi-behavior recommendation. The combination of multi-task learning and contrastive learning enables joint learning of the dependencies and similarities between behaviors, making full use of behavioral data to improve the quality of user and item embeddings. This approach also enhances the interpretability of the model and effectively alleviates the problem of data sparsity and cold start. We conduct extensive experiments on two real-life datasets, and the results demonstrate that the proposed CGCCN outperforms all state-of-the-art techniques on each datasets.

Acknowledgement. Dr. Xuyun Zhang is the recipient of an ARC DECRA (project No. DE210101458) funded by the Australian Government. This work is supported by the Open Research Project of State Key Laboratory of Novel Software Technology (No. KFKT2022B28).

References

1. Cheng, Z., Chang, X., Zhu, L., Kanjirathinkal, R.C., Kankanhalli, M.: MMALFM: explainable recommendation by leveraging reviews and images. ACM Trans. Inform. Syst. (TOIS) **37**(2), 1–28 (2019)
2. Liu, Z., Luo, X., Zhou, M.: Symmetry and graph bi-regularized non-negative matrix factorization for precise community detection. IEEE Trans. Autom. Sci. Eng. **21**(2), 1406–1420 (2023)
3. Mnih, A., Salakhutdinov, R.R.: Probabilistic matrix factorization. In: Advances in Neural Information Processing Systems, vol. 20 (2007)
4. : He, X., Deng, K., Wang, X., Li, Y., Zhang, Y., Wang, M.: Lightgcn: simplifying and powering graph convolution network for recommendation. In: Proceedings of the 43rd International ACM SIGIR Conference on Research and Development in Information Retrieval, pp. 639-648 (2020)
5. Liu, M., Yang, Z., Li, L., Li, Z., Xie, S.: Auto-weighted collective matrix factorization with graph dual regularization for multi-view clustering. Knowl.-Based Syst. **260**, 110145 (2023)
6. Jin, B., Gao, C., He, X., Jin, D., Li, Y.: Multi-behavior recommendation with graph convolutional networks. In: Proceedings of the 43rd international ACM SIGIR Conference on Research and Development in Information Retrieval, pp. 659–668 (2020)
7. Yan, M., et al.: Cascading residual graph convolutional network for multi-behavior recommendation. ACM Trans. Inform. Syst. **42**(1), 1–26 (2023)
8. Cheng, Z., Han, S., Liu, F., Zhu, L., Gao, Z., Peng, Y.: Multi-behavior recommendation with cascading graph convolution networks. In: Proceedings of the ACM Web Conference, pp. 1181–1189 (2023)
9. Xia, L., et al.: Knowledge-enhanced hierarchical graph transformer network for multi-behavior recommendation. In: Proceedings of the AAAI Conference on Artificial Intelligence Vol. 35, No. 5, pp. 4486–4493 (2021)
10. Loni, B., Pagano, R., Larson, M., Hanjalic, A.: Bayesian personalized ranking with multi-channel user feedback. In: Proceedings of the 10th ACM Conference on Recommender Systems, pp. 361–364 (2016)
11. Rendle, S., Freudenthaler, C., Gantner, Z., Schmidt-Thieme, L.: BPR: Bayesian personalized ranking from implicit feedback. arXiv arXiv:1205.2618 (2012)
12. Qiu, H., Liu, Y., Guo, G., Sun, Z., Zhang, J., Nguyen, H.T.: BPRH: Bayesian personalized ranking for heterogeneous implicit feedback. Inf. Sci. **453**, 80–98 (2018)
13. Chen, P., Chen, P., et al.: Cold-start based multi-scenario ranking model for click-through rate prediction. In: Wang, X., et al. (eds.) Database Systems for Advanced Applications: 28th International Conference, DASFAA 2023, Tianjin, China, April 17–20, 2023, Proceedings, Part II, pp. 64–79. Springer Nature Switzerland, Cham (2023). https://doi.org/10.1007/978-3-031-30672-3_5

14. Li, Q., Ma, H., Zhang, R., Jin, W., Li, Z.: Intra- and inter-behavior contrastive learning for multi-behavior recommendation. In: Wang, X., et al. (eds.) Database Systems for Advanced Applications: 28th International Conference, DASFAA 2023, Tianjin, China, April 17–20, 2023, Proceedings, Part II, pp. 147–162. Springer Nature Switzerland, Cham (2023). https://doi.org/10.1007/978-3-031-30672-3_10
15. Meng, C., Zhao, Z., Guo, W., Zhang, Y., Wu, H., Gao, C., ... Tang, R.: Coarse-to-fine knowledge-enhanced multi-interest learning framework for multi-behavior recommendation. ACM Trans. Inform. Syst. **42**(1), 1–27 (2023)
16. Chen, C., et al.: Graph heterogeneous multi-relational recommendation. In: Proceedings of the AAAI Conference on Artificial Intelligence, vol. 35, No. 5, pp. 3958–3966 (2021)
17. Xia, L., Xu, Y., Huang, C., Dai, P., Bo, L.: Graph meta network for multi-behavior recommendation. In: Proceedings of the 44th International ACM SIGIR Conference on Research and Development in Information Retrieval, pp. 757–766 (2021)
18. Yao, L., Mao, C., Luo, Y.: Graph convolutional networks for text classification. In: Proceedings of the AAAI Conference on Artificial Intelligence vol. 33(01), pp. 7370–7377 (2019)
19. Gutmann, M., Hyvärinen, A.: Noise-contrastive estimation: a new estimation principle for unnormalized statistical models. In: Proceedings of the Thirteenth International Conference on Artificial Intelligence and Statistics. JMLR Workshop and Conference Proceedings, pp. 297–304 (2010)
20. Gao, C., et al.: Learning to recommend with multiple cascading behaviors. IEEE Trans. Knowl. Data Eng. **33**(6), 2588–2601 (2019)
21. Schlichtkrull, M., Kipf, T.N., Bloem, P., van den Berg, R., Titov, I., Welling, M.: Modeling relational data with graph convolutional networks. In: Gangemi, A., et al. (eds.) The Semantic Web: 15th International Conference, ESWC 2018, Heraklion, Crete, Greece, June 3–7, 2018, Proceedings, pp. 593–607. Springer International Publishing, Cham (2018). https://doi.org/10.1007/978-3-319-93417-4_38
22. Xia, L., Huang, C., Xu, Y., Dai, P., Lu, M., Bo, L.: Multi-behavior enhanced recommendation with cross-interaction collaborative relation modeling. In: 2021 IEEE 37th International Conference on Data Engineering (ICDE), pp. 1931–1936 (2021)
23. Gao, C., et al.: Neural multi-task recommendation from multi-behavior data. In: 2019 IEEE 35th International Conference on Data Engineering (ICDE), pp. 1554–1557 (2019)
24. Gu, S., Wang, X., Shi, C., Xiao, D.: Self-supervised graph neural networks for multi-behavior recommendation. In: IJCAI, pp. 2052–2058 (2022)

Social Relation Enhanced Heterogeneous Graph Contrastive Learning for Recommendation

Jiaxi Wang, Bingce Wang, Liwen Zhang, Tong Mo, and Weiping Li(✉)

School of Software and Microelectronics, Peking University, Beijing, China
wangjxi@pku.edu.cn, {wangbingce,zhanglw0529}@stu.pku.edu.cn,
{motong,wpli}@ss.pku.edu.cn

Abstract. Recommender systems have emerged as pivotal components in numerous online services, facilitating the personalized discovery of items that align with users' interests. These systems have showcased their significance in diverse scenarios, with particular prominence observed in applications related to social networks. Heterogeneous Graph Neural Networks (HGNNs) have shown success in recommendation tasks by embedding rich semantics from different relations into latent representations. However, the representation power of existing HGNNs is often limited by sparse data availability, particularly for sparse interaction labels during optimization. To address these challenges, we propose a novel self-supervised learning model called Social Relation-based Heterogeneous Graph Contrastive Learning (SR-HGCL). Our approach merges the user-user social graph and the user-item interaction graph into a unified heterogeneous graph, creating the heterogeneous view. We also construct the social relation enhanced view by resampling the user-item interaction graph. In the learning process, we leverage metapath based graph learning and graph diffusion with attention to obtain multi-view embeddings for users and items. Additionally, we incorporate view-level contrastive learning to encourage distinct embeddings from different views, improving interpretability. We evaluate SR-HGCL on three benchmark datasets and demonstrate its superiority over state-of-the-art methods.

Keywords: Heterogeneous Graph Representation · Social Recommendation · Graph convolutional network · Contrastive Learning

1 Introduction

Recommender systems play a crucial role in many online services, enabling the discovery of items that are of interest to users. They have proven to be particularly valuable in various scenarios, especially in social network-related applications. Collaborative Filtering (CF) is a popular technique that utilizes user-item interactions to learn latent embeddings [9]. Despite its widespread use, CF-based

M. Onizuka et al. (Eds.): DASFAA 2024, LNCS 14855, pp. 19–34, 2024.
https://doi.org/10.1007/978-981-97-5572-1_2

methods still face challenges such as data sparsity and the cold start problem [11,15]. Traditional recommender systems make the assumption of independent and identically distributed data, overlooking the asymmetric distribution of users and the complexity of social connections. Researchers have endeavored to improve CF by incorporating auxiliary information, also known as side information, which goes beyond users' activities. Various types of side information have been employed to enhance the representation of users, items, and their interactions. Examples include item reviews [6], commodities metadata [16], user locations [25], as well as social relations [14].

One of the challenges in recommender systems is data sparsity, especially for users with limited behavior data. Models relying solely on user-item interaction history may suffer from inadequate information. However, the growth of social networks offers an opportunity to address this issue. Researchers have increasingly investigated social relationships and incorporated them as critical side information to enhance user and item embeddings.

An viable approach to address the challenge of data sparsity is through contrastive learning [22], which effectively addresses this issue by augmenting data from unlabeled sources. By integrating contrastive learning with graph neural networks, a method known as Graph Contrastive Learning (GCL) has proven effective in enhancing the robustness of learned representations, especially when dealing with limited observed labels within graph structures. The main idea behind GCL is to examine the alignment between embeddings encoded from different graph contrastive representation views. In GCL-based model, the objective is to maximize the agreement between representations of positive contrastive samples while pushing apart the embeddings of negative pairs.

Practical recommendation scenarios often encompass a multitude of heterogeneous relationships. Researchers are motivated by the success of GNNs in diverse recommendation tasks and are currently striving to develop Heterogeneous Graph Neural Networks capable of embedding rich semantics from various relations into latent representations. However, the representation power of most existing studies is often hampered by the limited availability of sparse data. Current Heterogeneous Graph Neural Networks heavily depend on labeled data for learning. Consequently, models may struggle to generate high-quality embeddings for users and items, especially when dealing with sparse interaction labels during the optimization process.

To tackle the aforementioned challenges, We propose a novel model based on the self-supervised learning paradigm, named Social Relation-based Heterogeneous Graph Contrastive Learning(SR-HGCL). Specifically, we merge the user-user social graph and the user-item interaction graph into a unified heterogeneous graph, creating the heterogeneous view. Next, we split the user-item interaction graph and create the social relation enhanced view by resampling the user-item interaction graph. During the learning process, we leverage meta-path based graph learning and perform graph diffusion with attention to obtain multi-view embeddings for users and items. Additionally, we incorporate view-level contrastive learning to regulate significant differences among embeddings from different views, enhancing interpretability. To evaluate the effectiveness of SR-HGCL, we conduct comparison experiments on three benchmark datasets. The results clearly demonstrate the superiority of SR-HGCL over state-of-the-art methods.

Our contributions are summarized as follows:

- To our best knowledge, this is the first attempt to study the heterogeneous graph neural networks based on the view-level contrastive learning with heterogeneous view and social relation enhanced view. By contrastive learning based on graph learning, the high-level factors can be captured, enabling better embeddings without label supervision.
- We propose a novel model based on the self-supervised learning paradigm. SR-HGCL innovatively employs heterogeneous view and social relation enhanced view to collaboratively supervise each other. Moreover, a multi-task learning mechanism is designed to further enhance the contrastive performance.
- We conduct diverse experiments on three public datasets and the proposed SR-HGCL outperforms the state-of-the-arts and even semi-supervised method, which demonstrates the effectiveness from various aspects.

2 Related Work

In this section, we mainly review social recommendation, heterogeneous graph learning and contrastive learning in GNN-based recommender system.

2.1 Social Recommendation

Before the advent of deep learning, social recommendation has been a subject of study since 1997 [15], primarily based on collaborative filtering. Typical regularization methods such as SocialMF [7] and Social Regularization [12] employ matrix factorization techniques to regulate user preferences.

Despite the considerable success of collaborative filtering in terms of efficiency and performance, it still suffers from data sparsity issues and cold start problems. Deep learning-based methods have emerged as a means to model the social embeddings of users and items, especially through Graph Neural Networks (GNNs). GraphRec [1] is a GNN-based model for social recommendation rating prediction, aggregating representations for items and users from their linked neighbors. Diffnet [20] creates a layer-by-layer influence propagation system to model how consumers' preferences develop as social influence spreads recursively. LightGCN [4] linearly propagates user and item embeddings with weighted sums. HOSR [10] proposes to iteratively "propagate" embeddings along the social network, effectively introducing the influence of high-order neighbors into user representation to overcome the issue of rapidly rising high-order neighbors with the order size. MHCN [23] employs hierarchical mutual information maximization to retrieve connection information, incorporating self-supervised learning into the training of the hypergraph convolutional network.

2.2 Heterogeneous Graph Learning

Heterogeneous graphs are prevalent in real-life applications, comprising various types of nodes and connections. The objective of representation learning over such graphs is to encode node embeddings that preserve rich semantics with relation heterogeneity effectively [13]. To accomplish this, heterogeneous graph neural networks have emerged as promising techniques, yielding state-of-the-art representation results. For instance, HAN [17] enhances the graph attention network by incorporating meta-path construction, allowing it to handle heterogeneous types of nodes and relations. Drawing inspiration from the transformer framework, HGT [5] introduces a graph transformer network, enabling heterogeneous message passing through self-attention to calculate propagation weights between nodes. Furthermore, MAGNN [2] considers both intra-metapath and inter-metapath aggregation to fuse information from diverse meta-paths within heterogeneous graphs. In unsupervised setting, HetGNN [24] samples a fixed size of neighbors, and fuses their features using LSTMs.

2.3 Contrastive Learning for Recommendation

Contrastive learning methods [3,18] learn discriminative node representations from unlabeled data by maximizing the distance between negative pairs while minimizing the distance between positive pairs. Recently, some has been done towards contrastive learning on recommendation. SGL [19] utilizes a multi-view approach to generate multiple perspectives of a node, aiming to maximize the agreement between different views of the same node while comparing them to other nodes. The method employs three operators to generate these views as node dropout, edge dropout, and random walk. KGCL [21] presents a novel knowledge graph augmentation schema with the purpose of mitigating noise during knowledge graph information aggregation. This approach facilitates the derivation of more robust and knowledge-aware representations for items. Moreover, it effectively utilizes the KG augmentation process to guide cross-view contrastive learning. KGIC [26] constructs both local and non-local graphs for users and items in the Knowledge Graph (KG) and performs layer-wise contrastive learning on these graphs. While all the aforementioned methods utilize contrastive learning at the graph view level and tend to overlook the more detailed level of user intents.

3 Methodology

In this section, we propose a novel heterogeneous graph contrastive learning model for online recommendation in social relation-based service, named SR-HGCL as Fig. 2 shows. SR-HGCL consists of three modules. (1) Cross-View Heterogeneous Graph Construction (2) View-Based Graph Learning. (3) View-level Contrastive Learning.

3.1 Definitions and Problem Formulation

Real-life recommender systems frequently involve heterogeneous relations that encompass diverse semantic information from both users and items. The context of recommendation basically contains two types of entities. let $U = \{u_1, u_2, \ldots, u_n\}$ and $V = \{v_1, v_2, \ldots, v_m\}$ be the set of users and items respectively where n and m are the numbers of users and items. The interaction graphs are formed from U and V. User-item interaction graph can be denoted as $\boldsymbol{G}_{ui}$, we can use the adjacency matrix to represent the interactions in the graph. $\mathbf{M}_{inter} \in \mathbb{R}^{n \times m}$ is the interaction matrix between users and items. If there is an interaction between the user u and the item i, y_{ui} is 1, otherwise it is 0. User-user social network can be denoted as $\boldsymbol{G}_{uu}$, $\mathbf{M}_{social} \in \mathbb{R}^{n \times n}$ is the interaction matrix between users. We formulate the studied recommendation task in this paper as:

Input: The User set U and the item set V, $\mathbf{M}_{inter} \in \mathbb{R}^{n \times m}$ the adjacency matrix representation of the user-item interaction graph $\boldsymbol{G}_{ui}$, $\mathbf{M}_{social} \in \mathbb{R}^{n \times n}$ the adjacency matrix representation of the user-user social relation network $\boldsymbol{G}_{uu}$.

Output: A predictive function $\text{Predict}(z_{ij}; \theta_P)$ that effectively forecasts the future user-item interaction.

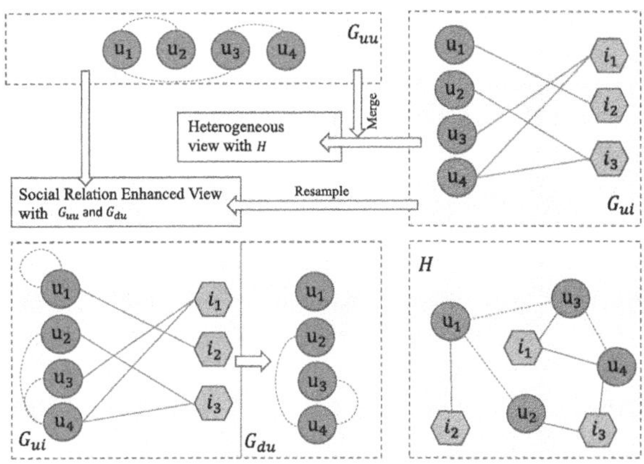

Fig. 1. Social relations across heterogeneous views: the heterogeneous view (user-item merged relation) and the social relation enhanced view (user-user relation and user-user dummy relation)

3.2 Cross-View Heterogeneous Graph Construction

To comprehensively investigate social relations across heterogeneous views, we decouple the different relationships into two views shown in Fig. 1: the heterogeneous view (user-item merged relation) and the social relation enhanced view (user-user relation and user-user dummy relation). The alignment of different views is based on the users that appear in both views simultaneously. Initially, we extract valuable social information from the homogeneous user-user relation and user-user dummy relation, and subsequently integrate it into the heterogeneous view for making the final predictions of user interests.

The heterogeneous view comprises the interacted user and item nodes, along with the original social relation. These two graphs are merged into a unified heterogeneous graph, encompassing two types of nodes and two types of edges. On the other hand, the social relation enhanced view consists of two sub-graphs: the original user-user social graph $\boldsymbol{G}_{uu}$ and the dummy social graph $\boldsymbol{G}_{du}$. In the dummy social graph, edges are created between two users if they have purchased the same items, based on the user-item interaction graph. While the dummy social relation does not represent real existing relationships, it reflects the potential interactions between users through user-item connections. It intuitively captures the shared purchasing habits among users. Considering that the strength of the dummy social relation is low and might introduce noise to graph learning, we filter the edges between two users by adding the restriction of social-friend relations. The adjacency matrix in dummy social graph is $\mathbf{M}_{du} = \left(\mathbf{M}_{inter} \cdot \mathbf{M}_{inter}^{T}\right) \odot \mathbf{M}_{social}$, where $\mathbf{M}_{inter}$ and $\mathbf{M}_{social}$ are the adjacency matrix of user-item interaction and user-user social relationship, $\odot$ is the Hadamard product. The edges in the social-friend view are established based on social friendship relations. However, we must consider that some friendships might form on an occasional basis and may not significantly contribute to modeling users' social information. On the other hand, the structure of friendships in ternary closures tends to be more stable. Therefore, we choose to retain only the edges with a ternary closure social structure in the social relation graph.The adjacency matrix of relation graph $\mathbf{M}_s$ is computed as $\mathbf{M}_s = (\mathbf{M}_d \cdot \mathbf{M}_d) \odot \mathbf{M}_d$

To harness the rich information shared across $\boldsymbol{G}_{ui}$ and $\boldsymbol{G}_{uu}$, we propose concatenating them to form a heterogeneous graph $\boldsymbol{H}$. This enables effective quantification of user similarity, which is crucial for social recommendation.

When dealing with large-scale real-world recommender systems modeled by heterogeneous graph, computational efficiency becomes a crucial challenge in identifying implicit friends while preserving the original network's embedded information. To address this challenge, we draw inspiration from successful network embedding models and propose using meta-paths to generate biased meta-path-based random walks to solve the problem.

A meta-path schema represents a composite relation between the start type and the end type of a path, characterized by a series of relation types. In this work, we define six carefully designed meta-paths to capture user similarities across the user-item bipartite network and the user-user social network. These meta-paths effectively identify similar entities that may be distant from each other on the networks.

To leverage meta-paths, we guide the random walks, resulting in a social corpus of node sequences. However, social relations in heterogeneous graph can be noisy, and thus it is essential to identify reliable sequences. To achieve this, we use UUU, UUI, UIU, and UUUI as heterogeneous graph meta-path investigation strategies.

Upon procuring the instances of metapaths, while meticulously preserving the multifarious relational structures inherent to both the user and item domains, we advocate the implementation of a graph-based neural architecture to systematically encode the inherent heterogeneity of relations. Within the ambit of our proposed framework, which is guided by metapath-facilitated message propagation, the embedding of each discrete node is derived through neural layer. This particular layer is adept at aggregating and assimilating feature information emanating from neighboring nodes, as defined by the metapath-based contextual linkages (Table 1).

Table 1. Meta-paths Description

Path	Schema	Detailed Information
#P1	UUU	Friend's friend,two-hop connection
#P2	UUI	short-term similar preference relationship
#P3	UIU	Users who have similar preference
#P4	UUUI	long-term similar preference relationship

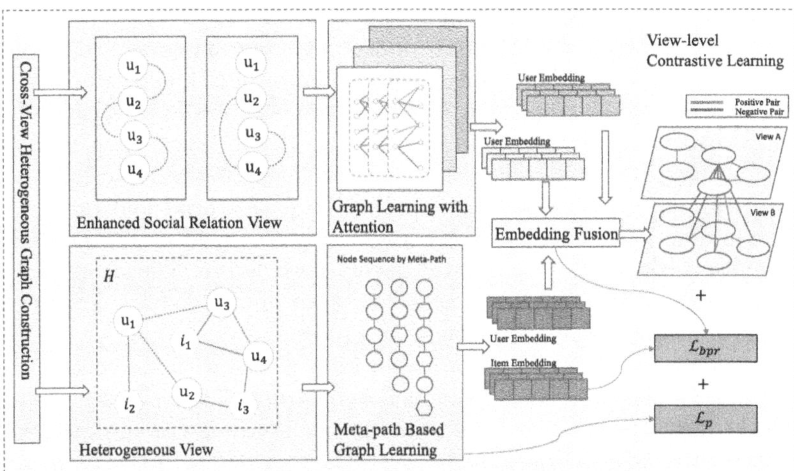

Fig. 2. Model overview. SR-HGCL consists of three modules. (1) Cross-View Heterogeneous Graph Construction (2) View-Based Graph Learning. (3) View-level Contrastive Learning

3.3 View-Based Graph Learning

We employ the attention network structure to aggregate adaptively the learning embeddings of different neighbors in the user-user social graph $\boldsymbol{G}_{uu}$ and the dummy social graph $\boldsymbol{G}_{du}$. Through iterative aggregation, information can be diffused and propagated in the network. Some high-order cooperative information can also be captured, so the model can model high-order collaborative signals. In our work, we simplify feature transformation and nonlinear activation to reduce model complexity. We aggregate all nodes in the heterogeneous information network, the graph convolution operation in the SR-HGCL is defined as:

$$e_a^* = e_a + \sum_{b \in \mathcal{N}(a)} \tilde{\alpha}_{ab} \cdot e_b \tag{1}$$

where e_a^* represents the embedding representation of node $a, \mathcal{N}(a)$ represents the set of neighbors of a, and $\tilde{\alpha}_{ab}$ is the attention value, which indicates that how strongly node a is influenced by node b. The attention value is specifically defined as follows:

$$\tilde{\alpha}_{ab} = \frac{\exp(\alpha_{ab})}{\sum_{b' \in \mathcal{N}(a)} \exp(\alpha_{ab,})} \tag{2}$$

$$\alpha_{ab} = (\boldsymbol{e}_a \odot \boldsymbol{e}_b)^T \tanh(\boldsymbol{r}_{ab} \odot \boldsymbol{e}_b) \tag{3}$$

where r_{ab} represents the interaction type between node a and node b. High-order correlations play a crucial role in encoding collaborative signals and estimating correlation scores between users and items. To achieve this, we iteratively execute the above aggregation process, enabling the encoding of high-order collaborative signals and exploration of high-order connectivity information.

In heterogeneous graph $\boldsymbol{H}$, entities with a relationship should exhibit a strong path dependency in the graph structure (or representation). Conversely, entities without a relationship should have a weak path dependency. With this understanding, we aim to reconstruct the path dependency between each entity pair from the learned graph representation. For the path's head and tail nodes, if there are interactions between the two nodes in the ground-truth, the path will have a higher score; otherwise, it will have a lower score. The main goal is to encourage users to focus more on nodes that are more likely to interact with themselves during the embedding learning process.

We calculate the dependency probabilities of paths with the following approach. Unlike which proposed to model reconstructed sequences as sequence generation, the sequence nature of paths in our study is not obvious. Specifically, since there is a relationship between the two nodes in the path, we use self-attention to calculate the hidden state $\boldsymbol{p}_l$ of each node $\boldsymbol{q}_{l-1}$ on the path $\phi(n)$, as follows:

$$\boldsymbol{p}_l = \text{self-attention } (\boldsymbol{p}_{l-1}, \boldsymbol{q}_{l-1}) \tag{4}$$

where $\boldsymbol{p}_l$ is the embedding of the node, and the initialization of $\boldsymbol{p}_0$ is the output of the aggregation layer. Then $\boldsymbol{p}_l$ is given to the softmax layer to calculate the probability of node v_{b_l} in the path:

$$P(v_l \mid v_{<l}) = \frac{\exp(\boldsymbol{p}_l \boldsymbol{W}_r \boldsymbol{q}_l)}{\sum_n \exp(\boldsymbol{p}_l \boldsymbol{W}_r \boldsymbol{q}_n)}, \tag{5}$$

where $\boldsymbol{W}_r \in \mathbb{R}^{d\times d}$ is the parameter matrix. Then we get the set of node probabilities for path $\phi(n)$ as $\{P(v_1 \mid v_{<1}), P(v_2 \mid v_{<2}), \cdots, P(v_L \mid v_{<L})\}$, finally, the probability of path $\phi(n)$ is calculated as:

$$N(\phi(n)) = \prod_{l=1}^{L} P(v_l \mid v_{<l}) \tag{6}$$

We use the reconstructed path probability to compute an additional reconstruction loss over the training data set:

$$\mathcal{L}_p = -\frac{1}{\sum_{t=0}^{T} N_t} \sum_{t=1}^{T} \sum_{n=1}^{N_t} \{r \log \mathcal{N}(\phi_n) + (1 - r_n^t) \log(1 - \mathcal{N}(\phi_n)\}, \tag{7}$$

where r is one of $\{0, 1\}$, we maximize the probability of the path $\mathcal{N}(\phi_n)$ if the entity pair has relation, and minimize the probability otherwise.

3.4 View-Level Contrastive Learning

In our recommender system, the acquired user-item relations and user-user dependencies are considered complementary to each other. To effectively integrate both relational contextual signals, we have devised a cross-view denoised self-supervised learning paradigm. This approach helps mitigate the noisy effects that may arise when transferring social knowledge into user-item interaction modeling. In real-life scenarios, passive social relations, like colleagues or classmates, may not significantly impact user interaction preferences due to their diverse shopping tastes. Over reliance on such irrelevant social ties to infer users' interests could adversely affect the performance of social recommendation models. To overcome this challenge, we apply a filtering mechanism to remove noisy social influences between dissimilar users, particularly regarding their interaction preferences. This filtering ensures unbiased self-supervision during the learning process.

Using our heterogeneous graph relation learning technique and user-user relation enhanced with user-user dummy relation, we derive two sets of embeddings for users and items: $\mathbf{E}_{uu}$, $\mathbf{E}_{du}$, and $\mathbf{E}_{Hu}$ for users, and $\mathbf{E}_{Hi}$ for items. These embeddings are obtained by encoding the user-item interaction data and the social relation enhanced view. Drawing inspiration from the successful application of recent contrastive learning in recommendation systems, we propose to leverage the user/item representation learning from our SR-HGCL method with an InfoNCE-based contrastive learning loss between view-wise embeddings, as follows:

$$\mathcal{L}_{cl}^{u} = \sum_{u \in \mathcal{V}_u} -\log \frac{\exp\left(s\left(\mathbf{e}_{uu} + (1-\alpha_u)\mathbf{e}_{du}, \mathbf{e}_{Hu}\right)/\tau\right)}{\sum_{u' \in \mathcal{V}_u} \exp\left(s\left(\mathbf{e}_{uu} + (1-\alpha_u)\mathbf{e}_{du}, \mathbf{e}'_{Hu}\right)/\tau\right)} \tag{8}$$

where $\mathbf{e}_{uu}^{M} \in \mathbb{R}^d, \mathbf{e}_{d_u} \in \mathbb{R}^d, \mathbf{e}_{Hu} \in \mathbb{R}^d$ are the embedding vectors from the matrices $\mathbf{E}_{uu}, \mathbf{E}_{du}$ and $\mathbf{E}_{Hu}$, respectively. α_u denotes the fusion ratio for the enhanced social relation user embeddings, $s(\cdot)$ denotes the similarity function, which can be inner product or cosine similarity. Here we use cosine similarity as our $s(\cdot).\tau$ represents the temperature coefficient, which is capable of automatically identifying difficult negative samples. u' indicates negative samples with different indices.

To optimize our SR-HGCL method for the recommendation task, we adopt the Bayesian Personalized Ranking (BPR) pairwise loss function, following recent works in the field. For each training sample, comprising a user u, a positive item i^+ (i.e., an item the user has interacted with), and a negative item i^- (i.e., an item the user has not interacted with), we aim to maximize the prediction score as follows:

$$\mathcal{L}_{bpr} = \sum_{(u,i^+,i^-)\in O} -\ln\left(\text{sigmoid}\left(\hat{y}_{u,i^+} - \hat{y}_{u,i^-}\right)\right) + \lambda\|\Theta\|^2 \tag{9}$$

where $\ln(\cdot)$ and sigmoid $(\cdot)$ denote the logarithm function and the sigmoid function, respectively. λ denotes a hyperparameter to determine the weight of the regularization term.

3.5 Multi-task Training

To improve recommendation with the CL task, we leverage a multi-task training strategy to jointly optimize the classic recommendation task and the self-supervised learning task:

$$\mathcal{L} = \mathcal{L}_p + \mathcal{L}_{bpr} + \lambda_1 \mathcal{L}_{cl}^u + \lambda_2\|\Theta\|_2^2, \tag{10}$$

where Θ is the set of model parameters in $\mathcal{L}_{b\ pr}, \mathcal{L}_p$ since $\mathcal{L}_{cl}^u$ introduces no additional parameters; λ_1 and λ_2 are hyperparameters to control the strengths of CL task and L_2 regularization, respectively.

4 Experiment

In this section, we evaluate the performance of SR-HGCL on a series of real-world datasets. Specifically, we choose Epinions, Yelp and Ciao dataset for experiment, to answer the following research questions:

- **RQ1**: Can SR-HGCL perform better than other baseline methods?
- **RQ2**: Does the incorporation of key components in our SR-HGCL lead to a noticeable improvement in the recommendation performance?
- **RQ3**: How do key hyperparameters affect model performance?

4.1 Experimental Setting

Dataset. In this study, we validate the performance of SR-HGCL through experiments conducted on three public datasets: Epinions, Yelp, and Ciao. Detailed descriptions of these datasets are provided below (Table 2).

Table 2. The statistics of the three datasets

Dataset	Epinions	Yelp	Ciao
#User	22164	17235	7375
#Item	296277	37378	105114
#Ratings	911442	207945	169730
#Relations	574202	169150	57544
#Rating density	0.014%	0.032%	0.035%
#Relation density	0.116%	0.087%	0.313%

- **Epinions:** The dataset comes from a general consumer review site Epinions.com which contains trust relationships interactions and review ratings information.
- **Yelp:** This dataset is a location-based review dataset which also contains social information and ratings.
- **Ciao:** This dataset contains rating information of users given to items, and also contain item category information.

Evaluation Metrics. To assess the performance of the methods, we utilize two widely accepted ranking-based metrics: Hit Ratio at rank k (Hit Ratio@k) and Normalized Discounted Cumulative Gain at rank k (NDCG@k). The recommended set is generated by reporting the top K (K = 5 and K = 10) items from the ranking list.

Baselines. We compare the proposed SR-HGCL with the following baseline models:

- **SocialMF** [7]: This method employs matrix factorization techniques with a trust propagation mechanism to reduce the cold start problem and improve prediction accuracy.
- **GraphRec** [1]: This approach introduces a novel Graph Neural Network (GNN)-based model that integrates both the user-user social graph and the user-item interaction graph. An attention network is employed to filter out high-influence users and items, enhancing the model's performance.
- **Diffnet** [20]: Diffnet adopts a layer-wise GNN diffusion architecture to effectively model users' preferences.
- **LightGCN** [4]: User and item embeddings are obtained through LightGCN using a linear propagation technique on the user-item interaction graph. The model's simplicity in implementation and training sets it apart from other approaches.
- **HAN** [17]: HAN introduces a novel heterogeneous graph neural network based on hierarchical attention, incorporating node-level and semantic-level attentions to enhance the model's representation learning capabilities.
- **HeCo** [18]: This approach presents a novel co-contrastive learning mechanism for HGNNs and employs cross-view contrastive learning, contributing to improved performance in heterogeneous graph neural networks.

- **MHCN** [23]: MHCN utilizes hierarchical mutual information maximization and incorporates self-supervised learning into the training of the hypergraph convolutional network, thereby retrieving valuable connection information and enhancing the model's performance.

Parameter Settings. We have implemented our proposed model using PyTorch. The optimization of the gradient descent processes is carried out using Adam [8]. To obtain the best results for all the baseline models, we conduct a learning rate search within the range of [0.001, 0.005, 0.1, 0.2, 0.5]. To mitigate overfitting, a dropout rate of 0.5 is employed. The embedding size is varied between 32, 64, 128, and 256. For the diffusion model, we experiment with layer numbers of 1, 2 and 3. During evaluations, we employ early stopping as a training termination strategy when the performance degrades for 5 consecutive epochs on the validation dataset.

4.2 Performance Comparision(RQ1)

To provide a comprehensive evaluation of the performance of the proposed model and compare it with other recommendation methods, we report the results in terms of HR@[5, 10] and NDCG@[5, 10] in Table 3.

Our proposed model consistently outperforms all baseline models across all datasets, Epinions, Yelp, and Ciao. It achieves the highest Hit Ratio at rank k (H@k) and Normalized Discounted Cumulative Gain at rank k (N@k) scores, indicating its effectiveness in generating superior recommendations. The model leverages multi-level contrastive learning, which enables it to explore user preferences from diverse intentional perspectives, leading to a more nuanced understanding of user behavior and resulting in more personalized and accurate recommendations. Among the baseline models, HeCo and MHCN also demonstrate competitive performance, exhibiting relatively higher H@k and N@k scores across all datasets. This performance suggests the effectiveness of incorporating heterogeneous graph-based information, including social relations from both the user and item sides, into the recommender system.

Table 3. Overall Comparison with Baseline. Best results are in boldface. Best baseline are marked with underline. Improv. denotes relative improvement over the best baseline

Dataset	Epinions				Yelp				Ciao			
	H@5	H@10	N@5	N@10	H@5	H@10	N@5	N@10	H@5	H@10	N@5	N@10
SocialMF	0.411	0.534	0.359	0.385	0.600	0.701	0.420	0.464	0.241	0.336	0.154	0.180
GraphRec	0.430	0.541	0.341	0.388	0.591	0.813	0.450	0.535	0.260	0.384	0.170	0.211
Diffnet	0.460	0.576	0.350	0.391	0.600	0.749	0.441	0.490	0.241	0.365	0.163	0.196
LightGCN	0.485	0.609	0.360	0.391	0.644	0.790	0.474	0.526	0.253	0.390	<u>0.172</u>	0.227
HAN	0.488	0.597	0.357	0.380	0.649	0.786	0.478	0.515	0.261	0.380	0.170	0.219
HeCo	<u>0.494</u>	0.607	0.367	0.381	0.653	0.791	0.480	0.519	<u>0.268</u>	0.387	0.171	0.221
MHCN	0.491	<u>0.611</u>	<u>0.372</u>	<u>0.420</u>	<u>0.667</u>	<u>0.809</u>	<u>0.511</u>	<u>0.556</u>	0.262	<u>0.397</u>	0.170	<u>0.231</u>
Ours	**0.530**	**0.647**	**0.395**	**0.453**	**0.684**	**0.838**	**0.563**	**0.581**	**0.290**	**0.421**	**0.190**	**0.274**
Improv.	7.29%	5.89%	6.18%	7.86%	2.55%	3.59%	10.18%	4.50%	8.21%	6.05%	10.46%	18.62%

Table 4. Overall Comparison with Different Version Model

Dataset	Epinions				Yelp				Ciao			
	H@5	H@10	N@5	N@10	H@5	H@10	N@5	N@10	H@5	H@10	N@5	N@10
Our-Normal	0.421	0.527	0.361	0.381	0.613	0.690	0.411	0.453	0.239	0.327	0.141	0.188
Ours-nHG	0.491	0.601	0.369	0.385	0.649	0.793	0.488	0.521	0.272	0.389	0.177	0.226
Ours-nSE	0.491	0.603	0.361	0.377	0.650	0.787	0.475	0.514	0.265	0.386	0.173	0.219
Ours-nCL	0.486	0.591	0.341	0.377	0.644	0.780	0.474	0.511	0.258	0.385	0.173	0.222
Ours	**0.530**	**0.647**	**0.395**	**0.453**	**0.684**	**0.838**	**0.563**	**0.581**	**0.290**	**0.421**	**0.190**	**0.274**

4.3 Experiment with Effectiveness(RQ2)

To evaluate the individual contributions of the main components in our model, we conducted ablation experiments by creating three variants that represent different perspectives:

- **Ours-nHG:** This variant omits the heterogeneous graph in our model and replaces it with the user-item interaction graph for model training.
- **Ours-nSE:** This variant omits the social graph enhancement part of our model and utilizes only the heterogeneous graph for model training.
- **Ours-nCL:** This variant omits the contrastive learning part of our model and uses the BPR (Bayesian Personalized Ranking) loss as the regulation loss.
- **Ours-Normal:** This version omits all key components of our model and only employs diffusion and aggregation for embedding, serving as the basic baseline for comparison.

The Table 4 present the results of ablation experiments conducted to assess the individual contributions of the main components in the proposed model. We created three variants (Ours-nHG, Ours-nSE, and Ours-nCL) by removing specific components from the original model and compared their performance with the complete model (Ours). The table displays the evaluation metrics (H@5, H@10, N@5, and N@10) for each variant and the complete model across three datasets (Epinions, Yelp, and Ciao).According to the results in the table, the complete model (Ours) consistently outperforms all three variants (Ours-nHG, Ours-nSE, and Ours-nCL) on all datasets. This indicates that each component plays a vital role in enhancing the recommendation performance. Specifically, by omitting the heterogeneous graph and using only the user-item interaction graph, the variant shows improved performance compared to the basic model (Ours-Normal) but still falls short compared to the complete model (Ours). This suggests that the heterogeneous graph is essential in providing additional information for more accurate recommendations. The omission of the social graph enhancement part results in slightly lower performance compared to the complete model (Ours). This indicates that the social graph enhances the model's ability to capture social influence and improve the recommendation accuracy. When contrastive learning is omitted and replaced with the BPR loss, the variant shows a decrease in performance compared to the complete model

(Ours). This demonstrates the significance of contrastive learning in improving the model's understanding of user preferences and generating more personalized recommendations.

4.4 Hyper-parameter Analysis(RQ3)

In this section, we delve into the influence of embedding dimension on the performance of the SR-HGCL model and validate its effectiveness, the experiment result shown in Fig. 3. To comprehensively assess the impact of the embedding dimension, we conduct a thorough study, varying the dimension in the range of [64, 128, 256, 512]. The objective is to identify the optimal embedding dimension that yields the best results. As depicted in Fig. 1, the experimental findings demonstrate that an embedding dimension of around 128 provides the most favorable outcomes in terms of recommendation performance. Another critical aspect of the SR-HGCL model that warrants examination is the diffusion layer in the graph neural network. To gain insights into the influence of the diffusion layer, we perform a dedicated study, evaluating different layer settings. Specifically, we explore the options of 1, 2 and 3 diffusion layers in the graph neural network. After thorough analysis, the results show that the optimal number of diffusion layers is 2. This finding highlights the importance of leveraging the diffusion layer to achieve optimal recommendation performance.

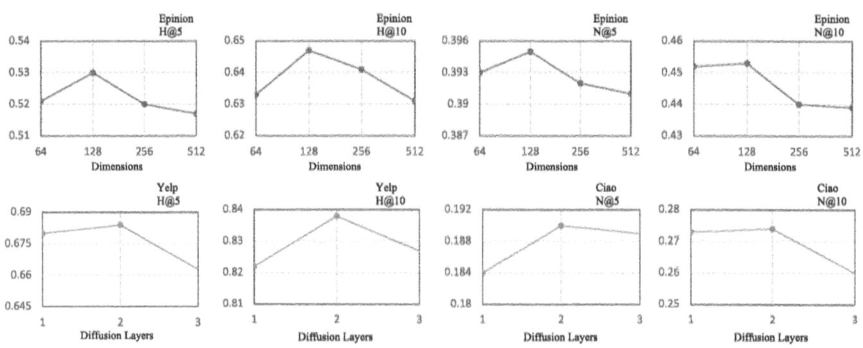

Fig. 3. The Embedding Dimension and Diffusion Layer Analysis

5 Conclusion

In this work, we present a novel recommendation framework called SR-HGCL, designed to address the issue of sparse interaction labels. To overcome these challenges, we merge the user-user social graph and the user-item interaction graph into a unified heterogeneous graph, thereby creating the heterogeneous view. Additionally, we construct the social relation enhanced view through resampling the user-item interaction graph. During the learning process, we utilize meta-path based graph learning and graph diffusion with attention to obtain multi-view embeddings for both users and items. Furthermore, we introduce view-level contrastive learning to encourage distinct embeddings from different views, thereby enhancing interpretability.

The SR-HGCL model we propose provides an innovative solution to tackle the complexities inherent in heterogeneous recommendation scenarios. By leveraging self-supervised learning combined with contrastive learning, we effectively capture diverse relationships and generate superior embeddings, thereby leading to enhanced recommendation performance.

Acknowledgement. This work was supported by the National Key R&D Program of China [2022YFF0902703, 2023YFC3304900].

References

1. Fan, W., et al.: Graph neural networks for social recommendation. In: The World Wide Web Conference, pp. 417–426 (2019)
2. Fu, X., Zhang, J., Meng, Z., King, I.: Magnn: metapath aggregated graph neural network for heterogeneous graph embedding. In: Proceedings of the Web Conference 2020, pp. 2331–2341 (2020)
3. Hassani, K., Khasahmadi, A.H.: Contrastive multi-view representation learning on graphs. In: International Conference on Machine Learning, pp. 4116–4126. PMLR (2020)
4. He, X., Deng, K., Wang, X., Li, Y., Zhang, Y., Wang, M.: Lightgcn: simplifying and powering graph convolution network for recommendation. In: Proceedings of the 43rd International ACM SIGIR Conference on Research and Development in Information Retrieval, pp. 639–648 (2020)
5. Hu, Z., Dong, Y., Wang, K., Sun, Y.: Heterogeneous graph transformer. In: Proceedings of the Web Conference 2020, pp. 2704–2710 (2020)
6. Huang, C., Jiang, W., Wu, J., Wang, G.: Personalized review recommendation based on users' aspect sentiment. ACM Trans. Internet Technol. **20**(4) (oct 2020)
7. Jamali, M., Ester, M.: A matrix factorization technique with trust propagation for recommendation in social networks. In: Proceedings of the Fourth ACM Conference on Recommender Systems, pp. 135–142 (2010)
8. Kingma, D.P., Ba, J.: Adam: a method for stochastic optimization. arXiv preprint arXiv:1412.6980 (2014)
9. Konstan, J.A., Miller, B.N., Maltz, D., Herlocker, J.L., Gordon, L.R., Riedl, J.: Grouplens: applying collaborative filtering to usenet news. Commun. ACM **40**(3), 77–87 (1997)
10. Liu, Y., Liang, C., He, X., Peng, J., Zheng, Z., Tang, J.: Modelling high-order social relations for item recommendation. IEEE Transactions on Knowledge and Data Engineering (2020)
11. Ma, H., Yang, H., Lyu, M.R., King, I.: Sorec: social recommendation using probabilistic matrix factorization. In: Proceedings of the 17th ACM Conference on Information and Knowledge Management, pp. 931–940 (2008)
12. Ma, H., Zhou, D., Liu, C., Lyu, M.R., King, I.: Recommender systems with social regularization. In: Proceedings of the fourth ACM International Conference on Web Search and Data Mining, pp. 287–296 (2011)

13. Shi, C., Li, Y., Zhang, J., Sun, Y., Philip, S.Y.: A survey of heterogeneous information network analysis. IEEE Trans. Knowl. Data Eng. **29**(1), 17–37 (2016)
14. o Song, W., Xiao, Z., Wang, Y., Charlin, L., Zhang, M., Tang, J.: Session-based social recommendation via dynamic graph attention networks. In: Proceedings of the Twelfth ACM International Conference on Web Search and Data Mining, pp. 555–563 (2019)
15. Tang, J., Hu, X., Liu, H.: Social recommendation: a review. Soc. Netw. Anal. Min. **3**(4), 1113–1133 (2013)
16. Wang, J., Huang, P., Zhao, H., Zhang, Z., Zhao, B., Lee, D.L.: Billion-scale Commodity Embedding for E-commerce Recommendation in alibaba. In: Proceedings of the 24th ACM SIGKDD International Conference on Knowledge Discovery Data Mining, pp. 839–848. KDD '18, Association for Computing Machinery, New York, NY, USA (2018)
17. Wang, X., Ji, H., Shi, C., Wang, B., Ye, Y., Cui, P., Yu, P.S.: Heterogeneous graph attention network. In: The World Wide Web Conference, pp. 2022–2032 (2019)
18. Wang, X., Liu, N., Han, H., Shi, C.: Self-supervised heterogeneous graph neural network with co-contrastive learning. In: Proceedings of the 27th ACM SIGKDD Conference on Knowledge Discovery and Data Mining, pp. 1726–1736 (2021)
19. Wu, J., et al.: Self-supervised graph learning for recommendation. In: Proceedings of the 44th International ACM SIGIR Conference on Research and Development in Information Retrieval, pp. 726–735 (2021)
20. Wu, L., Sun, P., Fu, Y., Hong, R., Wang, X., Wang, M.: A neural influence diffusion model for social recommendation. In: Proceedings of the 42nd International ACM SIGIR Conference on Research and Development in Information Retrieval, pp. 235–244 (2019)
21. Yang, Y., Huang, C., Xia, L., Li, C.: Knowledge graph contrastive learning for recommendation. In: Proceedings of the 45th International ACM SIGIR Conference on Research and Development in Information Retrieval, pp. 1434–1443 (2022)
22. You, Y., Chen, T., Sui, Y., Chen, T., Wang, Z., Shen, Y.: Graph contrastive learning with augmentations. Adv. Neural. Inf. Process. Syst. **33**, 5812–5823 (2020)
23. Yu, J., Yin, H., Li, J., Wang, Q., Hung, N.Q.V., Zhang, X.: Self-supervised multichannel hypergraph convolutional network for social recommendation. In: Proceedings of the Web Conference 2021, pp. 413–424 (2021)
24. Zhang, C., Song, D., Huang, C., Swami, A., Chawla, N.V.: Heterogeneous graph neural network. In: Proceedings of the 25th ACM SIGKDD International Conference on Knowledge Discovery and Data Mining, pp. 793–803 (2019)
25. Zheng, V.W., Zheng, Y., Xie, X., Yang, Q.: Towards mobile intelligence: learning from GPS history data for collaborative recommendation. Artif. Intell. **184**, 17–37 (2012)
26. Zou, D., Wei, W., Wang, Z., Mao, X.L., Zhu, F., Fang, R., Chen, D.: Improving knowledge-aware recommendation with multi-level interactive contrastive learning. In: Proceedings of the 31st ACM International Conference on Information and Knowledge Management, pp. 2817–2826 (2022)

Higher-Order Graph Contrastive Learning for Recommendation

ZhenZhong Zheng[1,2], Jianxin Li[3], Xiaoming Wu[1,2(✉)], Xiangzhi Liu[1,2], and Lili Pei[4]

[1] Key Laboratory of Computing Power Network and Information Security, Ministry of Education, Shandong Computer Science Center, Qilu University of Technology (Shandong Academy of Sciences), Jinan, China
z_zhong2021@163.com, xiaomingwu1981@126.com, liuxzh@sdas.org

[2] Shandong Provincial Key Laboratory of Computer Networks, Shandong Fundamental Research Center for Computer Science, Jinan, China

[3] Deakin University, Geelong, Australia
jianxin.li@deakin.edu.au

[4] Chang'an University, Xi'an, China
peilili@chd.edu.cn

Abstract. Collaborative filtering, as an effective recommendation technique, models user preferences by representing user interactions as a bipartite graph(user-item). However, the graph-based model struggles to mitigate the impact of data sparsity. Recent studies have attempted to tackle this problem by utilizing contrastive learning. Nevertheless, most of these methods rely on augmenting the data based on the original graph to construct contrastive views, neglecting the utilization of higher-order relationships(user-user, item-item). Alternatively, some methods erroneously compare higher-order relationships directly with the original graph. Such an asymmetric contrasting approach makes it challenging to fully exploit higher-order knowledge. To address the issues, we propose HoGCL, a novel contrastive learning paradigm. HoGCL enriches representation learning by mining higher-order dependencies from the original graph to enhance supervisory signals. Specifically, we construct two contrasting views: higher-order and general views. In the higher-order view, we devise a high-order symmetric contrastive scheme to better explore higher-order dependencies. For the general view, the objective is to fully leverage the learned higher-order knowledge and transfer it into node representations. Moreover, to remove redundant information from higher-order knowledge, HoGCL employs an adaptive fusion strategy to obtain the final representations. Experimental results demonstrate the effectiveness of HoGCL, which significantly outperforms existing models..

Keywords: Contrastive Learning · Collaborative Filtering · Recommendation

M. Onizuka et al. (Eds.): DASFAA 2024, LNCS 14855, pp. 35–51, 2024.
https://doi.org/10.1007/978-981-97-5572-1_3

1 Introduction

Recommendation systems aim to personalize knowledge and enhance users' experience by filtering out irrelevant content based on their historical behaviors. Collaborative filtering(CF) algorithms, as a fundamental recommendation means, initially employ matrix factorization techniques [11,17] to embed users/items into vector spaces. In the present scenario, Graph Neural Network(GNN) based techniques have emerged as the preferred approach due to their exceptional performance [9,23,29]. These methods model user interactions as a bipartite graph and employ GNN to gather neighbor information that is propagated across graph layers, enabling learning reliable node representations.

Although GNN-based CF methods have proven to be effective in various applications, they often encounter issues of data sparsity [25]. This is mainly because the explicit interactions in the data are limited compared to the overall dataset. Therefore, researchers have introduced contrastive learning(CL) to alleviate this issue by incorporating additional supervised signals [14,25,31]. CL-based methods focus on constructing effective contrastive views and encouraging consistent representations of the same nodes across multiple views, aiming to improve learning quality. In current research, data augmentation is a commonly used technique for constructing contrastive views [28], including structural enhancement [25] and feature enhancement [30]. However, such methods also come with inevitable drawbacks. Structural augmentation methods can disrupt the original dependencies within the graph, while feature augmentation methods introduce same-scale noise but overlook the distinctive features of individual nodes. Furthermore, these methods construct contrastive views solely based on the original interaction graph, neglecting higher-order dependencies(e.g., the similarity of user behaviors, and commonalities among items). These higher-order pieces of information are often implicit but crucial.

Recent research aims to use higher-order knowledge to improve node representation learning. However, studies such as NCL [14] and HCCF [26] leveraging such knowledge indirectly do not fully exploit its potential. While HGCL [1] and AdaMCL [31] explicitly construct higher-order interaction graphs (i.e., user-user and item-item graphs), it establishes an asymmetric contrastive that seeks consistency between the higher-order graph and the original graph. Importantly, the higher-order graph preserves one-sided higher-order dependencies (i.e., only user-side or item-side), whereas node representations in the original graph aggregate double-side content (both users and items). This asymmetric contrastive approach struggles to uncover higher-order semantic knowledge fully. Furthermore, in HGCL the construction of higher-order graphs relies on specific dataset formats, making it challenging to generalize to other datasets. Therefore, a more effective contrastive paradigm is necessary for mining and transferring higher-order knowledge.

In light of this, we introduce a novel CL paradigm in this paper for recommendation to address the issues above, named HoGCL[1] Specifically, two high-order

[1] HoGCL is available at https://github.com/zzDeepML/HoGCL.

interaction graphs, G_u and G_i, are constructed based on the original graph to aid in learning and transferring high-order knowledge. GNNs are utilized to iteratively aggregate information from neighboring nodes on the high-order graphs. HoGCL's contrastive strategy involves constructing a higher-order contrastive view and a general contrastive view. For the higher-order view, the primary objective is to address the previously mentioned issue of asymmetric contrast, aiming to enhance the learning effectiveness of higher-order knowledge while filtering out noise. And provides high-quality supervision signals for node representation learning. Figure 1 illustrates the distinction between the contrastive approach in HoGCL and the asymmetric contrastive approach. We group similar nodes in the semantic space using clustering to obtain corresponding cluster prototypes, which are considered higher-order neighbors. In simple terms, a prototype can be understood as the centroid of the neighbor cluster space, encapsulating the common characteristics of these nodes. In the high-order contrastive task, our objective is to pursue consistency between node representations learned from high-order interaction graphs and corresponding prototypes, continually updating and optimizing the results throughout the model's iterations. For the general view, its purpose is to transfer knowledge learned from high-order graphs while removing redundant information to enhance the reliability of node representations. The high-order graphs and the original graph have overlapping redundancies in their acquired knowledge, these redundancies not only increase the training cost of the model but also negatively impact model optimization. Furthermore, these negative impacts grow exponentially during the convolution process. Given this, we propose an adaptive fusion strategy determined by the convolution layers and degree of redundancy to control the receptiveness to higher-order information, thus filtering out redundancy. Finally, combining the fused representation with those in the original graph produces a contrasting view. This process enhances representation by transferring higher-order knowledge into node representations.

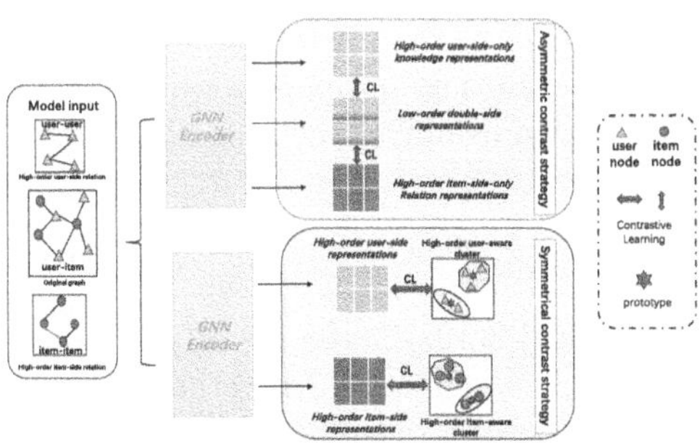

Fig. 1. Asymmetric Contrast and Symmetric Contrast

Extensive experiments on three public datasets have demonstrated the effectiveness of HoGCL. Our contributions can be summarized as follows:

- In this paper, we propose HoGCL, a self-supervised contrastive learning model that enhances node representation learning by explicitly uncovering relationship dependency knowledge within high-order interaction graphs.
- We address insufficient utilization of higher-order knowledge in existing research on asymmetric contrastive strategies by constructing a high-order contrastive view. The data sparsity problem is effectively mitigated by extracting supervised signals from high-order interactions.
- We propose an adaptive fusion strategy that utilizes higher-order knowledge and filters redundant information.
- Extensive experiments demonstrate the outstanding performance of HoGCL, which significantly outperforms existing neural CF models.

2 Preliminaries

In this section, we provide a brief overview of GNN-based Collaborative Filtering (CF) algorithms.

Adjacency Matrix Construction. Current GNN-based CF methods [8,22] typically models interactions between users and items as a bipartite graph $\mathcal{G} = \{\mathcal{U}, \mathcal{I}, R\}$, where $\mathcal{U}(|\mathcal{U}| = m)$ and $\mathcal{I}(|\mathcal{I}| = n)$ represents the set of user nodes and item nodes, respectively. and $R \in \mathbb{R}^{m \times n}$ is used to describe interactions between them. When there is an interaction between user u and item i, $R_{ui} = 1$; otherwise, 0. To facilitate matrix operations, we construct the adjacency matrix as described in NGCF [23].

$$A = \begin{pmatrix} \mathbf{0} & R \\ R^T & \mathbf{0} \end{pmatrix} \tag{1}$$

where R^T denotes the transpose matrix of R.

Propagation and Aggregation. The update of node representations is achieved by aggregating messages from neighboring nodes, and this process can be divided into two stages: message propagation and aggregation. Taking the update of the user node as an example, the representations e_u^l at the l-th layer can be expressed as follows:

$$e_u^l = f_{prop}(e_u^{l-1}, f_{agg}(\{e_i^{l-1} | i \in N_u\})) \tag{2}$$

where N_u represents the neighboring nodes set of u. The neighbors' representations at the $(l-1)$-th layer are aggregated together using $f_{agg}(\cdot)$. These aggregated representations, along with the representation of user u e_u^{l-1}, are then propagated to the l-th layer by the function $f_{prop}(\cdot)$, effectively completing the node update. To obtain a more comprehensive semantic representation, it is common to combine the representations produced by multiple layers of the network as the final representation of node u:

$$e_u = f_{readout}(e^0, e^1, \cdots, e^{l-1}, e^l) \tag{3}$$

$f_{readout}(\cdot)$ can be a simple weighted average. We can use a similar approach to obtain the item representation e_i. Let e_u and e_i denote representations of the user and item, respectively. By utilizing inner products, we can predict how much a user likes an item based on the learned node representations:

$$\hat{y}_{ui} = e_u^T e_i \tag{4}$$

In summary, the goal of GNN-based CF methods is to learn high-quality node representations for more accurate preference prediction.

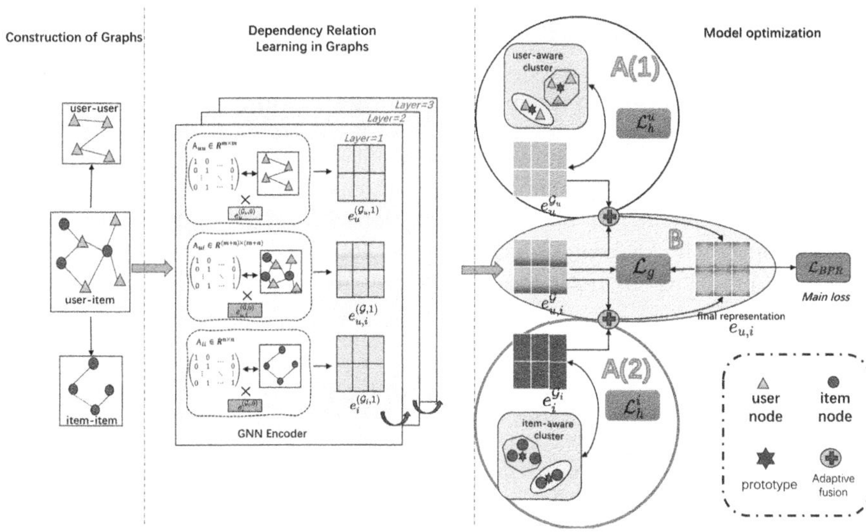

Fig. 2. Overall framework of HoGCL

3 The Proposed Method

This section provides a detailed description of HoGCL. Figure 2 shows its overall architecture.

3.1 Construction of High-Order Graphs

The problem of sparse supervision affects node representation learning. To obtain high-quality and reliable node representations, we conduct mining on the original graph to uncover higher-order interaction patterns that provide more supervision signals. There are various ways to construct higher-order graphs [1,12,31], and based on the most naive CF concept, we employ the Jaccard similarity coefficient $\mathcal{J}(\cdot)$ [7], as follows:

$$\mathcal{G}_u = \begin{cases} \mathcal{J}(a,b), & \text{if } \mathcal{J}(a,b) \geq \delta \\ 0, & otherwise. \end{cases} \tag{5}$$

$$\mathcal{J}(a,b) = \frac{|N_a| \cap |N_b|}{|N_a| \cup |N_b|} \tag{6}$$

where N_a and N_b denote neighbor node sets of user a and b, respectively. δ is the threshold used to filter out noise. $\mathcal{J}(\cdot)$ is used to calculate the similarity between two users. Item-side relationship $\mathcal{G}_i$ is constructed in the same way.

3.2 Message Propagation and Knowledge Fusion

Message Propagation. The GNN-based CF algorithm improves node representations by aggregating messages from neighboring nodes in the graph, which is critical for high-quality representations. Building upon previous works [9,25], we adopt the message propagation method used in LightGCN [9] to learn node representations in both the original graph $\mathcal{G}$ and the high-order graph $\mathcal{G}_h = \{\mathcal{G}_u \cup \mathcal{G}_I\}$:

$$e_u^{\mathcal{G},l} = \sum_{i \in N_{\mathcal{G}}(u)} \frac{1}{\sqrt{|N_{\mathcal{G}}(u)| \times |N_{\mathcal{G}}(i)|}} e_i^{\mathcal{G},(l-1)} \tag{7}$$

$$e_j^{\mathcal{G}_h,l} = \sum_{i \in N_{\mathcal{G}_h}(j)} \frac{1}{\sqrt{|N_{\mathcal{G}_h}(j)| \times |N_{\mathcal{G}_h}(i)|}} e_i^{\mathcal{G}_h,(l-1)} \tag{8}$$

In this message propagation process, we have eliminated the cumbersome feature transformations and activation functions. For the node representation $e_j^{\mathcal{G}_h,l}$ at the l-th layer of the graph $\mathcal{G}_h$, its receptiveness to neighbor information is controlled based on the number of neighboring nodes (i.e., $N_{\mathcal{G}_h}(j)$ and $N_{\mathcal{G}_h}(i)$).

Adaptive Knowledge Fusion. Learning dependencies in the high-order graph can effectively improve the quality of node representations. Since the high-order graph is generated based on the original graph, it tends to contain overlapping information. Directly integrating these representations would increase computational complexity and amplify popularity bias [31]. Thus, we propose an adaptive fusion strategy based on convolutional layers and similarity.

$$\begin{cases} e_i^l = e_i^{\mathcal{G},l} + \beta_i \cdot e_i^{\mathcal{G}_h,l} \\ \beta_i = \frac{1}{\sqrt{l^2 + s^2(e_i^{\mathcal{G},l}, e_i^{\mathcal{G}_h,l})}} \end{cases} \tag{9}$$

$s(\cdot)$ represents cosine similarity, and we use it to indicate the degree of information redundancy between two representations. As the convolutional layers deepen, redundant information increases, amplifying errors. Therefore, there is a negative correlation between the high-order relationship acceptance coefficient β and the network layer l and $s(\cdot)$. When the adaptive approach is used, it becomes possible to effectively eliminate redundant information within nodes, while also controlling popularity bias, allowing for more effective utilization of high-order dependency relationships.

After performing the multi-layer graph convolution operations, we obtain node representations $\{e^1, ..., e^l\}$ from each layer, which incorporates high-order dependency knowledge. These representations are then combined with the initial representation (e^0) through a weighted average to obtain the final node representation.

$$e_u = \frac{1}{L+1}\sum_{l=0}^{L} e_u^l,\ e_i = \frac{1}{L+1}\sum_{l=0}^{L} e_i^l \tag{10}$$

then we can predict the user's preference for an item by using vector inner products as mentioned in the preliminary.

3.3 Contrastive Learning for High-Order View

Compared to the original graph, interactions on the high-order graph are sparser, making it more challenging to learn node dependencies due to the lack of explicit supervision signals. In previous studies, such as HGCL [1], a direct contrast is made between the nodes obtained from the high-order graph and nodes from the original graph. Clearly, this approach does not fully leverage high-order knowledge because high-order interactions involve deeper homogenous relationships (i.e., contain only user-side or item-side information), whereas the original graph contains both user and item information. Such an asymmetrical contrastive view can't effectively explore high-order knowledge. To address the issue of high-order asymmetrical contrast, we propose a contrastive learning paradigm in the high-order view.

Figure 2 (Part A) demonstrates the high-order contrastive view. Taking the example of user-side contrast in Part A(1), we first form K_u neighbor clusters through clustering. Each cluster is associated with a prototype that effectively reflects the semantic characteristics of the corresponding class. Our goal is to create a symmetric contrastive by considering the user node and the class prototype as positive contrastive pairs, maximizing their consistency to uncover high-order relationships. Building on previous research [13,14,28], the EM algorithm is used during the clustering process to learn clustering prototypes by maximizing the following likelihood function:

$$\hat{\Theta}_k = argmax\sum_{u=1}^{m} logp(e_u; \Theta_k) = argmax\sum_{u=1}^{m} log\sum_{c_u \in C_u} p(e_u, c_u; \Theta_k) \tag{11}$$

where Θ_k represents the set of model parameters used to cluster, and c_u represents the prototype corresponding to the user e_u.

After obtaining K_u prototypes, we construct the high-order contrast task on the user side based on InfoNCE [16] as follows:

$$\mathcal{L}_h^u = \sum_{u \in \mathcal{U}} -log \frac{exp(\mathbf{e}_u{}^T \mathbf{c}_i / \tau)}{\sum_{C_j \in C} exp(\mathbf{e}_u{}^T \mathbf{c}_j / \tau)} \tag{12}$$

where C_j denotes the prototype set, τ represents the temperature hyperparameter, and c_i represents the prototype corresponding to u. Using a similar approach, we obtain the loss $\mathcal{L}_h^i$ on the item side.

Finally, we define the optimization objective for high-order contrastive learning as follows:

$$\mathcal{L}_h = \mathcal{L}_h^u + \alpha \mathcal{L}_h^i \tag{13}$$

α is the hyperparameter that balances the weight between the user side and the item side. By constructing symmetric contrastive learning tasks, we can effectively learn the dependency relationships on high-order graphs.

3.4 Contrastive Learning for General View

In existing research, the construction of contrastive learning tasks is typically based on observable user-item interactions, lacking the utilization of high-order interaction. To address this issue, we propose an approach similar to feature enhancement. By leveraging the dependency relationships learned from high-order graphs, we construct a general contrastive view as illustrated in Part B of Fig. 2. Specifically, we adaptively fuse the knowledge learned from high-order relationships with the nodes in the original graph. Subsequently, We consider the fused nodes and the nodes in the original graph as positive contrastive pairs. This provides supervisory signals for transferring high-order knowledge into node representations, thereby enhancing the quality of node representation learning.

Let $e_u^{\mathcal{G}}$ represent the node representations learned in the original graph $\mathcal{G}$:

$$e_u^{\mathcal{G}} = \frac{1}{L} \sum_{i=1}^{L} e_u^{\mathcal{G},l} \tag{14}$$

We form positive contrast pairs using $e_u^{\mathcal{G}}$ and the fused node representations e_u from Eq. 10, aiming to maximize the consistency between the two representations. our optimization objective on the user side can be formalized as follows:

$$\mathcal{L}_g^u = \sum_{u \in \mathcal{U}} -log \frac{exp(\mathbf{e}_u{}^T \mathbf{e}_u^{\mathcal{G}} / \tau)}{\sum_{j \in \mathcal{U}} exp(\mathbf{e}_u{}^T \mathbf{e}_j^{\mathcal{G}} / \tau)}, \tag{15}$$

We can obtain the optimization objective on the item side $\mathcal{L}_g^i$ in a similar way.

Similar to the approach in the high-order view, we can define the optimization objective for the contrastive learning task on the generic view as follows:

$$\mathcal{L}_g = \mathcal{L}_g^u + \alpha \mathcal{L}_g^i \tag{16}$$

3.5 Optimization

To better train the model, we first define the BPR loss function [18] as the primary loss function, which is commonly used in recommendation tasks:

$$\mathcal{L}_{bpr} = \sum_{(u,i^+,i^-)\in\mathcal{O}} -ln(sigmoid(\hat{y}_{ui^+} - \hat{y}_{ui^-})) \tag{17}$$

Table 1. Statistics of the datasets

Datasets	#Users	#Items	#Interactions	Density
Yelp	45,478	30,709	1,777,765	0.00127
Gowalla	29,859	40,989	1,027,464	0.00084
Alibaba	300,000	81,614	1,607,813	0.00007

where $\mathcal{O}$ represents the set of all nodes in the batch, i^+ denotes items interacted with u, and i^- represents items with no interactions.

Combining the two contrastive tasks in the general view and high-order view, the total loss function for the HoGCL model is defined as follows:

$$\mathcal{L} = \mathcal{L}_{bpr} + \lambda_1\mathcal{L}_g + \lambda_2\mathcal{L}_h + \lambda_3||\Theta||^2 \tag{18}$$

where Θ represents the set of trainable parameters in the model, λ_1, λ_2, and λ_3 are hyperparameters used to control the importance assigned to each part.

4 Experiments

4.1 Experimental Settings

Dataset. To evaluate the performance of HoGCL, we conduct extensive experiments on three datasets: Alibaba [3], Yelp[2], and Gowalla-merged [4]. Table 1 presents the statistical information for these datasets. We split the datasets into training, validation, and test sets in an 8:1:1 ratio.

Baseline Models. We compared our model with the following baseline models, which encompass a variety of traditional models, GNN-based models, and Graph CL-based models:

- **BPRMF** [18] employs matrix factorization to optimize the BPR loss, enabling the learning of latent representations for users and items.
- **NeuMF** [10] employs a multi-layer perceptron and matrix factorization approach to learn the relationships between users and items.

[2] https://www.yelp.com/dataset.

Table 2. Statistics of the hyperparameters

Hyperparameter	#Value
Batch_size	4096
Embedding_size	64
δ	0.8
λ_1&λ_2& λ_3	[1e-10, 1e-4]
k	[100,2000]
τ	[0.05, 0.2]

• **NGCF** [23] utilizes GNN to address collaborative filtering tasks by designing how messages are propagated and aggregated among nodes in the graph.
• **Multi-GCCF** [19] learns relationships between nodes from both the original graph and high-order graphs to obtain high-quality node representations.
• **DGCF** [24] improves node quality and model interpretability by decoupling complex user intents.
• **LightGCN** [9] eliminates the complex feature transformation and activation functions used in NGCF, resulting in improved model performance.
• **SGL** [25] introduces a self-supervised contrastive learning paradigm to enrich the supervisory signals. We adopted SGL-ED as the instantiation of SGL.
• **SimGCL** [30] introduces noise into the node embedding space as a substitute for data augmentation methods to create contrastive learning tasks.
• **NCL** [14] utilizes both the semantic space and graph structural neighbors to generate contrastive pairs.
• **CGCL** [8] improves recommendation effectiveness by learning relationships between user nodes and candidate items across different layers.
• **AdaMCL** [31] has proposed an adaptive multi-view fusion approach to leverage high-order knowledge.

Evaluation Metrics. We employ widely used evaluation metrics in the Top-N recommendation task, including Recall@N and NDCG@N, where N was set to 10, 20, and 50. Following NCL [14], we employed a full-ranking strategy, ranking all candidate items for users with whom they have had no interactions.

Implementation Details. HoGCL and the baseline models used in the experiments were implemented using RecBole. All models are optimized and fine-tuned using the Adam optimizer to find the best parameter settings. Parameters are initialized using Xavier [5]. Table 2 provides a summary of the hyperparameters used in the models.

4.2 Overall Performance Comparison

Table 3. The Overall Performance Comparison of Different Models. The best result is **bolded** and the second best is underlined.

Dataset	Metric	BPRMF	NeuMF	NGCF	MultiGCCF	DGCF	LightGCN	SGL	SimGCL	NCL	CGCL	AdaMCL	HoGCL	Improv.
Yelp	Recall@10	0.0643	0.0531	0.0630	0.0646	0.0723	0.0730	0.0833	0.0818	0.0920	0.0969	0.0939	**0.1374**	+41.80%
	NDCG@10	0.0458	0.0377	0.0446	0.0450	0.0514	0.0520	0.0601	0.0606	0.0678	0.0702	0.0712	**0.1105**	+55.20%
	Recall@20	0.1043	0.0885	0.1026	0.1053	0.1135	0.1163	0.1288	0.1229	0.1377	0.145	0.1400	**0.1896**	+30.76%
	NDCG@20	0.0580	0.0486	0.0567	0.0575	0.0641	0.0652	0.0739	0.0731	0.0817	0.085	0.0851	**0.1261**	+48.18%
	Recall@50	0.1862	0.1654	0.1864	0.1882	0.1989	0.2016	0.2140	0.1992	0.2247	0.2394	0.2297	**0.2836**	+18.46%
	NDCG@50	0.0793	0.0685	0.0784	0.0790	0.0862	0.0875	0.0964	0.0929	0.1046	0.1097	0.1085	**0.1506**	+37.28%
Gowalla	Recall@10	0.1158	0.1039	0.1005	0.1108	0.1252	0.1362	0.1465	0.1228	0.1500	0.1537	0.1613	**0.2344**	+45.32%
	NDCG@10	0.0833	0.0731	0.0713	0.0791	0.0902	0.0876	0.1048	0.0876	0.1082	0.1102	0.1179	**0.1749**	+48.35%
	Recall@20	0.1695	0.1535	0.1572	0.1626	0.1829	0.1976	0.2084	0.1732	0.2133	0.2212	0.2278	**0.3034**	+33.19%
	NDCG@20	0.0988	0.0873	0.0893	0.0940	0.1066	0.1152	0.1225	0.1022	0.1265	0.1298	0.1368	**0.1948**	+42.40%
	Recall@50	0.2756	0.2510	0.2546	0.2631	0.2877	0.3044	0.3197	0.2632	0.3259	0.3392	0.3438	**0.4089**	+18.94%
	NDCG@50	0.1450	0.1110	0.1130	0.1184	0.1322	0.1414	0.1497	0.124	0.1542	0.1588	0.1653	**0.2214**	+33.94%
Alibaba-iFashion	Recall@10	0.303	0.182	0.0371	0.0401	0.0447	0.0468	0.0461	0.0246	0.0477	0.055	0.0632	**0.2376**	+275.95%
	NDCG@10	0.0161	0.0092	0.0193	0.0207	0.0241	0.025	0.0248	0.0131	0.0259	0.0301	0.0351	**0.1857**	+429.06%
	Recall@20	0.0467	0.0302	0.0589	0.0634	0.0677	0.0708	0.0692	0.0038	0.0713	0.0811	0.0917	**0.2700**	+194.44%
	NDCG@20	0.0203	0.0123	0.0248	0.0266	0.0299	0.0311	0.0307	0.0165	0.0319	0.0367	0.0423	**0.1940**	+358.63%
	Recall@50	0.0799	0.0576	0.1035	0.1107	0.1120	0.1169	0.1141	0.0635	0.1165	0.1305	0.1413	**0.3194**	+126.04%
	NDCG@50	0.0269	0.0177	0.0337	0.0360	0.0387	0.0402	0.0396	0.0216	0.0409	0.0465	0.0522	**0.2039**	+290.61%

Table 3 shows a performance comparison between the proposed HoGCL and baseline models. HoGCL achieves outstanding results on all datasets and outperforms other models with absolute advantage, especially on sparser datasets like Alibaba, where the performance improvement is more pronounced(over 200%).

From the experimental results, it is clear that traditional methods like NeuMF perform significantly worse than GNN-based methods. The reason behind this is traditional methods only utilize low-order neighbor interactions, hence limiting the node representation capabilities significantly. Among all GNN-based methods, LightGCN achieved the best performance by eliminating complex feature transformations and activation functions that affect the quality of node learning. It is worth noting that although MultiGCCF utilizes higher-order relationships, its excessive redundancy amplifies errors, leading to poor performance.

Among the three types of models, Graph CL-based models exhibit significantly superior overall performance, thanks to the introduction of contrastive learning, which adds valuable supervised signals and greatly alleviates the issue of data sparsity. Within this category of models, NCL, CGCL, and AdaMCL outperform the others. This is because the data augmentation approach in SGL somewhat disrupts crucial node features. At the same time, NCL constructs the contrastive task by introducing higher-order features, preserving the original characteristics of the nodes. On the other hand, SimGCL introduces the same scale noise without considering the differences between nodes, leading to slightly lower performance compared to SGL.

HoGCL demonstrates the best performance among all models, especially on sparser datasets. This is attributed to the effective utilization of higher-order information. By constructing high-order contrastive views to unearth high-order dependencies on the graph, and with the aid of signals provided by general contrastive views, higher-order knowledge can be seamlessly incorporated into node representations. The utilization of the adaptive fusion strategy in combination leads to enhanced exploitation of information and improved recommendation performance.

4.3 Further Analysis of HoGCL

We carry out additional experiments in this section to analyze the individual components of the HoGCL model and examine their impact on the overall performance. We use the moderately sparse Gowalla-merged dataset and the excessively sparse Alibaba dataset for the experiments to ensure their reliability.

Table 4. Ablation study on high-order and general contrastive learning.

Model	Alibaba		Gowalla	
	Recall@20	NDCG@20	Recall@20	NDCG@20
HoGCL	0.2700	0.1940	0.3034	0.1948
w/o h	0.2575	0.1700	0.2603	0.1533
w/o g	0.2573	0.1698	0.2193	0.1269
LightGCN	0.0708	0.0311	0.1976	0.1152

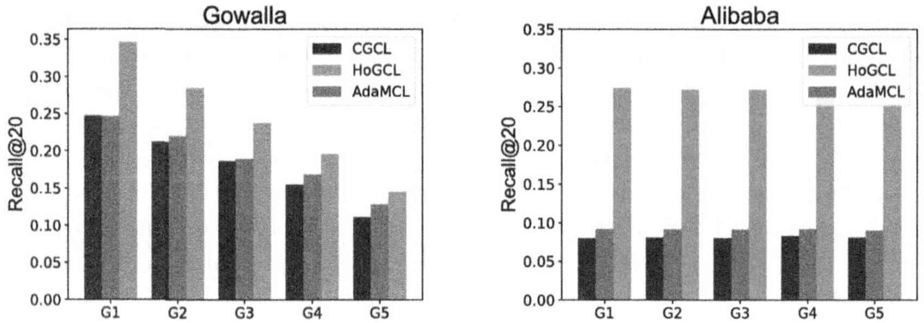

Fig. 3. Performance analysis for different sparsity-level users. $G1$-$G5$ denotes the different average number of interactions.

Ablation Study of HoGCL. In HoGCL, we construct two contrastive tasks: a general one and a higher-order one. To investigate the contributions of these two contrastive learning tasks, we conduct ablation experiments, and the results are shown in Table 4. Here, 'w/o g' and 'w/o h' represent the removal of the general contrastive task and the higher-order contrastive task, respectively. The results demonstrate the vital roles played by both higher-order contrast and general contrast in model performance. Removing any of these components results in a significant drop in effectiveness. Compared with LightGCN, it is evident that the introduction of higher-order dependency relationships can effectively enhance the learning quality of nodes in the graph. However, solely relying on individual contrastive tasks does not significantly improve performance. This is because the high-order contrast task captures high-order dependency relationships between nodes, while the general contrast task facilitates the transfer of this knowledge into node representations. By combining these two tasks, we can achieve the maximum improvement in model performance, as they are complementary to each other.

The Effect of Sparsity. We conduct experiments inspired by NCL to examine HoGCL's performance with sparse data. Users are grouped based on interaction sparsity, with equal total interactions per group. We compare HoGCL with CGCL and AdaMCL models, which also utilize high-order relationships. The results are presented in Fig. 3. The results show that there is minimal performance distinction among different groups in the Alibaba dataset due to its inherent sparsity, whereas the Gowalla dataset exhibits significant variations among groups. Compared with CGCL and AdaMCL, HoGCL exhibits a more pronounced advantage as interaction sparsity increases, especially with the highly sparse Alibaba dataset. This suggests that HoGCL effectively reveals higher-order dependencies in the original graph, crucial for addressing challenges posed by sparse data.

Symmetry vs Asymmetry. To enhance node representation and overcome limitations observed in previous studies [1,31], we introduce symmetric contrast in HoGCL. We validate its performance against asymmetric contrast in a controlled experiment, as shown in Table 5. In the experiment, asymmetric contrast is implemented in the form of considering the same node in the high-order graph and the original graph as positive pairs. We also include the AdaMCL [31] model, which has the asymmetric contrastive task. From the experimental results, we can observe that the proposed symmetric contrastive approach in HoGCL can better unearth high-order knowledge. It outperforms asymmetric contrast in all metrics, particularly on the sparser Alibaba dataset. This shows that symmetric comparison can provide a higher quality Supervisory signal, which can better solve the problem of data sparsity than asymmetric.

Hyperparameters Sensitivity Impact of k. Figure 4 shows the influence of k on the HoGCL, where $k = 0$ indicates the removal of the high-order contrastive objective. We can observe that when $k < 100$, relative to the thousands of nodes, having too few user (or item) clusters makes it challenging to separate nodes of different types, resulting in poor model performance. However, this does not necessarily imply that a higher cluster number is better. Larger k implies increased clustering computation costs without enhancing the model's performance. Through experiments, we find that the k around 500 yields satisfactory model results.

Table 5. Performance comparison w.r.t. Symmetry vs Asymmetry.

Model	Alibaba			Gowalla		
	Symmetry(HoGCL)	Asymmetry	AdaMCL	Symmetry(HoGCL)	Asymmetry	AdaMCL
Recall@10	**0.2376**	0.2183	0.0623	**0.2344**	0.2055	0.1613
Recall@20	**0.2700**	0.2366	0.0917	**0.3034**	0.2811	0.2278
Recall@50	**0.3194**	0.2989	0.1413	**0.4089**	0.4010	0.3438
NDCG@10	**0.1857**	0.1482	0.0351	**0.1749**	0.1504	0.1179
NDCG@20	**0.1940**	0.1580	0.0423	**0.1948**	0.1719	0.1368
NDCG@50	**0.2039**	0.1622	0.0522	**0.2214**	0.2016	0.1653

Impact of τ. The τ defined in Eqs. (12) and (15), plays a crucial role in mining hard negative samples. To identify the appropriate τ, we conduct the experiments with τ ranging from 0.04 to 0.2, and the results are depicted in Fig. 4. Results indicate that the model performs relatively well when τ is in [0.08, 0.12]. When τ is smaller, it amplifies the differences in softmax outputs, increasing the discriminative capacity of samples may also lead to a situation where a small number of negative samples dominate the optimization. Conversely, if τ is too large, it may not provide sufficient discriminative power. Therefore, for HoGCL, it is appropriate to control τ at 0.1 ± 0.02.

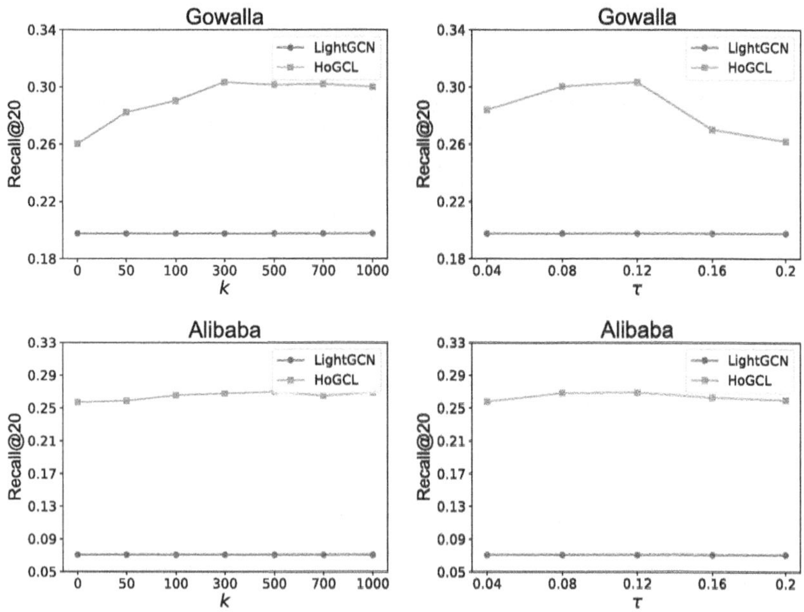

(a) (b)

Fig. 4. Performance comparison w.r.t. different τ and k.

5 Related Work

Graph Neural Network in Recommendation. Early matrix factorization-based recommendation models [6,11,27] combine multilayer perceptrons with matrix factorization to capture collaborative signals. However, such an approach has significant limitations as it relies solely on first-order neighbor information. The advent of GNN has addressed this issue. GNN-based models like NGCF [23] iteratively aggregate knowledge from neighboring nodes to introduce high-order signals. DGCF [24] disentangles complex user intentions to obtain fine-grained embeddings, enhancing recommendation performance. However, these methods often require multiple feature transformations, which can increase the cost of training without providing corresponding benefits. Therefore, LightGCN [9] was introduced, removing feature transformations and activation functions, and proposing a lightweight and more effective recommendation model. IPM-GCN [15] further enhances LightGCN by clustering similar nodes. Although these methods have shown good results, they still face difficulties in dealing with data sparsity. As a solution, CL has been incorporated into recommendation tasks in recent years to address the issue of data sparsity.

Contrastive Learning in Recommendation. Thanks to the success of computer vision [2,21], contrastive learning(CL) has been introduced into recommendation [8,20,25] to address the issue of sparse supervision. It maximizing consistency of positive pairs provides supervised signals to learn high-quality node representations. Among these efforts, the impact of SGL [25] is the most significant. As a result, research on CL, such as simGCL [30] and HCCF [26], has advanced. NCL [14] builds contrastive tasks from both structural and semantic neighbor perspectives, further advancing the study of graph CL. Subsequently, the work of HGCL [1] introduces heterogeneous relationships to construct contrastive views, explicitly leveraging higher-order dependencies. However, in above models, HGCL has certain limitations. Its construction of heterogeneous relationships relies on specific datasets, and there are asymmetric pair issues during the CL process. Other models also lack a clear exploration of high-order dependency relationships. To address these issues and improve the effectiveness of recommendation models, we propose HoGCL, which explicitly explores high-order relationships in the original graph and combines CL to enhance the quality of node representation learning.

6 Conclusion

This paper introduces a novel contrastive learning paradigm called HoGCL for recommendation. In HoGCL, we use higher-order knowledge to improve the quality of node representations. This is achieved by utilizing high-order contrastive view and general contrastive view. We employ an adaptive fusion strategy to effectively transfer knowledge obtained from higher-order graphs into node representations. This enhances the quality of node embeddings and improves the recommendation performance of the model. Extensive experiments demonstrate the effectiveness of HoGCL.

Acknowledgements. This work is supported by the National Key R&D Program of China (Grant No. 2021YFF0901300), "20 New Universities" Project of Jinan City (No. 202228077), Major innovation projects of the pilot project of science, education and industry integration (2022JBZ01-01), Taishan Scholars Program (tsqn202211203) and Taishan Industrial Experts Program(No. tscy20231203).

References

1. Chen, M., Huang, C., Xia, L., Wei, W., Xu, Y., Luo, R.: Heterogeneous graph contrastive learning for recommendation. In: WSDM, pp. 544–552. ACM (2023)
2. Chen, T., Kornblith, S., Norouzi, M., Hinton, G.E.: A simple framework for contrastive learning of visual representations. In: ICML Proceedings of Machine Learning Research, vol. 119, pp. 1597–1607. PMLR (2020)
3. Chen, W., et al.: POG: personalized outfit generation for fashion recommendation at Alibaba ifashion. In: KDD, pp. 2662–2670. ACM (2019)
4. Cho, E., Myers, S.A., Leskovec, J.: Friendship and mobility: user movement in location-based social networks. In: KDD, pp. 1082–1090. ACM (2011)
5. Glorot, X., Bengio, Y.: Understanding the difficulty of training deep feedforward neural networks. In: AISTATS JMLR Proceedings, vol. 9, pp. 249–256. JMLR.org (2010)
6. Guo, H., Tang, R., Ye, Y., Li, Z., He, X.: Deepfm: a factorization-machine based neural network for CTR prediction. In: IJCAI, pp. 1725–1731. ijcai.org (2017)
7. Hamers, L., et al.: Similarity measures in scientometric research: Jaccard index versus Salton's cosine formula. Inf. Process. Manag. **25**(3), 315–318 (1989)
8. He, W., Sun, G., Lu, J., Fang, X.S.: Candidate-aware graph contrastive learning for recommendation. In: SIGIR, pp. 1670–1679. ACM (2023)
9. He, X., Deng, K., Wang, X., Li, Y., Zhang, Y., Wang, M.: Lightgcn: simplifying and powering graph convolution network for recommendation. In: SIGIR, pp. 639–648. ACM (2020)
10. He, X., Liao, L., Zhang, H., Nie, L., Hu, X., Chua, T.: Neural collaborative filtering. In: WWW, pp. 173–182. ACM (2017)
11. Koren, Y., Bell, R.M., Volinsky, C.: Matrix factorization techniques for recommender systems. Computer **42**(8), 30–37 (2009)
12. Li, B., Guo, T., Zhu, X., Li, Q., Wang, Y., Chen, F.: SGCCL: siamese graph contrastive consensus learning for personalized recommendation. In: WSDM, pp. 589–597. ACM (2023)
13. Lin, S., et al.: Prototypical graph contrastive learning. IEEE Transactions on Neural Networks and Learning Systems (2022)
14. Lin, Z., Tian, C., Hou, Y., Zhao, W.X.: Improving graph collaborative filtering with neighborhood-enriched contrastive learning. In: WWW, pp. 2320–2329. ACM (2022)
15. Liu, F., Cheng, Z., Zhu, L., Gao, Z., Nie, L.: Interest-aware message-passing GCN for recommendation. In: WWW, pp. 1296–1305. ACM/IW3C2 (2021)
16. van den Oord, A., Li, Y., Vinyals, O.: Representation learning with contrastive predictive coding. ArXiv:1807.03748 (2018)

17. Rendle, S.: Factorization machines. In: 2010 IEEE International Conference on Data Mining (2010)
18. Rendle, S., Freudenthaler, C., Gantner, Z., Schmidt-Thieme, L.: Bpr: Bayesian personalized ranking from implicit feedback. arXiv:1205.2618
19. Sun, J., et al.: Multi-graph convolution collaborative filtering. In: ICDM, pp. 1306–1311. IEEE (2019)
20. Tang, H., Zhao, G., Wu, Y., Qian, X.: Multisample-based contrastive loss for top-k recommendation. IEEE Transactions on Multimedia (2021)
21. Wang, F., Liu, H.: Understanding the behaviour of contrastive loss. In: CVPR, pp. 2495–2504. Computer Vision Foundation/IEEE (2021)
22. Wang, H., et al.: Knowledge-adaptive contrastive learning for recommendation. In: WSDM, pp. 535–543. ACM (2023)
23. Wang, X., He, X., Wang, M., Feng, F., Chua, T.: Neural graph collaborative filtering. In: SIGIR, pp. 165–174. ACM (2019)
24. Wang, X., Jin, H., Zhang, A., He, X., Xu, T., Chua, T.: Disentangled graph collaborative filtering. In: SIGIR, pp. 1001–1010. ACM (2020)
25. Wu, J., et al.: Self-supervised graph learning for recommendation. In: SIGIR, pp. 726–735. ACM (2021)
26. Xia, L., Huang, C., Xu, Y., Zhao, J., Yin, D., Huang, J.X.: Hypergraph contrastive collaborative filtering. In: SIGIR, pp. 70–79. ACM (2022)
27. Xue, H., Dai, X., Zhang, J., Huang, S., Chen, J.: Deep matrix factorization models for recommender systems. In: IJCAI, pp. 3203–3209. ijcai.org (2017)
28. Yang, Y., et al.: Generative-contrastive graph learning for recommendation. In: SIGIR, pp. 1117–1126. ACM (2023)
29. Ying, R., He, R., Chen, K., Eksombatchai, P., Hamilton, W.L., Leskovec, J.: Graph convolutional neural networks for web-scale recommender systems. In: KDD, pp. 974–983. ACM (2018)
30. Yu, J., Yin, H., Xia, X., Chen, T., Cui, L., Nguyen, Q.V.H.: Are graph augmentations necessary?: Simple graph contrastive learning for recommendation. In: SIGIR, pp. 1294–1303. ACM (2022)
31. Zhu, G., Lu, W., Yuan, C., Huang, Y.: Adamcl: adaptive fusion multi-view contrastive learning for collaborative filtering. In: SIGIR, pp. 1076–1085. ACM (2023)

FNDPro: Evaluating the Importance of Propagations during Fake News Spread

Herun Wan[1], Ningnan Wang[2], Xiang Zhao[3], Rui Li[4], Hui Yang[5], and Minnan Luo[1](✉)

[1] School of Computer Science and Technology, Xi'an Jiaotong University, Xi'an, China
minnluo@xjtu.edu.cn

[2] College of Artificial Intelligence, Xi'an Jiaotong University, Xi'an, China
wanherun@stu.xjtu.edu.cn

[3] National University of Defense Technology, Changsha, China

[4] School of Continuing Education, Xi'an Jiaotong University, Xi'an, China

[5] The 30th Research Institute of China Electronics Technology Group Corporation, Chengdu, China

Abstract. Existing fake news detection models fall into two categories: content-based and graph-based. Content-based models identify fake news depending on news content, which may fail to determine fake news with disguised content. Graph-based models adopt extra media to construct graphs, which provide social context to identify fake news. However, existing graph-based models treat each media equally, neglecting the echo chamber phenomenon where most media have the same opinion or similar content. A model that can dynamically evaluate the contribution of each propagation and find critical ones during news spread is needed. To this end, we proposed FNDPro, which models the news propagation process as a heterogeneous dynamical graph. The key is that it models news as the first propagation and ℓ-hop neighbors as the $(\ell + 1)$-th propagation. FNDPro contains a multi-modality encoder to encode each media and a propagation encoder to encode each propagation. FNDPro then employs a propagation transformer module to make every propagation embedding interact and obtain the importance score of each propagation. FNDPro achieves the best performance on three real-world datasets. Further experiments show the propagation transformer is helpful. Notably, FNDPro shows great generalization capabilities and can detect fake news even when news media are limited and manipulated. (Resources are available at https://github.com/whr000001/FNDPro.)

Keywords: Fake News Detection · Social Network · Graph Neural Networks

1 Introduction

With the development of social media, the spread of misinformation and disinformation is at an all-time high. Fake news contains intentional false information

M. Onizuka et al. (Eds.): DASFAA 2024, LNCS 14855, pp. 52–67, 2024.
https://doi.org/10.1007/978-981-97-5572-1_4

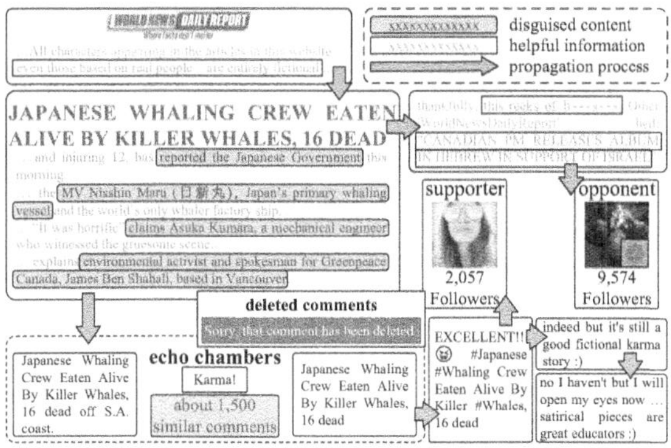

Fig. 1. An example of a fake news propagation process. It is disguised as real news by adding made-up information, which escapes detection from content-based models. Meanwhile, fake news publishers reinforce echo chambers by deleting debunking comments and using social bots, which confuses traditional graph-based models.

and can disrupt social order. Since widespread fake news has harmful societal effects, there is an urgent need to develop fake news detection methods. Traditional methods focus on news content, including texts [9] or pictures [40], to identify fake news with feature engineering [27] or neural networks [30]. These methods believe that word choice, language style, or sentiment of fake news is distinct from real ones, which leads to content-based models that can identify fake news. But, to confuse people, fake news publishers often make content and pictures of fake news similar to the real one [33]. Fake news publishers tend to disguise fake news as authentic reporting by adding made-up sources such as "CNN" or "Rachel Maddow" [24]. They may also post authentic but misleading images of artwork or ones from outdated news to confuse people [40]. Meanwhile, there may be noisy correspondences between the text and images of news articles [4]. As a result, content-based models fail to identify fake news disguised as real news. Besides disguised news content, people could identify fake news through different media containing sources, comments, and users during the spread. To this end, models that adopt social context information are proposed, which usually construct graphs to extract and fuse features from different media [30,33]. These models usually are called graph-based models, which employ graph neural networks to tackle topological information from graphs [5,25].

Existing graph-based models treat each node equally, ignoring the echo chamber phenomenon, where most media have the same opinion or similar content. Meanwhile, fake news publishers may reinforce echo chambers by deleting debunking comments [15] or leveraging social bots [10,18] to spread repeated fake news in large numbers. As a result, existing graph-based models treating the comments individually and homogeneously neglect the possibility that there are missing comments because of opinion manipulation.

Figure 1 illustrates an example of a fake news spread, where the news content is disguised, and publishers reinforce echo chambers to confuse people. This results in very little information during the propagation process to help identify fake news. To correctly identify fake news in this situation, graph-based models need to find out what matters during the propagation of fake news, namely, which propagation is helpful. Overall, a model that can identify disguised fake news, dynamically evaluate the contribution of related media, and point out which propagation processes are critical for fake news detection is needed.

In this paper, we propose **FNDPro**, a Fake News Detection framework that evaluates the importance of **Pro**pagations during fake news spread. FNDPro models the news spread process as a heterogeneous dynamic graph dubbed news propagation network consisting of news, source, user, comment media, and different edges among the media. As a result, FNDPro can extract heterogeneous information from the network to identify disguised fake news. When news is published, it spreads across social networks through the edges between different media. To this end, FNDPro treats news as the first propagation and ℓ-hop neighbors of it as $(\ell+1)$-th propagation, so that finding what matters during the spread process of news is to evaluate the contribution of each propagation. FNDPro designs a multi-modality encoder to extract and fuse node multi-modality information as the first propagation embedding. FNDPro adopts time encoding and graph neural networks to extract temporal and spatial information during propagation. Note that, after ℓ GNN layers, a node can aggregate the first to $(\ell+1)$-hop neighbor features, so that FNDPro treats the output of ℓ-th GNN layer as the $(\ell+1)$-th propagation embedding. To dynamically evaluate the contribution of different propagation, FNDPro designs a propagation transformer module, which can make each propagation embedding attend to the information at each propagation and then process its representation. The attention weights reflect the importance of each propagation, where the higher the weight, the more critical the propagation. Our main contributions are summarized as follows:

- We model the news propagation process as a heterogeneous dynamic graph, which provides additional information to identify disguised fake news. Meanwhile, we define news propagation and find critical propagation during news spread to solve echo chamber problems.
- We propose a novel fake news detection framework, which exploits an end-to-end neural network to encode each node in the news propagation network and learn representations of each news propagation. FNDPro dynamically evaluates the contribution of each propagation, which enables it to find out what matters during the news propagation process.
- We conduct extensive experiments on real-world datasets to prove the effectiveness of FNDPro, which demonstrates that FNDPro is better than existing models. Ablation studies show every module of FNDPro helps to identify fake news. Further experiments prove that FNDPro can identify fake news in more realistic situations.

2 Related Work

Existing fake news detectors fall into content-based [9] and graph-based [41].

2.1 Content-Based Models

Previous works on fake news detection mainly focus on news content. Castillo et al. [3] use features from users' posting and retweeting behavior, tweet content, and URLs. Popat et al. [28] propose a classifier that uses factors to assess the credibility of the claim from different sources. Karimi and Tang [16] utilize automatic document structure learning for fake news detection. Ma et al. [20] integrate structure and content semantics based on tree-structured recursive neural networks. Tan et al. [36] introduce the task of detecting machine-generated news that includes images. Dun et al. [6] propose a knowledge-aware attention network for fake news detection. Pelrine et al. [26] improve traditional NLP baselines to achieve better detection performance. However, content-based fake news detection methods can not handle fake news which disguises itself as genuine news. Graph-based models that leverage additional information from media during the news propagation process are proposed.

2.2 Graph-Based Models

Graph-based methods adopt graph neural networks to extract information from social contexts [39] to detect fake news. Monti et al. [23] propose a novel fake news detection based on deep learning on the graph. Hu et al. [12] and Feng et al. [7] utilize external knowledge through entities for fake news detection. Han et al. [8] use the propagation pattern of news on social media and focus on a propagation-based approach for fake news detection. Mehta and Goldwasser [21] incorporate cross-media consistency checking to detect the fine-grained knowledge elements and make news articles misinformation. Though existing graph-based fake news detection methods have made great achievements in detecting fake news through graph structure, they treat each node the same, ignoring the possibility of the echo chamber phenomenon. There is a need for a model that can dynamically evaluate the contribution of media during news dissemination.

3 Methodology

In this section, we present FNDPro for fake news detection as Fig. 2 shows. FNDPro is mainly faced with three problems: (A) How to construct heterogeneous information from various media and what information is needed; (B) How to define news propagation and obtain spatial-temporal information from each propagation; (C) How to evaluate the importance of each propagation to find out what matters during the propagation process.

3.1 News Propagation Network

FNDPro builds a heterogeneous news propagation network, dubbed the news propagation network, that contains various media information to solve (A). Formally, we define the news propagation network as $\mathcal{G} = (\mathcal{V}, X, \mathcal{E}, \mathcal{C}^v, \mathcal{R}^e, \varphi, \psi, \mathcal{T})$

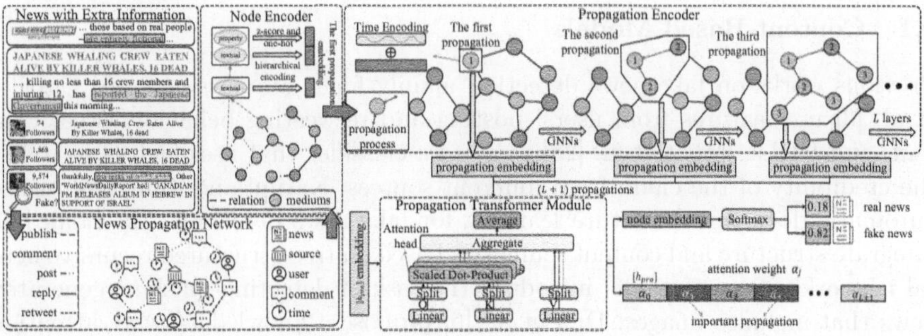

Fig. 2. The overview of FNDPRO. FNDPRO first constructs a heterogeneous dynamic **News Propagation Network**, including news, source, user, and comment nodes and multiple relations. FNDPRO then employ a **Propagation Encoder** containing a heterogeneous information encoder and a time encoder to capture the spatial and temporal information of each propagation. FNDPRO finally proposes a **Propagation Transformer Module** to make each propagation interact, capturing important propagation during the news spread process to relieve the echo chamber effect.

with news node $\mathcal{V}_n \subset \mathcal{V}$ and associated news label $\mathcal{Y}$, where $\mathcal{V} = \{u_i\}$ is node set, X is node attribution set, $\mathcal{E} = \{(u_i, v_i)\}$ is edge set, $\mathcal{C}^v$ is node type set, $\mathcal{R}^e$ is edge type set, $\varphi : \mathcal{V} \to \mathcal{C}^v$ is node mapping function, $\psi : \mathcal{E} \to \mathcal{R}^e$ is edge mapping function, and $\mathcal{T} : \mathcal{V} \to \mathbb{R}$ is time stamp mapping function. Specifically, $\mathcal{C}^v$ contains four kinds of nodes, namely news media:

- **News.** The object to be detected in the task. FNDPRO needs the title and content of a piece of news. A title leads the outline and attracts the audience, which can help detect fake news [11]. Fake news content may have a language style different from real ones [24].
- **Source.** The source is an organization or online website publishing news like BBC, the credibility of which cannot be easily changed [17]. FNDPRO needs source descriptions containing rich textual information.
- **User.** Users are involved in news dissemination, which plays an important role in spreading fake news [34]. FNDPRO needs the description on the homepage and property information.
- **Comment.** The comment expresses user opinions. Comments with rebuttals may help to identify fake news [30]. FNDPRO needs the content and property information like support count to generate initial features.

For $\mathcal{R}^e$, there are five types of edges between media, which are the *publish* relation between source and news, *comment* relation between comment and news, *post* relation between user and comment, *reply* relation between comment and comment, and *retweet* relation between comment and comment.

3.2 Propagation Encoder

When news is published, it spreads across social networks through edges between different media. For example, some users might comment on this news, and other users get this news through the comments. Users post comments, and comments are replied to or retweeted by other users, and other users post new ones, and so on, which leads to propagation. Inspired by this, for a piece of news, FNDPro treats itself as the first propagation, neighbors as the second propagation, two-hop neighbors as the third propagation, and so on, which solves the first half of (B). FNDPro encodes each node by extracting and concatenating features from various information, which is divided into textual and property information.

- **Textual information** includes news title, news content, source description, user description, and comment content. FNDPro adopts hierarchical encoding, which can extract different levels of semantic information [42], for textual content A. FNDPro leverages NLTK [2] to tokenize A into sentences $\{a_i\}_{i=1}^{Q}$, where a_i denotes the i-th sentence and Q is sentence count. FNDPro then employs the pre-trained language model RoBERTa [19] as text encoder $\mathrm{LM}(\cdot)$ to encode tokens in each sentence, *i.e.*,

  $$\{\overline{a}_{i,j}\} = \mathrm{LM}(\{a_{i,j}\}_{j=1}^{Q_i}),$$

  where $a_{i,j}$ is j-th word in i-th sentence and Q_i is the token count. After that, a mean pooling is adopted to obtain sentence-level embedding, *i.e.*,

  $$\overline{a}_i = \frac{1}{Q_i}\sum_{j=1}^{Q_i} \overline{a}_{i,j}.$$

  Finally, FNDPro feeds the text-level embedding, which is adopted by a mean pooling, into a linear layer to obtain the textual embedding:

  $$\overline{a} = W_a \cdot (\frac{1}{Q}\sum_{i=1}^{Q} \overline{a}_i) + b_a,$$

  where W_a and b_a are learnable parameters.
- **Property information** includes user property, and comment property. FNDPro employs z-score normalization and one-hot encoding to normalize numerical properties and encode categorical properties. FNDPro feeds processed properties into a linear layer to obtain the property embedding.

FNDPro obtains the first propagation embedding of a node i, represented by $\overline{r}_i$, by concatenating textual embedding and property embedding. Note that news and source nodes only have textual embedding. To extract temporal information from propagations as (B) mentions, FNDPro adopts time encoding, which reflects the impact of the time difference between source node u and target node v, where $(u,v) \in \mathcal{E}$. Note that time encoding parameters for each edge type are different because the time difference distribution is different, *i.e.*,

$$r_v = \overline{r}_v + \underset{(u,v)\in\mathcal{E}}{\mathrm{mean}}(\mathrm{enc}_{\psi(u,v)}(\mathcal{T}(u) - \mathcal{T}(v))),$$

where mean$(\cdot)$ is the mean pooling and $\text{enc}_{\psi(s,e)}(\cdot)$ is the time encoding function, which transforms a time difference to an embedding with the same dimension as $\overline{r}_v$. In this paper, we choose the RTE module [13] as $\text{enc}_{\psi(s,e)}(\cdot)$.

To extract spatial information from propagations mentioned by (B), FNDPro employs message-passing graph neural networks, where the representation h_u of each node u is updated iteratively in each layer by collecting messages from its neighbors denoted as $\mathcal{N}_{\mathcal{R}}(u)$ among each edge type $\mathcal{R} \in \mathcal{R}^e$. For the ℓ-th layer, $h_u^{(\ell)}$ is obtained from the previous layer representation $h_u^{(\ell-1)}$ ($h_u^{(0)} = r_u$) via a propagation operation Prop$(\cdot)$ and an aggregation operator Aggr$(\cdot,\cdot)$, *i.e.*,

$$h_{\mathcal{N}_{\mathcal{R}}(u)}^{(\ell)} = \text{Prop}(\{h_v^{(\ell-1)} : v \in \mathcal{N}_{\mathcal{R}}(u)\}), \mathcal{R} \in \mathcal{R}^e,$$
$$h_u^{\ell} = \text{Aggr}(\{h_{\mathcal{N}_{\mathcal{R}}(u)}^{\ell} : \mathcal{R} \in \mathcal{R}^e\}, h_u^{(\ell-1)}).$$

Note that $h_u^{(\ell)}$ contains information of ℓ-hop neighbors so that FNDPro treats it as $(\ell+1)$-th propagation embedding. FNDPro is adapted for arbitrary GNNs, without loss of generality, we adopt GAT [38] as the graph neural network. For a news node u, we adopt L layers of graph neural networks so that we can obtain $L+1$ propagation embeddings $\{h_u^{\ell}\}_{\ell=0}^{L}$.

3.3 Propagation Transformer Module

FNDPro dynamically evaluates the importance of each propagation with the propagation transformer module to solve (C). The attention weights reflect the related importance of each propagation embedding. Another advantage of transformer architecture is the global receptive field, where each propagation embedding can attend to the information at each propagation and then process its representation. For a node u, FNDPro concatenates a learnable parameter as a special propagation embedding $[h_{pro}]$ shared with all nodes, and the final embedding of news u would be the embedding of $[h_{pro}]$. FNDPro adopts multi-head attention [37] and calculates query, key, and value for the c-th attention head,

$$q_{c,i} = W_{c,q} \cdot x_i + b_{c,q},$$
$$k_{c,i} = W_{c,k} \cdot x_i + b_{c,k},$$
$$v_{c,i} = W_{c,v} \cdot x_i + b_{c,v},$$

where x_i is the i-th embedding of the input data $[h_{pro}] \| \{h_u^{\ell}\}_{\ell=0}^{L}$, q, k and v are query, key, and value of the attention mechanism, and all W and b are learnable parameters with regard to attention heads. After that, FNDPro models the importance of each propagation by calculating attention weights, *i.e.*,

$$\alpha_{c,ij} = \frac{\langle q_{c,i}, k_{c,j} \rangle}{\sum_{k=0}^{\ell} \langle q_{c,i}, k_{c,k} \rangle},$$

where $\alpha_{c,ij}$ denotes the attention weight between propagation i and j, $\langle q, k \rangle = \exp(\frac{q^T k}{\sqrt{d}})$ is the exponential scale dot-product function, where d refers to the hidden dimension of each attention head. FNDPro then aggregates over attention heads to obtain propagation representation, *i.e.*,

$$z_i = \frac{1}{C}\sum_{c=1}^{C}\sum_{j}\alpha_{c,ij}\cdot v_{c,j},$$

where C is the number of attention heads. Note that z_0 is the output of $[h_{pro}]$, namely, the final embedding z_u of news u. Meanwhile, we define a shorthand notation $\alpha_j = \frac{1}{C}\sum_{j=1}^{C}\alpha_{c,0j}$, which reflect the importance of j-th propagation. The higher α_j is, the more important j-th embedding is.

3.4 Learning and Optimization

We identify fake news u according to the news embedding z_u. The goal is to minimize the following loss function:

$$\mathcal{L}(\theta) = -y\log(\hat{y}) - (1-y)\log(1-\hat{y}) + \lambda\sum_{\omega\in\theta}\omega^2,$$

where θ are learnable parameters of FNDPro, λ is a hyperparameter, y refers to the ground truth of this news, and $\hat{y}$ is the predicted probability of fake news.

4 Experiments

We make use of a widely used dataset, FakeNewsNet [31] dataset and its publicly available version at the official GitHub repository[1]. We use the official GitHub repository to obtain a slightly different, copy of the dataset. This dataset consists of 28,433 news, 2,165 sources, 2,446,885 comments, 805,842 users, and 4,966,149 relations. To obtain general results, we experiment with the following three settings: (i) **Merged**, the whole dataset that contains news with different types; (ii) **Politifact**, politics-related news; (iii) **Gossipcop**, entertainment-related news.

We compare FNDPro with 13 state-of-the-art baselines including 6 content-based approaches and 7 graph-based approaches. For content-based, we employ dEFEND [30], RoBERTa [19], UPF [34], NEP [29], CMTR [9] and FinerFact [14]. For graph-based, we employ HPN [32], Bi-GCN [1], FANG [25], UPFN [5], Cross-D [35], InfOp [22] and GET [41]. We re-implement these baselines under the three settings to have a fair comparison. Note that our modified dataset cannot provide the data needed for some baselines. Therefore, for these baselines, we directly delete relevant news items.

4.1 Main Results

We present the performance of FNDPro and state-of-the-art baselines in Table 1, where we could conclude as follows:

[1] https://github.com/KaiDMML/FakeNewsNet

Table 1. Performance of FNDPro and baselines under three settings. We conduct experiments in five runs and report the average values and standard deviation. The † mark denotes that the model cannot obtain all the required data from the dataset. **Bold** and underline indicate the highest and second highest performance. We conduct an unpaired t-test with a 95% confidence interval as the significant test and report the p-value (* for $p < 0.0005$, ** for $p < 0.005$, *** for $p < 0.05$).

Type	Models	Merged		Politifact		Gossipcop	
		Accuracy	F1-score	Accuracy	F1-score	Accuracy	F1-score
Content-based	dEFEND [30]	84.1 (±4.1)*	67.8 (±6.9)*	84.4 (±1.0)*	78.5 (±2.5)*	87.4 (±2.8)*	73.0 (±6.5)*
	RoBERTa [19]	85.5 (±0.2)*	67.6 (±1.1)*	88.8 (±0.3)*	86.4 (±0.3)*	86.3 (±0.3)*	69.7 (±1.0)*
	UPF† [34]	94.4 (±0.2)*	86.0 (±0.4)*	86.8 (±0.6)*	84.4 (±0.7)*	95.2 (±0.0)*	88.0 (±0.1)*
	NEP [29]	81.8 (±0.1)*	58.8 (±1.5)*	87.5 (±0.4)*	84.0 (±0.5)*	82.1 (±0.1)*	59.9 (±1.4)*
	CMTR [9]	91.8 (±0.7)*	82.1 (±3.0)*	83.4 (±1.9)*	77.3 (±3.3)*	93.3 (±0.6)*	86.2 (±0.5)*
	FinerFact† [14]	95.5 (±0.1)*	91.6 (±0.2)*	86.2 (±0.1)*	88.1 (±0.2)*	95.9 (±0.1)*	93.1 (±0.2)*
Graph-based	HPN† [32]	87.3 (±0.2)*	82.4 (±0.3)*	78.1 (±0.7)*	81.2 (±0.6)*	87.1 (±0.1)*	81.4 (±0.2)*
	Bi-GCN [1]	90.4 (±0.5)*	90.0 (±0.7)*	77.1 (±1.4)*	74.7 (±2.2)*	90.8 (±0.3)*	90.4 (±0.3)*
	FANG [25]	81.3 (±0.2)*	53.7 (±1.5)*	80.3 (±1.5)*	75.4 (±3.1)*	82.7 (±0.1)*	51.2 (±1.2)*
	UPFN† [5]	93.6 (±1.0)*	93.7 (±0.8)***	89.5 (±2.1)*	87.2 (±2.6)*	95.8 (±0.8)*	**95.8** (±0.8)***
	Cross-D† [35]	89.8 (±0.3)*	86.1 (±0.4)*	88.3 (±0.9)*	87.8 (±0.9)*	90.3 (±0.3)*	87.1 (±0.5)*
	InfOp [22]	95.8 (±0.1)*	91.5 (±0.2)*	85.8 (±1.3)*	81.7 (±1.6)*	96.1 (±0.1)*	92.2 (±0.1)*
	GET [41]	91.3 (±0.4)*	81.5 (±1.7)*	88.5 (±0.2)*	83.8 (±0.7)*	91.1 (±1.4)*	82.9 (±0.5)*
	FNDPro	**97.0** (±0.2)	**93.9** (±0.3)	**92.9** (±0.5)	**90.8** (±0.6)	**97.4** (±0.1)	94.7 (±0.2)

- FNDPro ***archives state-of-the-art performance.*** FNDPro achieves the best accuracy with very statistically significant under all settings. It improves by 1.3% under **Merged** setting, 3.8% under **Politifact**, and 1.4% under **Gossipcop**. For the f1-score, it outperforms all baselines except under the **Gossipcop** setting. It has only minor inadequacies of 1.1% compared to UPFN, where we speculate that this may be due to missing news items.
- ***Graph-based methods could detect disguised fake news better.*** Under the most comprehensive setting **Merged**, graph-based baselines are generally more effective, where they improve the average accuracy by 2.1% compared with content-based. It proves that content-based models suffer from disguised content and extra information can improve detection performance.
- FNDPro ***could mitigate the echo chamber effect.*** Compared to other graph-based models, FNDPro achieves a 1.3% improvement from InfOp in accuracy. Note that compared to FANG, a model that adopts the same media as FNDPro, FNDPro achieves a significant improvement. We speculate it benefits from the propagation transformer model, which could dynamically evaluate propagation distributions and find critical propagation.

4.2 Propagation Transformer Study

FNDPro fuses propagation embeddings by the propagation transformer, evaluating the importance of each propagation simultaneously. We compare this module with the following six modules which can also fuse embeddings: (i) **First**: We consider the first propagation representation as the final news representation. (ii) **Last**: Similar to (i), while we adopt the last. (iii) **Mean**: We average

Table 2. Performance of the Embedding Fusion Modules. FNDPRO achieves the best performance, showing the ability to fuse the propagation features.

Module	Accuracy	Difference	F1-score	Difference
First	84.9 (±0.3)*	−12.1	64.9 (±1.3)*	−29.0
Last	96.2 (±0.5)**	−0.9	92.3 (±0.7)**	−1.6
Mean	85.7 (±1.0)*	−11.3	76.0 (±1.2)*	−17.9
Max	95.9 (±0.4)*	−1.1	91.8 (±0.7)*	−2.1
MLP	96.0 (±0.6)**	−1.0	92.0 (±1.1)**	−1.9
GRU	94.3 (±0.9)*	−2.7	89.2 (±1.4)*	−4.7
FNDPRO	**97.0** (±0.2)	0	**93.9** (±0.3)	0

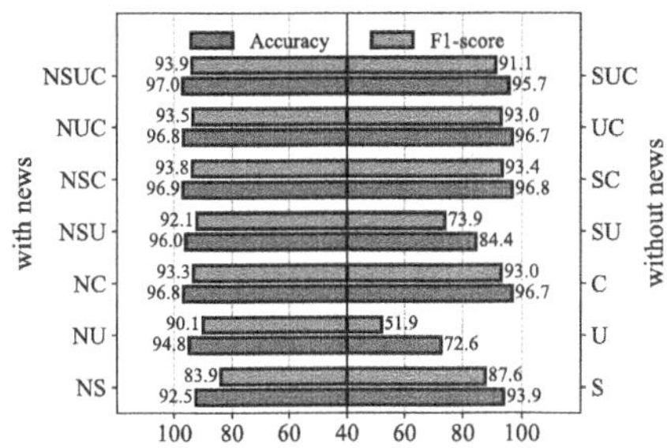

Fig. 3. The performance of FNDPRO media ablation study under the **Merged** setting. FNDPRO could keep performance without specific media.

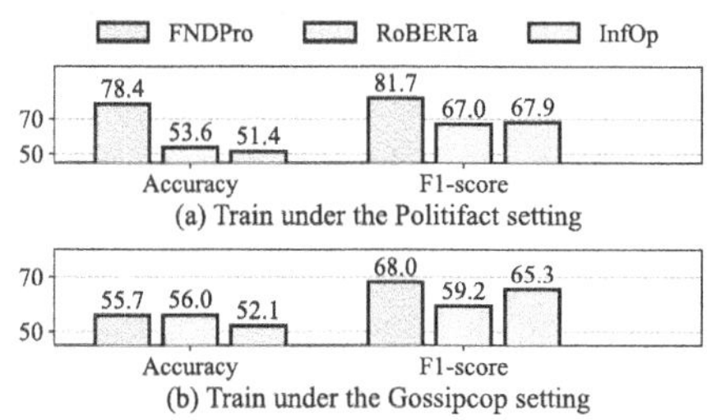

Fig. 4. Train FNDPRO and baselines under one setting and test under another. FNDPRO can identify different types of fake news even if not trained on them.

each representation. (iv) **Max**: The max operation is replaced with mean. (v) **MLP**: We stack each representation and apply a linear layer. (vi) **GRU**: We apply a GRU layer to get the representation. The results in Table 2 illustrate that the module we proposed achieves the best performance, showing a significant advantage in accuracy and f1-score against first, mean, and max. Compared to MLP and GRU which aggregate features in a deep way, our proposed module outperforms as well. It illustrates that the proposed module can effectively fuse the propagation embeddings, alleviating the echo chamber effect.

4.3 Discussion

In realistic detection situations, fake news detectors suffer from echo chamber effects, which would be enhanced by missing related information, manipulation by fake news publishers, etc. We conduct further experiments to explore whether FNDPRO could better solve echo chamber problems.

Media Ablation Study. FNDPRO leverages multiple media during news propagation, however, these media may not be accessed in a realistic detection situ-

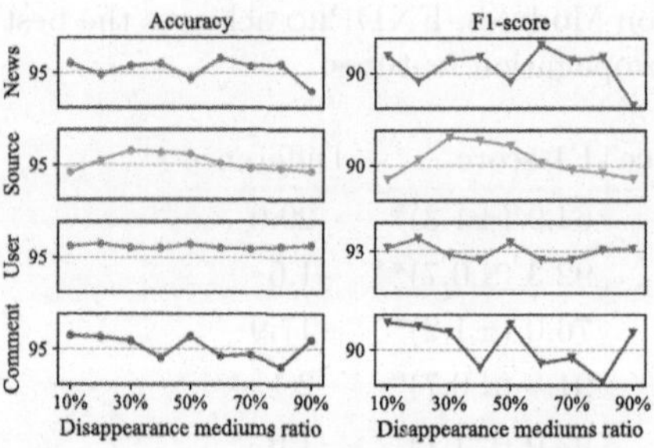

Fig. 5. Performance of FNDPRO when a portion of the specific media is gradually removed. FNDPRO could keep performance, showing great robustness to the related media in real situations.

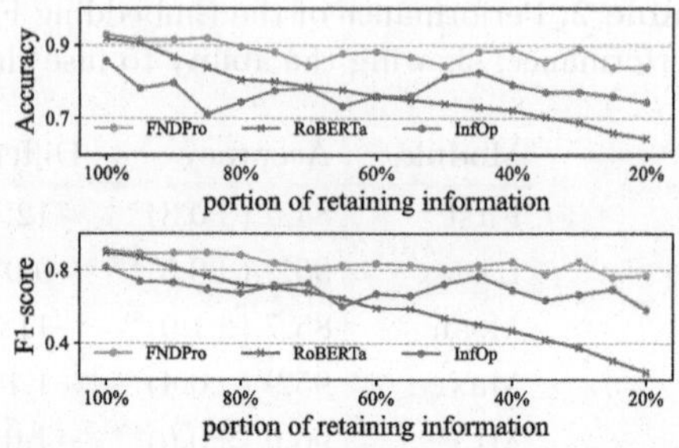

Fig. 6. Performance of FNDPRO and baselines when a portion of the whole graph information is gradually removed. FNDPRO keeps the performance with only 20% portion of the information.

ation. We remove kinds of media by deleting the node and related edges of the networks to evaluate whether FNDPRO can keep performance. Figure 3 (Left) illustrates that removing any media in the news propagation network leads to a drop in performance, which proves it is sensible to leverage these media. On the other hand, the drop in performance is acceptable. Under the NS setting, the accuracy drops to 92.5%, which still outperforms most of the baselines. It proves that FNDPRO could identify fake news when the media are missing. We speculate it benefits from the propagation transformer, which finds the critical propagation during detection. Meanwhile, we also remove the news itself and the results are shown in Fig. 3 (Right). It illustrates that even without news content information, FNDPRO can still detect fake news with extra information.

Transfer Study. There are various types of fake news, such as political-related or entertainment-related. To check if detectors can detect different types of fake news, we train them on one news type and test them on the others. We compare FNDPRO with ROBERTA and INFOP. We sample 314 political-related news from **Politifact** and 5,464 entertainment-related from **Gossipcop**, where the fake and real news is balanced. Figure 4 illustrates that FNDPRO can better identify other types of news even when not trained on them. It establishes the claim that FNDPRO can leverage extra information and benefit from extractions of critical media information thanks to the propagation transformer module.

Robustness to Limited Media Information. Fake news publishers might manipulate the related media, which means that we may not access all information from the media during news dissemination. We gradually remove media information under the **Merged** setting by setting the initial embedding to zero. Figure 5 illustrates that FNDPRO can maintain performance when the portion of a specific media, even the news, gradually disappears. Note that the accuracy drops to a minimum of 93.8%, but FNDPRO still outperforms several baselines. We speculate the remained node can fuse helpful spatial-temporal information

(a) FNDPro (V-Measure: 0.869) (b) UPFN (V-Measure: 0.656) (c) RoBERTa (V-Measure: 0.204)

Fig. 7. 2D t-SNE plot of the news representations of FNDPro, UPFN, and RoBERTa and the corresponding V-Measure. FNDPro achieves the best score and shows excellent collocation for the group of fake and real news.

thanks to time encoding and graph neural networks. Then the propagation transformer module aggregates critical propagation embeddings to obtain the final news embedding, which can maintain the performance.

Robustness to the Limited Whole Information. In the early dissemination of news, there may be little related information. Meanwhile, fake news publishers may delete related media to create echo chambers. Namely, fake news detectors need to be robust to information missing. We simulate this situation by randomly gradually removing a portion of the graph. We compare FNDPro with RoBERTa and InfOp under a more challenging setting, **Politifact**. Figure 6 illustrates that FNDPro outperforms other baselines with any portion of retaining information. Besides, when retaining 20% or more of the graph information, the performance of FNDPro does not decrease significantly. It may benefit from the propagation transformer module, which can dynamically evaluate the importance of remaining media. Meanwhile, FNDPro successfully identifies inconsistencies in the remaining media, which makes it maintain performance.

Representation Learning Study. The news representations learned by FNDPro fuse heterogeneous information and propagation embeddings. High-quality representation means that the model has good adaptability. The representations can be adopted for other downstream tasks like stance detection. To examine the quality of the learned representation, we cluster representation using k-means and calculate the V-Measure, an external cluster evaluation. We compare FNDPro with two models, UPFN and RoBERTa, that can generate representation. Figure 7 visualizes the representations and we can obviously observe that the representations of FNDPro show excellent collocation for the group of fake and real news, while UPFN and RoBERTa show little. Quantitatively, FNDPro's clusters achieve the best V-Measure score of 0.869, compared to a 0.656 score for the UPFN clusters and 0.204 for the RoBERTa.

4.4 Case Study

We study a piece of specific news and related media with the propagation attention weight in Fig. 8. For this fake news, other models make the wrong detection

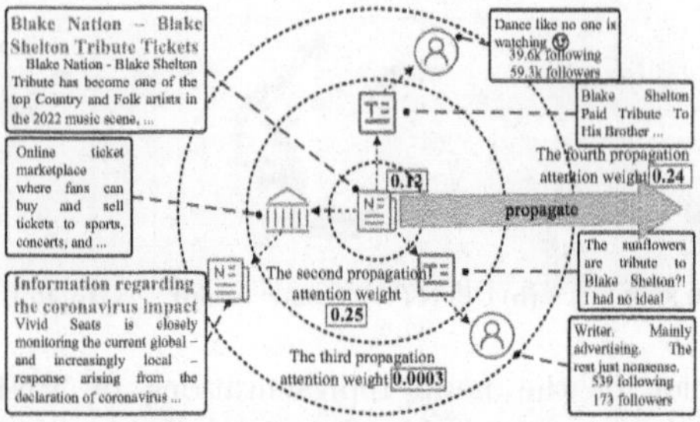

Fig. 8. Example propagation network of a piece of fake news and its propagation attention weight. The second and fourth propagation are important during news spread.

but FNDPRO succeeds. It is illustrated that the second and the fourth propagation have a more considerable attention weight (0.25 and 0.24), but the third one (0.0003) has a more minor weight in this specific fake news. Moreover, content-based models can hardly identify this fake news depending on its content (with an attention weight of 0.12), but FNDPRO can identify it according to the related comments. The rest of the media, the two users and one news, do not help judge the veracity of this news. It shows that FNDPRO could evaluate the importance of each propagation and find what matters during news propagation.

5 Conclusion

We propose FNDPRO, a fake news detection framework that identifies disguised fake news with additional media information and dynamically evaluates which media are critical to identifying fake news. Specifically, FNDPRO models the propagation process of news as a heterogeneous dynamic graph dubbed news propagation network, which contains source, news, comment, and user as media. FNDPRO then designs a node encoder to extract and fuse multi-modality information, which helps identify disguised fake news. To evaluate which media are critical, FNDPRO considers each node as the first propagation and ℓ-hop neighbors as the $(\ell + 1)$-th propagation. FNDPRO then adopts time encoding and graph neural networks to fuse spatial-temporal features and obtains propagation embeddings. After that, a propagation transformer module is adopted to make each propagation embedding interact and measure the importance scores, so that FNDPRO can find what matters during the news propagation process. We conduct extensive experiments on real-world datasets, which show that FNDPRO outperforms all the baselines. The ablation study shows that our proposed propagation transformer can effectively improve detection performance. Further discussions prove that FNDPRO can identify fake news in a realistic situation.

Acknowledgement. This work was supported by the National Key Research and Development Program of China (No. 2022YFB3102600), National Nature Science Foundation of China (No. 62272374, No. 62192781, No. 62202367, No. 62250009, No.

62137002), Project of China Knowledge Center for Engineering Science and Technology, and Project of Chinese academy of engineering "The Online and Offline Mixed Educational Service System for 'The Belt and Road' Training in MOOC China". We would like to express our gratitude for the support of K. C. Wong Education Foundation.

References

1. Bian, T., et al.: Rumor detection on social media with bi-directional graph convolutional networks. In: Proceedings of the AAAI Conference on Artificial Intelligence, vol. 34, pp. 549–556 (2020)
2. Bird, S., Klein, E., Loper, E.: Natural Language Processing with Python: Analyzing Text with the Natural Language Toolkit. O'Reilly Media, Inc. (2009)
3. Castillo, C., Mendoza, M., Poblete, B.: Information credibility on twitter. In: Proceedings of the 20th International Conference on World Wide Web, pp. 675–684 (2011)
4. Dang, Z., Luo, M., Jia, C., Dai, G., Chang, X., Wang, J.: Noisy correspondence learning with self-reinforcing errors mitigation. In: Proceedings of the AAAI Conference on Artificial Intelligence, vol. 38, pp. 1463–1471 (2024)
5. Dou, Y., Shu, K., Xia, C., Yu, P.S., Sun, L.: User preference-aware fake news detection. In: Proceedings of the 44th International ACM SIGIR Conference on Research and Development in Information Retrieval (2021)
6. Dun, Y., Tu, K., Chen, C., Hou, C., Yuan, X.: KAN: knowledge-aware attention network for fake news detection. In: Proceedings of the AAAI Conference on Artificial Intelligence, vol. 35, pp. 81–89 (2021)
7. Feng, S., Tan, Z., Zhang, W., Lei, Z., Tsvetkov, Y.: Kalm: knowledge-aware integration of local, document, and global contexts for long document understanding. In: The 61st Annual Meeting of the Association For Computational Linguistics (2023)
8. Han, Y., Karunasekera, S., Leckie, C.: Graph neural networks with continual learning for fake news detection from social media. CoRR abs/2007.03316 (2020)
9. Hartl, P., Kruschwitz, U.: Applying automatic text summarization for fake news detection. In: Proceedings of the Thirteenth Language Resources and Evaluation Conference, pp. 2702–2713. European Language Resources Association (2022)
10. Heidari, M., et al.: Bert model for fake news detection based on social bot activities in the covid-19 pandemic. In: 2021 IEEE 12th Annual Ubiquitous Computing, Electronics & Mobile Communication Conference (UEMCON), pp. 0103–0109. IEEE (2021)
11. Horne, B.D., Adali, S.: This just in: Fake news packs a lot in title, uses simpler, repetitive content in text body, more similar to satire than real news. In: Eleventh International AAAI Conference on Web and Social Media (2017)
12. Hu, L., et al.: Compare to the knowledge: Graph neural fake news detection with external knowledge. In: Proceedings of the 59th Annual Meeting of the Association for Computational Linguistics and the 11th International Joint Conference on Natural Language Processing (Volume 1: Long Papers), pp. 754–763 (2021)
13. Hu, Z., Dong, Y., Wang, K., Sun, Y.: Heterogeneous graph transformer. In: Proceedings of The Web Conference 2020, pp. 2704–2710 (2020)
14. Jin, Y., et al.: Towards fine-grained reasoning for fake news detection. In: Proceedings of the AAAI Conference on Artificial Intelligence, vol. 36, pp. 5746–5754 (2022)

15. Jung, A.K., Ross, B., Stieglitz, S.: Caution: rumors ahead-a case study on the debunking of false information on twitter. Big Data Soc. **7**(2), 2053951720980127 (2020)
16. Karimi, H., Tang, J.: Learning hierarchical discourse-level structure for fake news detection. In: NAACL, pp. 3432–3442. Association for Computational Linguistics (2019)
17. Kulkarni, V., Ye, J., Skiena, S., Wang, W.Y.: Multi-view models for political ideology detection of news articles. In: Proceedings of the 2018 Conference on Empirical Methods in Natural Language Processing, pp. 3518–3527 (2018)
18. Lei, Z., et al.: BIC: twitter bot detection with text-graph interaction and semantic consistency. In: Proceedings of the 61st Annual Meeting of the Association for Computational Linguistics (Volume 1: Long Papers), pp. 10326–10340 (2023)
19. Liu, Y., et al.: Roberta: a robustly optimized BERT pretraining approach. CoRR abs/1907.11692 (2019)
20. Ma, J., Gao, W., Wong, K.F.: Rumor detection on twitter with tree-structured recursive neural networks. In: Proceedings of the 56th Annual Meeting of the Association for Computational Linguistics (Volume 1: Long Papers), pp. 1980–1989 (2018)
21. Mehta, N., Goldwasser, D.: Tackling fake news detection by interactively learning representations using graph neural networks. In: Proceedings of the First Workshop on Interactive Learning for Natural Language Processing, pp. 46–53 (2021)
22. Mehta, N., Pacheco, M.L., Goldwasser, D.: Tackling fake news detection by continually improving social context representations using graph neural networks. In: Proceedings of the 60th Annual Meeting of the Association for Computational Linguistics (Volume 1: Long Papers), pp. 1363–1380 (2022)
23. Monti, F., Frasca, F., Eynard, D., Mannion, D., Bronstein, M.M.: Fake news detection on social media using geometric deep learning. CoRR abs/1902.06673 (2019)
24. Narwal, B.: Fake news in digital media. In: 2018 International Conference on Advances in Computing, Communication Control and Networking (ICACCCN), pp. 977–981. IEEE (2018)
25. Nguyen, V.H., Sugiyama, K., Nakov, P., Kan, M.Y.: Fang: leveraging social context for fake news detection using graph representation. In: Proceedings of the 29th ACM International Conference on Information & Knowledge Management, pp. 1165–1174 (2020)
26. Pelrine, K., Danovitch, J., Rabbany, R.: The surprising performance of simple baselines for misinformation detection. In: Proceedings of the Web Conference 2021, pp. 3432–3441 (2021)
27. Pennebaker, J.W., Boyd, R.L., Jordan, K., Blackburn, K.: The development and psychometric properties of liwc2015. Technical report (2015)
28. Popat, K., Mukherjee, S., Strötgen, J., Weikum, G.: Credibility assessment of textual claims on the web. In: Proceedings of the 25th ACM International on Conference on Information and Knowledge Management, pp. 2173–2178 (2016)
29. Sheng, Q., Cao, J., Zhang, X., Li, R., Wang, D., Zhu, Y.: Zoom out and observe: news environment perception for fake news detection. In: Proceedings of the 60th Annual Meeting of the Association for Computational Linguistics, pp. 4543–4556. Association for Computational Linguistics (2022)
30. Shu, K., Cui, L., Wang, S., Lee, D., Liu, H.: Defend: explainable fake news detection. In: Proceedings of the 25th ACM SIGKDD International Conference on Knowledge Discovery & Data Mining, pp. 395–405 (2019)

31. Shu, K., Mahudeswaran, D., Wang, S., Lee, D., Liu, H.: Fakenewsnet: a data repository with news content, social context, and spatiotemporal information for studying fake news on social media. Big data **8**(3), 171–188 (2020)
32. Shu, K., Mahudeswaran, D., Wang, S., Liu, H.: Hierarchical propagation networks for fake news detection: investigation and exploitation. In: Proceedings of the International AAAI Conference on Web and Social Media, vol. 14, pp. 626–637 (2020)
33. Shu, K., Wang, S., Liu, H.: Beyond news contents: the role of social context for fake news detection. In: Proceedings of the Twelfth ACM International Conference on Web Search and Data Mining, pp. 312–320 (2019)
34. Shu, K., Zhou, X., Wang, S., Zafarani, R., Liu, H.: The role of user profiles for fake news detection. In: Proceedings of the 2019 IEEE/ACM International Conference on Advances in Social Networks Analysis and Mining, pp. 436–439 (2019)
35. Silva, A., Luo, L., Karunasekera, S., Leckie, C.: Embracing domain differences in fake news: Cross-domain fake news detection using multi-modal data. In: Proceedings of the AAAI Conference on Artificial Intelligence, vol. 35, pp. 557–565 (2021)
36. Tan, R., Plummer, B.A., Saenko, K.: Detecting cross-modal inconsistency to defend against neural fake news. In: Proceedings of the 2020 Conference on Empirical Methods in Natural Language Processing, pp. 2081–2106. Association for Computational Linguistics (2020)
37. Vaswani, A., et al.: Attention is all you need. In: Advances in Neural Information Processing Systems, vol. 30 (2017)
38. Velickovic, P., Cucurull, G., Casanova, A., Romero, A., Lio, P., Bengio, Y.: Graph attention networks. Stat **1050**, 20 (2017)
39. Wan, H., Feng, S., Tan, Z., Wang, H., Tsvetkov, Y., Luo, M.: Dell: generating reactions and explanations for LLM-based misinformation detection. arXiv preprint arXiv:2402.10426 (2024)
40. Wang, Y., et al.: EANN: event adversarial neural networks for multi-modal fake news detection. In: Proceedings of the 24th ACM SIGKDD International Conference on Knowledge Discovery & Data Mining, pp. 849–857 (2018)
41. Xu, W., Wu, J., Liu, Q., Wu, S., Wang, L.: Evidence-aware fake news detection with graph neural networks. In: Proceedings of the ACM Web Conference 2022, pp. 2501–2510 (2022)
42. Yang, Z., Yang, D., Dyer, C., He, X., Smola, A., Hovy, E.: Hierarchical attention networks for document classification. In: Proceedings of the 2016 conference of the North American Chapter of the Association for Computational Linguistics: Human Language Technologies, pp. 1480–1489 (2016)

Leveraging Homophily-Augmented Energy Propagation for Bot Detection on Graphs

Bradley Ashmore and Lingwei Chen(✉)

Wright State University, Dayton, OH 45435, USA
{ashmore.3,lingwei.chen}@wright.edu

Abstract. As the developers of malware continuously evolve their attacks and infection methods, so to must bot detection methods advance. Graph Neural Networks (GNNs) have emerged as a promising detection method. However, in most cases communications graphs reflecting bot-infected networks are plagued with class imbalance and a high level of heterophily. Graph oversampling techniques employed to tackle class imbalance on graphs have drawbacks, such as introducing noisy topological structures or exacerbating heterophily within the graph. Out-of-distribution detection (ODD) is considered as an alternative solution to address data imbalance issues, but when applied to graphs, it assumes that the underlying graph structure does not interfere with the learning of data distributions. In this paper, we present the first application of ODD methods for bot detection in a network. We propose a new energy-based ODD model, which surpasses existing ODD methods, including those tailored for ODD on graph data, and effectively mitigates performance degradation caused by graph heterophily. We substantiate our claims through extensive experiments on the TON IoT dataset, which comprises real captured bot data. The experimental results demonstrate that our model achieves state-of-the-art performance in bot detection on graphs with high graph heterophily and extreme class imbalance.

Keywords: Bot Detection · Out-of-distribution Detection · Class Imbalance · Graph Heterophily

1 Introduction

Malicious bots are compromised victim computers or IoT devices, infiltrated by automated malware programs that perform predefined assignments, which can infect a system, steal data, or commit other malicious activities [5,28]. As these bots generally reside in a network by communicating in peer-to-peer (P2P) structures [37], graph neural networks (GNNs) have emerged as one of the most effective bot detection methods [22,36,39], due to their graph learning capabilities [7], where GNN models [17] are devised to perform neighborhood aggregation

---

Supplementary Information The online version contains supplementary material available at https://doi.org/10.1007/978-981-97-5572-1_5.

M. Onizuka et al. (Eds.): DASFAA 2024, LNCS 14855, pp. 68–83, 2024.
https://doi.org/10.1007/978-981-97-5572-1_5

and label propagation through graph structure. For example, Zhao et al. [39] customized GNNs to detect botnet nodes within Internet communication graphs by identifying their topological patterns. Lo et al. [22] built a graph isomorphism network (GIN) with grouped reversible residual connections to handle performance degradation on botnet communication graphs.

Communication graphs often exhibit a natural imbalance among labeled nodes due to the inherent rarity of bots. We calculate the imbalance ratios (as defined in Sect. 2) over TON IoT networks [1,26], a dataset widely used for bot detection, where in most cases the imbalance ratios are below 0.03. This indicates a significant disparity where malicious bots are greatly outnumbered by benign devices. When the GNN model is trained on such class-imbalanced graphs, it is difficult to accurately identify bots [2,14]. Unfortunately, though class imbalance is prevalent in communication graphs for bot detection, existing graph-based bot detection methods have rarely explored this issue. To mitigate the impact of class imbalance on GNN models, oversampling has recently been generalized to imbalanced graphs [2,38]. This line of research synthesizes nodes in the embedding space and then generates edges to connect them with existing nodes, or removes selected edges to refine the node embedding to exclusively synthesizes nodes that balance class distribution. Either way may significantly disrupt graph structural semantics, resulting in noisy or incomplete neighborhoods for message passing. For instance, GraphSMOTE [38] generates edges using weighted inner products of node embeddings, which may not represent real-world relations between devices only enabled by TCP; HOVER [2] removes edges that satisfy heuristic criteria to prevent node oversmoothing, which may in turn conceal bots from their exploitation of other devices.

Out-of-distribution detection (ODD) [8] has been introduced to empower models with an introspective capability to recognize samples that deviate from the training data used for model preparation, and reject data for which they exhibit low confidence [18]. This becomes particularly advantageous in security-critical applications where anomalous or malicious samples are often limited in representation [4]. This feasibility inspires us to cast bot detection as an ODD problem, which aims at discriminating bots (out-of-distribution) from benign devices (in-distribution). While a surge of ODD methods have been developed [3,12,30], their effectiveness hinges on the assumption of independently sampled inputs which disables them from generalizing with graph data. Such a premise propels ODD exploration extended to graph data [24,32,34]. Addressing distribution shifts in an inter-dependence graph is more challenging, which often requires graph-specific technical innovations. For example, Li et al. [16] introduced a generative framework that incorporates node features, labels, and graph structures to derive a posterior distribution for ODD. Stadler et al. [29] explored uncertainty quantification by proposing Graph Posterior Network (GPN) to perform Bayesian posterior updates for predicting out-of-distribution nodes.

In this paper, we investigate how to leverage ODD for bot detection on graphs. More specifically, our model proceeds by learning node embeddings that encode both node features and graph structure for representation and prediction; these node embeddings are then fed to energy-based supervised classification loss that is regulated by semi-supervised energy propagation for optimization. Recent work GNNSafe [32] utilizing energy function to handle ODD focuses on energy formulation and transformation across different nodes, which, however, is encumbered with the same drawbacks that plague GNNs - over-smoothing and reliance on the presence of homophily [40]. As discussed in Section 3.1, GNNSafe surprisingly underperforms on graphs when imbalance exists between in-distribution and out-of-distribution nodes, due to its extracted energies being degraded by the GNN's message passing and energy propagation. In communication graphs for bot detection, class imbalance is often accompanied by a high degree of heterophily [2], where bots tend to massively connect with benign devices. To address this issue, we propose a simple yet effective method to mitigate over-smoothing on node embeddings caused by heterophily. Our method treats node neighborhood as structural features of nodes, embedding both structural and semantic features separately, which are then integrated to learn the final node embeddings. These embeddings enable subsequent homophily-augmented energy propagation to amplify the extracted energies for effective out-of-distribution (i.e., bot) detection. In summary, our major contributions are listed as follows:

- We cast bot detection on graphs as an ODD problem to discriminate bots (out-of-distribution) from benign devices (in-distribution).
- We design a new energy-based ODD framework, which can better deal with graph data with extreme low imbalance ratio and high heterophily level.
- Extensive experiments are conducted on TON IoT networks, demonstrating our model can achieve state-of-the-art bot detection performance on graphs.

2 Preliminaries and Problem Statement

Notations. We represent a communication graph for bot detection as $\mathcal{G} = (\mathcal{V}, \mathcal{E}, \mathbf{X})$, where $\mathcal{V}(n = |\mathcal{V}|)$ is the set of devices, $\mathcal{E}$ is the set of edges indicating communication between devices, and $\mathbf{X} \in \mathbb{R}^{n \times d}$ is the feature matrix. Edges E can be organized as an adjacency matrix $\mathbf{A} \in \mathbb{R}^{n \times n}$ and $\mathbf{A}_{ij} = \{0, 1\}$ [15]. Each labeled node has a ground truth $y \in \mathcal{Y} = \{benign, bot\}$. Bot detection learns a model $f_{\mathbf{W}} : (\mathbf{A}, \mathbf{X}) \rightarrow \mathbf{y}$ in a semi-supervised manner. The node embeddings $\mathbf{Z} = f_{\mathbf{W}}(\mathbf{A}, \mathbf{X})$ are used to formulate loss function for model optimization.

Class Imbalance. We define imbalance ratio to quantify the nature of class imbalance in communication graphs as follows: given a graph $\mathcal{G}$ where $\#m$ is the number of malicious bots and $\#b$ is the number of benign devices, the imbalance ratio is represented as $r_i = \frac{\#m}{\#b}$.

Homophily and Heterophily. This paper focuses on homophily in class labels, where a graph with high homophily suggests that connected nodes share the same label with a high probability across an entire graph. Homophily ratio can thus be defined: given a graph $\mathcal{G}$, the homophily ratio is measured as a ratio of homophilic edges to all edges $r_h = \frac{1}{|\mathcal{E}|}\sum_{(v_i,v_j)\in\mathcal{E}} 1(y_{v_i} = y_{v_j})$. Heterophily is the opposite of homophily to describe the status of heterophilic edges that connect nodes belonging to different labels. Graphs with high homophily have $r_h \to 1$, while graphs with high heterophily exhibit low homophily with $r_h \to 0$.

Energy-based Out-of-Distribution Detection. In this paper, we explore out-of-distribution detection to identify bots using energy-based models (EBMs) [20]. An EBM defines an energy function that maps any input $\mathbf{x}$ to a single scalar value $E(\mathbf{x}) : \mathbb{R}^D \to \mathbb{R}$. This non-probabilistic scalar is called *energy*, which can be expressed as the negative of the log partition function:

$$E(\mathbf{x}) = -\log \sum_{y'} e^{-E(\mathbf{x},y')} \tag{1}$$

A collection of energy values can be converted to a probability density $p(\mathbf{x})$ through Gibbs distribution as:

$$p(y|\mathbf{x}) = \frac{e^{-E(\mathbf{x},y)}}{\sum_{y'} e^{-E(\mathbf{x},y')}} = \frac{e^{-E(\mathbf{x},y)}}{e^{-E(\mathbf{x})}} \tag{2}$$

When connecting the EBM with a discriminative neural classifier $f(\mathbf{x})$ that maps the input $\mathbf{x}$ to k-dimensional logits (k is the number of classes), we can define an energy for a given input $\mathbf{x}$ with a given class label y as $E(\mathbf{x}, y) = -f_y(\mathbf{x})$. In this respect, the energy function can be further derived as:

$$E(\mathbf{x}; f) = -\log \sum_{i}^{k} e^{f_i(\mathbf{x})} \tag{3}$$

Due to the intrinsic characteristics of the energy $E(\mathbf{x}; f)$ that the extracted energy values for in-distribution samples tend to be lower than those of out-of-distribution samples, $E(\mathbf{x}; f)$ can be effectively used to facilitate ODD [20,32] with a pre-specified threshold τ:

$$g(\mathbf{x}; \tau, f) = \begin{cases} 0, & \text{if } E(\mathbf{x}; f) \le \tau, \\ 1, & \text{if } E(\mathbf{x}; f) > \tau \end{cases} \tag{4}$$

In this paper, we designate positive (1) as the prediction of out-of-distribution (bots), while negative (0) signifies the declaration for in-distribution (benign).

3 Proposed Model

In this section, we present our proposed model to leverage ODD for bot detection. We first analyze our motivation to better understand the impacts of graph structure on in-distribution learning before diving into technical details. Figure 1 depicts the overview of our complete model.

3.1 Impacts of Graph Structure on In-Distribution Learning

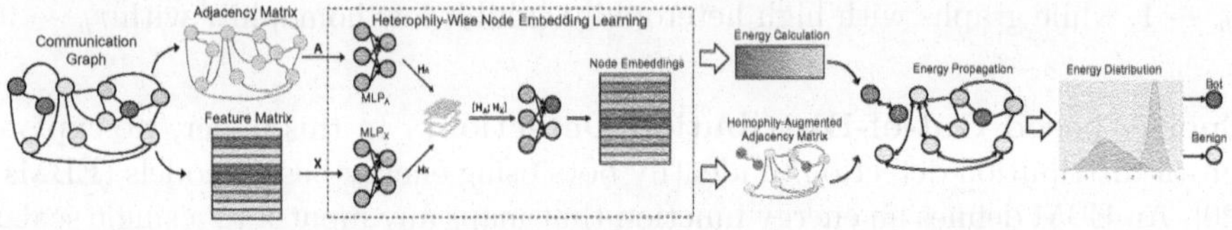

Fig. 1. Overview of our proposed model.

When extending energy-based ODD to graph data, the reliance on GNN backbones may result in these models inheriting heterophiliy, oversmoothing, and imbalance problems that are known to degrade GNN performance [33]. To demonstrate the impacts less-than-favorable graph conditions can have on graph-based ODD, we replicate GNNSafe [32], which is a recent state-of-the-art ODD model for graphs, with out-of-distribution exposure and belief propagation to work on imbalanced versions of the Cora dataset [35] by varying the number of in-distribution classes between six and three, and inversely adjusting the number of out-of-distribution classes between one and three. A more detailed description of the data configurations can be found in Table 1. Each adjustment to the dataset impacts which edges are homophilic, which edges are heterophilic, and the number of homophilic edges present in the in-distribution classes. This is significantly relevant due to the presence of the graph structure nullifies the independent and identically distribution (i.i.d.) assumption that regular machine learning-based ODD models (defined in Sect. 2) rely on, making the learning process sensitive to topological features of the graph [24].

Table 1. Data configurations on Cora dataset for graph structure analysis.

In-dist classes	In-dist Nodes	OOD Classes	OOD Nodes
6	2,140	1	568
5	1,722	2	986
4	904	3	1,804

We report the learned energy distributions in Fig. 2, with in-distribution nodes highlighted in blue and out-of-distribution nodes in orange. Notably, there is an observable trend of increased overlap between in-distribution and out-of-distribution energy values as the class imbalances between them grow, which implies a greater number of misclassifications. This phenomenon can be attributed to the fact that while shifting a single Cora class does not notably affect the overall number of heterophilic edges throughout the entire graph,

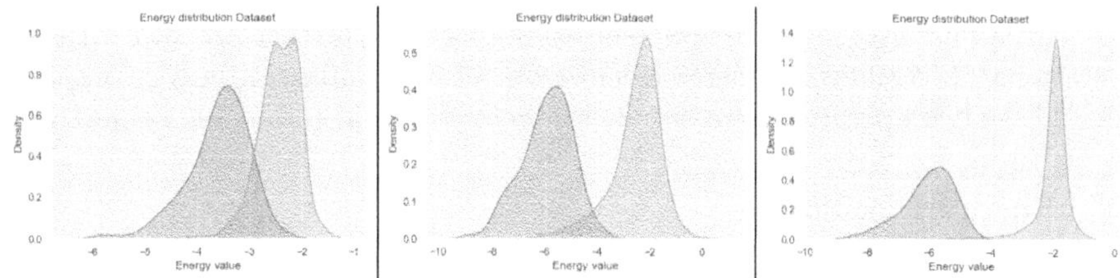

2,140 In-distribution Nodes | 1,722 In-distribution Nodes | 904 In-distribution Nodes

Fig. 2. Impacts of varying the number of in-distribution classes on energy distribution.

it does influence specific edges, potentially leading to the formation of heterophilic neighborhoods around individual nodes. Nodes with heterophilic neighborhoods tend to contribute to their over-smoothed embeddings and predicted logits; this results in less differentiable energy values that impact on subsequent energy propagation, ultimately leading to the observed overlap. Figure 2 provides insight into the graph-based ODD detection capability and its potential improvement by mitigating the impacts of heterophilic graph structure.

3.2 Heterophily-Wise Node Embedding Learning

To generalize GNNs to heterophilic graphs, the prevalent methods attend to refine neighborhood aggregation to adaptively exploit homophilic and heterophilic information [21,23,27], which improve the learning performance. However, these approaches not only complicate the models with extra computational cost, but also unstable generalization with high variance across diverse graphs [19]. In this paper, we take a different direction to formulate a learning-effective yet cost-efficient node embedding method to mitigate the impact of heterophily. Considering that heterophily impacts on node embeddings through neighborhood aggregation mechanism [25], we treat graph structure encoded as adjacency matrix $\mathbf{A}$ as structural features of nodes indicating inter-dependence with other nodes and their label distribution, instead of using it as low-pass filter to perform neighborhood aggregations. More specifically, we extract structural information from adjacency matrix $\mathbf{A}$ and semantic information from feature matrix $\mathbf{X}$, and then integrate them to learn the final node embeddings, which encode both node features and graph structure without excessive smoothing or compromise.

Formally, given a communication graph $\mathcal{G} = (\mathcal{V}, \mathcal{E}, \mathbf{X})$, we train two separate multi-layer perceptions (MLPs) on feature matrix $\mathbf{X}$ and adjacency matrix $\mathbf{A}$ to capture higher-level representations $\mathbf{H}_{\mathbf{X}}$ and $\mathbf{H}_{\mathbf{A}}$. Both representations are then concatenated as $[\mathbf{H}_{\mathbf{X}}; \mathbf{H}_{\mathbf{A}}]$ and fed to a fully-connected layer to fine-tune the presentation importance, deriving the final node embeddings $\mathbf{H}$:

$$\mathbf{H} = \sigma([\mathrm{MLP}_{\mathbf{X}}(\mathbf{X}), \mathrm{MLP}_{\mathbf{A}}(\mathbf{A})]\mathbf{W}), \quad \mathbf{H} \in \mathbb{R}^{n \times d} \tag{5}$$

The learned heterophily-wise node embeddings $\mathbf{H}$ will further proceed with two operations: (1) facilitating energy calculation; and (2) improving energy propagation. These two operations are detailed in Sect. 3.3 and Section 3.4, respectively.

3.3 Energy Calculation

To map the node embeddings $\mathbf{H}$ to the logit matrix $\mathbf{Z}$, we train another MLP to perform $\mathbf{Z} = \text{MLP}_{\mathbf{Z}}(\mathbf{H}), \mathbf{Z} \in \mathbb{R}^{n\times k}$, where $k = 2$ in our application scenario since we only have two classes to predict in bot detection. In this way, given an input node $\mathbf{x} \in \mathbf{X}$ from the graph $\mathcal{G}$ and its logit $\mathbf{z} \in \mathbf{Z}$, the energy function can be calculated as follows:

$$E(\mathbf{x}, \mathcal{G}_{\mathbf{x}}; f) = -\log \sum_{i}^{k} e^{\mathbf{z}[i]} \tag{6}$$

Since $\mathbf{z}$ is derived from node embeddings $\mathbf{h} \in \mathbf{H}$ that encodes the graph structure, the energy function $E(\mathbf{x}, \mathcal{G}_{\mathbf{x}}; f)$ indicates the energy information of the node itself and other nodes that share potential dependencies within the graph structure.

3.4 Homophily-Augmented Energy Propagation

Inspired by label propagation, our model further elaborates a non-parametric energy propagation to aggregate energy values from neighboring nodes and takes advantage of unlabeled nodes across graph structure to enhance the generalization ability. A general edge propagation can be implemented as follows:

$$\mathbf{E}^{(l)} = \alpha\mathbf{E}^{(l-1)} + (1-\alpha)\mathbf{D}^{-1}\mathbf{A}\mathbf{E}^{(l-1)} \tag{7}$$

where $E_{\mathbf{x}}^{(l)} = E^{(l)}(\mathbf{x}, \mathcal{G}_{\mathbf{x}}; f) \in \mathbf{E}^{(l)}$ at step l, and α is a balance parameter in the range of $[0, 1]$ to adjust the relative weight on the energy of the node itself and its neighboring nodes. Such a formulation implicitly relies on a high degree of graph homophily, where connected nodes tend to be sampled from similar distributions, such that the propagation will push energy towards the majority of neighbored nodes. Unfortunately, the given assumption does not hold under heterophilic conditions, leading to the failure of the propagation in Eq. (7). Based on this observation, we assert the following lemma:

Lemma 1. *Nodes with a heterophilic neighborhood, identified by a homophily ratio $r_h < 0.5$, will not trend towards the mean energy level of the distribution within which the node is contained.*

Proof of Lemma 1. For ODD to succeed, two distributions must be considered, one containing in-distribution data, $\mathcal{D}_{\text{in}}$, and another containing out-of-distribution data, $\mathcal{D}_{\text{out}}$. Each distribution has an expected value denoted as $\mathbb{E}_{\text{in}}$ and $\mathbb{E}_{\text{out}}$ respectively, representing the average energy value of in-distribution or out-of-distribution nodes, which can be specified as:

$$\mathbb{E}_{\text{type}} = \frac{\sum_{i}^{|\mathcal{V}_{\text{type}}|} E_i}{|\mathcal{V}_{\text{type}}|}, \quad \text{type} \in \{\text{in}, \text{out}\} \tag{8}$$

The expected energy values change during the training process. The effectiveness of ODD depends on a sufficiently large distance ψ between these two expected values, ensuring significant separation to minimize overlap in the tails of the learned distributions:

$$\psi < ||\mathbb{E}_{\text{in}} - \mathbb{E}_{\text{out}}|| \tag{9}$$

A node's neighborhood can be naturally divided into homophilic edges $\mathcal{E}_{\text{homo}}$ and heterophlic edges $\mathcal{E}_{\text{heter}}$. During a given training epoch l, $E_i^{(l)}$ is expected to approach a weighted average of the expected value of its neighbors. For an in-distribution node, this approximation can be expressed as:

$$E_i^{(l)} \approx \frac{|\mathcal{E}_{\text{homo},i}|\mathbb{E}_{\text{in}} + |\mathcal{E}_{\text{heter},i}|\mathbb{E}_{\text{out}}}{\sum_j \mathbf{A}_{ij}} \tag{10}$$

In homophilic neighborhoods, $|\mathcal{E}_{\text{homo},i}| > |\mathcal{E}_{\text{heter},i}|$ holds true and $E_i^{(l)}$ approaches the current $\mathbb{E}_{\text{in}}$, maintaining the desired relationship with ψ. In heterophilic neighborhoods where $|\mathcal{E}_{\text{homo},i}| < |\mathcal{E}_{\text{heter},i}|$, $E_i^{(l)}$ approaches the current $\mathbb{E}_{\text{out}}$. Adjustments to E_i impact the expected value of $\mathcal{D}_{in}$, potentially reducing the distance between expected values and violating the desired relationship with ψ.

To counter the negative impact of heterophily on energy propagation, we propose to empower our model to identify homophilic and heterophilic edges. This capability facilitates adjusting the energy propagation accordingly, ensuring a sufficiently large distance between $\mathbb{E}_{\text{in}}$ and $\mathbb{E}_{\text{out}}$.

Identification of Homophilic and Heterophilic Edges. We assess the likelihoods of edges being homophilic by examining the similarity of the connected node embeddings $\mathbf{H}$, as learned in Sect. 3.2. This is achieved through the application of two similarity measures:

- *Cosine Similarity.* Cosine similarity is used to evaluate the angle between two vectors, or in this case, node embeddings, which is calculated as:

$$\cos(\mathbf{h}_i, \mathbf{h}_j) = \frac{\mathbf{h}_i \cdot \mathbf{h}_j}{||\mathbf{h}_i||||\mathbf{h}_i||} \tag{11}$$

  To predict homophilic and heterophilic edges, we consider a cosine similarity > 0 as homophilic and ≤ 0 as heterophilic.
- *Euclidean Distance.* We first calculate Euclidean distance between the connecting node embeddings to obtain distances for all edges:

$$d_{i,j} = ||\mathbf{h}_i - \mathbf{h}_j||_2, \quad \forall (v_i, v_j) \in \mathcal{E} \tag{12}$$

  Subsequently, each distance $d_{i,j}$ is normalized as follows:

$$\text{norm}(d_{i,j}) = 1 + \frac{d_{\min} - d_{i,j}}{d_{\max} - d_{\min}} \tag{13}$$

  where $d_{\min}$ and $d_{\max}$ are the minimal and maximal distances among all edges. The normalized output falls within the range $[0, 1]$, where we consider $\text{norm}(d_{i,j}) \leq 0.5$ as heterophilic and $\text{norm}(d_{i,j}) > 0.5$ as homophilic.

Both similarity metrics can serve as either a binary or a fractional belief assessment of homophily. Here we use a binary assessment to improve energy propagation. Their effectiveness in edge identification will be evaluated in Sect. 4.2.

Use of Homophily-Augmented Adjacency Matrix. To incorporate the identification of heterophilic and homophilic edges into downstream tasks, we create an altered adjacency matrix $\hat{\mathbf{A}}$, which is defined as the edges believed to be homophilic in the graph, with belief determined by one of the two similarity measures described in Sect. 3.4. $\hat{\mathbf{A}}$ replaces the original adjacency matrix in the energy propagation, explicitly discouraging the propagation of energy across heterophilic edges. The energy propagation can be accordingly updated as:

$$\mathbf{E}^{(l)} = \alpha \mathbf{E}^{(l-1)} + (1-\alpha)\hat{\mathbf{D}}^{-1}\hat{\mathbf{A}}\mathbf{E}^{(l-1)} \tag{14}$$

The original normalization term $\mathbf{D}^{-1}$ is also replaced in our improved belief scheme with $\hat{\mathbf{D}}^{-1}$, which is the inverse diagonal degree matrix of $\hat{\mathbf{A}}$. This replacement is required to account for instances where a high-degree node contains a very small number of homophilic edges, ensuring energy being propagated as a proportion of homophilic edges and not restricted by a high degree. After L-step propagation, the energy values will be fed to the loss function for optimization, or used to make final out-of-distribution (bot) prediction using Eq. (4).

3.5 Loss Function

Our model utilizes both a supervised loss and a regularization loss for optimization. The supervised loss $\mathcal{L}_{\text{sup}}$ is a modified negative log-likelihood of the labeled training data designed to accommodate energy values:

$$\mathcal{L}_{\text{sup}} = \mathbb{E}_{(\mathbf{x},\mathcal{G}_{\mathbf{x}},y)\sim\mathcal{D}_{\text{in}}}(-\log p(y|\mathbf{x},\mathcal{G}_x)) = \sum_{i\in\mathcal{V}_{\text{in}}}(-\mathbf{z}_{i,[y_i]} + \log\sum_{j}^{k} e^{\mathbf{z}_{i,[j]}}) \tag{15}$$

where $\mathcal{D}_{\text{in}}$ is the in-distribution data where the labeled nodes (i.e., benign nodes) of the training data are sampled. Inspired by the bounding constraints for absolute energy [20], we further add a regularization loss $\mathcal{L}_{\text{reg}}$ to constrain the energy gap, bounding the energy for in-distribution data [32], and also expose a fraction of out-of-distribution data (i.e., bot nodes) to training that gives the model the opportunity to learn how to discriminate against known out-of-distribution nodes and assists in fine-tuning the model:

$$\begin{aligned}\mathcal{L}_{\text{reg}} &= \frac{1}{|\mathcal{V}_{\text{in}}|}\sum_{i\in\mathcal{V}_{\text{in}}}(\text{ReLU}(E(\mathbf{x}_i,\mathcal{G}_{\mathbf{x}_i};f)-t_{\text{in}}))^2 \\ &+ \frac{1}{|\mathcal{V}_{\text{out}}|}\sum_{j\in\mathcal{V}_{\text{out}}}(\text{ReLU}(t_{\text{out}}-E(\mathbf{x}_j,\mathcal{G}_{\mathbf{x}_j};f)))^2\end{aligned} \tag{16}$$

$\mathcal{L}_{\text{reg}}$ will push energy values in $[t_{\text{in}}, t_{\text{out}}]$ to be lower for in-distribution and higher for out-od-distribution nodes. The final loss function for optimizing our model is $\mathcal{L} = \mathcal{L}_{\text{sup}} + \lambda\mathcal{L}_{\text{reg}}$, where λ is a balance parameter.

Table 2. Statistics of bot detection datasets.

File	1	2	3	4	5	6	7	8	9	10
#Bots	19	35	32	38	41	39	49	57	57	56
#Benigns	1,534	1,563	1,604	1,623	1,638	1,654	1,688	2,304	2,043	2,048
Imb. Ratio	0.012	0.022	0.020	0.023	0.024	0.023	0.028	0.027	0.027	0.027
#Edges	8,264	9,296	9,492	9,644	9,716	9,812	10,016	11,816	11,880	11,928
Homo. Ratio	0.300	0.264	0.651	0.270	0.269	0.660	0.272	0.668	0.668	0.668

4 Experimental Results and Analysis

4.1 Experimental Setup

Datasets. We perform our experiments using the TON IoT dataset which is an assortment of recorded Industry 4.0/IoT and IIoT network traffic [1,26]. Each network is recorded in an individual capture file that is translated into an undirected communication graph, where protocol used, packet size, and amount of data transmitted and received are used as node features. To represent unique graph constructs, we select capture files with high heterophily ($r_h < 0.4$), high homophily ($r_h > 0.6$), and extreme class imbalance ($r_i < 0.03$) respectively. Table 2 details the data statistics. We separate nodes in each graph into an 80-10-10 split for training, testing, and validation.

Baselines. We select 11 different models as our baselines, which include three traditional GNNs (GCN [13], GAT [31], and GraphSAGE [9])), three graph-based oversampling models (SMOTE [6], GraphSMOTE [38], and HOVER [2]), and five ODD models (MSP [10], MSP with outlier exposure (OE) [11], and and three variants of GNNSafe [32]).

Implementation Details. We implement our model as described in Sect. 3 using the following hyperpameter settings: hidden size in $\text{MLP}_\mathbf{A} = \{128, 256\}$; hidden size in $\text{MLP}_\mathbf{X} = \{24, 48, 64, 128\}$; output size is $\{1x, 2x, 4x, 8x, 64x\}$ per number of classes x. In addition to these adjustable parameters, we use the following fixed configurations: layers in $\text{MLP}_\mathbf{A} = \{2\}$ and $\text{MLP}_\mathbf{X} = \{2\}$; propagation layer number $L = 2$; balance parameter $\alpha = 0.5$; we select the values $t_{in} = -5$ and $t_{out} = -1$ used in the regularization term of our loss function.

4.2 Effectiveness of Edge Prediction

We initiate our experimentation by assessing our model's ability to differentiate between homophilic and heterophilic edges. We perform this experiment by training our model on capture files 1, 3, 5, and 6 from TON IoT, due to their inclusion of both high and low levels of heterophily. During every training epoch, we generate and store improved node embeddings, and feed them into our edge

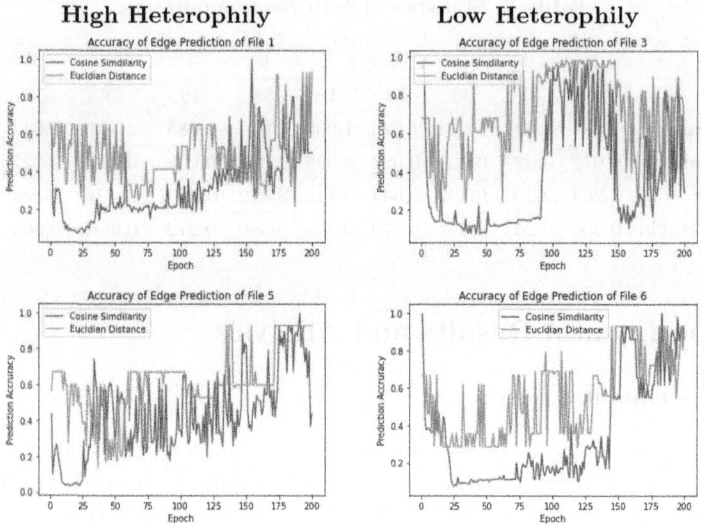

Fig. 3. Effectiveness of edge identification by comparing cosine similarity and Euclidean distance across different training epochs.

prediction module. The module assesses if an edge is homophilic or heterophilic according to cosine similarities and normalized Euclidean distances. We compare the edge prediction outputs with the true edge type as determined by node labels, and calculate the accuracy of both predictions at every epoch. Our model learns only from a loss function quantifying the quality of node classifications.

The edge prediction performance is recorded in Fig. 3. From this figure, we can see that our model's edge identification approach is capable of making highly accurate assessments of edge predictions, exceeding 80% accuracy during the training process. Both cosine similarity and Euclidean distance similarity approaches display generally stable behaviour. However, Euclidean distance outperforms cosine similarity almost universally. As such, we conclude that we can identify homophilic edges in heavily heterophilic graphs using heterophily-wise embeddings learned from node features and graph structure.

4.3 Comparison with Baselines for Bot Detection

We compare our model against multiple baselines, including basic GNN models: GCN [13], GAT [31], and GraphSage [9], oversampling approaches: SMOTE (applied on node embeddings learned by GCN) [6], GraphSMOTE [38], and HOVER [2], and ODD methods: MSP [10], MSP with outlier exposure [11], and three variants of GNNSafe [32]. We report the performance of each model and our proposed model as measured by F-1 (%) in Table 3.

Our model shows substantial performance over existing baselines with limited exceptions. Traditional GNN models fall inline with the expected behavior under heterophilic and extremely class-imbalanced conditions, which cannot gen-

Table 3. Comparison of our model with different baselines in terms of F-1 (%).

Model	File 1	File 2	File 3	File 4	File 5	File 6	File 7	File 8	File 9	File 10
GCN	00.75	06.09	03.95	04.85	08.51	04.62	01.46	07.37	09.95	01.19
GAT	00.00	01.09	03.75	04.75	04.68	08.51	08.84	07.37	01.01	08.82
GraphSage	03.13	01.54	05.41	08.70	02.22	05.41	02.58	01.69	01.80	07.92
SMOTE	02.28	02.13	01.23	01.57	07.75	01.10	08.04	02.89	01.24	02.81
GraphSMOTE	66.74	55.10	49.50	64.21	54.28	58.10	57.00	56.00	56.00	49.33
HOVER	60.45	33.33	**80.00**	74.16	66.67	66.67	60.00	69.21	71.78	66.67
MSP	07.27	06.00	06.67	06.15	05.46	05.71	04.00	04.00	05.00	04.29
OE	06.67	05.46	06.67	05.72	05.46	05.33	03.81	03.81	04.44	04.44
GNN-ODD w/o Reg	75.00	66.67	60.00	61.54	57.14	50.00	53.33	35.29	31.58	42.11
GNNSafe	85.71	66.67	60.00	66.67	61.54	54.55	57.14	35.30	35.29	42.11
GNNSafe++	02.14	01.60	01.31	01.47	01.53	01.39	01.73	01.70	02.90	01.65
Our model	**88.89**	**72.22**	71.43	**80.00**	**88.89**	**85.71**	**75.00**	**83.33**	**73.68**	**70.00**

erate enough discriminative information to separate bots in the learned embedding space. Oversampling approaches provide a good improvement over the traditional GNN models but all have their flaws: SMOTE does not synthesize structural information, resulting in oversampled nodes being disjoint from the rest of the graph with very limited contribution to the learning process; GraphSMOTE addresses the structural limitations of SMOTE by creating synthetic edges, which, however, does not consider the impact of heterophily present in the graph, limiting its classification performance; HOVER periodically shows very good performance but performance is inconsistent, and it requires multiple training cycles, modifies the semantics of the graph structure, and is sensitive to hyperparameter selection. ODD methods provide far more completion for our model. An interesting point is the negative impact that belief propagation has when graph heterophily is high. A comparison of GNNSafe and GNNSafe++ quantifies this observation, where in all data cases, the addition of belief propagation to GNNSafe hurts the detection performance. The cause of this is the nature of bot graphs. The out-of-distribution nodes, or bots, have a high local heterophily ratio regardless of the heterophily ratio of the entire graph. Naively, applying belief propagation to heterophilic graphs will pull bot samples toward the mean of non-bot samples. Bot nodes are then masked by benign nodes after energy values are calculated and propagated.

Our model outperforms all models in the tradition and ODD categories, with only one exception that our model performs second to HOVER. A direct comparison of our model to the various flavors of GNNSafe demonstrates the benefit of our heterophily-wise node embedding learning and homophily-augmented energy propagation scheme. The addition of edge prediction to energy propagation reduces the influence heterophily has on ODD.

4.4 Case Study: ODD for Bot Detection

We dig deeper into the performance of our model and the performance of a selection of baselines with a case study. Specifically, in this section, we look at

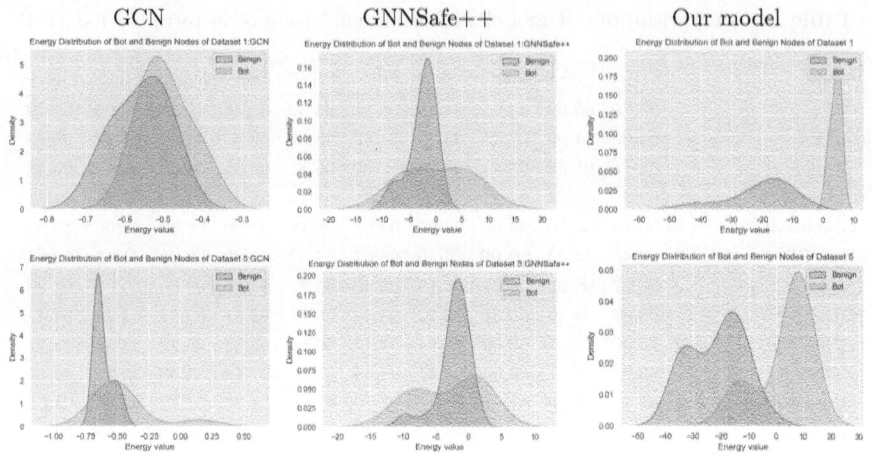

Fig. 4. Energy distributions learned by GCN, GNNSafe++, and our model. The upper row contains the distributions learned on capture file 3, while the lower row contains the distributions learned on capture file 5.

the learned energy distributions from GCN, GNNSafe++, and our model. We select two capture files for use in this case study: File 3 has low heterophily, File 5 has high heterophily, and both files has extremely low imbalance ratio.

GCN struggles to learn separated distributions for bot and benign samples. The impacts of heterophily can be seen by comparing the GCN results of each capture file. Capture file three, the top left of Fig. 4, contains mostly homophilic edges. This allows the GCN to constrain bot samples to energy levels less than the most extreme values of benign samples. In contrast, capture file five, the bottom left of Fig. 4, contains mostly heterophilic edge which causes the benign samples to attract bot samples. GNNSafe++ fails to separate or even constrain bot samples. GNNSafe++ does utilize OE and regularization during training which causes both learned distributions to be centered at approximately the same group about the same energy value. However, regularization alone is not capable of restraining the bot distribution. The bot distribution overlaps the entire learned benign distribution, creating a situation when no useful division between distributions can be found.

Our model is capable of separating the densest portions of the learned distributions. Moreover, our model impacts the extremes of both distributions. The energy values of lower tail of benign samples is pushed to extremely low energy values, extending below the -20 and -30 values that competing models reach on files three and five, respectively. The upper tail of the bot distribution displays similar behavior extending into positive energy values in both instances. This behavior allows our model to clearly separate many bots from benign samples.

4.5 Ablation Study

We conclude our experimentation with an ablation study to verify that all aspects of our model contribute in a manner that are both beneficial and nec-

Table 4. Ablation study to evaluate different model components in terms of F-1 (%).

Model	File 1	File 2	File 3	File 4	File 5	File 6	File 7	File 8	File 9	File 10
No Independent Embedding	02.14	01.15	01.24	14.24	01.60	01.38	02.88	01.58	01.59	01.58
No Propagation	75.00	**72.73**	60.00	66.67	61.53	61.53	57.14	15.38	22.48	15.38
No Edge Prediction	40.00	25.00	66.67	40.00	77.78	79.99	57.14	73.68	43.90	15.85
Our model	**88.89**	72.22	**71.43**	**80.00**	**88.89**	**85.71**	**75.00**	**83.33**	**73.68**	**70.00**

essary to the learning and classification processes. We evaluate the necessity of three elements to our model: (1) **Independent node and structural embedding**: we remove the MLPs from our model and replace them with a GCN. We refer to this experiment as "No independent embedding" in Table 4; (2) **Energy propagation**: we remove energy propagation entirely, which is denoted as "No Propagation" in Table 4, where we make a determination if a data sample is in or out of distribution immediately following the energy calculation; (3) **Edge prediction**: we repeat the experiment without using the homophily-augmented $\hat{\mathbf{A}}$ adjacency matrix, but use the original adjacency matrix $\mathbf{A}$, which is denoted as "No Edge Prediction" in Table 4.

The experiment results in Table 4 allows us to draw several conclusions. First, the fact that our model outperforms the reduced models on almost all capture files shows that all elements of our model contribute to the overall performance. File 2 in the "No Propagation" case is the lone instance where our complete model does not exceed the performance of a reduced model. As the difference between the complete and reduced model is negligible, we do not consider this undermines the utility of energy propagation. Moreover, as the "No Edge Prediction" case shows dramatically reduced performance, we assess that the anomaly speaks more to the benefit of independent representation embedding than it implies frivolous use of energy propagation. Our model is outperformed by "No Propagation" implying that the edge classification scheme is lacking in this instance. We consider this a motivation to explore improved edge prediction schemes in the future. A column-wise comparison of results highlights the individual contribution of each element of our model. The addition of edge prediction making arguable the biggest impact, improving performance by up to 48% on extremely heterophilic graphs and a more modest by still respectable improvement of 10% to 30% on the homophilic graphs (Files 6, 8, 9, and 10).

5 Conclusion

In this paper, we generalize the utility of ODD to the bot detection problem on graphs. To address over-smoothing and graph heterophily introduced by class imbalance, we propose a new energy-based ODD model to better deal with graph data. Our model formulates a simple yet effective node embedding method to encode node features and graph structure without the need for neighborhood aggregation. These node embeddings not only enhance energy calculation, but also enable homophily-augmented energy propagation, which significantly separate energy distributions between bots and benign samples, and thus improve the detection performance. Our experiments validate our proposed model and demonstrate that independent node embedding can overcome extreme heterophily, and the addition of edge prediction to existing energy propagation approaches advances the existing state-of-the-art in graph ODD and bot detection.

Acknowledgments. This work is partially supported by the NSF under grant CNS-2245968. The authors would also like to thank the reviewers for their valuable feedback.

References

1. Alsaedi, A., Moustafa, N., Tari, Z., Mahmood, A., Anwar, A.: Ton_iot telemetry dataset: a new generation dataset of iot and iiot for data-driven intrusion detection systems. IEEE Access **8**, 165130–165150 (2020)
2. Ashmore, B., Chen, L.: Hover: Homophilic oversampling via edge removal for class-imbalanced bot detection on graphs. In: CIKM, pp. 3728–3732 (2023)
3. Bitterwolf, J., Meinke, A., Augustin, M., Hein, M.: Breaking down out-of-distribution detection. In: ICML, pp. 2041–2074 (2022)
4. Cen, J., Yun, P., Cai, J., Wang, M.Y., Liu, M.: Deep metric learning for open world semantic segmentation. In: ICCV, pp. 15333–15342 (2021)
5. Chatterjee, M., Namin, A.S., Datta, P.: Evidence fusion for malicious bot detection in iot. In: IEEE International Conference on Big Data, pp. 4545–4548 (2018)
6. Chawla, N.V., Bowyer, K.W., Hall, L.O., Kegelmeyer, W.P.: Smote: synthetic minority over-sampling technique. JAIR **16**, 321–357 (2002)
7. Chen, L., Li, X., Wu, D.: Enhancing robustness of graph convolutional networks via dropping graph connections. In: ECML PKDD, pp. 412–428 (2021)
8. Cui, P., Wang, J.: Out-of-distribution (ood) detection based on deep learning: a review. Electronics **11**(21), 3500 (2022)
9. Hamilton, W., Ying, Z., Leskovec, J.: Inductive representation learning on large graphs. In: Advances in Neural Information Processing Systems, vol. 30 (2017)
10. Hendrycks, D., Gimpel, K.: A baseline for detecting misclassified and out-of-distribution examples in neural networks. In: ICLR (2017)
11. Hendrycks, D., Mazeika, M., Dietterich, T.: Deep anomaly detection with outlier exposure. In: ICLR (2019)
12. Hsu, Y.C., Shen, Y., Jin, H., Kira, Z.: Generalized odin: detecting out-of-distribution image without learning from out-of-distribution data. In: CVPR, pp. 10951–10960 (2020)
13. Kipf, T.N., Welling, M.: Semi-supervised classification with graph convolutional networks. In: ICLR (2017)
14. Li, Q., Chen, L., Cai, Y., Wu, D.: Hierarchical graph neural network for patient treatment preference prediction with external knowledge. In: PAKDD, pp. 204–215 (2023)
15. Li, Q., Li, X., Chen, L., Wu, D.: Distilling knowledge on text graph for social media attribute inference. In: SIGIR, pp. 2024–2028 (2022)
16. Li, Z., Wu, Q., Nie, F., Yan, J.: Graphde: a generative framework for debiased learning and out-of-distribution detection on graphs. Adv. Neural. Inf. Process. Syst. **35**, 30277–30290 (2022)

17. Li, Z., Liu, F., Yang, W., Peng, S., Zhou, J.: A survey of convolutional neural networks: analysis, applications, and prospects. TNNLS **33**(12), 6999–7019 (2021)
18. Liang, S., Li, Y., Srikant, R.: Enhancing the reliability of out-of-distribution image detection in neural networks. In: ICLR (2018)
19. Lim, D., Hohne, F., Li, X.: Large scale learning on non-homophilous graphs: new benchmarks and strong simple methods. NeurIPS **34**, 20887–20902 (2021)
20. Liu, W., Wang, X., Owens, J., Li, Y.: Energy-based out-of-distribution detection. Adv. Neural. Inf. Process. Syst. **33**, 21464–21475 (2020)
21. Liu, Y., Zheng, Y., Zhang, D., Lee, V.C., Pan, S.: Beyond smoothing: Unsupervised graph representation learning with edge heterophily discriminating. In: AAAI, pp. 4516–4524 (2023)
22. Lo, W.W., Kulatilleke, G., Sarhan, M., Layeghy, S., Portmann, M.: Xg-bot: an explainable deep graph neural network for botnet detection and forensics. Internet Things **22**, 100747 (2023)
23. Luan, S., Hua, C., Lu, Q., Zhu, J., Zhao, M.: Revisiting heterophily for graph neural networks. NeurIPS **35**, 1362–1375 (2022)
24. Ma, J., Deng, J., Mei, Q.: Subgroup generalization and fairness of graph neural networks. NeurIPS **34**, 1048–1061 (2021)
25. Ma, Y., Liu, X., Shah, N., Tang, J.: Is homophily a necessity for graph neural networks? In: ICLR (2022)
26. Moustafa, N.: A new distributed architecture for evaluating ai-based security systems at the edge: Network ton_iot datasets. Sustainable Cities and Society (2021)
27. Pei, H., Wei, B., Chang, K.C.C., Lei, Y., Yang, B.: Geom-gcn: geometric graph convolutional networks. In: ICLR (2020)
28. Rahal, B.M., Santos, A., Nogueira, M.: A distributed architecture for ddos prediction and bot detection. IEEE Access **8**, 159756–159772 (2020)
29. Stadler, M., Charpentier, B.: Graph posterior network: Bayesian predictive uncertainty for node classification. NeurIPS **34**, 18033–18048 (2021)
30. Sun, Y., Guo, C., Li, Y.: React: Out-of-distribution detection with rectified activations. Adv. Neural. Inf. Process. Syst. **34**, 144–157 (2021)
31. Veličković, P., Cucurull, G., Casanova, A., Romero, A., Lio, P., Bengio, Y.: Graph attention networks. In: ICLR (2018)
32. Wu, Q., Chen, Y., Yang, C., Yan, J.: Energy-based out-of-distribution detection for graph neural networks. In: ICLR (2023)
33. Yan, Y., Hashemi, M., Swersky, K., Yang, Y., Koutra, D.: Two sides of the same coin: Heterophily and oversmoothing in graph convolutional neural networks. In: ICDM, pp. 1287–1292 (2022)
34. Yang, N., Zeng, K., Wu, Q., Jia, X., Yan, J.: Learning substructure invariance for out-of-distribution molecular representations. NeurIPS **35**, 12964–12978 (2022)
35. Yang, Z., Cohen, W., Salakhudinov, R.: Revisiting semi-supervised learning with graph embeddings. In: ICML, pp. 40–48 (2016)
36. Zhang, B., Li, J., Chen, C., Lee, K., Lee, I.: A practical botnet traffic detection system using gnn. In: CSS, pp. 66–78 (2022)
37. Zhang, J., Perdisci, R., Lee, W., Luo, X., Sarfraz, U.: Building a scalable system for stealthy p2p-botnet detection. TIFS **9**(1), 27–38 (2013)
38. Zhao, T., Zhang, X., Wang, S.: Graphsmote: imbalanced node classification on graphs with graph neural networks. In: WSDM, pp. 833–841 (2021)
39. Zhou, J., Xu, Z., Rush, A.M., Yu, M.: Automating botnet detection with graph neural networks. In: MLSys (2020)
40. Zhu, J., et al.: Graph neural networks with heterophily. In: AAAI, pp. 11168–11176 (2021)

Multi-level Contrastive Learning on Weak Social Networks for Information Diffusion Prediction

Zihan Feng[1], Rui Wu[1], Yajun Yang[1(✉)], Hong Gao[2], Xin Wang[1], Xueli Liu[1], and Qinghua Hu[1]

[1] College of Intelligence and Computing, Tianjin University, Tianjin, China
{zihanfeng,circle,yjyang,wangx,xueli,huqinghua}@tju.edu.cn

[2] College of Mathematics and Computer Science, Zhejiang Normal University, Jinhua, China
honggao@zjnu.edu.cn

Abstract. Information diffusion prediction, as a fundamental task in social network analysis, aims to identify potential users who are likely to participate in an information diffusion process. Most existing works learn user representations based on the collected social network data and then complete downstream prediction tasks. However, due to data privacy protection and low data quality, these methods are always limited by weak information issues of the social network data. For example, incomplete network structure, sparse labels, and insufficient features severely obstruct user representation learning. To mitigate these issues, we design an effective two-stage method MGCL. In the first stage, an enhanced representation is learned for every user even though the social network is with weak information. A multiplex heterogeneous network is adaptively constructed to enrich social network information. To facilitate user representation learning under sparse labels and insufficient features, we further propose self-supervised training specifically tailored for social networks with weak information. In the second stage, the cascade representations are learned using the multi-head self-attention network for information diffusion prediction. Extensive experiments on four real-world datasets validate that MGCL always outperforms state-of-the-art methods.

Keywords: Social networks analysis · Information diffusion prediction · Graph representation learning · Contrastive learning

1 Introduction

With the rapid development of the mobile Internet, social network platforms have become the most important place for people to obtain and share information. Information diffusion prediction is a fundamental task in social networks, which can assist in understanding the evolution of trending topics and has many

M. Onizuka et al. (Eds.): DASFAA 2024, LNCS 14855, pp. 84–100, 2024.
https://doi.org/10.1007/978-981-97-5572-1_6

important applications, such as public opinion analysis [1], fake news control [2], and online social marketing [3].

Information diffusion prediction is to identify potential users who are likely to participate in information sharing, e.g., re-tweets. Existing methods can be classified into two categories: predefined model-based methods and cascade learning-based methods. Predefined model-based methods artificially design various propagation models [4] (e.g., SIR model) to predict which users will be involved in the information diffusion. However, these methods may not be strictly consistent with information propagation in the real world. To address this limitation, cascade learning-based methods have been proposed recently. An information diffusion cascade is a sequence of tuples (v_x, t_x), where v_x is a user who participates in this propagation and t_x is her/his participation time. On the one hand, some works focus on extracting the macro-level features, such as user statistics or information contents, to predict the overall growth of cascade size [5,6]. On the other hand, some methods aim to predict the next user that will participate in this cascade based on representation learning. These methods utilize various techniques such as GNNs and RNNs to learn users' social relationships and sequential cascade interaction [10–13] for information diffusion prediction.

Despite the improvements made in describing the user social homogeneity and cascade interaction, existing methods still suffer from three critical limitations: (1) Existing works are designed on the fundamental assumption that observed social network data contains complete and sufficient information. However, in the real world, these data suffer from extremely weak information issues, manifested by incomplete structure, sparse labels, and insufficient user features. Such low-quality data severely obstructs learning user representations. (2) All existing methods only utilize cascade sequences as the prediction-oriented objective to jointly learn social networks and information diffusion cascades, which is insufficient for learning reliable user representations of the social network and further results in suboptimal prediction performance. (3) Existing end-to-end methods all use the full-graph training pipeline for the entire social network, which is straightforward but significantly restricts practical flexibility.

To address these limitations, we propose a novel two-stage method MGCL for information diffusion prediction. In the first stage, an enhanced representation can be learned for every user even though the social network only has weak information. On the one hand, by leveraging the relationships of users' follower-followee and historical participation behavior, a multiplex heterogeneous social network is constructed to enrich the structure information in the original social network and then an adaptive early-fusion strategy is well-designed to ensure a reliable adjacency matrix for graph learning. On the other hand, to facilitate user representation learning under sparse labels and insufficient features, we propose self-supervised training specifically tailored for social networks with weak information and further design contrastive learning tasks at the user and preference level respectively. Our method can effectively mitigate the challenges posed by weak information issues to user preference modeling and thereby improve the performance of downstream prediction. Moreover, we employ an inductive graph learning method with the mini-batch setting that can cope with the graph structure changing and is more flexible for model training. In the second stage, the cascade representations are learned using the multi-head self-attention network for the downstream information diffusion prediction task.

The main contributions of this paper are summarized as follows:

(1) We propose an effective method to enhance the original social network by integrating users' historical behavior into a multiplex heterogeneous social network and designing an adaptive early-fusion strategy to ensure a reliable user adjacency matrix for graph learning.
(2) We design a novel multi-level graph contrastive learning method that can provide more sufficient self-supervised signals for graph learning. In this way, our method can obtain more expressive user representations even if it does not have user features and labels.
(3) We conduct extensive experiments on four real-world datasets. The results show that: (i) MGCL can enable more effective supervision for graph learning and further effectively improve prediction performance. (ii) In various scenarios, MGCL consistently outperforms the state-of-the-art methods.

2 Preliminaries

Definition 1. (**Information Diffusion Cascade**). *An information diffusion cascade C_k with k users is a sequence of k tuples, i.e., $C_k = ((v_1, t_1), (v_2, t_2), \ldots, (v_k, t_k))$, where $v_x \neq v_y$ and $t_x \leq t_{x+1}$. Each tuple (v_x, t_x) is called a "participation" indicating that user v_x is the x-th participant in information diffusion and the participation time is t_x.*

The information diffusion process can be regarded as a communication chain. An information diffusion cascade records the complete journey of a specific piece of information spreading through the social network. In information diffusion prediction, C_k is used as an input to predict the next participation (v_{k+1}, t_{k+1}).

Definition 2. (**Ideal Social Network**). *We define the ideal social network as social network data that contains sufficient information. Let the ideal social network be $\hat{\mathcal{D}} = (\hat{\mathcal{G}}, \hat{\mathbf{Y}}_L) = ((\mathcal{V}, \hat{\mathcal{E}}, \hat{\mathbf{X}}), \hat{\mathbf{Y}}_L)$, where $\hat{\mathcal{E}}$ is an ideal edge set containing all relevant links between users, $\hat{\mathbf{X}}$ is an ideal feature matrix containing all informative features for each user, and $\hat{\mathbf{Y}}_L$ is an ideal label matrix containing sufficient labels (with number $\hat{n}_L$) for learning tasks with balanced distribution.*

In real-world scenarios, there often is insufficient data for model training and deployment [14]. Specifically, the structure can be incomplete with an *incomplete edge set* $\check{\mathcal{E}} \subsetneq \hat{\mathcal{E}}$ that contains limited edges to provide sufficient information. Meanwhile, some critical elements in the feature matrix are missing, which can be represented by an *incomplete feature matrix* $\check{\mathbf{X}} = \mathbf{M} \odot \hat{\mathbf{X}}$, where $\mathbf{M} \in \{0, 1\}^{n \times d}$ is the missing mask matrix. Besides, the available labels for training can be scarce, resulting in an *insufficient label matrix* $\check{\mathbf{Y}}_L$ with training number $\check{n}_L \ll \hat{n}_L$.

Definition 3. (**Weak Social Network**). *Due to privacy security issues, weak social networks usually do not have user features and labels, and observed relationships are always incomplete, further exacerbating the challenges. Let social network data with weak information be* $\mathcal{D}_x = ((\mathcal{V}, \check{\mathcal{E}}, \check{\mathbf{X}}), \check{\mathbf{Y}}_L)$. *The target is to learn the representation of users* $\mathcal{V}$ *with* $\mathcal{D}_x$ *for model training.*

In this paper, we confront a more challenging scenario wherein the structure, features, and labels exhibit serious deficiencies. Specifically, manual labels are absent for all users, initial user features are randomized, and observed user relationships are extremely incomplete. The sole additional information available is users' historical participation in information diffusion processes.

Definition 4. (**Multiplex Heterogeneous Network**). *A multiplex heterogeneous network can be denoted as* $G = (V, E)$, *where* V *and* E *are the sets of nodes and edges respectively.* $\phi : V \to O$ *is a node type mapping function and* $\psi : E \to R$ *is an edge type mapping function, where* O *and* R *represent the sets of node types and edge types respectively, and they satisfy* $|O| + |R| > 2, |R| > 1$. *Different from traditional heterogeneous networks, there exist multiple edges with different types between two nodes, i.e.* $E \subseteq V \times V \times R$.

In the real world, a social network inherently involves multiple types of complex relationships between users and thus it can be modeled by a multiplex heterogeneous network. For instance, follower-followee relationships on platforms such as Twitter constitute one type of edge. Moreover, users engaging in the same information diffusion events can be viewed as another type of edge, signifying shared interests in specific trending topics. This multiplex heterogeneous network, comprising multiple types of relationships, can offer more comprehensive information conducive to mitigating weak information issues.

Problem Formulation. The task of information diffusion prediction is to predict the next participation (v_{k+1}, t_{k+1}) based on the current information diffusion cascade C_k. Formally, the prediction task is to select an optimal v_x from $V \backslash V_k$, to maximize the following conditional likelihood, as the next participant v_{k+1}.

$$\hat{v}_x = \arg\max P(v_x | G, C_k), v_x \in V \backslash V_k \tag{1}$$

where V_k is the set of vertices in C_k, G is the social network and C_k is the information diffusion cascade sequence.

3 Methodology

Overview. The overall architecture of MGCL is shown in Fig. 1. Specifically, MGCL is a two-stage model designed for the information diffusion prediction task. In the first stage, a multiplex heterogeneous social network is constructed adaptively, taking into account both the original relationships of users' follower-followee connections and their historical participation. This enhanced structure mitigates weak information issues present in the original social network. Subsequently, we design three self-supervised tasks that can contribute to adaptive graph construction and effective graph learning. In this way, we can obtain better user representations for the downstream prediction even if it does not have user features and labels. In the second stage, we further exploit the multi-head self-attention network to learn the information diffusion cascade for prediction.

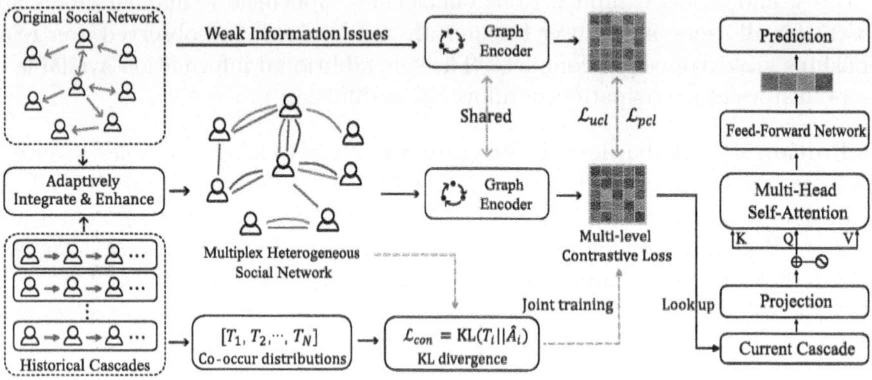

Fig. 1. The overall architecture of the proposed method MGCL.

3.1 Multiplex Heterogeneous Graph Learning

Adaptive Graph Construction. An ideal graph structure is the basis for graph learning. However, the weak social network hinders the graph encoder's ability to capture crucial dependencies between users and leads to biased user representations [14]. By leveraging the assumption that user dependencies mirror real-world information diffusion patterns, we propose an adaptive graph construction method to enhance the original social network. On the one hand, users' historical participation relationship indicates which users prefer to participate in the same information diffusion and prediction task can benefit from it. Given an information diffusion cascade C_k and an integer h, an edge (v_x, v_{x+i}) will be built for every $i \leq h$ and every v_x and v_{x+i} in C_k. For example of $h = 3$, three edges (v_x, v_{x+1}), (v_x, v_{x+2}), and (v_x, v_{x+3}) are created for every v_x in C_k, where $1 \leq$ x $\leq k-3$. We can select an appropriate h to state users' historical participation. On the other hand, the follower-followee relationships can be treated as two types of edges due to the limitation of graph convolution on directed graphs. For example, user v_y is a follower of user v_x, thus (v_y, v_x) and (v_x, v_y) are created to represent the follower and followee relationship respectively. The intuitive meaning behind that is followers are usually influenced by their followees but not vice versa and it reveals the different influential roles of users.

For each edge type $r \in R$, a "***projection***" of a multiplex heterogeneous network G on edge type r, denoted as $G_r = (V, E_r)$, is a subgraph of G consisting of all the nodes in V and all the edges in E_r, where $E_r = \{e|e \in E \wedge \psi(e) = r\}$. Each G_r $(1 \leq r \leq |R|)$ only focuses on one type of relationship in G. We use $\boldsymbol{A}_r$ to denote the adjacency matrix of G_r. To adaptively integrate and enhance multiplex heterogeneous relationships between users, we utilize the relation-aware weights β_r to aggregate the adjacency matrix of different projections as follows:

$$\hat{\boldsymbol{A}} = \sum_{r=1}^{|R|} \beta_r \boldsymbol{A}_r \tag{2}$$

where β_r $(1 \leq r \leq |R|)$ is a learnable relationship weight. We use the *softmax* function to ensure $\sum_{r=1}^{|R|} \beta_r = 1$. In this way, the weights imply the importance of different types of relationships, and thus the aggregated adjacency matrix $\hat{\boldsymbol{A}}$ can characterize the multiplex heterogeneous relationships among users.

Inductive Graph Encoder. In this paper, we use the inductive GraphSAGE [16] as the basic graph encoder to propagate the node message exploiting the multiplex heterogeneous social network $\hat{\boldsymbol{A}}$. The message propagation and aggregation at the t-th layer of the graph neural network are as follows:

$$\alpha_{v_i}^{(t)} = \text{Aggregate}^{(t)}(\{h_{v_j}^{(t-1)} : v_j \in N_{v_i}\}) \tag{3}$$

$$h_{v_i}^{(t)} = \text{Update}^{(t)}(\alpha_{v_i}^{(t)}, h_{v_i}^{(t-1)}) \tag{4}$$

where N_{v_i} denotes the set of v_i's neighbors in $\hat{\boldsymbol{A}}$, $h_{v_i}^{(t)}$ denotes the representation of v_i at the t-th GNN layer. $h_{v_i}^{(0)}$ is the initial user embedding randomly generated by the normal distribution. Aggregate$(\cdot)$ is the function aggregating the neighborhood information of a central node v_i, and Update$(\cdot)$ is the function that combines neighborhood information to update the node embedding. In this paper, we use weighted aggregation as the neighborhood aggregation function based on the weight of relationships in the adjacency matrix $\hat{\boldsymbol{A}}$. We denote the embedding at the last layer as the final user representation, which can be used as an input for subsequent information diffusion prediction.

3.2 Self-supervised Graph Training

Previous works entangle graph training and the downstream prediction task for end-to-end learning, but this joint training pipeline has two serious limitations: (1) The supervision signal of joint training predominantly focuses on downstream prediction, overlooking crucial supervision for the social network itself. (2) The joint training pipeline necessitates preloading the entire social network to acquire extensive representations of all users and subsequently perform lookups during downstream tasks. This unnecessary overhead constrains the practical flexibility of GNN-based methods. To achieve more efficient and effective graph learning, we design mini-batch self-supervised objectives to supervise graph construction and user representation learning. Subsequently, these representations can be projected into the downstream model for improved prediction.

Graph Structure Regularization. A critical foundation of graph learning is to generate a more reliable adjacency matrix $\hat{\boldsymbol{A}}$ that requires effective learning of β_r in Eq. (2). However, without proper supervision, the adaptive strategy may result in sub-optimal solutions, i.e., the quality of the constructed graph may not be consistent with the preferences exhibited by users in actual information diffusion participation. Inspired by [18], we leverage co-occur patterns in historical cascades as supervision to conduct graph structure regularization. This method capitalizes on the reasonable observation that frequently co-occurring users in historical cascades often share similar preferences, making it a natural and effective supervision for constructing a reliable graph structure.

We first precalculate the co-occur pattern $P_i \in N^N$ for each user v_i, where the co-occur weight to the j-th user equals the total frequency that v_i and v_j participate in the same information diffusion in all historical cascades. Furthermore, we transform P_i to the discrete distribution $T_i \in N^N$ by row normalization and softmax operation. We formalize the regularization of graph structure as the KL divergence between the approximate distribution of each user from the adjacency matrix $\hat{\boldsymbol{A}}$ and its corresponding distribution T_i of co-occur regularity, which indicates the discrepancy between these two distributions. In practice, we transform each row $\hat{A}_i$ of $\hat{\boldsymbol{A}}$ to the approximate distribution with softmax operation and compute the regularization term $\mathcal{L}_{con}$ as:

$$\mathcal{L}_{con} = \frac{1}{N} \sum_{i=1}^{N} KL(T_i || softmax(\hat{A}_i)) \tag{5}$$

This graph structure regularization objective ensures a refined graph structure, consequently mitigating the challenges posed by weak information issues. Moreover, the learned graph structure $\hat{\boldsymbol{A}}$ explicitly captures the pertinent dependencies between users in real-world scenarios, which is more beneficial for prediction.

Multi-level Contrastive Learning. Then we can conduct graph encoder training on the constructed multiplex heterogeneous social network. Motivated by the principle that user representations should be consistent with their inherent preferences in information diffusion, we introduce a novel multi-level graph contrastive learning method with the mini-batch setting, which operates on two different levels: user-level and preference-level.

As shown in Fig. 2, denote the distance between the representations of v_i in different views as d_u, the distance between v_i and another user v_j with similar preferences as d_s, and the distance between the user and another user v_k with quite different preference as d_f. Note that the generated user representations can be different due to input graph structure differences, i.e. $0 < d_u$. Meanwhile, excessively increasing d_u will result in overlapping between users with the other and therefore lead to suboptimal results, i.e. $d_u < \{d_s, d_f\}$. Furthermore, user representations with similar preferences should have smaller distances than those with larger differences, i.e. $d_s < d_f$. To further refine our distance considerations, we introduce d_δ representing the approximate boundary of the preference decision area of v_i. Ideally, the representation of user v_i should be distributed at

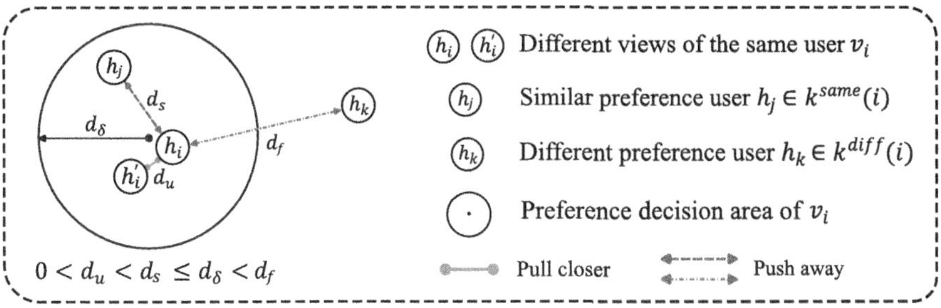

Fig. 2. Here is an example of multi-level contrastive learning.

the center of the area, and then d_s can be approximated as the distance from h_j to the center of the area. Therefore, we strive to maintain $d_s \leq d_\delta$ (or $d_s \approx d_\delta$), ensuring h_j resides within the appropriate boundary.

In summary, the total goal can be further described as:

$$0 < d_u < d_s \leq d_\delta < d_f \tag{6}$$

Based on the guidance of this goal, we can generate positive samples and negative samples within a mini-batch. Then we take user v_i as an example to introduce the detailed generation strategy. Given the representation h_i of user v_i learned from the constructed multiplex heterogeneous social network as a ***query***.

Positive Key Generation. In Sect. 3.1, we enhance the original social network by integrating the original structure and historical participation. Although the graph structure is not the same, the user representations in the original social network and the enhanced social network should be similar. Therefore, we also learn the original social network using the shared graph encoder. In this way, we can obtain the original user representation h_i^* as the positive key.

Negative Key Generation. Initially, we treat all representations of other in-batch samples as its negative keys, where $k^{neg}(i) = \{h_j, h_j^*\}_{j=1...n, j \neq i}$. However, this does not represent the crucial link between user representations and their actual information diffusion behavior. Furthermore, $k^{neg}(i)$ can be divided into two groups according to the co-occur pattern $P_i \in N^N$:

$$k^{diff}(i) = \{k \in k^{neg}(i) : P_{i,k} < \delta\} \tag{7}$$

$$k^{same}(i) = \{k \in k^{neg}(i) : P_{i,k} \geq \delta\} \tag{8}$$

where $k^{diff}(i)$ is the group of users who do not frequently co-occur with user v_i in this batch, and $k^{same}(i)$ is the opposite. In this way, these two groups of users can respectively represent two types of users with different and similar preferences as user v_i in information diffusion. This in-batch generation of positive and negative examples ensures the scalability of our method. Now given the query h_i with its positive key h_i^* and negative keys $k^{neg}(i) = \{k^{diff}(i), k^{same}(i)\}$, we can introduce the user-level and preference-level contrastive loss as follows.

User-level Contrastive Loss. We aim to guarantee $0 < d_u < \{d_s, d_f\}$ at the user level. To achieve this, we design an InfoNCE-based loss as follows:

$$\mathcal{L}_{ucl} = -\frac{1}{N}\sum_{i=1}^{N}\left(\log\frac{\mathrm{f}(h_i, h_i^*)}{\mathrm{f}(h_i, h_i^*) + \sum_{j\neq i}\mathrm{f}(h_i, h_j^*)}\right) \tag{9}$$

where $\mathrm{f}(\mathbf{a}, \mathbf{b}) = \mathrm{e}^{\mathrm{sim}(\mathbf{a},\mathbf{b})/\tau}$, $\mathrm{sim}(\cdot,\cdot)$ is the cosine similarity, and τ is temperature coefficient. Note that taking $\mathrm{f}(h_i, h_i^*)$ as the denominator can ensure $0 < d_u$.

Preference-level Contrastive Loss. At the preference level, our goal is to satisfy $0 < \{d_s, d_\delta\} < d_f$. If depict the exact preference boundary of user v_i, we should calculate for each user in $k^{same}(i)$, but this is quite time-consuming. Therefore, we introduce the preference prototype. For each user v_i, we acquire its preference prototype $p_i, p_i^* = \mathbf{MEAN}(k^{same}(i))$, which are correlated to the preference boundary d_δ. The preference-level contrastive loss can be defined as:

$$\mathcal{L}_{pcl} == -\frac{1}{N}\sum_{i=1}^{N}\left(\log\frac{\mathrm{f}(h_i, p_i)}{\sum_{j\in k_-^{diff}(i)}\mathrm{f}(h_i, h_j)} + \log\frac{\mathrm{f}(h_i, p_i^*)}{\sum_{j\in k_-^{diff}(i)}\mathrm{f}(h_i, h_j^*)}\right) \tag{10}$$

where $\mathrm{f}(\mathbf{a}, \mathbf{b})$ is the same as Eq.(9). $\mathcal{L}_{pcl}$ increases the agreement between the user and its similar users, which explicitly incorporates the crucial link between user representations and their actual information diffusion behavior.

Overall Training Objective. *Graph Structure Regularization* and *Multi-level Contrastive Learning* are all formulated for a single node, eliminating the need to preload the entire graph. So the overall graph training objective is:

$$\mathcal{L}_{graph} = \gamma_1\mathcal{L}_{con} + \gamma_2\mathcal{L}_{ucl} + \gamma_3\mathcal{L}_{pcl} \tag{11}$$

where $\gamma_1, \gamma_2, \gamma_3$ are the hyper-parameter. In each mini-batch, we randomly sample n nodes to calculate $\mathcal{L}_{graph}$. In this way, we can simultaneously achieve better graph construction and better graph learning, which provides an effective and comprehensive solution for learning the weak social network.

3.3 Information Diffusion Prediction

In this section, we aim to convert the user representation sequence into the cascade representation for information diffusion prediction. Here we employ the basic multi-head self-attention mechanism [17] and eschew more complex encoders, which emphasizes the effectiveness of our proposed graph learning method.

Cascade Interaction Encoder. We first look up user representations $\boldsymbol{H}_k = \{\boldsymbol{h}_x | v_x \in V_k\}$ from multiplex heterogeneous graph encoder for the cascade C_k. Then we encode position information to obtain $\boldsymbol{H}_k^{'} = \{\boldsymbol{h}_x^{'} | v_x \in V_k\}$, where $\boldsymbol{h}_x^{'} =$

$[\boldsymbol{h}_x||\boldsymbol{p}_x]$ and the $\boldsymbol{p}_x$ is a learnable position embedding. $\boldsymbol{H}_k^{'}$ is user representations in C_k and preserves the relative sequential information of users.

We further employ the basic Multi-Head Attention (MHA) to model cascade interaction. The MHA of the i-th attention head can be calculated as follows:

$$\boldsymbol{Q}_i = \boldsymbol{H}_k^{'}\boldsymbol{W}_i^Q, \boldsymbol{K}_i = \boldsymbol{H}_k^{'}\boldsymbol{W}_i^K, \boldsymbol{V}_i = \boldsymbol{H}_k^{'}\boldsymbol{W}_i^V \tag{12}$$

$$\text{MHA}_i(\boldsymbol{Q}_i, \boldsymbol{K}_i, \boldsymbol{V}_i) = \text{softmax}(\frac{\boldsymbol{Q}_i\boldsymbol{K}_i^{\top}}{\sqrt{d^{'}}} + \boldsymbol{M})\boldsymbol{V}_i \tag{13}$$

where d is the dimension of the embedding, B represents the number of heads, $d^{'} = d/B$, $\boldsymbol{W}_i^Q$, $\boldsymbol{W}_i^K$, and $\boldsymbol{W}_i^V$ are learnable projection matrices. To avoid label leakage, we introduce a mask matrix $\boldsymbol{M} \in \mathbb{R}^{n \times n}$ to block out future information and achieve this via setting the upper triangle of the attention map with $-\infty$. We concatenate the outputs of B attention heads to form the output $\boldsymbol{S}$ as:

$$\boldsymbol{S} = [\text{MHA}_1||\dots||\text{MHA}_B]\boldsymbol{W}^O \tag{14}$$

where $||$ is the concatenation operation, and $\boldsymbol{W}^O$ is the transformation matrix.

To introduce non-linearity to the MHA, we use a position-wise feed-forward network, allowing it to model complex dependencies within the cascade:

$$\boldsymbol{Z} = (\text{ReLU}(\boldsymbol{S}\boldsymbol{W}_t^1 + \boldsymbol{b}_t^1))\boldsymbol{W}_t^2 + \boldsymbol{b}_t^2 \tag{15}$$

where $\boldsymbol{W}_t^1$, $\boldsymbol{W}_t^2$, $\boldsymbol{b}_t^1$, and $\boldsymbol{b}_t^2$ are learnable parameters, $\boldsymbol{Z} \in \mathbb{R}^{k \times d}$ is the final information diffusion cascade representation.

Prediction. After that, we calculate the probability $\hat{y} \in \mathbb{R}^{|C_k| \times |V|}$ of the next participant using a softmax function:

$$\hat{y} = \text{softmax}(\boldsymbol{W}_p\boldsymbol{Z} + Mask) \tag{16}$$

where $\boldsymbol{W}_p$ maps $\boldsymbol{Z}$ to user-specific space, $Mask$ is used to mask users who have participated in information diffusion before prediction. Finally, we can use the cross-entropy function to supervise the training of cascade prediction:

$$\mathcal{L}(\theta) = -\sum_{j=2}^{|C_k|}\sum_{i=1}^{|V|} y_{ij}\log(\hat{y}_{ij}) \tag{17}$$

in which θ denotes all parameters needed to be learned in the model, if the user v_i participates in cascade C_k at the step j, $y_{ij} = 1$, otherwise $y_{ij} = 0$.

4 Performance Evaluation

In this section, we conduct extensive experiments to evaluate the performance of our MGCL model and answer the following Research Questions (RQs):

RQ1: How does the proposed model perform compared to baseline models?
RQ2: How do the critical components affect the model performance?
RQ3: How do the different hyperparameter settings affect the performance?
RQ4: How does the proposed model perform under different practical scenarios?

Table 1. Statistics of the four used datasets.

Datasets	Twitter	Douban	Android	Christianity
# Users	12,627	12,232	9958	2897
# Links	309,631	396,580	48,573	35,624
# Cascades	3442	3475	679	589
Avg. Len.	32.60	21.76	33.30	22.90

4.1 Experimental Settings

Datasets. We study the model performance using four real-world social network datasets, i.e. **Twitter** [27], **Douban** [28], **Android** [10], and **Christianity** [10]. These datasets are widely used and each contains the social network and information diffusion cascades. The descriptive statistics are in Table 1.

Evaluation Metrics. According to the problem formulation, the prediction task can be regarded as a retrieval problem. Therefore, we choose the widely used *Hits rate* on top-k (Hits@k) and *Mean Average Precision* on top-k (MAP@k) to evaluate the performance, where k ={10, 50, 100}.

Compared Methods. We compare our method with seven recent information diffusion prediction methods. **Topo-LSTM** [7] and **NDM** [8] utilize RNNs and CNNs to effectively capture the dependencies inherent in the diffusion cascade. **SNIDSA** [9], **Inf-VAE** [10] and **FOREST** [11] employ GNNs to learn users' social relationships and utilize RNN to explore the context of cascades for prediction. **DyHGCN** [12] and **MS-HGAT** [13] focus on integrating various user relationships and then designing novel GNNs for user representation learning.

Implementation Details. The experiments are conducted using PyTorch on a 24GB NVIDIA TITAN RTX GPU. For each dataset, we split these cascades by 8:1:1 for training, validation, and testing. The maximum cascade length is set to 200. To avoid information leakage, we use the training data as the historical cascades. We set the embedding dimension $d = 128$ and the attention head $B = 6$. For the model training, the AdamW optimizer is chosen, initialized with a learning rate of 5e-4 and a weight decay coefficient set to 0.01. The training configuration includes a batch size of 32 and a dropout rate of 0.3. The maximum number of training epochs in the downstream task is set to 50, and we early stop after 3 consecutive epochs without observed improvement on the validation set. The number of training epochs in the graph training is chosen from 4 to 8. The settings for the baselines remain consistent with the original papers.

Table 2. The performance on **Hits@k** metrics (%), scores are the higher the better.

Models	Twitter			Douban			Android			Chiristianity		
	@10	@50	@100	@10	@50	@100	@10	@50	@100	@10	@50	@100
TopoLSTM	8.45	15.80	25.42	8.57	16.53	21.47	4.56	12.63	16.53	12.28	22.63	31.52
NDM	15.21	28.23	32.30	10.00	21.13	30.14	4.85	14.24	18.97	15.41	31.36	45.86
SNIDSA	25.37	36.64	42.89	16.23	27.24	35.59	5.63	15.22	20.93	17.74	34.58	48.76
Inf-VAE	14.85	32.72	45.72	8.94	22.02	35.72	5.98	14.70	20.91	18.38	38.50	51.05
FOREST	28.67	42.07	49.75	19.50	32.03	39.08	9.68	17.73	24.08	24.85	42.01	51.28
DyHGCN	31.88	45.05	52.19	18.71	32.33	39.71	9.10	16.38	23.09	26.62	42.80	52.47
MS-HGAT	33.50	49.59	58.91	21.33	35.25	42.75	10.41	20.31	27.55	28.80	47.14	55.62
MGCL	**34.28**	**51.79**	**62.47**	**21.37**	**35.44**	**43.27**	**10.79**	**21.43**	**29.35**	**31.70**	**50.22**	**60.17**

Table 3. The performance on **MAP@k** metrics (%), scores are the higher the better.

Models	Twitter			Douban			Android			Chiristianity		
	@10	@50	@100	@10	@50	@100	@10	@50	@100	@10	@50	@100
TopoLSTM	8.51	12.68	13.68	6.57	7.53	7.78	3.60	4.05	4.06	7.93	8.67	9.86
NDM	12.41	13.23	14.30	8.24	8.73	9.14	2.01	2.22	2.93	7.41	7.68	7.86
SNIDSA	15.34	16.64	16.89	10.02	11.24	11.59	2.98	3.24	3.97	8.69	8.94	9.72
Inf-VAE	19.80	20.66	21.32	11.02	11.28	12.28	4.82	4.86	5.27	9.25	11.96	12.45
FOREST	19.60	20.21	21.75	11.26	11.84	11.94	5.83	6.17	6.26	14.64	15.45	15.58
DyHGCN	20.87	21.48	21.58	10.61	11.26	11.36	6.09	6.40	6.50	15.64	16.30	16.44
MS-HGAT	22.49	23.17	23.30	11.72	12.52	12.60	6.39	6.87	6.96	17.44	18.27	18.40
MGCL	**22.96**	**23.55**	**23.79**	**11.76**	**12.64**	**12.81**	**6.76**	**7.20**	**7.31**	**18.81**	**19.64**	**19.80**

4.2 Overall Performance (RQ1)

To evaluate the effectiveness of our proposed MGCL, we compare the overall prediction performance of MGCL with the seven recent baselines. The experimental results are summarized in Table 2 and Table 3.

Specifically, we have the following observations: First, compared to the best baseline MS-HGAT, our MGCL adaptively constructs a multiplex heterogeneous social network to mitigate weak structure present in the original social network and conduct the multi-level contrastive learning that can provide more self-supervised signals for graph learning. As a result, MGCL achieves more expressive user representations and consistently outperforms all baseline models in Hits and MAP scores. Second, these methods that incorporate multiple user relationships (DyHGCN, MS-HGAT, and MGCL) show better performance than social-only methods (SNIDSA, FOREST, Inf-VAE), while cascade-only methods (Topo-LSTM, NDM) show the worst performance. This underscores the importance of sufficient social network information for downstream prediction.

4.3 Ablation Study (RQ2)

To investigate the effect of each component on the MGCL, we compare different variants with the original model. Specifically, **w/o GE** only uses the original social network without enhancing the graph. **w/o GC** replaces adaptive relationships aggregation with average aggregation. Furthermore, we remove our graph training objectives to obtain **w/o** $\mathcal{L}_{con}$, **w/o** $\mathcal{L}_{ucl}$, and **w/o** $\mathcal{L}_{pcl}$.

Table 4. Performance of MGCL with its variants.

Models	Twitter		Android	
	H@100	M@100	H@100	M@100
MGCL	**62.47**	**23.79**	**29.35**	**7.31**
w/o GE	59.82	21.03	28.11	6.97
w/o GC	61.45	20.82	29.18	7.24
w/o $\mathcal{L}_{con}$	60.59	19.23	28.50	7.02
w/o $\mathcal{L}_{ucl}$	61.46	21.12	29.11	7.25
w/o $\mathcal{L}_{pcl}$	60.45	23.01	28.74	7.13

As shown in Table 4, MGCL achieves the best performance compared to any of its variants. Specifically, we can observe the following: First, utilizing the original social network leads to a significant degradation in performance, highlighting the detrimental impact of an incomplete social network on user learning. Second, employing the average aggregation of various user relationships also diminishes performance, underscoring the complexity of relationships. Our structure regularization can ensure that the enhanced structure aligns more closely with real-world scenarios. Third, there is a significant performance degradation upon the removal of either the user-level or preference-level contrastive learning objective, confirming the crucial role of self-supervised signals in graph learning.

4.4 Hyperparameter Analysis (RQ3)

We study four important hyperparameters $\gamma_1, \gamma_2, \gamma_3$ in Eq. (11) and δ in Eq. (7), which determine the effectiveness of the three self-supervised graph learning objectives we proposed in this paper.

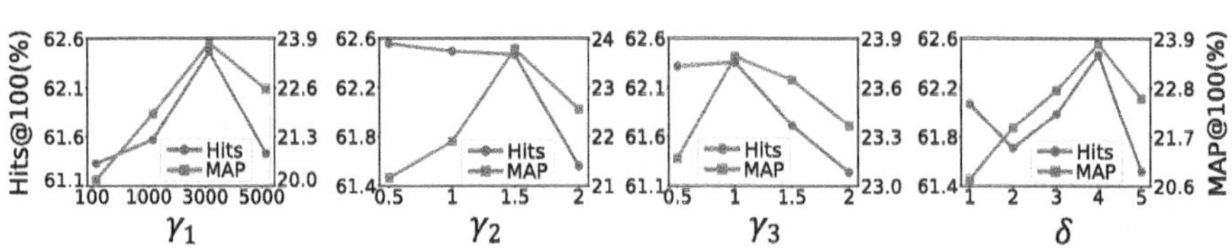

Fig. 3. Impact of different hyperparameters on Twitter.

Figure 3 shows the effect of hyperparameters on the Twitter dataset, the weight γ_1 of the graph regularization is vital to enhance social network structure. The weight γ_2 determines the distance between the user and others, which is more important for the prediction precision. γ_3 and δ determine the user's preference dependence, and excessively bringing similar users closer will lead to an overly dense distribution of user representations, resulting in suboptimal performance.

4.5 Performance in Different Scenarios (RQ4)

The Performance of High-ranking Position. In some specific scenarios, where precise identification of the evolution of information diffusion is crucial, there is a heightened requirement for more accurate prediction. To address this need, we investigate the performance of high-ranking position prediction.

Table 5. The performance of High-ranking position.

Datasets	Models	H/M@1	H@3	M@3	H@5	M@5
Twitter	MGCL	**15.91**	**24.77**	**19.88**	**28.69**	**20.63**
	MS-HGAT	15.46	23.40	18.95	26.81	19.73
Android	MGCL	**5.38**	**7.30**	**6.12**	**8.31**	**6.42**
	MS-HGAT	5.24	6.94	5.95	7.97	6.18

The results in Table 5 consistently demonstrate that MGCL outperforms MS-HGAT in high-ranking position prediction. We attribute this superiority to MGCL's ability to effectively learn the real behavior of users in social networks. In particular, multi-level contrastive learning of user representations greatly enhances the understanding of users' preferences in actual information diffusion, leading to a substantial improvement in the precision of high-ranking position.

Impact of Cascade Interaction Data. We conduct comparative experiments on Twitter and Android datasets under different training proportions and cascade lengths to further prove the stability of our model.

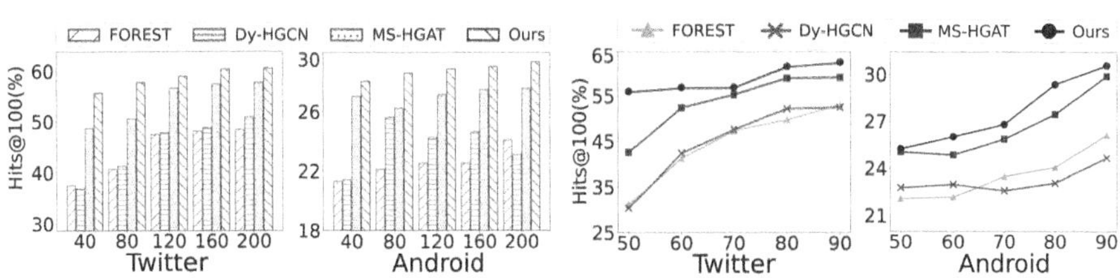

(a) Maximum length of cascade (b) Proportion of training cascade(%)

Fig. 4. Impact of cascade interaction data.

Referring to Fig. 4(a), we can observe that our model consistently achieves the best performance across different cascade lengths. Notably, for shorter cascade lengths, there is minimal degradation in performance compared to other models. Furthermore, as shown in Fig. 4(b), MGCL achieves comparable performance to the state-of-the-art model trained with less data. In conclusion, these results demonstrate the effectiveness of MGCL learning from user relationships even with extremely weak information issues.

5 Related Work

Graph Contrastive Learning. Graph data, often lacking labels, poses a challenge for current learning methods. Recently, contrastive learning has emerged as a promising solution, which utilizes self-supervised strategies to achieve effective graph learning. DGI [21] extends InfoMax [22] to graphs and maximizes MI between the global graph and local node. GraphCL [23] explores the impact of various combinations of different data augmentations. GCA [24] proposes an adaptive augmentation scheme at both topology and attributes. MVGRL [25] introduces graph diffusion to create another view. However, applying to social networks has certain challenges. For example, inappropriate data augmentation introduces noise and distorts user dependencies, and existing methods lack consideration of specific downstream tasks, limiting their effectiveness in practice.

Information Diffusion Prediction. Information diffusion prediction aims to predict future participants based on the current cascade and relevant knowledge, such as social networks, information content, and user profiles [19,20]. Recent methods focus on learning user representations from sequential or structured cascades using extended RNNs. For example, [7] extends the standard LSTM to model the cross-dependency of cascades. [9] introduces structural information into sequential information using RNN. However, these methods overlook social relationships among users. Some works [10,11] further attempt to embed social relationships to enhance prediction using GNNs. However, learning solely from original social network data is too weak to capture users' relationships. In existing datasets, users lack manual labels, initial user features are random, and the observed user relationships are incomplete. To fill this gap, we innovatively enhance the social network structure and propose multi-level contrastive tasks to obtain better user representations, which can benefit downstream prediction.

6 Conclusion

In this paper, we design a novel and effective two-stage framework named MGCL. Different from existing works, we construct the multiplex heterogeneous social network to enhance the original structure, and further design three self-supervised training objectives to improve graph learning. This method mitigates weak information issues of social network data and enables higher-quality user representations to improve the performance of the downstream task. Extensive experiments are conducted on four real-world datasets to validate that our MGCL method can consistently outperform state-of-the-art methods.

Acknowledgments. This work was supported by the State Key Laboratory of Communication Content Cognition Funded Project No. A32003, National Natural Science Foundation of China No. U22A2025 and No. 61972275.

References

1. Li, J., Yang, Y., Hu, Q., Wang, X., Gao, H.: Public opinion field effect fusion in representation learning for trending topics diffusion. In: NeurIPS (2023)
2. Sun L., Rao, Y., Wu, L., Zhang, X., Lan, Y., Nazir, A.: Fighting false information from propagation process: a survey. In: ACM Computing Surveys, vol. 55(10), pp. 1–38 (2023)
3. Chen, J., Hoops, S., Marathe, A., Mortveit, H., Lewis, B., Venkatramanan, S., et al.: Effective social network-based allocation of COVID-19 vaccines. In: KDD, pp. 1675–1683 (2022)
4. Broekaert, J. B., La Torre, D., Hafiz, F.: Competing control scenarios in probabilistic SIR epidemics on social-contact networks. In: ArXiv./abs/2108.13714 (2021)
5. Cheng, J., Adamic, L., Dow, P., Kleinberg, J. M., Leskovec, J.: Can cascades be predicted? In: WWW, pp. 925–936 (2014)
6. Gao, S., Ma, J., Chen, Z.: Effective and effortless features for popularity prediction in microblogging network. In: WWW, pp. 269–270 (2014)
7. Wang, J., Zheng, V. W., Liu, Z., Chang, K. C.: Topological recurrent neural network for diffusion prediction. In: ICDM, pp. 475–484 (2017)
8. Yang, C., Sun, M., Liu, H., Han, S., Liu, Z., Luan, H.: Neural diffusion model for microscopic cascade study. TKDE **33**(3), 1128–1139 (2021)
9. Wang, Z., Chen, C., Li, W.: A sequential neural information diffusion model with structure attention. In: CIKM, pp. 1795–1798 (2018)
10. Sankar, A., Zhang, X., Krishnan, A., Han, J.: Inf-VAE: a variational autoencoder framework to integrate homophily and influence in diffusion prediction. In: WSDM, pp. 510–518 (2020)
11. Yang, C., Tang, J., Sun, M., Cui, G., Liu, Z.: Multi-scale information diffusion prediction with reinforced recurrent networks. In: IJCAI, pp. 4033–4039 (2019)
12. Yuan, C., Li, J., Zhou, W., Lu, Y., Zhang, X., Hu, S.: DyHGCN: a dynamic heterogeneous graph convolutional network to learn users' dynamic preferences for information diffusion prediction. In: ECML/PKDD, pp. 347–363 (2020)
13. Sun, L., Rao, Y., Zhang, X., Lan, Y., Yu, S.: MS-HGAT: memory-enhanced sequential hypergraph attention network for information diffusion prediction. In: AAAI, pp. 4156–4164 (2022)
14. Liu, Y., Ding, K., Wang J., Lee, V., Liu, H., Pan, S.: Learning strong graph neural networks with weak information. In: KDD, pp. 1559–1571 (2023)
15. Yu, P., Fu, C., Yu, Y., Huang, C., Zhao, Z., Dong, J.: Multiplex heterogeneous graph convolutional network. In: KDD, pp. 2377–2387 (2022)
16. Hamilton, W. L., Ying, R., Leskovec, J.: Inductive representation learning on large graphs. In: NeurPIS (2017)
17. Vaswani, A., et al.: Attention is all you need. In: NeurIPS, pp. 5998–6008 (2017)

18. Wang, Z., Zhu, Y., Wang, C., Ma, W., Li, B., Yu, J.: Adaptive graph representation learning for next POI recommendation. In: WWW, pp. 393–402 (2023)
19. Zhang, H., Yang, Y., Wang, X., Gao, H., Hu, Q.: MLI: A multi-level inference mechanism for user attributes in social networks. In: TOIS, vol. 41(2), 1–30 (2022)
20. Wang, H., Yang, C., Shi, C.: Neural information diffusion prediction with topic-aware attention network. In: CIKM, pp. 1899–1908 (2021)
21. Velickovic, P., Fedus, W., Hamilton, W.L., Li'o, P., Bengio, Y., Hjelm, R.D.: Deep graph infomax. In: ICLR (Poster) (2019)
22. Hjelm, R.D., et al.: Learning deep representations by mutual information estimation and maximization. In: arXiv preprint arXiv:1808.06670 (2018)
23. You, Y., Chen, T., Sui, Y., Chen, T., Wang, Z., Shen, Y.: Graph contrastive learning with augmentations. In: NeurIPS, pp. 5812–5823 (2020)
24. Zhu, Y., Xu, Y., Yu, F., Liu, Q., Wu, S., Wang, L.: Graph contrastive learning with adaptive augmentation. In: WWW, pp. 2069–2080 (2021)
25. Hassani, K., Khasahmadi, A. H.: Contrastive multi-view representation learning on graphs. In: ICML, pp. 4116–4126 (2022)
26. An, W., Tian, F., Chen, P., Tang, S., Zheng, Q., Wang, Q.: Fine-grained category discovery under coarse-grained supervision with hierarchical weighted self-contrastive learning. In: EMNLP, pp. 1314–1323 (2022)
27. Hodas, N.O., Lerman, K.: The simple rules of social contagion. In: Scientific Reports, pp. 1–7 (2014)
28. Zhong, E., Fan, W., Wang, J., Xiao, L., Li, Y.: ComSoc: adaptive transfer of user behaviors over composite social network. In: KDD, pp. 696–704 (2012)

BiasRec: A General Bias-Aware Social Recommendation Model

Chunkai Zhang[1,2(✉)] and Guoqing Li[1]

[1] Harbin Institute of Technology, Shenzhen, China
ckzhang812@gmail.com, 21S151134@stu.hit.edu.cn

[2] Guangdong Provincial Key Laboratory of Intelligent System Research Center, Guangzhou, China

Abstract. The social recommendation aims to alleviate data sparsity problems and improve recommendation performance by incorporating user social data. And the recent popularity of graph neural networks (GNNs) has further advanced the development of social recommendation. However, most previous research erroneously assumed that user-item interaction data, such as ratings, directly reflects user preferences for items. In reality, various non-preference factors can also influence user ratings, referred to as biases in this paper. For example, users may still rate popular or high-quality items highly even if they are not interested in them, or some users may tend to give lower ratings to most items even if they like the item. Furthermore, existing research also lacks a general method for capturing and leveraging biases. To this end, we propose BiasRec, a bias-aware social recommendation model. BiasRec initially constructs a bias matrix for each user and item, calculates bias scores, and removes them from the raw rating data. Subsequently, the debiased data is fed into a GNN to learn users' genuine preferences. Last, it reasonably combines biases and preferences to make predictions. We performed experiments on three real-world datasets and attained state-of-the-art results, showcasing the efficacy of our model.

Keywords: Social recommendation · rating prediction · bias · graph neural network

1 Introduction

Recommendation systems leverage historical data to infer users' preferences and provide personalized content in the face of information overload on the internet. However, many existing recommendation systems grapple with the challenge of data sparsity, wherein there is insufficient interaction data (such as user ratings) to effectively infer each user's preferences. Some recommendation systems seek to incorporate additional information to alleviate this problem, such as social recommendation, which harnesses social network data to enhance the efficacy of recommendations. The fundamental rationale behind the social recommendation is that a user's preferences are influenced by or share similarities with those

M. Onizuka et al. (Eds.): DASFAA 2024, LNCS 14855, pp. 101–116, 2024.
https://doi.org/10.1007/978-981-97-5572-1_7

of individuals in their social environment [3]. Moreover, as social media gains popularity and user participation continue to rise, these platforms amass valuable data on users' social relationships and preferences, forming a crucial foundation for both business marketing and academic research [5,19,21,24].

To learn users' preferences, early collaborative filtering-based recommendation systems were mainly implemented by Matrix Factorization. And popularized by the Netflix competition, those matrix factorization-based recommendation models have attracted a great deal of attention, and many milestones have been achieved in this period [8,9,13]. Nonetheless, the linear nature of factorization models limits their suitability for managing extensive and intricate data. In recent years, deep learning has gained widespread adoption in recommendation systems due to its robust expressiveness and model fidelity [6,16]. More recently, Graph Neural Networks (GNNs), effective for learning graph-structured data, have propelled recommendation system development, given that recommendation data, including user-item interactions and user-user relations, naturally fit the graph structure. Many researchers have leveraged GNNs for social recommendation, yielding promising outcomes [1,7,14,19].

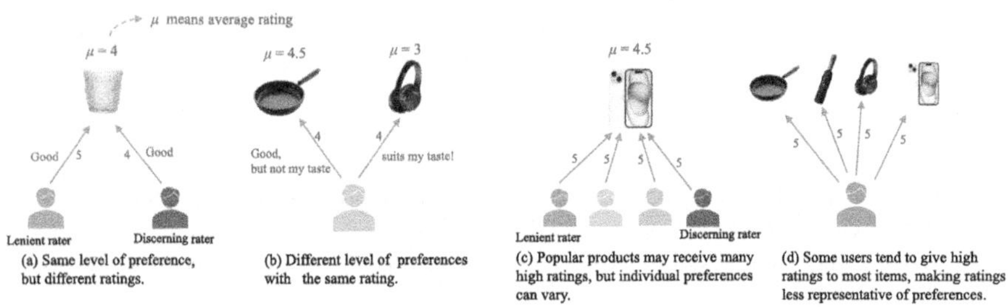

Fig. 1. Some examples: Why ratings cannot accurately represent user preferences. In the context of these four subplots, users can rate items on a scale of 1 to 5 and express their opinions.

However, most previous research erroneously assumed that user interaction behavior, such as ratings, directly reflects their preferences [1]. In reality, user ratings can be influenced by various non-preference factors, such as user-specific rating criteria and tendencies, inherent item quality, and popularity, among others, which are referred to as biases in this paper [9,12]. Figure 1 presents specific examples to elucidate why this assumption is incorrect. Before learning user preferences, mitigating the impact of bias information present in the raw rating data is crucial. This practice holds significant importance for two key reasons: Firstly, removing bias information aids in capturing more accurate user preferences and enhances interpretability [2,9]. Secondly, it prevents the model from directly equating high ratings with high preference, thereby alleviating the issue of over-recommending popular items and subsequently improving the diversity of the recommendation system [22,23]. To capture biases, BiasSVD [9] and SVD++

[8] incorporate bias as a trainable scalar applied to rating predictions, leading to a noticeable improvement in performance. However, this representation is overly simplistic and may not fully capture the richness of bias information. GDSRec [1] treats bias as a trainable vector and attempts to learn bias from simple preprocessed rating data. Nevertheless, the learned bias information actually contains a mixture of preference information. In summary, existing research lacks effective methods to accurately capture bias and preference information and a general approach to effectively integrating both for joint prediction purposes.

To this end, we propose BiasRec, a general bias-aware GNN-based model for social recommendation. Before model training, BiasRec constructs bias matrices for each user and item based on user-item rating data. These bias matrices are then used to calculate user bias scores regarding items. Then, BiasRec obtains the users' genuine preference levels for items by subtracting these bias scores from the original ratings. Meanwhile, we have distinguished the social connection strength among different users to assess their influence level better. During the training phase, we utilize a graph neural network to integrate user-item interaction data and user-user social network information to obtain more accurate embeddings to represent user and item preferences. During the prediction phase, we reasonably incorporate bias and preference data to make rating predictions.

The main contributions of this paper are as follows:

- We formulated a bias matrix to represent user and item bias information. Additionally, we introduced a bias network (BiasNet), which computes corresponding bias scores based on the user and item bias matrices.
- We have proposed a bias-aware GNN-based framework called BiasRec for social recommendation. It can effectively combine preference and bias information to make accurate rating predictions.
- We validated the effectiveness of BiasRec on three real-world datasets. Experimental results show BiasRec outperforms state-of-the-art baselines.
- We conducted additional experiments to compare the roles of preference and bias in predictions, confirming that bias information is more easily captured and plays a greater role in scenarios with particularly sparse data, such as social recommendations.

2 Related Work

2.1 Bias In Recommendation System

A wide variety of biases are prevalent in recommender systems, and understanding and dealing with them is important for recommendations' accuracy, diversity, and interpretability [2,22]. To capture the effects of biases, biasSVD [9] models user and item biases as trainable scalars to influence rating predictions, combining user and item biases, the inner product of user embedding and item embedding, and the average rating for the final prediction. This simple yet effective approach has inspired many subsequent methods [8,15]. DENC [10] presents a compelling approach to effectively address select bias, particularly those arising from the missing-not-at-random (MNAR) problem in recommendation systems. In GDSRec [1], biases are treated as vectors and integrated into the user and item representation learning process through decentralized neighborhood aggregation to capture statistical bias offsets. The PDA [22] introduces popularity-bias deconfounding and adjusting in recommender systems, a new approach that mitigates popularity bias in training and adjusts it during inference to enhance recommendation accuracy. Last, the study [2] provides a systematic survey of existing work on biases in recommendation systems, summarizing seven types of biases along with their definitions, characteristics, and solutions. While many methods targeting biases have been proposed, they mostly focus on a specific type of bias or a particular scenario (such as DENC and PDA). We aim to analyze the impact of all biases collectively, yet existing research still lacks a universal method to capture and address these biases comprehensively.

2.2 Social Recommendation

Incorporating social network data has proven to enhance the effectiveness and interpretability of recommendation systems. The matrix factorization-based model SoRec [12] firstly injects the user's social network information into rating prediction and achieves significant performance gains. The SUCP [15] method improves the recommendation performance by integrating social information and spatio-temporal information of user activities into matrix factorization. Deep learning-based model NeuMF [6] proposed a neural collaborative filtering framework to capture nonlinear relations between users and items. Recently, the application of graph neural networks has further advanced the development of social recommendations, leading to the success of various GNN-based models [1,17–19,21]. For example, GraphRec [5] utilizes GNNs to jointly capture interactions and rating scores for improved user and item representations, incorporating attention mechanisms to assess the strength of relationships between users and their various social network neighbors. Considering the diversity of social relationships, MPSR [11] further explores user preferences by constructing Multi-perspective social networks. SI-GAN [18] inherently combines adversarial learning-enhanced social network features and graph interaction features to tackle the issues of over-smoothing and digging latent feature representations. However, most social recommendation models have not considered the impact of biases, resulting in limited performance.

3 Proposed Method

3.1 Preliminaries and General Framework

Preliminaries. Let $U = \{u_1, u_2, \ldots, u_N\}$ represent the set of users and $V = \{v_1, v_2, \ldots, v_M\}$ represent the set of items, where N and M are the number of users and items respectively. In social recommendation, we have two networks: user-user social network $G_S = (U, T)$ and user-item rating network

$G_R = (U, V, R)$. In G_S, the social matrix $T \in \{0,1\}^{N \times N}$. T_{ij} is 1 if there is a social connection between user u_i and user u_j, otherwise it is 0. In G_R, $R \in R^{N \times M}$ is the rating matrix, where the r_{ij} in R is the rating score of user u_i on item v_j. $N_S(u_i)$ is the neighbors/friends set of user u_i in G_S, $N_R(v_j)$ is the users set who rate item v_j in G_R, and $N_R(u_i)$ is the items set of user u_i in G_R.

The objective of this paper is rating prediction, which involves learning user preference information from historical data to predict user ratings for new items.

General Framework. The overview of BiasRec is illustrated in Fig. 2. The BiasRec system comprises three key modules: the data transformation module, the encoder module, and the predictor module. The data transformation module converts raw graph data into a suitable format that enhances user and item modeling. In the encoding module, user and item embeddings are learned. Finally, the rating prediction module utilizes user preferences and bias information to predict ratings.

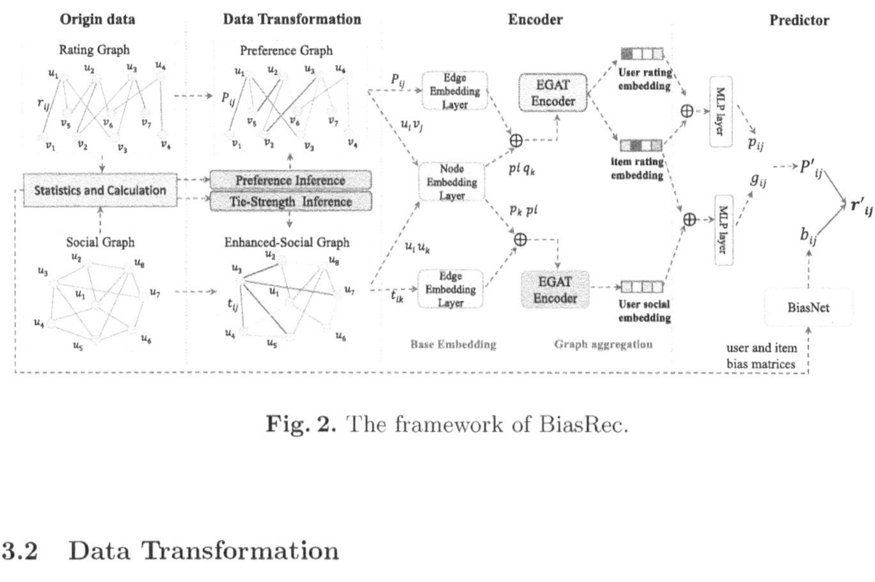

Fig. 2. The framework of BiasRec.

3.2 Data Transformation

Transformation of User-Item Rating Data. Raw ratings are jointly influenced by preferences and biases, which can lead to misguidance when used directly. A more accurate representation of user and item preferences can be learned by converting raw rating scores into preference scores and constructing a user-item preference graph. Preference scores are obtained by deducting the bias score from the rating score. The bias score represents the score a user might assign to an item when not considering preferences, in a state of neither liking nor disliking. The calculation of the bias score is illustrated in Fig. 3 and is detailed in the following:

First, we compute a bias matrix for each user and item separately to represent their respective inherent bias information. Taking the bias matrix M_{u_i} for user

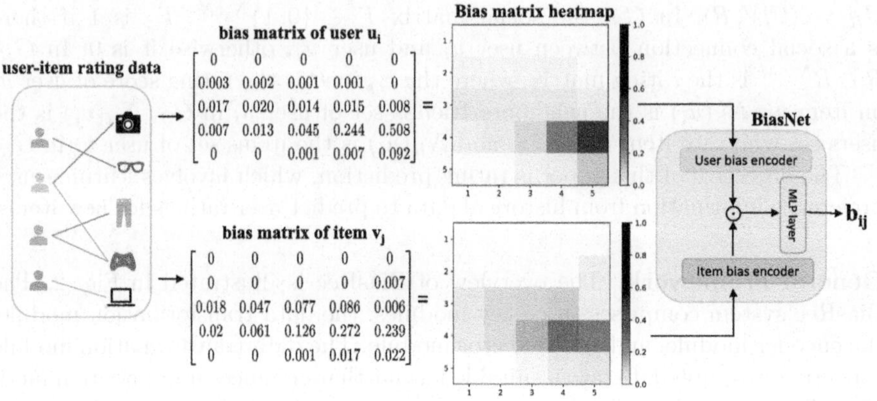

Fig. 3. The process of obtaining bias scores.

u_i as an example, $M_{u_i}[x][y]$ represents the probability of user u_i selecting an item with an average rating around x and giving it a rating of y. The sum of the x-th row of M_{u_i} represents the probability of the user selecting items with an average rating around x, while the sum of the y-th column of Correspondingly, in the bias matrix M_{v_j} for item v_j, $M_{v_j}[x][y]$ represents the probability that item v_j being selected by users with an average rating around x and receiving a rating of y. Considering that most rating scales in the dataset range from 1 to 5, we set the size of the bias matrix as 5×5 in this paper. Each element in the bias matrix is obtained by counting the relevant rating records and normalizing them. Considering the issue of data sparsity, where some users and items have limited rating records, making their bias matrices less reflective of accurate bias information, a global smoothing operation is required:

$$\tilde{M}_{u_i} = \frac{M_{u_i} \cdot \log(|N_R(u_i)| + c) + M_u}{\log(|N_R(u_i)| + c) + 1} \tag{1}$$

where $|N_R(u_i)|$ represents the number of ratings by user u_i; c is a constant, and M_u is the global user bias matrix, which is calculated by considering all ratings as one user's ratings to obtain M_u; the fewer the ratings, the higher the proportion of M_u in the smoothing process. Similarly, the item bias matrix M_{v_j} is also smoothed using the global item bias matrix M_v to obtain $\tilde{M}_{v_j}$. These bias matrices solely capture the rating bias characteristics of users or items and do not involve any preferences between users and items. To enhance understanding of this rating matrix, we can visualize it as a heatmap (refer to Fig. 3). For example, in the heatmap of users' matrices, a concentration of colors towards the bottom would indicate that users tend to select items with higher average ratings, while a concentration towards the right side would suggest that users tend to give higher ratings.

After obtaining all bias matrices, we input the smoothed bias matrices of user u_i and item v_j into a simple dual-tower model neural network(BiasNet), which aims to predict their bias scores b_{ij} without considering the influence of preferences:

$$b_{ij} = BiasNet(\tilde{M}_{u_i}, \tilde{M}_{v_j}) = W_1(F_u(\tilde{M}_{u_i}) \odot F_v(\tilde{M}_{v_j})) + b_1 \tag{2}$$

$$F_u(\tilde{M}_{u_i}) = \sigma(W_3 \cdot \sigma(W_2 \cdot vec(\tilde{M}_{u_i}) + b_2) + b_3) \tag{3}$$

$$F_v(\tilde{M}_{v_j}) = \sigma(W_5 \cdot \sigma(W_4 \cdot vec(\tilde{M}_{v_j}) + b_4) + b_5) \tag{4}$$

where F_u and F_v represent the bias encoders for users and items; they encode the bias matrix into a bias vector. Due to the small size of the input data, F_u and F_v can be either basic multi-layer perceptron (MLP) or small convolutional networks, both effective in this context; $\odot$ means the element-wise product; $vec(.)$ is used to flatten a matrix into a vector; $W_1 \sim W_5$ and $b_1 \sim b_5$ are the weights and biases of the linear layers; σ is the ReLU activation function.

It is worth noting that the bias matrix actually has the same semantics on different datasets, and the logic of BiasNet is consistent across datasets, so training BiasNet can be done naturally by using transfer learning or multi-source learning (putting training data from multiple datasets together to train). Additionally, we emphasize once again that the final learned bias scores are entirely independent of the preference relationship between users and items since neither the bias matrix nor BiasNet incorporates user and item ID information.

After getting the bias score b_{ij}, we use an operation similar to normalization to calculate the preference score:

$$P_{ij} = \lfloor \frac{r_{ij} - b_{ij}}{\sigma(v_j)} \rceil \tag{5}$$

where $\sigma(v_j)$ is the rating standard deviation of the item v_j, and the reason for employing it lies in that the rating distributions for different items vary, and it is necessary to eliminate distribution differences to obtain a more consistent standard for preference score. Inspired by the 3σ rule in statistics, we first clip the P_{ij} into [–3, 3] since most preference scores fall within this range. Subsequently, we apply rounding to P_{ij} for ease of further procedures. Now, we can use P_{ij} instead of r_{ij} to construct a preference graph $\tilde{G}_R$ for user-item interactions.

Transformation of User-User Social Data. In the raw user-user social network data, only information about whether users are connected or not is available, making it difficult to distinguish which connected neighbors are more important. Here, we use the numbers of common neighbors and common items(with similar preferences) to calculate the relationship strengths between user u_i and user u_j:

$$t_{ij} = |N_S(u_i) \cap N_S(u_j)| + \sum_{v_k \in N_R(u_i) \cap N_R(u_j)} I(|P_{ik} - P_{jk}| < \delta) \tag{6}$$

where $N_S(u_i)$ is the neighbors set of user u_i in G_S, I(x) = 1 when x is satisfied and 0 otherwise, and δ is a threshold value. Then we convert t_{ij} into a reasonable range and perform rounding operations:

$$t_{ij} = \lfloor log_2(t_{ij} + 2) \rceil \tag{7}$$

Later, by clipping $t_{ij}$to the range of [1, 10], we can obtain 10 levels of social connections with varying strengths. Finally, we can use t_{ij} instead of T_{ij} to construct an enhanced user-user social connection graph $\tilde{G}_S$.

The aforementioned operations occur within the Data Transformation module of Fig. 2. Through this module, the models can mitigate the influence of bias and subsequently learn more accurate representations of preferences.

3.3 Representation Learning

In this subsection, we will obtain embedding representations of users and items from the user-item preference graph $\tilde{G}_R$ and the enhanced user-user social connection graph $\tilde{G}_S$.

Modeling in Preference Graph. First, we use two embedding layers (MLP) to obtain the initial user representation p_i and initial item representation q_j. At first, all node embeddings are randomly initialized. The embedding representation of a user is also influenced by the items they have interacted with. When calculating these influences, we need to consider not only which items the user has interacted with but also their preference scores for these items. And the same is true for items. So in $\tilde{G}_R$, we use an edge aggregated graph attention layer (EGAT Encoder) to learn the representation of users and items. When aggregating information from neighbors, we will use p_i, q_j, and P_{ij} simultaneously to calculate the attention score between user u_i and item v_j. The calculation of the EGAT layer is as follows:

$$\tilde{p_i} = \sum_{v_j \in N_R(u_i)} \alpha_{i,j} \cdot W \cdot q_j \tag{8}$$

$$\tilde{q_j} = \sum_{u_i \in N_R(v_j)} \alpha_{j,i} \cdot W \cdot p_i \tag{9}$$

where W is the matrix to compute the information diffusion in the graph attention network, $\alpha_{i,j}$ is the attention score after normalize, $N_R(u_i)$ is the items set which user u_i rates and $N_R(v_j)$ is the users set who rates item v_j.

To better calculate attention scores, we use an embedding layer $f : R^1 \rightarrow R^{1 \times d}$ to map different preference scores into trainable vectors. For P_{ij}, its corresponding vector is $V_{P_{ij}}$. (Similarly, we map t_{ij} to $V_{t_{ij}}$). Then, we can get attention score by user embedding, item embedding, and preference vector:

$$\alpha_{i,j} = softmax_j(e_{i,j}) = \frac{\exp(e_{i,j})}{\sum_{v_k \in N_R(u_i)} \exp(e_{i,k})} \tag{10}$$

$$e_{i,j} = \sigma(W_e \left[p_i \oplus V_{P_{ij}} \oplus q_j\right]) \tag{11}$$

where $\oplus$ is the vector concatenation operation, and W_e is a weight matrix. Last, we get the user preference representation h_i and the item representation z_j by:

$$h_i = p_i \oplus \tilde{p_i}, \qquad z_j = q_j \oplus \tilde{q_j} \tag{12}$$

Modeling in Enhanced Social Graph. In a social network, users' choices and ratings of items may be influenced by the people around them, and we consider this influence as a type of preference since the preferences of people around a user may also become the potential preferences of that user. Similarly, we use another EGAT layer to aggregate influences from different neighbors for user u_i:

$$g_i = \sum_{u_j \in N_S(u_i)} \alpha_{i,j} \cdot W_I \cdot h_j \tag{13}$$

$$\alpha_{i,j} = softmax_j\left(e_{i,j}\right) = \frac{\exp\left(e_{i,j}\right)}{\sum_{u_k \in N_S(u_i)} \exp\left(e_{i,k}\right)} \tag{14}$$

$$e_{i,j} = \sigma(W_u \left[h_i \oplus V_{t_{ij}} \oplus h_j\right]) \tag{15}$$

where $\alpha_{i,j}$ is the attention score between user u_i and user u_j, W_I and W_u is the weight matrix, and $V_{t_{ij}}$ is the trainable vector corresponding to t_{ij}.

3.4 Rating Prediction

To predict a user's rating for an item, we need to predict the user's preference score for that item and then combine it with bias information to get the final rating prediction.

Preference Prediction. With the user embedding h_i and item embedding z_j, we can use a two-layer MLP to get predicted individual preference p_{ij} by:

$$p_{ij} = W_7 \cdot \sigma(W_6 \cdot [h_i \oplus z_j] + b_6) + b_7 \tag{16}$$

Similarly, we can get the predicted preference of people around the user on the item:

$$g_{ij} = W_9 \cdot \sigma(W_8 \cdot [g_i \oplus z_j] + b_8) + b_9 \tag{17}$$

After getting p_{ij} and g_{ij}, we need to combine them to predict user u_i's final preference for item v_j. Considering that different users may have different sensitivities to the influence of people around them, we use a trainable parameter $\alpha_i \in [0, 1]$ to represent the sensitivities of user u_i to the influence of people around him. Then the predicted preference score $P^{'}_{ij}$ can be obtained by:

$$P^{'}_{ij} = (1 - \alpha_i) \cdot p_{ij} + \alpha_i \cdot g_{ij} \tag{18}$$

The final predicted preference $P^{'}_{ij}$ can be viewed as a weighted sum of the user's preference p_{ij} and his/her friends' preferences g_{ij}.

Rating Prediction. We combine the bias score and the predicted preference score to get the predicted rating of user u_i on item v_j:

$$r_{ij}^{'} = b_{ij} + \sigma(v_j) \cdot P_{ij}^{'} \tag{19}$$

In reality, Eq. 19 is the inverse process of Eq. 5, and they encompass the same underlying ideas.

3.5 Loss Function

Before we perform the data transformation, we need to train the BiasNet first, using the following loss function:

$$\mathcal{L}_{bias} = \frac{1}{|R_{train}|} \sum_{r_{ij} \in R_{train}} (b_{ij} - r_{ij})^2 + \beta_1(||\Phi_1||^2) \tag{20}$$

where R_{train} is training data, and Φ_1 represents the model parameters, β_1 represents the regularization coefficient. In the actual training of BiasNet, we put the training data from multiple datasets together for training, which is called Multi-Source Learning. Then, we put it on specific datasets for fine-tuning. After the completion of the data transformation, we employ a similar loss function to train the backbone GNN network of BiasRec:

$$\mathcal{L}_{pred} = \frac{1}{|R_{train}|} \sum_{r_{ij} \in R_{train}} (r_{ij}^{'} - r_{ij})^2 + \beta_2(||\Phi_2||^2) \tag{21}$$

4 Experiment

4.1 Experimental Settings

Datasets. We evaluate our model on three widely used benchmark datasets for social recommendation: Ciao, Epinions[1], and FilmTrust[2]. They all contain user-item rating data and user-user trust data, and the statistics for these datasets are presented in Table 1. These datasets have frequently been used in social recommendation studies, making them ideal for comparisons with related research.

Evaluation Metrics. To assess the efficacy of our model in the rating prediction task, we employ two widely recognized metrics: root mean square error (RMSE) and mean absolute error (MAE). Smaller RMSE and MAE mean higher accuracy of the model. We also conducted experiments on the item ranking task to evaluate the performance of our model. For the predicted rating scores and real rating scores in the test set, we transform the rating score to 1 (positive) if it $\geq$ F, otherwise 0 (negative). Note that we only use the observed data to evaluate

[1] https://www.cse.msu.edu/~tangjili/trust.html.

[2] https://doi.org/10.7910/DVN/AKVGJ9.

Table 1. Dataset Statistics

Dataset	Ciao	Epinions	FilmTrust
Number of Users	7317	18069	1508
Number of Items	104975	261246	2071
Rating Class	1-5	1-5	0.5-4
Number of Ratings	281867	754939	35497
Number of Relations	111781	355503	1853

item ranking. Then, we use Recall@K and Normalized Discounted Cumulative Gain (NDCG@K) for the top-N ranking evaluation. Higher values of Recall and NDCG mean better ranking performance. Due to the limited observed data for validation, we only utilized recall@5 and NDCG@5 for testing.

Baselines. To evaluate the performance of our proposed model, we compare BiasRec with three other groups of methods, including:

- Matrix factorization-based recommendation models: **PMF** [13], matrix factorization methods, only use rating data; **SoRec** [12] and **TrustMF** [20], which start to exploit social information for recommendation;
- Deep learning-based social recommendation: **NeuMF** [6], A model that combines the neural network structure and matrix factorization. **DeepSoR** [4], a classical deep social recommendation method.
- Graph neural network-based social recommendation models: **GraphRec** [5], **Diffnet++** [19] **HGSR** [21], **MPSR** [11], **GDSRec** [1].

Parameter Settings. Each dataset is randomly partitioned into three subsets: training (80%), validation (10%), and testing (10%). We set the batch size to 128 and the embedding size to 64. We search learning rate in { 0.0001, 0.001, 0.01, 0.1}. In Eq. 6, we set δ to 3; For β_1 and β_2 in the loss function, We search them in { 0.0001, 0.001, 0.01, 0.1}; In the item ranking task, we set F to 4 for Epinions and Ciao while setting F to 3 for FilmTrust (The value of F depends on the average ratings in the dataset).

4.2 Experimental Results

Rating Prediction Task. Table 2 shows the performance of different models on the rating prediction task. Our model gets the best result on all three datasets, substantiating its effectiveness. The PMF, which only uses rating data, performs worst among the matrix factorization-based models, while SoRec and TrustMF consider the social network information and achieve better results. This indicates the user's social information is helpful for the improvement of the performance of the recommendation system. In the deep learning-based social recommendation models, NeuMF harnesses neural network architecture to address the problem, surpassing the performance of PMF. This underscores the efficacy of neural network models in recommendation algorithms. Last, the performance of graph neural network-based social recommendation models is better than previous models. We attribute this improvement to the ability of the graph neural network to iteratively aggregate information from user-item and user-user relations.

Table 2. Performance of different models on the task of rating prediction

Dataset	Ciao		Epinions		FilmTrust	
Metric	MAE	RMSE	MAE	RMSE	MAE	RMSE
PMF	0.9021	1.1238	0.9952	1.2128	0.7252	0.9680
SoRec	0.8410	1.0652	0.8961	1.1437	0.6661	0.8512
TrustMF	0.7690	1.0479	0.8410	1.1395	0.6410	0.8350
NeuMF	0.8062	1.0617	0.9072	1.1476	0.6480	0.8445
DeepSoR	0.7739	1.0316	0.8383	1.0972	0.6367	0.8342
GraphRec	0.7399	0.9766	0.8120	1.0604	0.6333	0.8229
Diffnet++	0.7398	0.9774	0.8201	1.0635	0.6301	0.8211
MPSR	0.7346	0.9731	0.8090	1.0511	0.6192	0.8326
HGSR	0.7324	0.9708	0.8039	1.0459	0.6201	0.8317
GDSRec	0.7323	0.9740	0.8047	1.0566	0.6147	0.8166
BiasRec (bias only)	*0.7302*	*0.9782*	*0.8055*	*1.0491*	*0.6722*	*0.8731*
BiasRec	**0.7162**	**0.9533**	**0.7859**	**1.0321**	**0.6071**	**0.7921**

Surprisingly, our model achieved competitive results(the second row from the bottom of Table 2) even without preference information (i.e., directly using bias score b_{ij} as predicted rating), indicating the effectiveness of the bias information we extracted. More detailed explanations and further analysis will be provided in Subsect. 4.4. After training and obtaining users' preference scores for items, the model performance was further improved, leading to state-of-the-art results.

Item Ranking Task. The experimental results of the item ranking task are shown in Fig. 4. From the experimental results, we can observe that the difference in recall rates between different models is insignificant, as the recall rate is a relatively rough indicator that cannot reflect the accuracy of item ranking. On the other hand, NDCG is an improved version of the recall rate that considers whether recommended items appear in the recommendation list and reflects the accuracy of item ranking and user preference toward different items. Therefore, NDCG can better reflect the performance differences between different methods. Although our method is not specifically optimized for item ranking tasks, it still achieved competitive results on NDCG@5, proving our approach's efficacy.

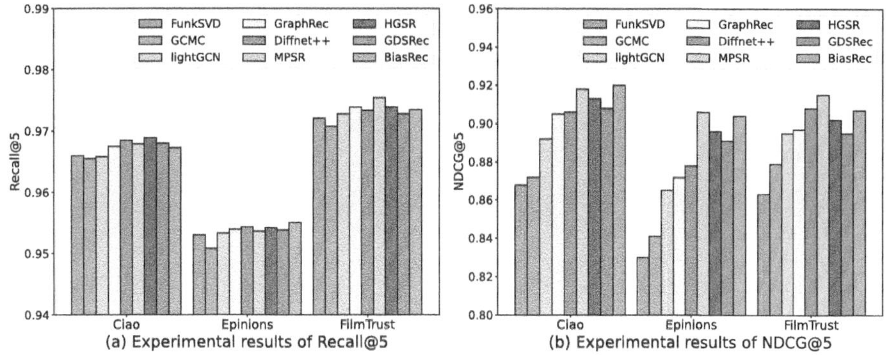

Fig. 4. Performance of different models on the task of item ranking.

4.3 Ablation Experiment

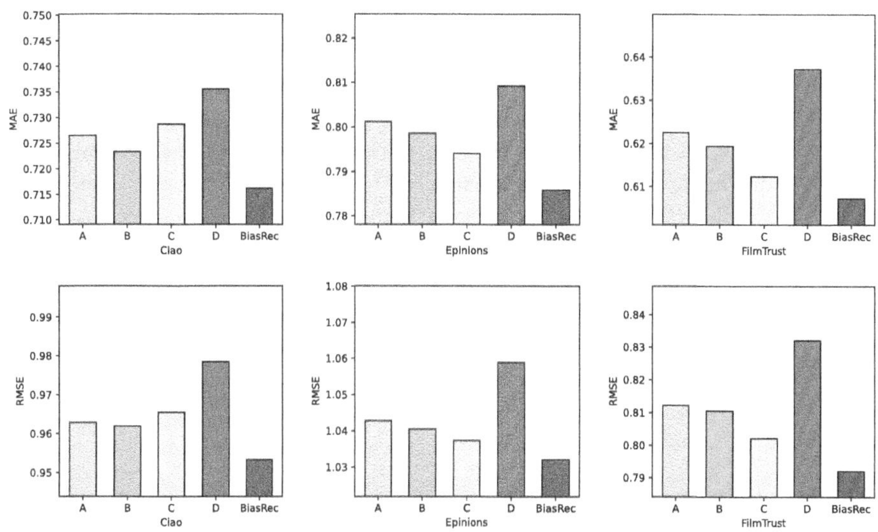

Fig. 5. Ablation study of BiasRec.

We conducted ablation experiments to evaluate different operations in BiasRec. We considered the following four variants of BiasRec: (A) without user-item rating graph transformation, (B) without user-user relationship graph transformation, (C) Training BiasNet without multi-source learning, (D) without distinguishing between preference and rating, directly using the user's preference score for the item as the predicted rating score to train.

The results of the ablation experiment are shown in Fig. 5. We can see that BiasRec performs the best against all variants, showing that each operation is essential for achieving optimal results. Among all variants, variant D exhibited the poorest performance, highlighting the necessity of distinguishing between ratings and preferences and underscoring the importance of utilizing biases. In addition, data transformation and multi-source learning also yield positive effects, with no clear superiority between them, the extent of improvement depends on the specific dataset.

4.4 Bias vs. Preference

As mentioned earlier, our model achieved competitive results by utilizing only the bias information. We attribute the above results to the fact that preferences represent a complex binary relationship between users and items, often necessitating substantial data for precise characterization. Conversely, biases are intrinsic attributes of users and items, demanding less data for effective capture. So, in cases of extremely sparse data, it becomes challenging for the model to learn preference information between users and items, while obtaining biased information is comparatively easier, leading to the rating prediction task primarily relies on bias information. Only when there is sufficient rating data for each user and item can the model learn effective preference information, allowing preferences to play a more significant role. To verify the above viewpoint, we performed experiments on the more densely populated Movielens 1M dataset[3] and Movielens 10M[4]. We use training data with varying proportions (ranging from 1% to 90%) to control the degree of data sparsity and observe the effects of bias and preference on rating prediction performance.

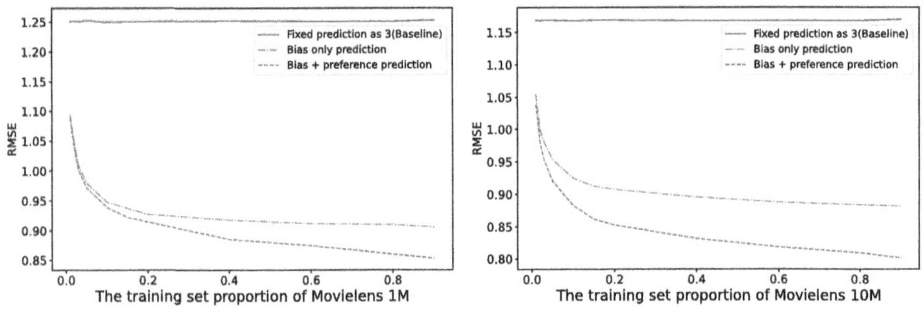

Fig. 6. The impact of bias and preference on model performance at different data sparsity levels. "Fixed predict as 3" means predicting all ratings as 3, which is used as a baseline. "Bias only prediction" means using only bias information for prediction.

The experimental results shown in Fig. 6 indicate that when the data is very sparse, the model performance mainly relies on bias information, which can bring significant performance improvement even with limited data. However, as more data is introduced, the impact of bias information on further enhancing model performance tends to saturate, as it has already reached a stable state.

[3] https://grouplens.org/datasets/movielens/1m/.

[4] https://grouplens.org/datasets/movielens/10m/.

In contrast, preference information cannot significantly enhance the model performance with limited data. However, as the data volume increases, preference's role in improving model performance becomes increasingly noticeable.

Social recommendation datasets are generally sparser than conventional recommendation datasets. For example, in Ciao and Epinions, there are only 3 ratings per item on average, which is an extremely sparse data situation. This is why we emphasize the use of bias in social recommendation.

5 Conclusion

This paper introduces BiasRec, a general bias-aware social recommendation model. BiasRec constructs bias matrices for each user and item to represent their bias information and introduces BiasNet to predict bias scores. Then, data transformation is applied before model training to learn more accurate preference representations. Finally, the model combines bias and preference information for prediction. The extensive experiments conducted on three real-world datasets and two tasks have demonstrated the effectiveness of BiasRec. Additional experiments have also demonstrated the necessity of using bias in social recommendation.

Acknowledgements. This work was supported by Guangdong Provincial Key Laboratory of Novel Security Intelligence Technologies, China (Grant NO. 2022B1 212010005), Shenzhen Research Council (Grant NO. Grant NO. WD2022081117025 3002) and Natural Science Foundation of Guangdong Province, China (Grant NO. 2024A1515010242).

References

1. Chen, J., Xin, X., Liang, X., He, X., Liu, J.: Gdsrec: graph-based decentralized collaborative filtering for social recommendation. IEEE Trans. Knowl. Data Eng. **35**(5), 4813–4824 (2023)
2. Chen, J., Dong, H., Wang, X., Feng, F., Wang, M., He, X.: Bias and debias in recommender system: a survey and future directions. ACM Trans. Inf. Syst. **41**(3), 1–39 (2023)
3. Cialdini, R.B., Goldstein, N.J.: Social influence: compliance and conformity. Annu. Rev. Psychol. **55**, 591–621 (2004)
4. Fan, W., Li, Q., Cheng, M.: Deep modeling of social relations for recommendation. In: Proceedings of the AAAI Conference on Artificial Intelligence vol. 32 (2018)
5. Fan, W., Ma, Y., Li, Q., He, Y., Zhao, E., Tang, J., Yin, D.: Graph neural networks for social recommendation. In: The World Wide Web Conference, pp. 417–426 (2019)
6. He, X., Liao, L., Zhang, H., Nie, L., Hu, X., Chua, T.S.: Neural collaborative filtering. In: Proceedings of the 26th International Conference on World Wide Web, pp. 173–182 (2017)
7. Huang, J., Xie, R., Cao, Q., Shen, H., Zhang, S., Xia, F., Cheng, X.: Negative can be positive: signed graph neural networks for recommendation. Inf. Process. Manage. **60**(4), 103403 (2023)

8. Koren, Y.: Factorization meets the neighborhood: a multifaceted collaborative filtering model. In: Proceedings of the 14th ACM SIGKDD International Conference on Knowledge Discovery and Data Mining, pp. 426–434 (2008)
9. Koren, Y., Bell, R., Volinsky, C.: Matrix factorization techniques for recommender systems. Computer **42**(8), 30–37 (2009)
10. Li, Q., Wang, X., Wang, Z., Xu, G.: Be causal: de-biasing social network confounding in recommendation. ACM Trans. Knowl. Discov. Data **17**(1), 1–23 (2023)
11. Liu, H., Zheng, C., Li, D., Zhang, Z., Lin, K., Shen, X., Xiong, N.N., Wang, J.: Multi-perspective social recommendation method with graph representation learning. Neurocomputing **468**, 469–481 (2022)
12. Ma, H., Yang, H., Lyu, M.R., King, I.: Sorec: social recommendation using probabilistic matrix factorization. In: Proceedings of the 17th ACM Conference on Information and Knowledge Management, pp. 931–940 (2008)
13. Mnih, A., Salakhutdinov, R.R.: Probabilistic matrix factorization. Adv. Neural. Inf. Process. Syst. **20**, 1257–1264 (2007)
14. Peng, S., Sugiyama, K., Mine, T.: Svd-gcn: a simplified graph convolution paradigm for recommendation. In: Proceedings of the 31st ACM International Conference on Information & Knowledge Management, pp. 1625–1634 (2022)
15. Seyedhoseinzadeh, K., Rahmani, H.A., Afsharchi, M., Aliannejadi, M.: Leveraging social influence based on users activity centers for point-of-interest recommendation. Inf. Process. Manage. **59**(2), 102858 (2022)
16. Tang, J., Wang, K.: Personalized top-n sequential recommendation via convolutional sequence embedding. In: Proceedings of the Eleventh ACM International Conference on Web Search and Data Mining, pp. 565–573 (2018)
17. Tu, K., Wu, Z., Hu, B., Zhang, Z., Cui, P., Li, X., Zhou, J.: A scalable social recommendation framework with decoupled graph neural network. In: International Conference on Database Systems for Advanced Applications, pp. 519–531. Springer (2023)
18. Wang, J., Li, H., Mo, T., Li, W.: A scalable social recommendation framework with decoupled graph neural network. In: International Conference on Database Systems for Advanced Applications, pp. 357–372. Springer (2023). https://doi.org/10.1007/978-3-031-30678-5_39
19. Wu, L., Li, J., Sun, P., Hong, R., Ge, Y., Wang, M.: Diffnet++: a neural influence and interest diffusion network for social recommendation. IEEE Trans. Knowl. Data Eng. **34**(10), 4753–4766 (2020)
20. Yang, B., Lei, Y., Liu, J., Li, W.: Social collaborative filtering by trust. IEEE Trans. Pattern Anal. Mach. Intell. **39**(8), 1633–1647 (2016)
21. Yang, Y., et al.: Hyperbolic graph learning for social recommendation. IEEE Trans. Knowl. Data Eng. (2023)
22. Zhang, Y., Feng, F., He, X., Wei, T., Song, C., Ling, G., Zhang, Y.: Causal intervention for leveraging popularity bias in recommendation. In: Proceedings of the 44th International ACM SIGIR Conference on Research and Development in Information Retrieval, pp. 11–20 (2021)
23. Zhao, Z., Chen, J., Zhou, S., He, X., Cao, X., Zhang, F., Wu, W.: Popularity bias is not always evil: Disentangling benign and harmful bias for recommendation. IEEE Trans. Knowl. Data Eng. (2022)
24. Zhou, W., Haq, A.U., Qiu, L., Akbar, J.: Multi-view social recommendation via matrix factorization with sub-linear convergence rate. Expert Syst. Appl. **237**, 121687 (2024)

Beyond the Known: Novel Class Discovery for Open-World Graph Learning

Yucheng Jin[1], Yun Xiong[1(✉)], Juncheng Fang[1], Xixi Wu[1], Dongxiao He[2], Xing Jia[1], Bingchen Zhao[3], and Philip S. Yu[4]

[1] Shanghai Key Laboratory of Data Science, School of Computer Science, Fudan University, Shanghai, China
yunx@fudan.edu.cn
[2] College of Intelligence and Computing, Tianjin University, Tianjin, China
[3] School of Informatics, University of Edinburgh, Edinburgh, UK
[4] Department of Computer Science, University of Illinois at Chicago, Chicago, USA

Abstract. Node classification on graphs is of great importance in many applications. Due to the limited labeling capability and evolution in real-world open scenarios, novel classes can emerge on unlabeled testing nodes. However, little attention has been paid to novel class discovery on graphs. Discovering novel classes is challenging as novel and known class nodes are correlated by edges, which makes their representations indistinguishable when applying message passing GNNs. Furthermore, the novel classes lack labeling information to guide the learning process. In this paper, we propose a novel method Open-world gRAph neuraL network (ORAL) to tackle these challenges. ORAL first detects correlations between classes through semi-supervised prototypical learning. Inter-class correlations are subsequently eliminated by the prototypical attention network, leading to distinctive representations for different classes. Furthermore, to fully explore multi-scale graph features for alleviating label deficiencies, ORAL generates pseudo-labels by aligning and ensembling label estimations from multiple stacked prototypical attention networks. Extensive experiments on several benchmark datasets show the effectiveness of our proposed method.

Keywords: Novel class discovery · Graph neural network · Open-world learning

1 Introduction

Graphs are ubiquitously used to reveal the interactions among various entities. One fundamental learning task on graphs is the node classification [11,12,14, 20,21], which has many applications such as predicting the areas of publications in academic networks [12,20], identifying the functions of proteins in biology graphs [10,21]. To ease the annotation burden, researchers have recently paid

M. Onizuka et al. (Eds.): DASFAA 2024, LNCS 14855, pp. 117–133, 2024.
https://doi.org/10.1007/978-981-97-5572-1_8

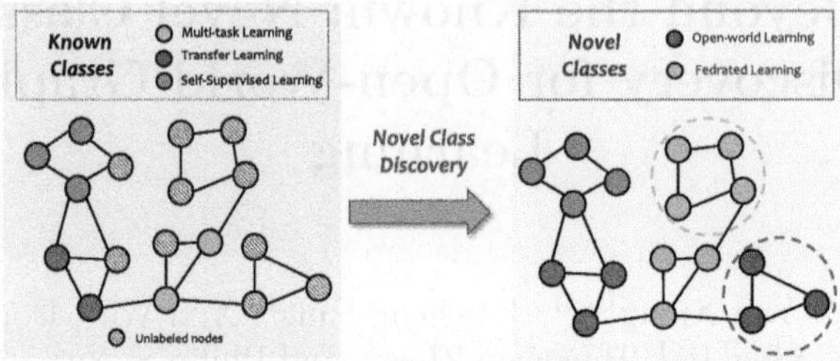

Fig. 1. Illustration of novel class discovery for open-world learning on an academic graph. Nodes represent papers, edges represent citation relationships, and each paper belongs to a certain research field (node class).

much attention to the semi-supervised setting [4,6,14,16], where only a small proportion of nodes on the graph is labeled.

However, existing node classification methods [2,11,12,14,20–22] assume a closed-world setting [3], where unlabeled testing nodes belong to classes seen on the labeled training nodes. This assumption rarely holds for graphs in the open world, since novel classes can frequently emerge on unlabeled nodes. Take the toy academic graph in Fig. 1 as an example, with nodes and edges denoting papers and citation relationships respectively. In addition to nodes labeled with well-known research topics, there are newly emerging unlabeled papers with unknown themes. The emergence of new research papers can introduce novel research topics. A crucial problem is to learn distinguishable representations for both known and novel classes, and facilitates novel class discovery along the way. We term this problem as **Open-world Graph Learning** in this paper.

While the open-world graph learning naturally suits many real-world applications, it poses unique challenges: (1) Novel and known classes are correlated by edges, which makes learning distinguishable representations difficult. For example, papers proposing new research areas will cite existing articles in related fields on the academic graph. Consequently, novel and existing research areas are correlated by edges representing citation relationships. Edges between novel and known classes can make their representations indistinguishable, since typical GNNs [2,11,12,14,22] tend to learn similar representations for connected nodes through message passing. (2) Only unlabeled nodes of novel classes are provided, lacking labeling information to guide representation learning and novel class discovery during the training process. Generating pseudo-labeling on novel class nodes is challenging due to the absence of labeled novel class nodes for learning the labeling mechanism. Given known class labeling information and rich features on graphs, how can we obtain novel class pseudo-labels and incorporate the generated label information for open-world graph learning?

In this paper, we propose a novel method Open-world gRAph neuraL network (ORAL) to tackle these challenges. ORAL first employs the prototypical attention network to detect and eliminate correlations between novel and known classes. By leveraging labeling granularity mined from known classes, the prototypical attention network first approximates the ground-truth label with semi-supervised prototypical clustering. Distinctive representations for different classes can be obtained by eliminating message passing over inter-class edges detected from label estimations. ORAL stacks multiple prototypical attention networks to explore graph features at different scales. To alleviate the lack of novel class labeling information, label estimations from different layers are aligned and ensembled to generate reliable pseudo-labels. Pseudo-labeling information is incorporated into the input graph data to guide the novel class discovery by recovering intra-class edges and removing inter-class ones. Confidence filtering and perturbation-robust consistency training are further introduced to combat pseudo-labeling noise during the learning process.

In summary, our contributions can be summarized as,

- Our approach constitutes the first attempt to investigate open-world graph neural network which can automatically discover novel classes on graphs.
- We propose the prototypical attention network to detect and eliminate inter-class correlations for distinguishing novel classes from others during representation learning.
- To ease the lack of novel class label information, we generate pseudo-labels according to multi-scale graph features and labeling granularity mined from known classes.
- We perform extensive experiments and analyze the results, proving the effectiveness of our method.

2 Problem Formulation

Notations. A graph is represented as $G = (\mathcal{V}, \mathcal{E}, \mathbf{X})$, where $\mathcal{V} = \{v_i\}_{i=1,\dots,N}$ is a node set representing nodes in a graph, $e_{ij} = (v_i, v_j) \in \mathcal{E}$ is an edge indicating the relationship between two nodes and $\mathbf{x}_i \in \mathbf{X}$ indicates content features associated with each node v_i.

Open-world Graph Learning. The node set $\mathcal{V}$ is composed of labeled training nodes $\mathcal{V}_L$ and unlabeled testing nodes $\mathcal{V}_U$, which belong to class set $\mathcal{Y}^L$ and $\mathcal{Y}^U$ respectively. In open-world scenarios, novel class can emerge on the unlabeled nodes, i.e., $\mathcal{Y}^L \subset \mathcal{Y}^U$. The aim of open-world graph learning is to discover novel classes automatically and classify the unlabeled nodes into known classes ($\mathcal{Y}^L$) or novel classes ($\mathcal{Y}^U \setminus \mathcal{Y}^L$).

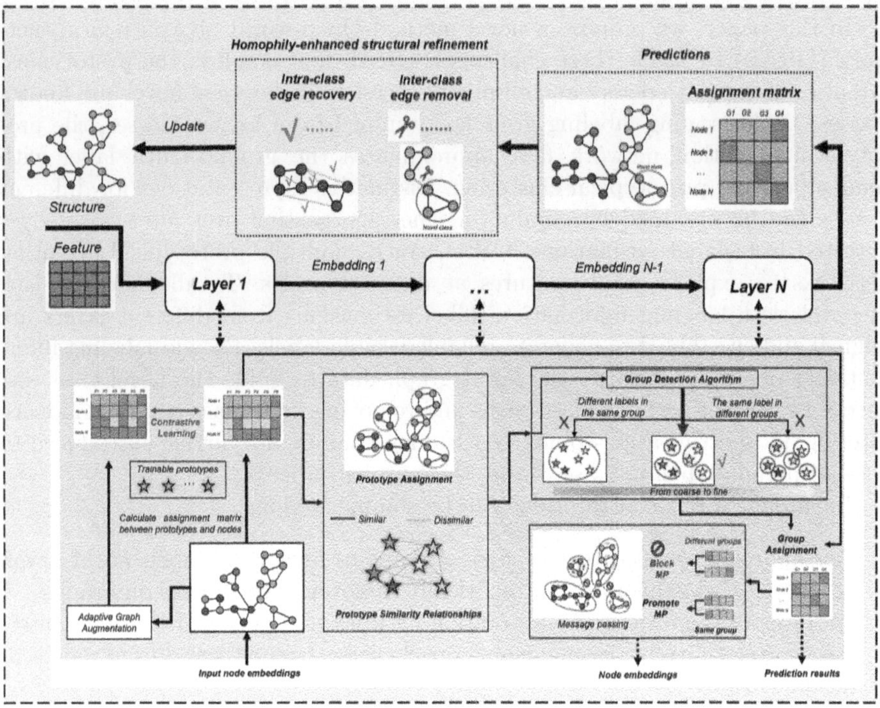

Fig. 2. An overview of the proposed model ORAL.

3 Methodology

Overview. The framework of our proposed method ORAL is shown in Fig. 2. The prototypical attention network (PAN) is first employed to eliminate inter-class correlations. Given representative nodes (prototypes) extracted by prototype learning, PAN performs semi-supervised clustering on the prototype graph of relatively small size, efficiently obtaining group assignments of prototypes and nodes. Message passing between groups are eliminated by the group-aware attention mechanism to generate separable representations. Furthermore, ORAL eases the lack of labeling information by the pseudo-label generator, which aligns and ensembles multiple group assignment estimations in the last stage, facilitating the open-world learning through structure refining.

3.1 Prototypical Attention Network

The prototypical attention network clusters nodes into groups and eliminates correlations between classes by reducing message passing between groups, where the clustering results are approximations of ground-truth labels. Nonetheless, the number of novel classes is unknown in the open world, we have to repeatedly perform clustering on graphs to search for the optimal clustering granularity by fitting the labeling information, which can be time-consuming. To ease the computation burden, we extract representative nodes through prototype learning and conduct semi-supervised clustering on the small prototype graph instead of the large whole graph. The node clustering result is approximated from that of prototypes, which is employed to eliminate message passing through the group-aware attention mechanism.

Representative nodes are modeled as a set of trainable prototypes $\mathcal{C} = \{\mathbf{c}_1, \mathbf{c}_2, \cdots \mathbf{c}_{N_{pro}}\}$, where N_{pro} denotes the number of prototypes. Given a node with representation $\mathbf{h}_i$ which is initialized as node feature $\mathbf{x}_i$, we compute the representativeness score of each prototype $\mathbf{c}_j$ to the node as follows,

$$r_{ij} = \frac{\exp(\mathbf{h}_i^T \mathbf{c}_j)}{\sum_{k=1}^{N_{pro}} \exp(\mathbf{h}_i^T \mathbf{c}_k)}. \tag{1}$$

To ensure each prototype is representative of a set of nodes, we regularize the marginal distribution of representativeness scores to be close to a balanced prior,

$$\mathcal{L}_{reg} = \mathrm{KL}[(\frac{1}{N_{pro}} \cdot \mathbf{1})||(\frac{1}{|\mathcal{V}|}\sum_{i=1}^{|\mathcal{V}|} \mathbf{r}_i)], \tag{2}$$

where $\mathbf{r}_i = [r_{i1}, \ldots, r_{iN_{pro}}] \in \mathbb{R}^{N_{pro}}$, $\mathbf{1} \in \mathbb{R}^{N_{pro}}$ represents the all-ones vector and KL represents the Kullback-Leibler divergence. Each node is regarded as an associated instance of prototypes with the top k largest representativeness score. The similarity relationships between prototypes $\mathbf{s} \in \mathbb{R}^{N_{pro} \times N_{pro}}$ can be calculated as the jaccard index of their associated node set, i.e., $s_{ij} = \mathrm{Jaccard}(\Gamma_i, \Gamma_j)$, with Γ_i denotes the associated node set of the i-th prototype.

Semi-supervised Prototypical Clustering. Applying a clustering algorithm with polynomial time complexity on the small prototype graph $(\mathcal{C}, \mathcal{E}^p)$ instead of the whole graph can substantially ease the computational burden, where $\mathcal{E}^p = \{(\mathbf{c}_i, \mathbf{c}_j, s_{ij})\}$. Given the groups $\mathcal{C}^g = \{\mathcal{C}_1^g, \cdots, \mathcal{C}_{N_{clu}}^g\}$ discovered by clustering on prototypes, the assignment probability of a node v_i to a group $\mathcal{C}_k^g$ can be approximated according to their relationship scores with prototypes,

$$p_{ik} = \sum_{\mathbf{c}_j \in \mathcal{C}_k^g} r_{ij}. \tag{3}$$

Despite prototype-based clustering can improve efficiency significantly, the labeling granularity which decides the class (cluster) number is still unknown. However, overly fine-grained clustering can cluster nodes of the same class into different clusters, while overly coarse-grained clustering can cluster nodes of different classes into the same cluster. Both cases lead to drops in clustering accuracy. Therefore, we search for the optimal clustering granularity by maximizing the accuracy on labeled known class nodes,

$$\text{ACC} = \max_{f \in \mathcal{P}(\mathcal{Y}^L)} \sum_{v_i \in \mathcal{V}_L} \frac{1}{|\mathcal{V}_L|} \mathbb{1}(\hat{y}_i = f(y_i)),$$
$$\hat{y}_i = \arg\max_k p_{ik}, \tag{4}$$

where $\mathbb{1}$ is the indicator function and $\hat{y}_i$ is the predicted group ID. Considering the potential misalignment between class and groups IDs, we find the alignment function from the set of all permutations $\mathcal{P}(\mathcal{Y}^L)$ by solving the maximization problem through the Hungarian algorithm [15]. Groups do not match with any known class are regarded as discovered novel classes.

Group-Aware Attention Mechanism. Besides local attribute information on nodes, the neighborhood information on graphs has also been shown to be important for node classification [2,11,12,14,22]. Despite graph attention networks can assign different weights to different neighbors, which are not guaranteed to contain knowledge about how to distinguish each novel class from others.

To learn compact representations for novel class discovery, we propose the group-aware attention mechanism. Group assignments predicted by semi-supervised prototypical learning are approximations of the ground-truth labels, which can be utilized to eliminate correlations between novel and known classes during neighborhood aggregation. The group-aware attention mechanism calculates the attention scores for neighbors according to the group assignment similarities $e_{ik} = \cos(\mathbf{p}_i, \mathbf{p}_k)$, where $\mathbf{p}_i = [p_{i1}, \ldots, p_{iN_{clu}}]$,

$$\alpha_{ij} = \text{softmax}(e_{ij}) = \frac{\exp(e_{ij})}{\sum_{k \in N(i)} \exp(e_{ik})}. \tag{5}$$

The low group assignment similarities between novel and known class nodes lead to low attention weights, while the attention weights between nodes from the same class are high. Consequently, aggregating information according to the group assignments can eliminate correlations between classes due to the low attention weights on inter-class edges, obtaining compact representations,

$$\mathbf{h}_i^* = \sigma\Big(\sum_{v_j \in N(i) \cup \{v_i\}} \alpha_{ij} \mathbf{W} \mathbf{h}_j\Big). \tag{6}$$

where σ is the relu activation function. Multiple prototypical attention networks are stacked by using the representations outputted by the last layer as the input for the next layer. Consequently, we get access to a list of predictions as well as node representations, $\{(\mathbf{p}^{(1)}, \mathbf{h}^{(1)}), ..., (\mathbf{p}^{(L)}, \mathbf{h}^{(L)})\}$, calculated at different scales using different hops of neighborhood information, where L is the number of layers, $(\mathbf{p}^{(i)}, \mathbf{h}^{(i)})$ is the group assignments and representations generated by the i-th prototypical attention network. The prototypical attention networks are trained by the cross-entropy loss layer by layer,

$$\mathcal{L}_{ce} = \sum_{v_i \in \mathcal{V}} \sum_{l=1}^{L} -\log p_{iy_i}^{(l)}, \tag{7}$$

where y_i is the label of node v_i and $p_{iy_i}^{(l)}$ is the predicted probability of node i belonging to the y_i-th group (class) in the l-th layer.

3.2 Pseudo-Label Guided Open-World Learning

To mitigate the lack of novel class labeling information, we generate pseudo-labels to guide the open-world graph learning. The generation process of pseudo-labels is shown in the pseudo-label generator part, while the utilization of pseudo-labels is detailed in the pseudo-label enhanced structure refinement part.

Pseudo-Label Generator. ORAL introduces the pseudo-label generator to mitigate the lack of novel class labeling information. To fully leverage the rich information captured at different scales, ORAL ensembles prediction results estimated by multiple stacked prototypical attention networks. However, ensembling multiple group assignment results is non-trivial. Due to the unsupervised nature of novel class discovery, groups from different layers are not aligned, even the number of groups varies from layer to layer.

We overcome these challenges through the proposed padding, aligning and suppressing operation. Given the predictions $\{\mathbf{p}^{(1)}, ..., \mathbf{p}^{(L)}\}$, we first pad them with zero vectors such that they have the same number of prototype groups. Then we align and ensemble predictions from different layers by the Hungarian algorithm [15],

$$\hat{p}_{ij} = \frac{1}{L} \sum_{k=1}^{L} p_{ij_k}^{(k)}, \tag{8}$$

where j_k denotes the prototype group ID in the k-th layer that corresponds to the j-th group in the initial layer. $\hat{p}_{ij}$ is the probability of assigning node i to group j obtained from ensemble inference. Groups which only exist in a certain layer or associates with almost no nodes can be regarded as detection noise. We suppress neurons responding to these noisy groups by masking over the prediction results,

$$\hat{\mathbf{p}}_i = \hat{\mathbf{p}}_i \odot \mathbf{m} \tag{9}$$

where the mask $\mathbf{m}$ is calculated by comparing the group popularity with a pre-defined threshold η, $m_i = \mathbb{1}(\frac{1}{|\mathcal{V}|}\sum_i \hat{p}_{i,j} > \eta)$. The symbol $\odot$ stands for the element-wise product. For each known class and discovered novel class, we select reliable pseudo labels as the labeling information on confident node subset, which is the top γ percentage of unlabeled nodes with the highest prediction probability (confidence) of belonging to this label.

Algorithm 1: ORAL

Data: Graph $G = (\mathcal{V}, \mathcal{E}, \mathbf{X})$, Labeled training nodes $(\mathcal{V}_L, \mathcal{Y}^L)$
Result: Predictions of unlabeled nodes $\mathcal{V}_U$, where new classes can emerge.

Randomly initialize model parameters;
repeat
  for $l = 1$ *to* L **do**
    Calculate relevance scores r_{ij} between nodes and prototypes;
    Conduct semi-supervised grouping over prototypes $\{\mathcal{C}_1^g, \cdots, \mathcal{C}_{N_{clu}}^g\}$;
    Calculate group assignments of nodes $p_{ik} = \sum_{\mathbf{c}_j \in \mathcal{C}_k^g} r_{ij}$;
    Generate node representations by assignment similarity as in Eq. (3) ;
  Update model parameters according to loss defined in Eq. (13);
  Pseudo-labeling on nodes by ensemble inference in Eq. (9);
  Update the graph structure according to pseudo labels as in Eq. (11);
until *Convergence or iteration number reaches the threshold*;
Predict test node labels by ensemble inference in Eq. (9);

Pseudo-Label Enhanced Structure Refinement. Reliable pseudo-labels information is incorporated into the input graph data for guiding open-world learning. Specifically, we recover intra-class edges to increase similarities between nodes of the same pseudo-label, while eliminating inter-class edges to decrease inter-class similarities. Structure refining according to pseudo-labeling makes it easier to distinguish novel classes from others.

Recovering intra-class edges according to pseudo-labels can introduce information from unconnected nodes in the original graph for open-world learning. However, excessive edge recovery can result in over-smoothing. We select the top μ percent of intra-class node pairs with the lowest similarities as recovered edges $\mathcal{E}_+$, where the similarity is calculated according to their relationship scores to prototypes in Eq. 1,

$$s_{ij}^n = \frac{1}{L} \sum_{l=1}^{L} \cos(\mathbf{r}_i^{(l)}, \mathbf{r}_j^{(l)}), \tag{10}$$

where $\mathbf{r}_i^{(l)}$ is the relationship between nodes and prototypes in the l-th layer. Consequently, edges between intra-class nodes associated to different prototypes are recovered, which increases similarities between prototypes related to the same class and facilitates more accurate semi-supervised prototypical clustering for novel class discovery.

ORAL further eliminates inter-class edges $\mathcal{E}_-$ according to pseudo-labeling. Since ensemble pseudo-labeling incorporates knowledge from different layers, taking the refined structure as input for each prototypical attention layer can transfer knowledge to eliminate unwanted correlations between classes among layers. The final refined structure can be formulated as,

$$\hat{\mathcal{E}} = (\mathcal{E} \setminus \mathcal{E}_-) \cup \mathcal{E}_+. \tag{11}$$

Table 1. Statistics of the experiment datasets.

Dataset	# Nodes	# Edges	# Attributes	# Classes	# Known Classes
Cora	2708	5429	1433	7	5
AmazonPhoto	7650	238162	745	8	6
BlogCatlog	5196	343486	8189	6	4

To make ORAL robust to the noise introduced during the structure refining process, we introduce the permutation-consistent training loss, which augments the structure information with the adaptive augmentation technique [34] and regularizes the prediction consistency between different views,

$$\mathcal{L}_{con} = \sum_{v_i \in \mathcal{V}} \sum_{l=1}^{L} \mathrm{KL}(\mathbf{p}_i^{(l)} || \bar{\mathbf{p}}_i^{(l)}), \tag{12}$$

where $\bar{\mathbf{p}}_i^{(l)}$ is the prediction over the augmented graph. The overall training loss can be calculated as,

$$\mathcal{L} = \mathcal{L}_{ce} + \mathcal{L}_{reg} + \mathcal{L}_{con} \tag{13}$$

The final predictions on unlabeled nodes can be obtained based on Eq. (9). The training process is summarized in Algorithm 1.

4 Experiments

4.1 Experiment Settings

Datasets. We test our method on three commonly used datasets: (1) **Cora** [31] is the citation network where classes are subjects of scientific publications. (2) **AmazonPhoto** [24] is the co-purchase network with categories of goods as node labels. (3) **BlogCatlog** [18] is the blog network where node labels are topics of blogs. The dataset statistics is shown in Table 1. For each dataset, we sample 80% of classes for which we have labels during training, the remaining ones are novel classes emerged in the test set. We further sub-sample 70% and 15% of the nodes from known classes to constitute the labelled training set and validation set. The remaining nodes from known classes, along with all instances from novel classes, constitute the unlabeled test node set V_U.

Evaluation Metrics. Following the novel class discovery literature [17,26] in the CV domain, we conduct experiments in scenarios where the number of classes is known as well as in scenarios where it is unknown. The model performance is measured with clustering accuracy, since the label IDs of ground-truth labeling and predictions may not be aligned. Besides the main metric indicating the classification performance over all testing nodes, we further report values for both known class subset (nodes in $\mathcal{V}_U$ belonging to classes in $\mathcal{Y}^L$) and novel classes subset (nodes in $\mathcal{V}_U$ belonging to classes in $\mathcal{Y}^U \setminus \mathcal{Y}^L$).

Table 2. Performance comparison of all methods, the experiment results are average over ten runs. "Prior" refers to the number of classes.

Dataset	Method	Without Prior					With Prior			
		GCN	GCD	OCD	GPC	ORAL	VGAE	GCD	OCD	ORAL
Cora	All	39.29	50.67	53.28	54.44	69.91	50.60	48.08	51.03	**71.39**
	Known	67.31	54.31	56.76	63.30	73.60	49.48	51.69	50.56	**76.97**
	Novel	0.00	45.50	48.41	42.03	64.72	52.22	43.03	51.67	**65.71**
BlogCatalog	All	26.04	37.79	42.83	39.33	70.23	38.18	37.91	42.73	**73.76**
	Known	70.28	26.00	34.55	49.79	87.39	25.03	27.13	34.84	**88.46**
	Novel	0.00	44.73	47.70	33.19	60.14	45.92	44.25	47.37	**65.11**
Amazon Photo	All	56.69	70.66	67.98	73.66	84.09	66.35	70.52	67.41	**88.05**
	Known	93.44	71.98	68.58	88.85	92.71	65.89	71.98	66.75	**95.36**
	Novel	0.00	68.44	67.04	50.22	70.79	67.06	68.27	68.43	**76.77**
Average	All	40.67	53.04	54.69	55.81	74.74	51.71	52.17	53.73	**77.73**
	Known	77.01	50.76	53.29	67.31	84.57	46.76	50.23	50.72	**86.93**
	Novel	0.00	52.89	54.38	41.81	65.21	54.96	51.85	55.82	**69.20**

Baselines. We choose the following methods as baselines, which can be classified into the three families. (1) **Semi-supervised Approach**. GCN [14] is a popular graph convolution network for semi-supervised node classification, which is unable to discover novel classes and will mistakenly classify novel class nodes into known classes.(2) **Unsupervised Approach**. VGAE [13] is an unsupervised method to learn graph embeddings. We extend this method for open-world graph learning by combining VGAE with the non-parametric kmeans classifier, where the number of classes is assumed to be a known prior. (3) **Open-world Semi-supervised Approaches.** GCD [26] is the most popular method for discovering novel categories in the CV domain, which estimates the number of novel classes by maximizing the clustering accuracy on labeled instances. OpenNCD [17] replaces the kmeans classifier with learnable prototypes and trains the network with a novel progressive bi-level contrastive learning method. We abbreviate OpenNCD as OCD in this article. GPC [32] proposes a semi-supervised variant of the gaussian mixture model which alternates between representation learning and class number estimation. We extend these methods to open-world graph learning by replacing the encoding module with graph convolution layers [14]. GCD, OCD and ORAL can additionally take the number of novel classes as a prior. In this scenario, these methods search for the clustering granularity where the number of clusters (classes) matches with the prior knowledge.

4.2 Main Results

Table 2 illustrates the experiment results of baselines and our work. From the results, we have the following observations: **(1)** Our proposed method

Table 3. Experiment results of class number estimation, where EST and MAE represents the estimated class number and mean absolute estimation error.

Method	Cora		Amazon		BlogCatalog		Average
	EST	MAE	EST	MAE	EST	MAE	MAE
GCD	6.0	1.0	7.1	0.9	6.3	<u>0.3</u>	0.7
OCD	7.4	**0.4**	8.8	<u>0.8</u>	5.6	0.4	<u>0.5</u>
GPC	9.0	2.0	10.0	2.0	8.0	2.0	2.0
ORAL	7.7	<u>0.7</u>	8.0	**0.0**	5.7	**0.3**	**0.3**

Table 4. Ablation study, where PAN stands for prototypical attention network and POL stands for pseudo-label guided open-world learning. L_{reg} is the regularization loss defined in Eq. 2, L_{con} is the contrastive loss defined in Eq. 12.

Modules		Losses		Without prior				With prior			
PAN	POL	L_{reg}	L_{con}	Cora		BlogCatalog		Cora		BlogCatalog	
				All	Novel	All	Novel	All	Novel	All	Novel
✓	✓	✓	✓	**69.91**	**64.72**	<u>70.23</u>	<u>60.14</u>	**71.39**	**65.71**	**73.76**	**65.11**
×	×	✓	✓	53.28	48.41	42.83	47.70	51.03	51.67	42.73	47.27
✓	×	✓	✓	<u>65.97</u>	55.27	**70.44**	**60.38**	<u>64.53</u>	49.10	<u>73.10</u>	<u>63.80</u>
✓	✓	×	✓	52.58	<u>60.34</u>	55.86	51.56	56.15	<u>58.68</u>	56.31	51.47
✓	✓	✓	×	60.41	48.96	66.50	54.18	60.83	49.54	69.97	60.00

consistently shows strong performance on both known and novel classes across all datasets. The improvements can be attributed to that ORAL distinguishes novel classes from others by utilizing node similarities discovered by semi-supervised prototypical clustering and generates reliable pseudo-labels to mitigate lack of novel class labeling information. We further show performance under varying numbers of novel classes in Fig. 3 to validate the generalization of the proposed method. **(2)** Despite existing semi-supervised method GCN performs well on known classes, they are unable to discover novel classes and mistakenly classify novel class nodes into known classes. Consequently, its accuracy over novel classes is zero. **(3)** Given the number of classes, we can approximate the ground-truth labels by combining clustering and unsupervised representation learning method VGAE, which results in low performance on known classes due to the lack of supervision information. **(4)** With the help of labeled known class nodes, open-world semi-supervised learning methods can discover novel classes automatically without knowing the number of classes in advance. ORAL achieves the lowest mean absolute error in the estimated class number as shown in Table 3. **(5)** The performance of open-world semi-supervised learning baselines is still unsatisfactory since they are unable to eliminate correlations between classes on graphs. As shown in Fig. 7, ORAL can learn a more separable embedding compared with baselines. We further compare the performance of baselines equipped with

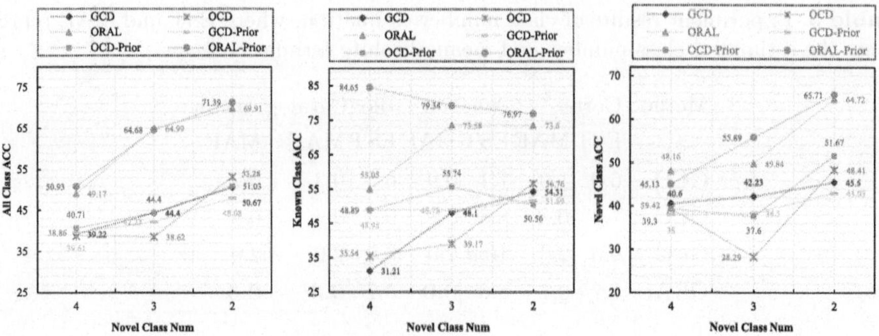

Fig. 3. Performance under varying numbers of novel classes on the Cora dataset, with methods suffixed by ‘prior’ being informed about the number of novel classes.

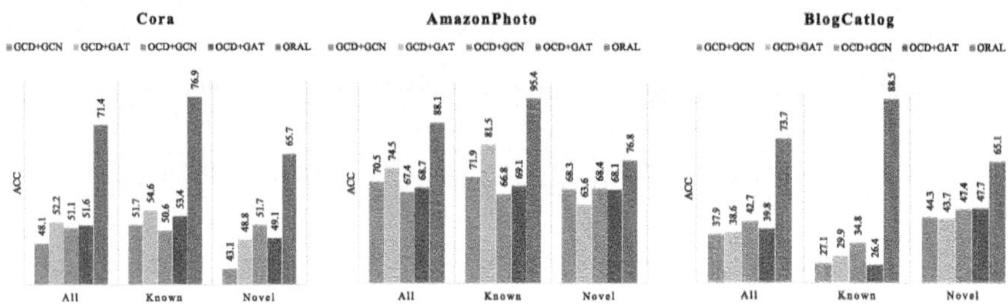

Fig. 4. Performance of baselines with different graph convolution networks. Our ORAL consistently outperforms all variants of existing open-world baseline methods, showing its superiority.

different graph convolution methods in Fig. 4. Although graph attention network [27] can assign learnable scores to edges, it still lacks the knowledge to distinguish novel classes from known ones. **(6)** By guiding the models to search for the proper labeling granularity during training, the number of novel classes can improve the novel class accuracy of OpenNCD and ORAL by 1.5% and 4.0% separately. Nonetheless, the prior information is simply incorporated into the clustering during testing in GCD, which can not guide the learning process and improve performance [28]. **(7)** Due to the lack of novel class labeling information, open-world semi-supervised baselines generally perform better on known classes than novel ones, except for the BlogCatlog dataset. Baseline methods tend to cluster different classes into the same group on BlogCatlog due to the strong inter-class correlations. To achieve best overall performance, the clustering accuracy is biased towards classes with more testing nodes, which constitute the novel classes in BlogCatlog.

4.3 Abaltion Study

We investigate the contribution of two main components as well as different loss terms in our model by ablation experiments. Specifically, we ablate the

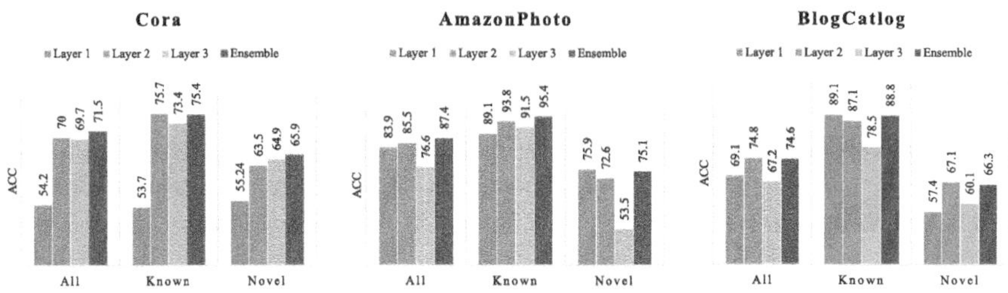

Fig. 5. Performance of different prototypical attention network layers.

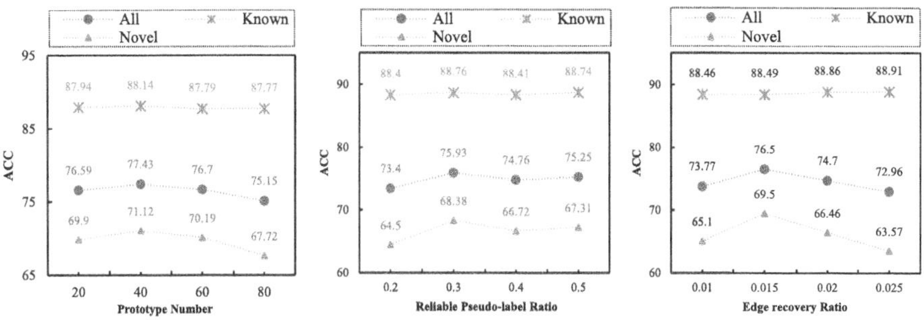

Fig. 6. Impact of key hyper-parameters on performance of ORAL on the BlogCatlog dataset when the class number is known.

prototypical attention network (PAN) by replacing it with the graph convolution layer [14], where the number of novel classes is estimated by OpenNCD [17]. The ablation of the pseudo-label guided open-world learning (POL) module is achieved by directly taking the model trained on the original graph for prediction, where the incorporation of pseudo-labeling information is ablated. We also ablate the regularization loss L_{reg} in Eq. (2) and contrastive loss in Eq. (12) by ignoring these loss terms in the learning process. The results of these variants are shown in Table 4, we have the following observations: **(1)** The improvement of PAN module is more significant on BlogCatlog which has more inter-class edges (60%) compared with Cora (19%). This observation shows the effectiveness of PAN in eliminating inter-class correlations on graphs. **(2)** Adding the POL module can further improve accuracy over novel classes, indicating the effectiveness of POL in mitigating the lack of novel class labeling information. **(3)** The regularization loss enhances the representativeness of prototypes for nodes, leading to a better approximation of node group assignments by the group assignments of prototypes. Ablation of the regularization loss results in less accurate group assignments of nodes obtained by semi-supervised prototypical clustering. **(4)** The contrastive loss enhances the robustness of ORAL against noise introduced during structure refining. Since the pseudo-labels on novel classes are less accurate compared to known classes, the structure refining guided by pseudo-labels for novel classes tends to be more noisy. Consequently, ablating contrastive loss results in more pronounced performance drops on novel classes.

We also conduct experiments to show the effectiveness of the proposed ensemble inference technique in Eq. 9. The performance of different prototypical attention network layers is shown in Fig. 5, we have the following observations: **(1)** The layer which achieves the best performance varies across datasets and classes. For instance, the second layer exhibits optimal performance for known classes classification in the AmazonPhoto dataset, whereas it is surpassed by the third layer on novel classes in the Cora dataset. **(2)** The ensemble inference consistently achieves strong performance over different classes, even surpassing any individual layer. Moreover, it alleviates the challenge of manually searching for the optimal layer for prediction, particularly in cases where labeled validation data for novel classes is unavailable.

4.4 Impact of Hyper-Parameter Settings

Next, we conduct experiments to investigate the impact of key hyper-parameters on the performance of ORAL:

Impact of Prototype Number N_{pro}**.** Prototypes are representative nodes on graphs, which are modeled as trainable vectors. We select {20, 40, 60, 80} as N_{pro} respectively, and show experiment results in Fig. 6. The performance improves as N_{pro} increases and peaks at 40. Then the performance decreases as N_{pro} becomes larger. Insufficient number of prototypes are not representative of nodes on graphs, while setting N_{pro} too high can lead to over-fitting, both results in performance drops.

Impact of Reliable Pseudo-Label Ratio γ**.** The pseudo-label generator selects the labeling information on top γ percentage of unlabeled nodes with the highest prediction probability (confidence) as reliable pseudo-labels. We range γ within {0.2, 0.3, 0.4, 0.5} and plot the results. In the beginning, performance increases as γ increases, indicating the effectiveness of utilizing pseudo-labels to guide open-world learning. The performance declines when γ is larger than 0.3, which may be caused by noise in pseudo-labels with low prediction confidence.

Impact of Recovery Edge Ratio μ**.** ORAL selects top μ percent of intra-class node pairs with the lowest similarities as recovered edges to guide open-world learning. We select μ as {0.01, 0.015, 0.02, 0.025} respectively. When μ is 0.015, ORAL achieves the best performance. This is because too many edges can lead to over-smoothing, while recovering too few edges can not incorporate enough information into the input graph to guide open-world learning.

5 Related Work

In this section, we briefly introduce the relevant research lines of our work, namely graph neural networks and open-world learning.

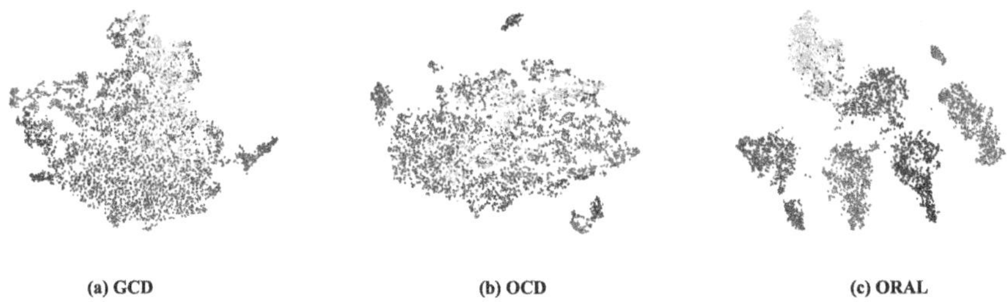

Fig. 7. Visualize embeddings of different models on BlogCatalog by t-SNE.

Graph Neural Networks. Graph Neural Networks (GNNs) are deep learning architectures on graph-structured data [8,23,29] and node classification is one of the most important downstream tasks of GNNs [2,11,12,14,20–22]. Early methods mainly focus on classifying the nodes within a set of fixed classes with abundant labelling [2,22]. To ease the annotation burden, researchers have paid much attention to semi-supervised setting [11,14] and few-shot setting [5,19]. However, these methods are unable to automatically discover novel classes emerged in the unlabeled test data where the method is required to estimate the number of novel classes and classify the unlabeled nodes into either known or novel classes.

Open-World Learning. Open-world learning methods [17,30] are proposed to handle the unknown novel classes during testing. Early works [7,25,30] propose out-of-distribution detection methods to reject samples not belonging to any known classes based on the statistics of prediction logits, which are unable to further discover the novel classes among the rejected samples. Novel class discovery methods [1,9,33] solve this problem by transferring knowledge from known classes. However, these methods share the strong assumption that unlabeled data all come from novel classes. Recently, generalized category discovery methods [17,26,32] relax this assumption by utilizing knowledge entailed in the pretrained models. Nonetheless, all these methods face challenges in distinguishing novel classes from others on graphs due to the unwanted correlations between classes caused by edge connections.

6 Conclusion

In this paper, we propose ORAL to discover novel classes for open-world graph learning. As far as we known, this is the first work to deal with this challenging yet practical problem. ORAL first employs the prototypical attention network to distinguish novel classes from others during representation learning. ORAL further generates pseudo-labels through clustering ensembles to exploit unlabeled data and mitigate the lack of novel class labeling information. Extensive experiments on public datasets show the superiority of our proposed method.

Acknowledgement. This work is supported in part by Ant Group, NSF under grant III-2106758, National Natural Science Foundation of China (62276187), and the Shanghai Science, Technology Development Fund No. 22dz1200704.

References

1. Brbić, M., Zitnik, M., Wang, S., et al.: Mars: discovering novel cell types across heterogeneous single-cell experiments. Nat. Methods **17**, 1200–1206 (2020)
2. Bruna, J., Zaremba, W., Szlam, A., et al.: Spectral networks and locally connected networks on graphs. arXiv preprint arXiv:1312.6203 (2013)
3. Cao, K., Brbic, M., Leskovec, J.: Open-world semi-supervised learning. In: Proceedings of ICLR (2022)
4. Chen, D., Jacob, L., Mairal, J.: Convolutional kernel networks for graph-structured data. In: Proceedings of ICML (2020)
5. Ding, K., Wang, J., Li, J., et al.: Graph prototypical networks for few-shot learning on attributed networks. In: Proceedings CIKM (2020)
6. Gao, X., Hu, W., Guo, Z.: Exploring structure-adaptive graph learning for robust semi-supervised classification. In: Proceedings of ICME (2020)
7. Geng, C., Huang, S.J., Chen, S.: Recent advances in open set recognition: a survey. In: TPAMI (2020)
8. Gori, M., Monfardini, G., Scarselli, F.: A new model for learning in graph domains. In: Proceedings of IJCNN. IEEE (2005)
9. Han, K., Vedaldi, A., Zisserman, A.: Learning to discover novel visual categories via deep transfer clustering. In: Proceedings of ICCV (2019)
10. Hetzel, L., Fischer, D.S., Günnemann, S., et al.: Graph representation learning for single-cell biology. Current Opinion in Systems Biology (2021)
11. Huang, W., Zhang, T., Rong, Y., et al.: Adaptive sampling towards fast graph representation learning. In: Proceedings of NeurlIPS (2018)
12. Izadi, M.R., Fang, Y., Stevenson, R.L., et al.: Optimization of graph neural networks with natural gradient descent. In: Proceedings of IEEE Big Data (2020)
13. Kipf, T.N., Welling, M.: Variational graph auto-encoders. In: Proceedings of NeurIPS (2016)
14. Kipf, T.N., Welling, M.: Semi-supervised classification with graph convolutional networks. In: Proceedings of ICLR (2017)
15. Kuhn, H.W.: The hungarian method for the assignment problem. In: NRL (1955)
16. Liao, R., Brockschmidt, M., Tarlow, D., et al.: Graph partition neural networks for semi-supervised classification. arXiv preprint arXiv:1803.06272 (2018)
17. Liu, J., Wang, Y., Zhang, T., et al.: Open-world semi-supervised novel class discovery. In: Proceedings of IJCAI (2023)
18. Liu, Y., Li, Z., Pan, S., et al.: Anomaly detection on attributed networks via contrastive self-supervised learning. In: IEEE TNNLS (2021)
19. Lu, B., Gan, X., Yang, L., et al.: Geometer: Graph few-shot class-incremental learning via prototype representation. In: Proceedings of SIGKDD (2022)
20. Luan, S., Hua, C., Lu, Q., et al.: Is heterophily a real nightmare for graph neural networks on performing node classification? arXiv preprint arXiv:2109.05641 (2021)
21. Muzio, G., O'Bray, L., Borgwardt, K.: Biological network analysis with deep learning. In: Briefings in bioinformatics (2021)
22. Niepert, M., Ahmed, M., Kutzkov, K.: Learning convolutional neural networks for graphs. In: Proceedings of ICML (2016)

23. Scarselli, F., Gori, M., Tsoi, A.C., et al.: The graph neural network model. In: IEEE Trans. Neural Netw. (2008)
24. Shchur, O., Mumme, M., Bojchevski, A., et al.: Pitfalls of graph neural network evaluation. arXiv preprint arXiv:1811.05868 (2018)
25. Sun, X., Yang, Z., Zhang, C., et al.: Conditional gaussian distribution learning for open set recognition. In: Proceedings of CVPR (2020)
26. Vaze, S., Han, K., Vedaldi, A., et al.: Generalized category discovery. In: Proceedings of CVPR (2022)
27. Velickovic, P., Cucurull, G., Casanova, A., et al.: Graph attention networks. In: Proceedings of ICLR (2018)
28. Wen, X., Zhao, B., Qi, X.: Parametric classification for generalized category discovery: a baseline study. In: Proceedings of ICCV (2023)
29. Wu, M., Pan, S., Zhou, C., et al.: Unsupervised domain adaptive graph convolutional networks. In: Proceedings of WWW (2020)
30. Wu, M., Pan, S., Zhu, X.: Openwgl: open-world graph learning. In: Proceedings of ICDM (2020)
31. Yang, Z., Cohen, W., Salakhudinov, R.: Revisiting semi-supervised learning with graph embeddings. In: Proceedings of ICML (2016)
32. Zhao, B., Wen, X., Han, K.: Learning semi-supervised gaussian mixture models for generalized category discovery. In: Proceedings of ICCV (2023)
33. Zhong, Z., Zhu, L., Luo, Z., et al.: Openmix: Reviving known knowledge for discovering novel visual categories in an open world. In: Proceedings of CVPR (2021)
34. Zhu, Y., Xu, Y., Yu, F., et al.: Graph contrastive learning with adaptive augmentation. In: Proceedings of WWW (2021)

Robust Graph Recommendation via Noise-Aware Adversarial Perturbation

Jiakai Tang, Zuxu Sun, Wei Yao, and Xu Chen(✉)

Gaoling School of Artificial Intelligence, Renmin University of China, Beijing, China
{tangjiakai5704,sunzexu21,wei.yao,xu.chen}@ruc.edu.cn

Abstract. Graph Neural Networks (GNNs) have achieved impressive success in recommender systems. However, the performance of GNNs is greatly affected by noisy data because of their distinct message propagation mechanism, which enlarges the adverse impact of noisy information on the representation learning. Although adversarial perturbation-based technique has been used to improve the resilience of models against noise, they still face several challenges when applied to GNN-based recommenders. Firstly, GNNs iteratively aggregate information from neighboring nodes, which amplifies the negative impact of noisy data on model training. Secondly, the uniform magnitude of perturbations to nodes potentially leads to the decreased or even polluted task-relevant semantic information for the generated user-item pairs (referred as adversarial examples). To address these challenges, we propose a simple but effective approach called RAP, which employs a two-stage learning framework. Specifically, in the first stage, we construct a weighted bipartite graph to model interaction's confidence-score, which effectively blocks the spread of noise information in GNN. Furthermore, in the second stage, RAP introduces noise-aware adversarial perturbations for different nodes. It accurately captures the intrinsic robustness at the instance level to help mitigate task-relevant semantic discrepancies between the original and adversarial examples. We conduct extensive experiments on three datasets to demonstrate the effectiveness of our method.

Keywords: Robust Recommendation · Adversarial Perturbation · Graph Neural Network

1 Introduction

Recommender systems (RS) have been successfully applied to various service platforms [11,24,25], such as Twitter, TikTok and Airbnb. Graph Neural Network (GNN) methods are an important branch of recommender algorithms because it enhances representation learning by incorporating high-order relationships among users and items [5,10,12,15]. While GNN-based models have shown impressive performance, they are actually more susceptible to the negative impact of noisy data compared to traditional collaborative filtering (CF) algorithms [9,12,20].

More recently, with the rapid development of adversarial robustness, lots of works have proven that using adversarial perturbations to generate augmented

M. Onizuka et al. (Eds.): DASFAA 2024, LNCS 14855, pp. 134–150, 2024.
https://doi.org/10.1007/978-981-97-5572-1_9

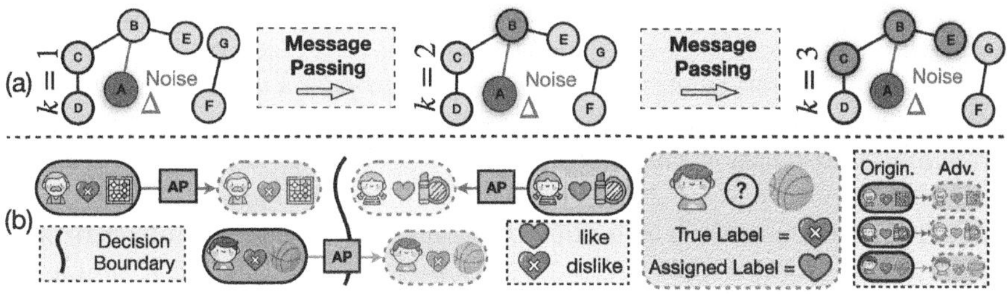

Fig. 1. An example to illustrate our motivation. (a) The message-passing mechanism of GNNs propagates the noise information from node 'A' to its connected neighboring nodes. (b) The uniform-magnitude adversarial perturbations may unintentionally decrease or even distort the potential semantic information of adversarial examples, leading to unexpected changes of the sample labels.

examples can significantly improve the resilience of models against noise perturbations [6,13,22,23]. For example, He *et al.* [6] propose to inject adversarial perturbations into user/item embedding vectors based on matrix factorization (MF) [8] algorithm. Wang *et al.* [13] propose an adversarial training framework based on adversarial gaussian perturbations to optimize the base recommenders. Despite their effectiveness, existing adversarial perturbation-based algorithms mainly focus on traditional recommender models, but lack research into their integration with GNN-based algorithms.

However, GNN-based recommenders are more vulnerable to noisy data compared to general DNN-based algorithms [10,14,21]. The unexpected sensitivity to noise information easily leads to the failure of the adversarial training framework in improving the robustness of GNN-based algorithms. The reasons can be given in twofold. *Firstly*, noise interactions are ubiquitous in real-world recommendation scenarios, such as accidental clicks and proxy purchasing. The adverse impact of noisy interactions is potentially amplified through GNN's message-passing mechanism, which iteratively aggregates the neighboring information. *Secondly*, existing adversarial perturbation-based methods adopt uniform magnitude of perturbations, overlooking the intrinsic variability in robustness across different instances. In special, the perturbations with uniform magnitude may unintentionally decrease or even destroy the task-specific semantic information for weak-robust adversarial examples. On the other hand, for strong-robust instances, the perturbations are too small to generate challenging adversarial examples that could help enhance robustness against noises.

To provide a better explanation, we take Fig. 1 as an example. As depicted in the upper part of Fig. 1, as the layer number of GNN increases, noise features progressively spread to connected neighboring representations through the message-passing mechanism. Furthermore, the lower part of Fig. 1 highlights that the uniform magnitude of adversarial perturbations, when applied to instances with different intrinsic robustness, may unexpectedly destroy task-relevant semantic information. What's worse, it results in changes to the real labels of adversarial examples for the weak-robust instances, ultimately misleading the model's learning process.

To address the aforementioned challenges, we propose a novel two-stage learning recommender framework: **R**obust Graph Recommendation via Noise-**A**ware Adversarial **P**erturbation (named as RAP). Specifically, *in the first stage*, we model the confidence score between each user-item interaction to construct a weighted bipartite graph, which aims to mitigate the negative impact of noise on the message-passing process. *In the second stage*, we develop a simple but effective method for generating adversarial perturbations that can adapt to different levels of intrinsic robustness in the training examples. Our algorithm's intuition is that data points closer to the decision boundary tend to have weaker tolerance for noise (namely weak-robust examples). In contrast, data points further away from the decision boundary are less affected by noise and thus have better robustness (namely strong-robust examples). Building on this idea, the well-designed noise-aware adversarial perturbations can dynamically and adaptively adjust the adversarial perturbations based on the distances between the input samples and the decision boundary of the recommender model. On three benchmark recommendation datasets, we conduct extensive experiments to validate the effectiveness of our proposed RAP.

In summary, the main contributions of this paper can be presented as follows:

- We study the reasons that GNN-based recommenders struggle to achieve ideal performance gains under the adversarial perturbation-based framework. Our work uncovers two important problems: (1) the negative influence caused from noise information is amplified through the message-passing mechanism of GNN algorithms. (2) Applying uniform-magnitude adversarial perturbations to the original data overlooks the different intrinsic robustness of data samples, leading to potential decreases or distortions of task-relevant semantic information in adversarial examples.
- We propose a novel robust GNN-based method for recommendation, which effectively improves the model robustness by incorporating confidence scores to interaction bipartite graph and introducing noise-aware adversarial perturbations. To the best of our knowledge, we are the first to explore graph adversarial perturbation for denoising in the field of collaborative filtering.
- We conducted extensive experiments on three real-world datasets to demonstrate the superiority of our proposed method. The source code for this work is publicly available at: https://github.com/TangJiakai/RAP.

2 Preliminary

Given the user-item interaction matrix $\mathbf{R}$, it consists of M users and N items. If the entry $r_{u,i} = 1$, it indicates that user u has interacted with item i. Conversely, $r_{u,i} = 0$ suggests that there is either no interaction between user u and item i or that the interaction has not been observed. In GNN-based recommender models, the bipartite interaction graph $\mathcal{G}$ contains a set of nodes $\mathcal{V}$ and a set of edges $\mathcal{E}$, where $\mathcal{V}$ consists of both user nodes $\mathcal{U}$ and item nodes $\mathcal{I}$. The adjacency matrix $\mathbf{A}$ of graph $\mathcal{G}$ is defined as follows:

$$\mathbf{A} = \begin{bmatrix} \mathbf{0} & \mathbf{R} \\ \mathbf{R}^\top & \mathbf{0} \end{bmatrix}. \tag{1}$$

The representations of user u and item i, $\mathbf{e}_u$ and $\mathbf{e}_i$, are defined by learnable embedding matrices $\mathbf{E}_U$ and $\mathbf{E}_I$, respectively. GNNs aim to learn high-order user/item representations by aggregating messages from their neighbors, which can be formulated as: $\mathbf{H}^{(l)} = f_{\text{AGG}}(\mathbf{H}^{(l-1)}, \mathbf{A})$, where $\mathbf{H}^{(l)}$ denotes the node representations at the l-th layer and $f_{\text{AGG}}(\cdot)$ represents the aggregation function of graph neural network. Here, we ignore the subscript u and i for simplicity.

After obtaining feature representations from L layers, a readout function $f_{\text{READOUT}}(\cdot)$ is employed to generate the comprehensive embeddings:

$$\mathbf{H}^* = f_{\text{READOUT}}([\mathbf{H}^{(0)}, \mathbf{H}^{(1)}, \cdots, \mathbf{H}^{(L)}]). \tag{2}$$

Then the preference prediction between the user u and item i is then computed using the inner product operation: $\hat{y}_{u,i} = \langle \mathbf{h}_u^*, \mathbf{h}_i^* \rangle$, where $\mathbf{h}_u^*$ and $\mathbf{h}_i^*$ indicate the comprehensive embeddings, obtained from $\mathbf{H}_U^*$ and $\mathbf{H}_I^*$, respectively. To optimize the model parameters, we employ the pairwise Bayesian Personalized Ranking (BPR) loss [8], which necessitates the prediction scores of observed interactions are higher than those for unobserved interactions:

$$\mathcal{L}(\mathcal{D}; \boldsymbol{\theta}) = \sum_{(u,i^+,i^-)\in\mathcal{D}} -\ln \sigma(\hat{y}_{u,i^+} - \hat{y}_{u,i^-}), \tag{3}$$

where $\mathcal{D}$ represents the training data and $\boldsymbol{\theta}$ indicates the model parameters. $\sigma(x) = \frac{1}{1+e^{-x}}$ denotes the sigmoid function. i^+ denotes the positive item while i^- represents the negative item. For each positive item i^+, we randomly draw a negative item from the candidate item pool that user u has not interacted with.

3 Proposed Methods

In this section, we present the proposed two-stage learning algorithm RAP. The overall framework is illustrated in Fig. 2. Specifically, in the first stage, we utilize a weighted interaction graph based on confidence scores between user-item pairs, which aims to block the spread of noise information on interaction bipartite graph. Following that, in the second stage, we design novel noise-aware perturbations that are implicitly applied to both semantic and structure features of nodes. This approach is designed to improve robustness and generalization performance of GNN-based backbones.

3.1 Confidence-Score Weighted Interaction Graph

Noisy interactions are ubiquitous in recommendation implicit feedback [6,16,17]. Compared with the traditional collaborative filtering methods, GNN-based recommender models are particularly affected by noisy data. This is because GNNs

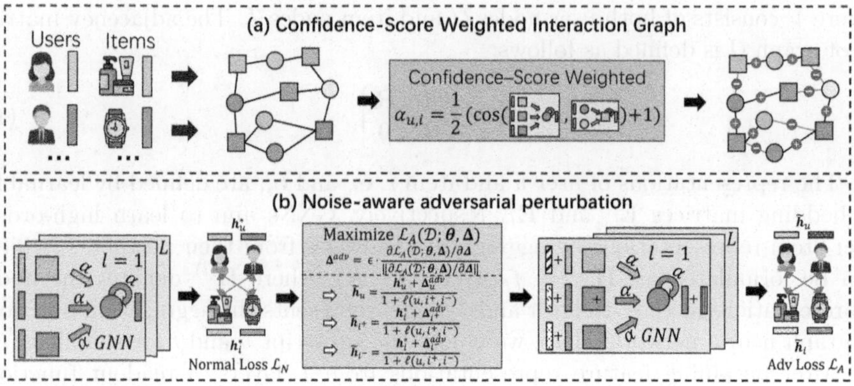

Fig. 2. Illustration of proposed RAP.

iteratively aggregate the information from neighboring nodes, which unintentionally amplify the negative impact of noise information on node's representation learning [10,14,17]. To overcome this challenge, we construct a weighted bipartite interaction graph based on confidence scores to purify the message propagation process of GNN-based recommenders.

Specifically, inspired by the homophily theory [7], which suggests that individuals with similar characteristics are more likely to have similar preferences. We leverage the structure features of user and item nodes to determine the confidence score for each user-item interaction pair. In specific, we calculate the structure representations of user u and item i as follows:

$$\mathbf{z}_u = \sum_{i \in \mathcal{N}_u} \frac{1}{\sqrt{|\mathcal{N}_u|\,|\mathcal{N}_i|}} \mathbf{e}_i,\ \mathbf{z}_i = \sum_{u \in \mathcal{N}_i} \frac{1}{\sqrt{|\mathcal{N}_i|\,|\mathcal{N}_u|}} \mathbf{e}_u, \tag{4}$$

where $|\mathcal{N}_u|$ and $|\mathcal{N}_i|$ represent the the number of neighboring nodes for user u and item i, respectively. Additionally, the structural features for user u and item i are denoted as $\mathbf{z}_u$ and $\mathbf{z}_i$, respectively. After that, We can obtain the confidence score $\alpha_{u,i}$ for the given user u and item i using the following method:

$$\alpha_{u,i} = \frac{1}{2}(\cos(\mathbf{z}_u, \mathbf{z}_i) + 1), \tag{5}$$

where $\cos(\mathbf{z}_u, \mathbf{z}_i) = \frac{\mathbf{z}_u^\top \mathbf{z}_i}{\|\mathbf{z}_u\|_2 \cdot \|\mathbf{z}_i\|_2}$. The purpose of normalizing the cosine similarity operation is to guarantee that the calculated confidence score is always a non-negative value. Furthermore, higher confidence scores indicate a higher reliability in the interactions, whereas lower scores signify a greater likelihood of the interactions being noise data.

Following that, based on the interaction bipartite graph weighted by the confidence scores, we can acquire the purified representations of users and items:

$$\mathbf{h}_u^{(l)} = \sum_{i \in \mathcal{N}_u} \frac{\alpha_{u,i} \cdot \mathbf{h}_i^{(l-1)}}{\sum_{j \in \mathcal{N}_u} \alpha_{u,j}},\ \mathbf{h}_i^{(l)} = \sum_{u \in \mathcal{N}_i} \frac{\alpha_{u,i} \cdot \mathbf{h}_u^{(l-1)}}{\sum_{j \in \mathcal{N}_i} \alpha_{j,i}}, \tag{6}$$

where $\mathbf{h}_u^{(l)}$ and $\mathbf{h}_i^{(l)}$ denote the l-th order graph representations of user u and item i, respectively. Note that we use $\mathbf{E}_U$ and $\mathbf{E}_I$ to initialize the user representations $\mathbf{h}_u^{(0)}$ and item representations $\mathbf{h}_i^{(0)}$ for the beginning layer, respectively. After propagating through L layers, the mean-pooling function is then employed as the readout function to compute the comprehensive user and item representations:

$$\mathbf{h}_u^* = \frac{1}{L+1}\sum_{k=0}^{L}\mathbf{h}_u^{(k)},\ \mathbf{h}_i^* = \frac{1}{L+1}\sum_{k=0}^{L}\mathbf{h}_i^{(k)}. \tag{7}$$

Finally, using the optimization objective (3), we can obtain the loss $\mathcal{L}_{\mathrm{N}} = \mathcal{L}(\mathcal{D};\Theta)$ under the standard training.

3.2 Noise-aware Adversarial Perturbation

The feature representations of users and items learned by the recommender models are highly dynamic and susceptible to contamination from noisy data. To overcome the above problem, we propose a novel noise-aware adversarial perturbation-based data augmentation in this subsection. Note that, unlike the first stage where we block the propagation of noisy information during message-passing process, in the second stage, we shift the focal point to the perturbations of the nodes themselves. The overall idea of noise-aware adversarial perturbation is to enhance the instance-level robustness capability by considering the distances between the data points and the decision boundary of the recommender model.

Specifically, we introduces adversarial perturbations to the high-order graph representations of user and item nodes. In contrast to the existing feature-based adversarial perturbation methods [6,13,23], our algorithm can implicitly achieve dual corruptions of both semantic and structure features of nodes. Consequently, such dual-role perturbations can more effectively align with the diverse and complicated noises present in real-world recommendation scenarios, which further strengthens the model robustness and generalization ability. In specific, the high-order graph adversarial perturbations $\boldsymbol{\Delta}^{adv}$ are defined as follows:

$$\begin{aligned} \boldsymbol{\Delta}^{adv} &= \arg\max_{\boldsymbol{\Delta},\|\boldsymbol{\Delta}\|\le\epsilon} \mathcal{L}_A(\mathcal{D};\boldsymbol{\theta},\boldsymbol{\Delta}), \\ \text{where}\quad \mathcal{L}_A(\mathcal{D};\boldsymbol{\theta},\boldsymbol{\Delta}) &= \sum_{(u,i^+,i^-)\in\mathcal{D}} -\ln\sigma\left(\left\langle\tilde{\mathbf{h}}_u,\tilde{\mathbf{h}}_{i^+}\right\rangle, -\left\langle\tilde{\mathbf{h}}_u,\tilde{\mathbf{h}}_{i^-}\right\rangle\right), \\ \tilde{\mathbf{h}}_u = \mathbf{h}_u^* + \boldsymbol{\Delta}_u,\ \tilde{\mathbf{h}}_{i^+} &= \mathbf{h}_{i^+}^* + \boldsymbol{\Delta}_{i^+},\ \tilde{\mathbf{h}}_{i^-} = \mathbf{h}_{i^-}^* + \boldsymbol{\Delta}_{i^-}, \end{aligned} \tag{8}$$

where ϵ denotes the magnitude of adversarial perturbations, and $\boldsymbol{\Delta}$ represents the perturbation matrix of graph representation, which is consisted of all user and item perturbation vectors, *i.e.*, $\boldsymbol{\Delta} = \{\{\boldsymbol{\Delta}_u\}_{u\in\mathcal{U}},\{\boldsymbol{\Delta}_i\}_{i\in\mathcal{I}}\}$. Note that $\boldsymbol{\Delta}$ can be initialized with either zeros or gaussian distribution. Similar to the previous adversarial perturbation-based works in recommender systems [6,13,23] we employ the Fast Gradient Sign Method (FGSM) [4] to calculate the adversarial perturbation $\boldsymbol{\Delta}^{adv}$, which can be formulated as follows:

$$\boldsymbol{\Delta}^{adv} = \epsilon\cdot\frac{\partial\mathcal{L}_A(\mathcal{D};\boldsymbol{\theta},\boldsymbol{\Delta})/\partial\boldsymbol{\Delta}}{\|\partial\mathcal{L}_A(\mathcal{D};\boldsymbol{\theta},\boldsymbol{\Delta})/\partial\boldsymbol{\Delta}\|_2}. \tag{9}$$

Furthermore, inspired by previous works [3,19], which suggests that samples closer to the decision boundary are more susceptible to noise, while samples farther away from the boundary are more insensitive. Building on this idea, we propose adjusting the adversarial perturbations based on the distance between data points and the decision boundary of recommender model. However, calculating accurate distances is extremely challenging because of the highly nonlinear nature of neural networks. Intuitively, the distance between data point and the decision boundary of the model is inversely proportional to the loss of input example. Therefore, we consider utilizing the normal loss of samples to guide the adjustment of the adversarial perturbation's magnitude for each sample. Specifically, we rewrite the adversarial loss $\mathcal{L}_A$ defined in Eq. (8) as follows:

$$\mathcal{L}_A(\mathcal{D};\boldsymbol{\theta},\boldsymbol{\Delta}^{adv}) = \sum_{(u,i^+,i^-)\in\mathcal{D}} -\ln\sigma\left(\left\langle\tilde{\mathbf{h}}_u,\tilde{\mathbf{h}}_{i^+}\right\rangle, -\left\langle\tilde{\mathbf{h}}_u,\tilde{\mathbf{h}}_{i^-}\right\rangle\right), \tag{10}$$

$$\text{where}\quad \tilde{\mathbf{h}}_u = \frac{\mathbf{h}_u^* + \boldsymbol{\Delta}_u^{adv}}{1+\ell(u,i^+,i^-)},\quad \tilde{\mathbf{h}}_{i^+} = \frac{\mathbf{h}_{i^+}^* + \boldsymbol{\Delta}_{i^+}^{adv}}{1+\ell(u,i^+,i^-)},\quad \tilde{\mathbf{h}}_{i^-} = \frac{\mathbf{h}_{i^-}^* + \boldsymbol{\Delta}_{i^-}^{adv}}{1+\ell(u,i^+,i^-)},$$

where $\ell(u,i^+,i^-) = -\ln\sigma(\hat{y}_{u,i^+} - \hat{y}_{u,i^-})$ represents the pair-wise loss with respect to the single instance (u,i,j) during standard training.

Algorithm 1. Learning algorithm for RAP

Input: Training Data $\mathcal{D} = \{(u,i^+,i^-)\}$;
Parameter: ϵ, λ, γ;
Output: Model parameters $\boldsymbol{\theta}$;
Initialize $\boldsymbol{\theta}$ with Xavier initialization;
while early stopping condition is not met **do**
  The First Stage:
  Calculate the confidence scores using Eq. (4) and (5);
  Construct a weighted bipartite graph;
  The Second Stage:
  $\mathcal{L}_N \leftarrow$ using Eq. (3) with original data;
  $\boldsymbol{\Delta}_{adv} \leftarrow$ using Eq. (8) and (9);
  Generate adversarial examples using $\boldsymbol{\Delta}_{adv}$ and $\mathcal{L}_N$;
  $\mathcal{L}_A \leftarrow$ using Eq. (10) with adversarial data;
  Model Optimization:
  $\mathcal{L} \leftarrow$ using Eq. (11);
  Update $\boldsymbol{\theta}$ using SGD;
end while
return $\boldsymbol{\theta}$

3.3 Optimization

In summary, the overall loss of RAP is formulated as follows:

$$\mathcal{L} = \mathcal{L}_N + \lambda\cdot\mathcal{L}_A + \gamma\|\boldsymbol{\theta}\|_2, \tag{11}$$

where λ and γ are hyper-parameters used to control the strength of adversarial term and regularization term, respectively, and $\|\cdot\|_2$ denotes ℓ_2 norm.

The complete training process is illustrated in Algorithm 1. In specific, *in the first stage*, we create a confidence-score weighted interaction graph based on the structure representations of users and items. Then, *in the second stage*, RAP calculates the normal loss $\mathcal{L}_N$ using the original data under standard training. Subsequently, the adversarial loss function Eq. (8) is maximized to obtain the noise-aware adversarial perturbations $\boldsymbol{\Delta}^{adv}$ through Eq. (9), which are utilized to generate adversarial exmaples. Following that, the adversarial loss $\mathcal{L}_A$ and final loss $\mathcal{L}$ are computed using Eq. (10) and (11), respectively. This process of solving the adversarial loss can be viewed as a minimax game. At last, we employ SGD approach to optimize the model parameters. The above two-stage learning algorithm is iteratively executed until the early stopping condition is met.

4 Experiments

We conduct extensive experiments to evaluate the effectiveness of our method by answering the following research questions:

- **RQ1**: How does RAP perform compared with the state-of-the-art traditional recommender models, non-adversarial and adversarial denoising models?
- **RQ2**: How does performance of the proposed method compare to baselines under different ratios of noisy interactions in training data?
- **RQ3**: How do different components impact the performance of RAP?
- **RQ4**: Can RAP effectively differentiate between clean and noisy interactions in terms of perturbation magnitude and training loss?
- **RQ5**: How do different settings of hyper-parameter influence the effectiveness of the proposed RAP algorithm?

4.1 Experiment Setup

Dataset. We evaluate our models on three public benchmark datasets: *MovieLens 1M*[1] (**ML-1M**), *Amazon Toys and Games*[2] (**Amazon**), and *2022-Yelp*[3] (**Yelp**). Their detailed statistics are shown in Table 1. Following by previous works [9], we regard the rating scores above three as the positive feedback. For ML-1M and Yelp datasets, we utilize 15-core operation (k-core means removing users and items with less than k interactions). For Amazon dataset, 10-core is employed to ensure both data quality and scale. For all datasets, we split data randomly into three sets: training, validation, and testing, with a ratio of 8:1:1.

Baseline Models. We compare RAP with different types of recommenders. Specifically, we classify the baseline models into three types as follows:

[1] https://grouplens.org/datasets/movielens/.
[2] https://jmcauley.ucsd.edu/data/amazon/.
[3] https://www.yelp.com/dataset/challenge.

Table 1. Dataset Statistics.

Dataset	# User	# Item	# Interaction	Sparsity
ML-1M	6,040	3,706	1,000,209	95.5316%
Amazon	4,204,994	624,792	8,201,231	99.9997%
Yelp	1,987,929	150,346	6,990,280	99.9977%

Traditional recommender baselines:

(1) **Pop** is a non-personalized recommender method which generates the recommendation list based on item popularity.
(2) **BPRMF** [8] is a popular matrix factorization model which utilizes Bayesian Personalized Ranking as the optimization objective function.
(3) **NGCF** [12] is a recommender framework that utilizes the user-item graph structure to incorporate collaborative signals into the learning process.
(4) **LGC** [5] is a well-known GNN-based recommender that improves performance by linearly propagating node representations on the interaction graph.

Non-adversarial denoising baselines:

(5) **CDAE** [16] is a denoising method for top-N recommendation using Denoising Auto-Encoders (DAE) structure as model backbone.
(6) **SGL** [14] is a self-supervised learning algorithm for improving recommendation accuracy, which leverages the graph contrastive learning technique to learn high-quality node representation.
(7) **RGCF** [10] is a collaborative filtering method that denoises unreliable interactions, which consists of a graph denoising module and a diversity preserving module under a multi-task learning framework.

Adversarial perturbation-based baselines:

(8) **AMF** [6] is the first approach to use adversarial perturbations for improving robustness of recommenders, which injects adversarial perturbations on the embeddings of users and items based on matrix factorization (MF) [8].
(9) **ACAE** [23] is an adversarial training method based on the CDAE backbone.
(10) **DBPR** [13] achieves robust recommendation by applying gaussian adversarial perturbations to enhance the randomness of the noise.
(11) **RocSE** [21] is a robust graph collaborative filtering model, which adds in-distribution perturbation to construct a contrastive view-generator, which aims to mimick the behaviors of adversarial attacks.

Evaluation Metrics. We adopt the popular ranking evaluation metrics to evaluate the recommendation performance, including Recall@K, NDCG@K, HR@K and MRR@K. In our experiments, the value of K is set to 10 and 20. All evaluation metrics are calculated over all items, *i.e.*, full ranking strategy.

Table 2. The overall performance comparison on three real-world datasets. The best result is **bolded** and the runner-up is underlined. '*' indicates the statistical significance for $p < 0.05$ compared to the best baseline by two-tailed t-test.

Dataset	Metric	Pop	BPRMF	NGCF	LGC	CDAE	SGL	RGCF	AMF	ACAE	DBPR	RocSE	RAP
ML-1M	Recall@10	0.0417	0.1868	0.1940	0.1981	0.1935	0.1931	0.1947	0.1866	0.1949	0.1879	0.1955	**0.2057***
	NDCG@10	0.0499	0.2213	0.2285	0.2298	0.2217	0.2213	0.2221	0.2207	0.2271	0.2268	0.2226	**0.2383***
	HR@10	0.3072	0.7330	0.7480	0.7530	0.7441	0.7477	0.7494	0.7312	0.7487	0.7340	0.7459	**0.7673***
	MRR@10	0.1053	0.4035	0.4139	0.4159	0.4077	0.4049	0.4023	0.4019	0.4131	0.4184	0.4037	**0.4284***
	Recall@20	0.0975	0.2793	0.2883	0.2911	0.2870	0.2847	0.2888	0.2783	0.2883	0.2778	0.2919	**0.2999***
	NDCG@20	0.0701	0.2421	0.2494	0.2516	0.2456	0.2433	0.2451	0.2412	0.2487	0.2455	0.2467	**0.2597***
	HR@20	0.5178	0.8395	0.8517	0.8567	0.8538	0.8519	0.8607	0.8379	0.8496	0.8443	0.8598	**0.8660***
	MRR@20	0.1201	0.4110	0.4212	0.4232	0.4154	0.4123	0.4101	0.4095	0.4202	0.4261	0.4118	**0.4353***
Amazon	Recall@10	0.0098	0.0750	0.0841	0.0849	0.0804	0.0815	0.0843	0.0747	0.0866	0.0801	0.0867	**0.0905***
	NDCG@10	0.0052	0.0505	0.0556	0.0558	0.0532	0.0541	0.0567	0.0491	0.0566	0.0532	0.0571	**0.0591***
	HR@10	0.0191	0.1329	0.1513	0.1570	0.1418	0.1472	0.1510	0.1317	0.1552	0.1445	0.1606	**0.1656***
	MRR@10	0.0053	0.0571	0.0631	0.0637	0.0596	0.0614	0.0654	0.0545	0.0640	0.0602	0.0659	**0.0676**
	Recall@20	0.0212	0.1105	0.1198	0.1268	0.1152	0.1184	0.1282	0.1076	0.1223	0.1161	0.1268	**0.1298***
	NDCG@20	0.0086	0.0615	0.0668	0.0688	0.0641	0.0656	0.0701	0.0593	0.0678	0.0643	0.0697	**0.0712**
	HR@20	0.0411	0.1957	0.2142	0.2279	0.2043	0.2088	0.2225	0.1909	0.2183	0.2064	0.2285	**0.2309***
	MRR@20	0.0068	0.0613	0.0675	0.0687	0.0639	0.0656	0.0703	0.0586	0.0682	0.0645	0.0707	**0.0720***
Yelp	Recall@10	0.0106	0.0573	0.0626	0.0573	0.0538	0.0630	0.0614	0.0646	0.0480	0.0687	0.0673	**0.0697***
	NDCG@10	0.0082	0.0435	0.0477	0.0439	0.0427	0.0475	0.0466	0.0489	0.0373	0.0527	0.0515	**0.0534**
	HR@10	0.0388	0.1892	0.2064	0.1865	0.1810	0.2064	0.2000	0.2098	0.1647	0.2228	0.2185	**0.2262***
	MRR@10	0.0145	0.0714	0.0788	0.0723	0.0728	0.0782	0.0760	0.0801	0.0633	0.0860	0.0839	**0.0870***
	Recall@20	0.0184	0.0927	0.1034	0.0940	0.0860	0.1036	0.0989	0.1065	0.0772	**0.1128**	0.1098	**0.1128**
	NDCG@20	0.0108	0.0559	0.0621	0.0568	0.0540	0.0617	0.0599	0.0637	0.0476	0.0681	0.0664	**0.0685**
	HR@20	0.0627	0.2854	0.3137	0.2873	0.2700	0.3123	0.3003	0.3186	0.2470	0.3351	0.3274	**0.3352**
	MRR@20	0.0162	0.0780	0.0861	0.0792	0.0789	0.0854	0.0829	0.0875	0.0689	0.0937	0.0914	**0.0945**

Implementation Details. We initialize the model parameters with Xavier. The Adam optimizer is used for all models. The learning rate is tuned in the range of {0.01,0.005,0.001,0.0005,0.0001}. The weight of regularization term γ is set to 1e-6. We tune ϵ in {1.0,0.5,0.1,...,0.001} and λ in {10.0,1.0,0.5,0.1,0.05,0.01}. The hyper-parameters of all baselines either follow the suggestions from the original papers or are carefully tuned. Moreover, early stopping strategy is used as the convergence criterion. Specifically, we evaluate the performance on the validation dataset every 10 epochs, and terminate the training process if there is no improvement in recommendation performance for 10 consecutive epochs.

4.2 Overall Performance (RQ1)

The overall comparison results are presented in Table 2. From the table, we can make the following observations:

– Among all baselines, Pop always performs worst on all datasets, which indicates the necessity of personalized recommendations. GNN-based recom-

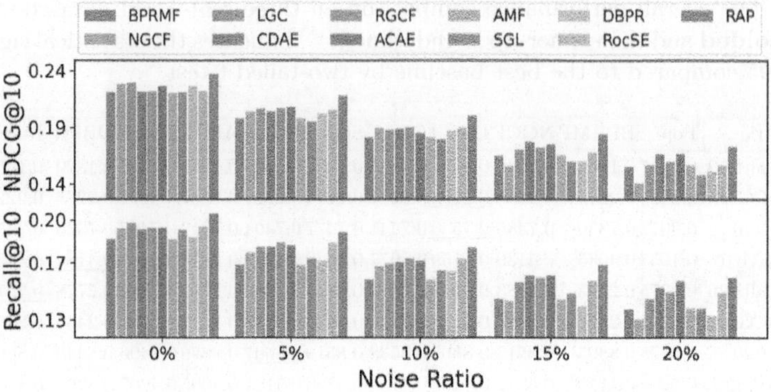

Fig. 3. Performance comparison under different noise ratios.

menders (such as LGC, RGCF, and RocSE) outperform traditional CF recommenders in most cases, especially on dense datasets (*e.g.*, ML-1M). This illustrates that leveraging the graph structure for recommendation task can obtain significant gains, which agrees with previous works [5,10].

- On extremely sparse datasets (*e.g.*, Amazon and Yelp), we can see that adversarial perturbation-based baselines (*e.g.*, DBPR) significantly outperform non-adversarial models. It suggests that adversarial examples can enhance the generalization ability when facing with noises and sparse data, overcoming the instability caused by the highly dynamic nature of real recommendation environment. However, all adversarial baselines fail to consistently achieve satisfactory performance across different datasets. We speculate the main reason is that these models assume the same magnitude of adversarial perturbations for all examples, overlooking the intrinsic variations in instance robustness.
- In contrast, our proposed RAP achieves the best performance on all evaluation metrics across different datasets. The advantage is brought by our designed two-stage learning algorithm that incorporates a confidence-score weighted interaction bipartite graph and noise-aware adversarial perturbations, which alleviate noise from both edge and node perspectives, respectively. More importantly, we carefully analyze the underlying principles of instance-level intrinsic robustness, thus achieving more fine-grained perturbation adjustments. Besides, we can find that the performance improvement at higher positions (*e.g.*, top-10 rank) is greater than that at lower positions (*e.g.*, top-20 ranks), suggesting that RAP tends to place the preferred items at the forefront.

4.3 Robustness Evaluation (RQ2)

To further investigate the robustness of RAP against noisy data, we randomly replace 5%, 10%, 15% and 20% of the observed interactions with noise interactions, respectively.[4] Specifically, for each observed interaction (u, i^+), we sample randomly an uninteracted item i^- of user u to replace the positive item i^+. In this experiment, we ignore the evaluation results of the Pop baseline due to its poor performance. Because of space limitations, we only show the results on ML-1M dataset based on NDCG@10 and Recall@10 metrics, but the same observations can be found on other datasets and evaluation metrics.

The results are reported in Fig. 3, we can observe that: (1) the recommendation performance consistently declines as the noise ratio increases for different recommenders. It indicates that an increased amount of noise data makes the recommenders more difficult to learn accurate user and item representations. Furthermore, these noisy data potentially misguide the models to capture the incorrect user preference and item characteristic, eventually resulting in poor generalization performance. (2) Our method achieves the best performance across different noise ratios, and the gap between our method and baselines becomes larger when the noise ratio increases from 0% to 20%. These observations confirm that RAP can effectively mitigate the negative impact of noisy data, and demonstrate the better robustness and denoising ability of our method.

Table 3. Ablation study on ML-1M dataset.

Model	Recall@10	Recall@20	NDCG@10	NDCG@20	HR@10	HR@20	MRR@10	MRR@20
w/o AT	0.1947	0.2888	0.2221	0.2451	0.7494	0.8607	0.4023	0.4101
w/o CW	0.2021	0.2928	0.2367	0.2560	0.7618	0.8558	0.4254	0.4320
w/o SP	0.1923	0.2819	0.2274	0.2470	0.7436	0.8482	0.4146	0.4220
w/o NA	0.2046	0.2993	0.2375	0.2587	0.7639	0.8641	0.4272	0.4343
RAP	**0.2057**	**0.2999**	**0.2383**	**0.2597**	**0.7673**	**0.8660**	**0.4284**	**0.4353**

4.4 Ablation Study (RQ3)

In this section, we analyze how the different components impact the overall recommendation performance. In specific, we compare RAP model against following ablated variants: (1) "w/o AT", where the entire adversarial training framework is removed; (2) "w/o CW", where the confidence-score weighted graph is replaced with the original 0-1 binary bipartite graph; (3) "w/o SP", where adversarial perturbations are only applied to the original feature instead of the high-order graph representations; (4) "w/o NA", where uniform-magnitude adversarial perturbations are assigned, similar to previous works [6,13,23].

[4] Some denoising works directly append noise interactions to original data for validating the model robustness. However, we argue that such experimental setting is unreasonable. This is because the data scale noticeably varies with different noise ratios, which may potentially mislead the analysis of experimental results.

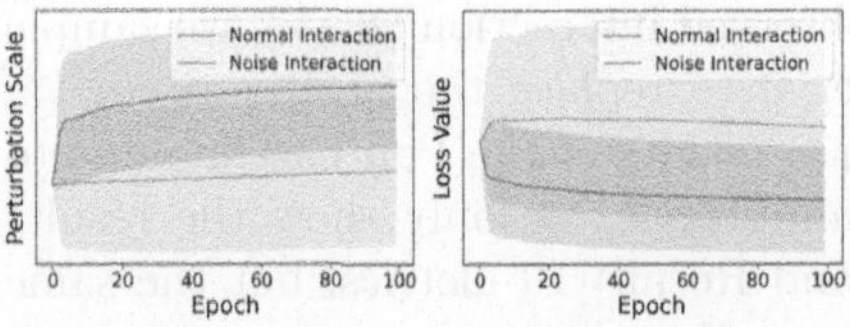

Fig. 4. Visualization of perturbation scale and loss during model training.

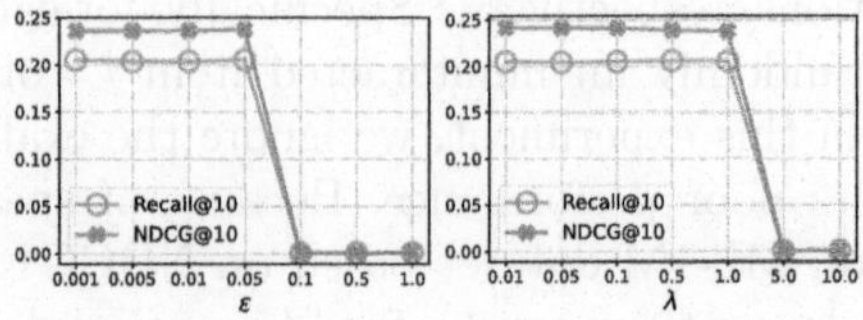

Fig. 5. Parameter Sensitivity Analysis.

The results are shown in Table 3. Based on the results, we can find:

- Comparing RAP with "w/o AT" variant, the significant performance improvement of our algorithm suggests that data augmentation based on adversarial perturbations can effectively enhance model robustness. As such, RAP has greater generalization ability in highly dynamic industrial environments.
- The consistently superior performance of RAP over "w/o CW" variant demonstrates the confidence-score weighted graph is effective in mitigating the pollution of noise interactions in the GNN's message propagation process, which facilitates purifying latent representations of users and items.
- The variant "w/o SP" demonstrates inferior performance compared to RAP. This suggests that perturbing shallow-layer representations does not yield ideal performance improvement. In contrast, our method targets higher-order graph representations, implicitly incorporating semantic and structural perturbations that could accurately mimic real-world recommendation scenarios.
- Our RAP significantly outperforms the variant "w/o NA". This suggests that taking into account the variations in the intrinsic robustness among original data is essential for improving the quality of generated adversarial examples.

4.5 Further Analysis (RQ4)

In this section, we compare the perturbation scales and loss values of normal interactions (*i.e.*, clean data) and noise interactions. Following the experimental setting in Section 4.3, we show the results in Fig. 4.

As shown in the left subgraph, we observe that as the training iteration increases, the perturbation scales of normal interactions become larger, while the perturbation scales of noisy interactions remain relatively small. These changing trends are in line with our assumptions. Specifically, the greater perturbations imposed on erroneous samples intensify the uncertainty in adversarial examples, which in turn would mislead the model's learning process. This is why RAP consistently assigns minor perturbations to noisy interactions. Conversely, clean data remain far from the decision boundary. Besides, as the model's capability improves during training, larger perturbations are produced to generate more challenging adversarial examples, which further strengthens the invariance of semantic features. Therefore, our method can effectively enhance the model robustness and achieve better generalization performance.

The right subgraph of Fig. 4 illustrates the changes in loss of normal and noisy interactions during the training process. It's obvious that the loss values of noisy samples always remain relatively high, and as the model continues training, the gap between the loss of clean and noisy data becomes significantly larger. This suggests that our RAP can accurately distinguish between noise and clean interactions, demonstrating superior denoising capability of our method.

4.6 Parameter Sensitivity (RQ5)

In this section, we evaluate the sensitivity of the model to hyper-parameters, including the initialized magnitude of adversarial perturbation ϵ and adversarial regularization strength λ. The experimental results are shown in Fig. 5, we can find that both hyper-parameters demonstrate similar patterns in the trends of model performance variations. More specifically, the performance remains relatively consistent when the parameter values are within reasonable ranges. It implies that our algorithm has good tolerance for the selection of hyper-parameters. However, it is important to note that when the parameter surpasses a certain threshold (*e.g.*, 0.05 for ϵ and 1.0 for λ), the accuracy of the recommender decreases rapidly. As such, we suggest that setting ϵ within the range of $[0.001, 0.05]$ and λ within the range of $[0.01, 1.0]$ can yield better model performance.

5 Related Work

We review prior works on denoising methods based on graph neural networks (GNNs) and adversarial perturbations in recommender systems.

GNNs-Based Denoising Recommendation. In recent years, Graph Neural Networks (GNNs) have shown remarkable performance in recommender systems [5,12]. However, the unique message-passing mechanism of GNNs enlarges the negative impact of noisy nodes and interactions on representation learning. Specifically, some methods aim to identify noises in user behavior by utilizing additional information such as auxiliary behaviors [17,18], explicit dislike signals [1], etc. Nevertheless, acquiring these additional data can be challenging and expensive. Instead, other works [10,14] adopt contrastive learning technique to expand the node views to enhance the robustness to noisy data. However, these methods are sensitive and vulnerable to the highly dynamic user preference and item characteristic in recommendation environments, *i.e.*, even imperceptible perturbations are sufficient to significantly degrade the model performance.

Adversarial Perturbation-Based Denoising Recommendation. Adversarial training becomes the de facto approach for improving model robustness [2].

The basic pipeline is to maximize the utility loss to obtain non-random perturbations, which are utilized to generate adversarial examples for data augmentation. For example, a large number of existing works exploit adversarial perturbations to different recommendation tasks [6,13,23]. Specifically, APR [6] utilizes adversarial perturbations based on matrix factorization (MF) [8] to generate adversarial user and item embedding vectors. Furthermore, DBPR [13] introduces adversarial Gaussian perturbations to enhance the randomness of noise. However, these methods overlook the fact that different samples have different intrinsic robustness. To this end, we propose a novel noise-aware adversarial perturbation-based recommender tailored for GNNs to improve the model robustness and effectiveness.

6 Conclusion

We propose a novel two-stage learning algorithm for GNN-based recommender, namely RAP. Specifically, in the first stage, to block the spread of noisy information, we construct a confidence-score weighted interaction graph, where we utilize the structure representations of users and items to obtain the confidence scores of interactions. In the second stage, we propose to adjust adversarial perturbations based on distances between data points and model decision boundary, which aims to considers different intrinsic robustness among training samples. The above two-stage training process mutually reinforces, effectively mitigating the impact of noise on the model's generalization ability. The expensive experimental results show that RAP significantly outperforms various types of recommender baselines, demonstrating the effectiveness of our proposed method.

Acknowledgement. This work is supported in part by National Key R&D Program of China (2023YFF0905402), National Natural Science Foundation of China (No. 62102420), Beijing Outstanding Young Scientist Program NO. BJJWZYJH012019100020098, Intelligent Social Governance Platform, Major Innovation & Planning Interdisciplinary Platform for the "Double-First Class" Initiative, Renmin University of China, Public Computing Cloud, Renmin University of China, fund for building world-class universities (disciplines) of Renmin University of China, Intelligent Social Governance Platform.

References

1. Bian, Z., et al.: Denoising user-aware memory network for recommendation. In: Proceedings of the 15th ACM Conference on Recommender Systems, pp. 400–410 (2021)
2. Deldjoo, Y., Noia, T.D., Merra, F.A.: A survey on adversarial recommender systems: from attack/defense strategies to generative adversarial networks. ACM Comput. Surv. (CSUR) **54**(2), 1–38 (2021)
3. Ding, G.W., Sharma, Y., Lui, K.Y.C., Huang, R.: Mma training: direct input space margin maximization through adversarial training. arXiv preprint arXiv:1812.02637 (2018)

4. Goodfellow, I.J., Shlens, J., Szegedy, C.: Explaining and harnessing adversarial examples. arXiv preprint arXiv:1412.6572 (2014)
5. He, X., Deng, K., Wang, X., Li, Y., Zhang, Y., Wang, M.: Lightgcn: simplifying and powering graph convolution network for recommendation. In: Proceedings of the 43rd International ACM SIGIR Conference on Research and Development in Information Retrieval, pp. 639–648 (2020)
6. He, X., He, Z., Du, X., Chua, T.S.: Adversarial personalized ranking for recommendation. In: The 41st International ACM SIGIR Conference on Research and Development in Information Retrieval, pp. 355–364 (2018)
7. McPherson, M., Smith-Lovin, L., Cook, J.M.: Birds of a feather: homophily in social networks. Ann. Rev. Sociol. **27**(1), 415–445 (2001)
8. Rendle, S., Freudenthaler, C., Gantner, Z., Schmidt-Thieme, L.: Bpr: Bayesian personalized ranking from implicit feedback. arXiv preprint arXiv:1205.2618 (2012)
9. Tang, J., Shen, S., Wang, Z., Gong, Z., Zhang, J., Chen, X.: When fairness meets bias: a debiased framework for fairness aware top-n recommendation. In: Proceedings of the 17th ACM Conference on Recommender Systems, pp. 200–210 (2023)
10. Tian, C., Xie, Y., Li, Y., Yang, N., Zhao, W.X.: Learning to denoise unreliable interactions for graph collaborative filtering. In: Proceedings of the 45th International ACM SIGIR Conference on Research and Development in Information Retrieval, pp. 122–132 (2022)
11. Wang, L., et al.: User behavior simulation with large language model based agents (2024)
12. Wang, X., He, X., Wang, M., Feng, F., Chua, T.S.: Neural graph collaborative filtering. In: Proceedings of the 42nd International ACM SIGIR Conference on Research and Development in Information Retrieval, pp. 165–174 (2019)
13. Wang, Z., Chen, X.: Robust recommendation with adversarial Gaussian data augmentation. In: Proceedings of the ACM Web Conference, pp. 897–905 (2023)
14. Wu, J., et al.: Self-supervised graph learning for recommendation. In: Proceedings of the 44th International ACM SIGIR Conference on Research and Development in Information Retrieval, pp. 726–735 (2021)
15. Shu, W., Tang, Y., Zhu, Y., Wang, L., Xie, X., Tan, T.: Session-based recommendation with graph neural networks. In: Proceedings of the AAAI Conference on Artificial Intelligence vol. 33, 346–353 (2019)
16. Wu, Y., DuBois, C., Zheng, A.X., Ester, M.: Collaborative denoising auto-encoders for top-n recommender systems. In: Proceedings of the Ninth ACM International Conference on Web Search and Data Mining, pp. 153–162 (2016)
17. Xin, X., et al.: Improving implicit feedback-based recommendation through multi-behavior alignment. In: Proceedings of the 46th International ACM SIGIR Conference on Research and Development in Information Retrieval, pp. 932–941 (2023)
18. Xu, J., et al.: Multi-behavior self-supervised learning for recommendation. arXiv preprint arXiv:2305.18238 (2023)
19. Xu, Y., Sun, Y., Goldblum, M., Goldstein, T., Huang, F.: Exploring and exploiting decision boundary dynamics for adversarial robustness. In: International Conference on Learning Representations, (2023)
20. Yang, H., Liu, Z., Zhang, Z., Zhuang, C., Chen, X.: Towards robust fairness-aware recommendation. In: Proceedings of the 17th ACM Conference on Recommender Systems, pp. 211–222 (2023)
21. Ye, H., Li, X., Yao, Y., Tong, H.: Towards robust neural graph collaborative filtering via structure denoising and embedding perturbation. ACM Trans. Inform. Syst. **41**(3), 1–28 (2023)

22. Yuan, F., Yao, L., Benatallah, B.: Adversarial collaborative auto-encoder for top-n recommendation. In: 2019 International Joint Conference on Neural Networks (IJCNN), pp. 1–8. IEEE (2019)
23. Yuan, F., Yao, L., Benatallah, B.: Adversarial collaborative neural network for robust recommendation. In: Proceedings of the 42nd International ACM SIGIR Conference on Research and Development in Information Retrieval, pp. 1065–1068 (2019)
24. Jingsen Zhang, X., Chen, J.T., Shao, W., Dai, Q., Dong, Z., Zhang, R.: Recommendation with causality enhanced natural language explanations. In: Proceedings of the ACM Web Conference, pp. 876–886 (2023)
25. Zhang, X., Dai, S., Xu, J., Dong, Z., Dai, Q., Wen, J.R.: Counteracting user attention bias in music streaming recommendation via reward modification. In: Proceedings of the 28th ACM SIGKDD Conference on Knowledge Discovery and Data Mining, pp. 2504–2514 (2022)

Learning Social Graph for Inactive User Recommendation

Nian Liu[1], Shen Fan[1], Ting Bai[2], Peng Wang[1], Mingwei Sun[1], Yanhu Mo[1], Xiaoxiao Xu[1], Hong Liu[1], and Chuan Shi[2(✉)]

[1] Alibaba Group Holding Limited, Hangzhou, China
{fanshen.fs,paulwong.wp,sunmingwei.smw,moyanhu.myh, xiaoxiao.xuxx,liuhong.liu}@alibaba-inc.com

[2] Peng Cheng Laboratory, Shenzhen, China
chuanshi1978@gmail.com

Abstract. Social relations have been widely incorporated into recommender systems to alleviate data sparsity problem. However, raw social relations don't always benefit recommendation due to their inferior quality and insufficient quantity, especially for inactive users, whose interacted items are limited. In this paper, we propose a novel social recommendation method called LSIR (**L**earning **S**ocial Graph for **I**nactive User **R**ecommendation) that learns an optimal social graph structure for social recommendation, especially for inactive users. LSIR recursively aggregates user and item embeddings to collaboratively encode item and user features. Then, graph structure learning (GSL) is employed to refine the raw user-user social graph, by removing noisy edges and adding new edges based on the enhanced embeddings. Meanwhile, mimic learning is implemented to guide active users in mimicking inactive users during model training, which improves the construction of new edges for inactive users. Extensive experiments on real-world datasets demonstrate that LSIR achieves significant improvements of up to 129.58% on NDCG in inactive user recommendation. Our code is available at https://github.com/liun-online/LSIR.

Keywords: Social recommendation · Graph structure learning

1 Introduction

The rise of social media has boosted interest in using social relations to enhance recommendation systems. Compared to general recommender systems [28], social recommendation [6,31] helps mitigate data sparsity issue, especially in cold-start setting [14,20]. Inactive users interact little with items on transaction platforms, making it challenging for precise recommendation.

Most existing social recommendation models operate solely on the raw social graph [6,31,34], with little effort to examine the quality of social relations on recommendation performance. *Are raw social relations adequately effective for inactive user recommendation?* Utilizing real social relations from one of China's

M. Onizuka et al. (Eds.): DASFAA 2024, LNCS 14855, pp. 151–167, 2024.
https://doi.org/10.1007/978-981-97-5572-1_10

largest e-commerce platforms, we analyze their impact on user behavior, particularly for inactive users. These analyses, uncover two flaws of raw social relations: (1) **Inferior quality**. The quality of raw social relations is generally inferior in real industrial scenario. Poor-quality social relations don't guarantee beneficial social information intake for users. (2) **Insufficient quantity**. Inactive users generally have fewer social relations than active users. This means less social information is available to inactive users. Combined with limited user-item interactions, inactive users receive very little information from both user and item aspects.

Above all, we realize that raw social relations are far from optimal for inactive users, requiring refinement for improved social recommendation, yet refining the social graph presents technical challenges. In the Graph Neural Network fidld, Graph Structure Learning (GSL) is employed to optimize graph structures [8,19,32]. GSL typically trains models to remove noisy edges and form better connections with distant nodes. Most GSL methods optimize relationships between all node pairs, leading to high computational demands [15]. In our case, applying GSL to the raw social graph involves eliminating irrelevant neighbors and incorporating high-quality users for inactive users, while also limiting the search range to address scalability issues.

In this paper, we explore the effect of social relations on inactive user recommendation, and incorporate GSL into social recommendation. We propose a novel model to ***L***earn ***S***ocial graph for ***I***nactive users ***R***ecommendation, named ***LSIR***. Specifically, users and items firstly recursively aggregate embeddings from each other, encoding collaborative signals. Using these aggregated embeddings, we refine the raw user-user social graph through delicately devised U2U deletion and U2C addition. U2U deletion retains high-quality existing social neighbors by removing noisy connections, while U2C addition adds connections between users and active interest clusters. These enhance inactive user representations with remaining purified relations and new connections to active interest clusters. Meanwhile, only generating new connections from representative active users avoids high computing complexity compared to traditional GSL settings requiring probability computation between every two nodes. Finally, mimic learning manipulates active users to 'mimic' the distribution of inactive users by maximizing similarity between pseudo inactive users and corresponding active interest clusters. Thus, the proposed LSIR can get more precise training on connecting inactive users and active users. Our contributions are summarized as follows:

- To the best of our knowledge, this is the first attempt to study the impact of social relations for social recommendation. We observe that inactive users tend to have limited social connections, some of which are low-quality, adversely affecting recommendation performance.
- We propose LSIR, incorporating graph structure learning into social recommendation to address this issue. By refining the raw user-user social graph, LSIR enhances social recommendation through noisy connection deletion and beneficial connection addition.

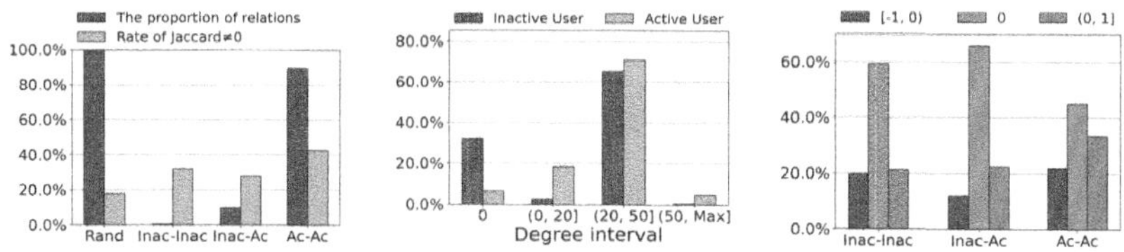

(a) Quality Analysis (b) Quantity Analysis (c) New Relation Effect

Fig. 1. The industrial observations on social relations based on Taobao.

– We validate the effectiveness and scalability of LSIR against state-of-the-art methods on public and industrial datasets. LSIR demostrates substantial improvements up to 129.58% on NDCG for inactive user recommendation.

2 Industrial Observations on Social Relation

In this section, we aim to investigate the impact of social relations on recommendation, especially for inactive users. We conduct our analysis based on real-world data from the Taobao e-commerce platform, part of Alibaba Group. Our observation includes over 100 million customers sampled uniformly from Taobao users. In this dataset, nearly 30% of users are labeled as inactive, with less than 10 monthly online behavior records. We also include items purchased by these selected customers over two months (August & September 2022)[1].

Observation 1: Raw Social Relations Generally Have Inferior Quality. We analyze the quality of social relations based on different cases. These cases involve interactions between inactive users and active users ('Inac-Ac'), interactions between two inactive users ('Inac-Inac'), and interactions between two active users ('Ac-Ac'). Additionally, we randomly pair users and consider it as a separate case ('Rand'), with the number of random pairs equal to the total number of actual relations. To evaluate the quality of each relation, we measure the rate of user pairs that have a non-zero Jaccard index for the items they have purchased. A non-zero Jaccard index indicates that the two users have at least one common interest within the observation window.[2] The results, shown in Fig. 1(a) as blue bars, indicate that for all three interaction types ('Inac-Inac', 'Inac-Ac', and 'Ac-Ac'), the rates of non-zero Jaccard index are no greater than 50%. These rates only slightly exceed that of the 'Rand' case. Therefore, only a partial connection between users is capable of reflecting each other's purchase

[1] In this section, we focus on dynamic data from Taobao. Public datasets are not fine-grained enough (lack of timestamps and 'inactive' definition). While the user population covered in this section is sufficiently large-scale, ensuring our observed phenomena are universally representative.

[2] This evaluation is focused on recommendation accuracy, as a higher Jaccard index signifies a stronger correlation between users.

preferences, while the assistance from the remaining connections is limited. This limited assistance is considered as an indication of inferior quality.

Observation 2: Social Connections for Inactive Users is Insufficient. We examine the degree distributions of active users and inactive users in the social graph, as depicted in Fig. 1 (b). The horizontal axis represents different intervals based on the social degrees of users. For instance, the interval (20, 50] includes users with more than 20 but no more than 50 social neighbors. The vertical axis represents the proportion of users falling into each interval. We observe that in the intervals (0, 20], (20, 50], and (50, Max), the proportions of active users are higher than those of inactive users. However, there is a notable disparity when it comes to isolated inactive users, which account for 31.9% of the total. In contrast, active users comprise only 6.2% of this category. This suggests that, on average, inactive users have fewer social connections and possess less social information compared to active users.

Observation 3: New-Founded Relations with Active Users Bring More Positive Effect. Our next observation focuses on the quality of newly established connections. Specifically, we examine the impact of these connections on user correlations. To do this, we calculate the difference in Jaccard indices between September and August for each pair of users who became connected during September. A higher positive difference indicates that the new connection has enhanced the correlations between these two users. We present these differences in Fig. 1 (c). To our surprise, the analysis reveals that connecting with active users ('Inac-Ac' and 'Ac-Ac') tends to yield more positive gains. On the other hand, the correlation changes between connected inactive users ('Inac-Inac') appear to be more random. This observation suggests that establishing connections with active users holds greater value for recommendation purposes.

Based on the above observations, we realize that the quality of social relations is probably inferior. On this basis, inactive users only own insufficient neighbors, which aggravates the challenges to recommend for them. To address this issue, it is crucial to consider importing high-quality information from active users and leveraging it to benefit inactive users. By doing so, we can enhance the overall recommendation process and improve the situation for inactive users.

3 Preliminary

Let $U = \{u_1, u_2, \ldots, u_m\}$ denotes the user set ($|U| = m$), and $I = \{i_1, i_2, \ldots, i_n\}$ denotes the item set ($|I| = n$). Users and items are assigned with original features, denoted as $\boldsymbol{X}_U = [\boldsymbol{x}_{u_1}, \boldsymbol{x}_{u_2}, \ldots, \boldsymbol{x}_{u_m}] \in \mathbb{R}^{m \times D_u}$ and $\boldsymbol{X}_I = [\boldsymbol{x}_{i_1}, \boldsymbol{x}_{i_2}, \ldots, \boldsymbol{x}_{i_n}] \in \mathbb{R}^{n \times D_i}$. $I(u)$ represents the set of items purchased by user u, and $U(i)$ means the set of users interacting with item i. As for social recommendation, we utilize N_u to denote user u's neighbors in raw social graph.

Definition 1. ***Inactive/Active user.*** *In a fixed transaction period, user u is defined as inactive user u_- if $|I(u)| < \epsilon$, where ϵ is a pre-defined threshold* *according to business consensus. And the rest users are defined as active user, denoted as u_+. In this case, the user set $U = \{u_+, u_-\}$.*

Definition 2. ***Enhanced social recommendation via refining social graph.*** *The probability of item $i \in I$ is purchased by an user $u \in \{u_+, u_-\}$ is:*

$$\widehat{r}_{u_i} = \mathcal{F}(i \in I | u, I(u), N'_u), \tag{1}$$

where the designed recommender $\mathcal{F}$ predicts the possibility as $\widehat{r}_{u_i} \in (0, 1)$. N'_u is the refined social neighbor set by solving the problems about raw relations uncovered in industrial observations. In this paper, we primarily enhance the social recommendation for these inactive users u_-.

4 The Proposed Model

In this section, we elaborate the proposed model LSIR, refining the social structure to improve the recommendation for inactive users. The overall architecture is shown in Fig. 2, including encoding user-item interactions, refining user-user social graph and mimic learning.

4.1 Encoding User-Item Interactions

To initially depict the characters of users and items, we focus on user-item interactions, and involve the collaborative filtering signals into their embeddings. Specifically, we need to firstly project features of users and items into a common latent vector space, since their features are composed in different ways. For an user u, we use a MLP with one hidden layer to transform his feature $\boldsymbol{x}_u$ into the common space:

$$\boldsymbol{h}_u = \boldsymbol{W}_U^{(1)} \sigma \left(\boldsymbol{W}_U^{(0)} \boldsymbol{x}_u + \boldsymbol{b}_U^{(0)} \right) + \boldsymbol{b}_U^{(1)}, \tag{2}$$

where $\boldsymbol{h}_u \in \mathbb{R}^{d \times 1}$ is the projected embedding of user u, σ is PReLU non-linear function, and $\{\boldsymbol{W}_U^{(1)}, \boldsymbol{W}_U^{(0)}, \boldsymbol{b}_U^{(1)}, \boldsymbol{b}_U^{(0)}\}$ are learnable parameters, respectively. Meanwhile, the feature $\boldsymbol{x}_i$ of item i can be mapped in a similar way:

$$\boldsymbol{h}_i = \boldsymbol{W}_I^{(1)} \sigma \left(\boldsymbol{W}_I^{(0)} \boldsymbol{x}_i + \boldsymbol{b}_I^{(0)} \right) + \boldsymbol{b}_I^{(1)}. \tag{3}$$

Next, we aim to enrich the user and item embeddings by aggregating interacted items and users, respectively. Considering potential degradation from further transformation and non-linear activation during aggregation [12], we merely adopt the simple weighted sum aggregator:

$$\boldsymbol{e}_u^{(k)} = \sum_{i \in I(u)} \frac{\boldsymbol{e}_i^{(k-1)}}{\sqrt{|I(u)|}\sqrt{|U(i)|}}, \quad \boldsymbol{e}_i^{(k)} = \sum_{u \in U(i)} \frac{\boldsymbol{e}_u^{(k-1)}}{\sqrt{|U(i)|}\sqrt{|I(u)|}}, \tag{4}$$

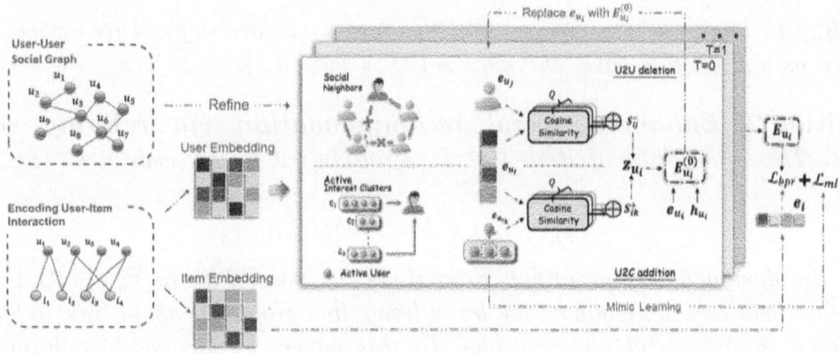

Fig. 2. The overview of our proposed LSIR, including 3 components: 1) Encoding User-Item Interactions, 2) Graph Structure Learning on Social Graph(U2U deletion & U2C addition) and 3) Mimic Learning.

where $\boldsymbol{e}_u^{(k)}$ and $\boldsymbol{e}_i^{(k)}$ represent embeddings of users and items in the kth layer, $\boldsymbol{e}_u^{(0)} = \boldsymbol{h}_u$ and $\boldsymbol{e}_i^{(0)} = \boldsymbol{h}_i$. $|\cdot|$ is the length of set. After K layers propagation, we further combine middle-layer results into the final embeddings of user u and item i under user-item interactions, denoted as $\boldsymbol{e}_u$ and $\boldsymbol{e}_i$:

$$\boldsymbol{e}_u = \frac{1}{K}\sum_{k=0}^{K}\boldsymbol{e}_u^{(k)}, \quad \boldsymbol{e}_i = \frac{1}{K}\sum_{k=0}^{K}\boldsymbol{e}_i^{(k)}. \tag{5}$$

4.2 Graph Structure Learning on Social Graph

After encoding user-item interactions, the embedding $\boldsymbol{e}_u$ of inactive user u_- still can not fully reflect his implicit interests, due to the limited consumption history. Thus, we need to supplement additional social information to boost the recommendation on u_-. Nevertheless, the raw social graph does not meet the requirements, pointed out in introduction. Hence, we refine it via two phases: U2U deletion and U2C addition. Next, we elaborate these two phases.

U2U Deletion. Observation 1 tells us that only few raw social relations are valid. These effective edges should be carefully discriminated and retained, while other noisy connections should be filtered out. Therefore, for user i and his one neighbor user j, we calculate the similarity between their embeddings to measure if the edge is effective. Without loss of generality, we adopt weighted cosine similarity [2] as their edge retention probability S_{ij-}:

$$S_{ij-} = \cos(\boldsymbol{W}_n \cdot \boldsymbol{e}_{u_i}, \boldsymbol{W}_n \cdot \boldsymbol{e}_{u_j}), \tag{6}$$

where $\boldsymbol{W}_n$ is learnable as a projection head to increase the flexibility of measurement. Furthermore, considering that there are multifaceted similarities between two users, we extend Eq. (6) to a multi-head version:

$$s_{ij-}^{(q)} = \cos(\boldsymbol{W}_n^{(q)} \cdot \boldsymbol{e}_{u_i}, \boldsymbol{W}_n^{(q)} \cdot \boldsymbol{e}_{u_j}), \quad S_{ij-} = \frac{1}{Q}\sum_{q=1}^{Q} s_{ij-}^{(q)}, \tag{7}$$

where we average results from Q aspects to get the comprehensive similarity. This multi-head extension can also stabilize the learning process [26,27]. With Eq. (7), we can get retention probabilities between u_i and all his neighbors. Next, we should consider how to adaptively select effective neighbors based on these probabilities for u_i. Please recall that inactive users absorb limited information from items (by definition) as well as social neighbors (Observation 2). Thus, the fewer items u_i interacts with, the more social information we should maintain for him to avoid uninformative initialization. Based on this principle, we devise the following tanh formula to calculate the deletion ratio for user u_i:

$$p_{del}(|I(u_i)|) = \frac{\exp(|I(u_i)|/r_1) - \exp(-|I(u_i)|/r_1)}{\exp(|I(u_i)|/r_1) + \exp(-|I(u_i)|/r_1)}, \tag{8}$$

where r_1 is a hyper-parameter to control the smoothness. If $|I(u_i)| \to 0$, then $p_{del}(|I(u_i)|) \to 0$. We sort previous retention probabilities between u_i and his neighbors in the descending order, and only retain top $1-p_{del}(|I(u_i)|)$ neighbors as the refined neighbor users set $N_{u_i}^U$.

U2C Addition. Enlightened by Observations 3, we realize that compared with randomly adding edges, linking inactive users to active ones is more efficient and reliable. However, active users still make up the vast majority (refer to Table 1). If we directly connect inactive user to every active user, the refined social graph will be very dense. To tackle this problem, we creatively devise a *U2C (user to cluster) addition* mechanism. Instead of employing all active users, we firstly mine interest clusters in active users (or active interest clusters), and then learn how to establish connections towards these clusters to reduce space complexity.

Mining Active Interest Clusters. The core insight is to bundle active users into several clusters, and select representative active user in each cluster as anchor connecting to inactive users. The whole process is sketched as follows:

- Cluster items via their raw features, and obtain the clusters set $\mathcal{C} = \{c_1, c_2, \dots, c_l\}$, where each $c_i = \{i_1^{c_i}, i_2^{c_i}, \dots, i_{|c_i|}^{c_i}\}$ represents one specific interest.
- For active user u_+, we calculate Jaccard index between $I(u_+)$ and c_i, denoted as $\mathcal{J}(I(u_+), c_i)$, and find the best-match cluster $\mathcal{C}(u_+)$:

$$\mathcal{C}(u_+) = \arg\max_{c_i} \mathcal{J}(I(u_+), c_i), \tag{9}$$

- For cluster c_i, we construct $\mathbb{U}(c_i) = \{u_+ | \mathcal{C}(u_+) = c_i\}$ via last step. In $\mathbb{U}(c_i)$, we choose the user u_{c_i} who owns the maximal Jaccard index with cluster c_i.

- Group these chosen users $U_C = \{u_{c_1}, u_{c_2}, \ldots, u_{c_l}\}$ to represent each corresponding item cluster.

Please note the above mining process only involves a hyper-parameter l, the cluster number. Next, we again resort to multi-head cosine similarity to define the probability S_{ik+} of connection between user u_i and one anchor user $u_{c_k} \in U_C$:

$$s_{ik+}^{(q)} = \cos(\boldsymbol{W}_n^{(q)} \cdot \boldsymbol{e}_{u_i}, \boldsymbol{W}_c^{(q)} \cdot \boldsymbol{e}_{u_{c_k}}), \quad S_{ik+} = \frac{1}{Q}\sum_{q=1}^{Q} s_{ik+}^{(q)}, \tag{10}$$

where $\boldsymbol{W}_n^{(q)}$ is shared with that in Eq. (7), but a new learnable matrix $\boldsymbol{W}_c^{(q)}$ is utilized to project anchor users specifically. Then, we still need to adaptively select the addition ratio for each user. Due to the inadequate social neighbors for inactive users (Observation 2), more information from high-quality active users should be involved into them as a supplementary. Thus, the addition ratio for user u_i is:

$$p_{add}(|I(u_i)|) = \frac{1}{1 + \exp(|I(u_i)|/r_2)}, \tag{11}$$

where r_2 is also a hyper-parameter similar to r_1. The less $|I(u_i)|$ is, the larger $p_{add}(|I(u_i)|)$ is. For u_i, we select top $p_{add}(|I(u_i)|)$ anchor users with maximal S_{ik+} as set $N_{u_i}^C$.

Now, we have got the refined neighbor users set $N_{u_i}^U$ and candidate anchor users set $N_{u_i}^C$ for u_i, then we fuse them into $\boldsymbol{z}_{u_i}$ under refined user-user social graph:

$$\boldsymbol{z}_{u_i} = \alpha\left(\sum_{j \in N_{u_i}^U} S_{ij-} \cdot \boldsymbol{e}_{u_j} + \sum_{k \in N_{u_i}^C} S_{ik+} \cdot \boldsymbol{e}_{u_{c_k}}\right) + (1-\alpha)\boldsymbol{e}_{u_i}, \tag{12}$$

where α is a balance parameter. Combining Eq. (2), (5) and (12), we utilize skip-connection to fuse these three user embeddings to get $\boldsymbol{E}_u^{(0)}$ at the first iteration:

$$\boldsymbol{E}_u^{(0)} = \boldsymbol{W}_f \cdot [\boldsymbol{h}_u || \boldsymbol{e}_u || \boldsymbol{z}_u] + \boldsymbol{b}_f, \tag{13}$$

where $\boldsymbol{W}_f$ and $\boldsymbol{b}_f$ are learnable parameters. Then, we replace $\boldsymbol{e}_u$ with $\boldsymbol{E}_u^{(0)}$, and redeploy U2U deletion and U2C addition to seek for a better social structure. After such T iterations, we get user final embedding $\boldsymbol{E}_u$.

Please note that although we mainly describe U2U deletion and U2C addition from the perspective of inactive users, this refinement process is also employed for active ones. Considering their large degree in social neighbors, we remove a large amount of useless social neighbors and very prudently involve new information for them according to Eq. (8) and (11). Therefore, the performance on active users can be guaranteed with improving inactive users.

To optimize the LSIR, BPR loss [23] is adopted as following:

$$\mathcal{L}_{bpr} = \sum_{(u,i,i') \in \mathcal{O}} -\log \sigma(\widehat{r}_{ui} - \widehat{r}_{ui'}) + \lambda ||\Theta||^2, \tag{14}$$

where Θ denotes the parameters of LSIR, $\widehat{r}_{ui} = \boldsymbol{E}_u^\top \boldsymbol{e}_i$ is the predicted score of u on i, and σ is sigmoid. Each triple instance in $\mathcal{O}$ includes target u, a ground-truth item i, and a negative item i' randomly chosen from unobserved items.

4.3 Mimic Learning

Although we have refined the raw user-user social graph, we notice that in U2C addition part, it is still difficult to match inactive users with suitable clusters. This is because in Eq. (10), only $\boldsymbol{e}_u$ (got from Eq. (5)) is utilized to measure the similarity, which only absorbs limited items for inactive users that cannot support them to connect towards true clusters. To solve this problem, we manipulate LSIR to know how to map users to corresponding clusters via the proposed mimic learning mechanism. Our intention is to generate pseudo inactive user embeddings $\hat{\boldsymbol{e}}_{u_+}$ for active user u_+. In this case, LSIR is expected to still match the changed $\hat{\boldsymbol{e}}_{u_+}$ to the real best-match $\mathcal{C}(u_+)$, given in Eq. (9).

MixUp [36] is proposed to efficiently improve results in supervised learning by adding arbitrary two samples to create a new one. MixGCF [13] introduces this strategy into graph-based recommendation, who involves positive samples into negative ones to generate harder negatives. Inspired by them, we inject inactive information into active embeddings, we call this method **Inactive Mixture**. Firstly, for each active user u_+, we randomly select η inactive users in one batch, and combine them with u_+ with weight $\beta \in (0, 1)$:

$$\hat{\boldsymbol{e}}_{u_+} = \beta \boldsymbol{e}_{u_+} + (1 - \beta)\boldsymbol{e}_{u_-}, \tag{15}$$

where $\boldsymbol{e}_{u_-}$ is embedding of one selected inactive user.

With the generated pseudo inactive user embeddings $\hat{\boldsymbol{e}}_{u_+}$, we manipulate encoder to make $\hat{\boldsymbol{e}}_{u_+}$ close to the anchor user who represents cluster $\mathcal{C}(u_+)$. To achieve this target, we adopt contrastive InfoNCE loss [22]:

$$\mathcal{L}_{ml} = -\sum_{u_+} \log \frac{\exp(\cos(\hat{\boldsymbol{e}}_{u_+}, \boldsymbol{e}_{u_{c_j}})/\tau)}{\sum_{k \neq j} \exp(\cos(\hat{\boldsymbol{e}}_{u_+}, \boldsymbol{e}_{u_{c_k}})/\tau)}, \tag{16}$$

where we assume $\mathcal{C}(u_+) = c_j$, and τ is a temperature parameter. With the optimization of Eq. (16), the encoder is trained to match pseudo inactive $\hat{\boldsymbol{e}}_{u_+}$ to the real $\mathcal{C}(u_+)$ and keep away from other clusters, even though u_+ has attempted to mimic other inactive users.

Combined with Eq. (14), the overall loss is:

$$\mathcal{L}_{\Omega} = \mathcal{L}_{bpr} + \xi \cdot \mathcal{L}_{ml}, \tag{17}$$

where ξ is a combination coefficient. We can optimize the proposed LSIR via back propagation with stochastic gradient descent.

4.4 Complexity Analysis

The complexity of each iteration inside LSIR mainly involves three parts: encoding user-item interactions, refining user-user social graph and mimic learning.

Table 1. The statistics of four datasets.

Dataset	#User	#Inactive User	#User-User	#Item	#User-Item	Inactive Cutoff
Flickr	8,358	2,386	187,273	82,120	327,815	≤ 2
Yelp	17,237	5,445	143,765	38,342	204,448	≤ 3
AliData *I*	40,428	12,071	323,608	115,502	1,016,122	≤ 7
AliData *II*	103,440	23,335	984,555	2,417,818	12,095,268	≤ 20

For the first part, the propagation between users and items consumes $\mathcal{O}(|\varepsilon|dK)$, where ε is the set of all user-item interactions, and d is the dimension of embedding. As for the refining process, we measure Q-head cosine similarity between target user and his neighbors or clusters. So, the complexity of this part is $\mathcal{O}(Q(|\zeta| + ml)d)$, where ζ is the set of all social relations and m is the number of users, and l is the number of active interest clusters. The cost of mimic learning is $\mathcal{O}(|B_+|\eta ld)$, where B_+ is the set of active users in one batch, and η is the number of selected inactive users. In summary, the complexity of LSIR is around $\mathcal{O}(|\varepsilon|dK + Q(|\zeta| + ml)d + |B_+|\eta ld)$, which is linear to $|\varepsilon|$ and $|\zeta|$. On one hand, the complexities of previous GNNs-based social recommendation models [31,34] are also linear to $|\varepsilon|$ and $|\zeta|$, which indicates that we do not drastically increase the model complexity. On the other hand, the complexities of traditional GSL based methods [8,15] are $\mathcal{O}(|\varepsilon|dK + m^2d)$, which is much higher than that of our proposed LSIR, considering $|\zeta| \ll m^2$.

5 Experiments

In this section, we conduct experiments to evaluate the performance of LSIR. Specifically, we aim to answer the following research questions:

- **RQ1**: How does LSIR perform compared to state-of-the-art baselines?
- **RQ2**: How does graph structure learning contribute to model performance?
- **RQ3**: How do key hyper-parameters affect the model?

5.1 Experimental Setup

Datasets. We employ four datasets, including two publicly available datasets [30] (i.e. Flickr, Yelp) and two real industrial datasets (i.e. AliData *I*, AliData *II*). AliData *I* represents a smaller-scale market with specific commodity categories and customer groups, where users have more sparse and concentrated behaviors. Contrarily, AliData *II* covers a broader range of commodities and serves diverse customers with different shopping needs, where interactions are denser. To provide transparency regarding our selection of inactive users, we present the cutoff criteria for inactive user behavior of each dataset in Table 1.

Baselines. We compare LSIR with three categories of baselines, including general recommendation methods {BPR [23], LightGCN [12], SimGCL [35]}, social recommendation methods {SBPR [38], DiffNet++ [30], S^2-MHCN [34], SEPT [33]} and cold-start recommendation methods {MeLU [17], TaNP [18]}.

Metrics. We focus on performing Top-K recommendation, where $K \in \{10, 20\}$. We choose two ranking based metrics *Normalized Discounted Cumulative Gain* (NDCG) [3] and *Hit Ratio* (HR) [3], and one relevancy-based metric *Precision* (PR) [34]. For each user, we randomly sample 1000 unrelated items as negative samples, and combine them with the ground-truth items in the ranking process to select top-k items for the final evaluation, which follows DiffNet++ [30].

Implementation Details. For a fair comparison, we set the dimension of embeddings as 64, and follow the settings in the original papers with carefully tuning. For LSIR, we use Glorot initialization [9] and Adam [16] optimizer. We tune the learning rate of Θ and Ω from 1e−4 to 1e−2. For temperature coefficients r_1 and r_2, we test ranging from {2, 10, 50, 100}. We search the number of attention heads from {1, 2, 3}. Finally, we tune ξ in {0.1, 0.01}.

5.2 Overall Recommendation Performance(RQ1)

In this subsection, we aim to validate the effectiveness of LSIR on recommendation. Specifically, we report the performances on Top-10/20 recommendation for

Table 2. Quantitative results on Top-10 and Top-20 recommendation for inactive users. (bold: best; underline: runner-up; '–': Out-Of-Memory on 32 GB GPU; 'Improv.' denotes the relative improvements over the best baseline)

Datasets	Metric	Top-K	BPR	LightGCN	SimGCL	SBPR	DiffNet++	S^2-MHCN	SEPT	MeLU	TaNP	**LSIR**	**Improv.**
Flickr	NDCG	10	0.0503	0.0797	0.0648	0.0659	0.1251	0.0990	0.0455	0.0660	0.0781	**0.2872**	**129.58%**
		20	0.0630	0.1042	0.0801	0.0808	0.1470	0.1137	0.0561	0.0994	0.1068	**0.3253**	**121.29%**
	HR	10	0.0731	0.1350	0.0951	0.0997	0.1752	0.1413	0.0675	0.1154	0.1391	**0.3927**	**124.14%**
		20	0.1112	0.2094	0.1444	0.1465	0.2462	0.2000	0.1028	0.2045	0.2176	**0.5164**	**109.75%**
	PR	10	0.0146	0.0270	0.0190	0.0199	0.0350	0.0283	0.0135	0.0231	0.0273	**0.0785**	**124.29%**
		20	0.0111	0.0209	0.0144	0.0147	0.0246	0.0200	0.0103	0.0204	0.0217	**0.0516**	**109.76%**
Yelp	NDCG	10	0.1437	0.1636	0.1847	0.1678	0.2261	0.1845	0.1480	0.1485	0.1562	**0.2669**	**18.05%**
		20	0.1725	0.1945	0.2152	0.2002	0.2605	0.2151	0.1834	0.1969	0.2126	**0.3038**	**16.62%**
	HR	10	0.2185	0.2439	0.2709	0.2525	0.3288	0.2742	0.2274	0.2492	0.2642	**0.3804**	**15.69%**
		20	0.3103	0.3409	0.3720	0.3555	0.4495	0.3754	0.3357	0.3971	0.4085	**0.4997**	**11.17%**
	PR	10	0.0437	0.0487	0.0542	0.0505	0.0658	0.0548	0.0455	0.0498	0.0531	**0.0761**	**15.65%**
		20	0.0310	0.0304	0.0372	0.0356	0.0450	0.0375	0.0336	0.0397	0.0408	**0.0500**	**11.11%**
AliData *I*	NDCG	10	0.1634	0.1629	0.1815	0.1644	0.1793	0.1817	0.1650	0.1648	0.1696	**0.2018**	**11.06%**
		20	0.1817	0.1766	0.1931	0.1812	0.1990	0.2019	0.1798	0.1784	0.1810	**0.2214**	**9.66%**
	HR	10	0.2201	0.2129	0.2224	0.2169	0.2515	0.2509	0.2122	0.2219	0.2223	**0.2748**	**9.26%**
		20	0.2859	0.2622	0.2638	0.2791	0.3223	0.3227	0.2655	0.2660	0.2724	**0.3530**	**9.39%**
	PR	10	0.0330	0.0301	0.0333	0.0325	0.0377	0.0376	0.0318	0.0317	0.0324	**0.0412**	**9.28%**
		20	0.0214	0.0187	0.0198	0.0209	0.0241	0.0242	0.0199	0.0202	0.0204	**0.0265**	**9.50%**
AliData *II*	NDCG	10	0.1126	0.1812	0.1757	0.1327	–	–	0.1796	0.1777	0.1808	**0.2060**	**13.69%**
		20	0.1251	0.1903	0.1821	0.1414	–	–	0.1897	0.1840	0.1866	**0.2208**	**16.03%**
	HR	10	0.1191	0.1638	0.1554	0.1368	–	–	0.1688	0.1610	0.1612	**0.2092**	**23.93%**
		20	0.1575	0.1955	0.1787	0.1664	–	-	0.2037	0.1812	0.1855	**0.2663**	**30.73%**
	PR	10	0.0329	0.0452	0.0430	0.0378	–	–	0.0467	0.0435	0.0444	**0.0578**	**23.77%**
		20	0.0218	0.0270	0.0247	0.0230	–	–	0.0282	0.0260	0.0262	**0.0368**	**30.50%**

Table 3. The NDCG@10 performances on active and overall users got from social recommenders. (bold: best; underline: runner-up; '–': Out-Of-Memory on 32 GB GPU)

Datasets	Flickr		Yelp		AliData *I*		AliData *II*	
Methods	Active	Overall	Active	Overall	Active	Overall	Active	Overall
SBPR	0.0728	0.0709	0.1687	0.1685	0.2128	0.1984	0.2964	0.2595
DiffNet++	<u>0.1011</u>	<u>0.1080</u>	<u>0.2193</u>	<u>0.2213</u>	0.2138	0.2035	–	–
S^2-MHCN	0.0918	0.0939	0.1991	0.1950	0.2274	0.2138	–	–
SEPT	0.0593	0.0554	0.1701	0.1638	**0.2362**	<u>0.2150</u>	**0.3415**	**0.3050**
LSIR	**0.1010**	**0.1542**	**0.2317**	**0.2418**	<u>0.2281</u>	**0.2203**	<u>0.3335</u>	<u>0.3048</u>

inactive users in Table 2, and NDCG@10 results for active and overall users in Table 3, where we focus on social recommenders. The conclusions are as follows:

- **LSIR consistently outperforms other baselines for inactive users.** Especially, LSIR achieves a significant improvement of 129.58% on NDCG of Flickr and 30.73% on HR of AliData *II*. This demonstrates the effectiveness of LSIR for inactive users. Besides, our model can successfully run on large-scale AliData *II*, while DiffNet++ and S^2-MHCN fail, showing the scalability of LSIR.
- **The results on Flickr stand out.** The reason is related to the average social neighbors per user. According to Table 1, the order follows Flickr (22.41) $\gg$ AliData *II* (9.52) $>$ Yelp (8.34) $>$ AliData *I* (8.00), which is in agreement with the order of performance improvements. Thus, higher number of neighbors leaves a larger room for LSIR to boost social relations.
- **Cold-start recommenders show weak competence in all the cases.** There are two possible causes. Firstly, these two cold-start models(MeLU and TaNP) are based on MAML rather than GNN, so superiority from message passing is neglected. Secondly, they do not absorb the useful knowledge from social relations, and only depend on user-item interactions.
- **LSIR is competitive constantly in both active and overall user cases.** As shown in Fig. 3, LSIR consistently achieves the best or second-best performance across all users in both academic and industrial datasets. This indicates that the proposed refinement process will not sacrifice the performance on active and overall users. LSIR successfully strikes a balance between enhancing recommendation quality for inactive users while maintaining or even improving performance for active and overall users.

5.3 Effects of Graph Structure Learning(RQ2)

Analysis of GSL Variants. To evaluate the effectiveness of designed GSL component in LSIR, we replace this part with two other GSL methods IDGL [2] and GAUG [37], denoted as LSIR+IDGL and LSIR+GAUG. In addition, we have a baseline variant, LSIR+UU, where the original social graph remains unchanged. We compare these variants with our model, and report *NDCG@10* and *NDCG@20* on inactive users on Yelp and AliData *I* in Table 4. From the table, we observe that LSIR+GAUG fails to run on AliData *I* due to its high space complexity. LSIR+IDGL performs even worse than LSIR+UU because it neglects some important users in its random anchors. Additionally, we explore the effectiveness of U2C addition and U2U deletion individually. The results, as shown in the Table 4, indicate that the absence of the U2C addition component leads to a larger drop in performance compared to the absence of the U2U deletion component. This indicates that for inactive users, high-quality supplement is more valuable than deleting limited neighbors.

Table 4. Test on the variants of GSL. ('–': Out-Of-Memory on 32GB GPU)

Datasets	Yelp		AliData *I*	
Metric	N@10	N@20	N@10	N@20
LSIR+UU	0.2211	0.2551	0.1984	0.2186
LSIR+IDGL	0.2066	0.2404	0.1883	0.2062
LSIR+GAUG	0.1974	0.2317	–	–
LSIR w/o U2C	0.2590	0.2903	0.1991	0.2187
LSIR w/o U2U	0.2622	0.2994	0.2009	0.2200
LSIR	**0.2669**	**0.3038**	**0.2018**	**0.2214**

Analysis of Mimic Learning. We test the effectiveness of mimic learning to enhance GSL. The results are given in Fig. 3, where Flickr and Yelp results are reported. In the figure, 'None' means no mimic learning. As shown in figure, the mimic learning performs better than 'None', with relative improvements around 6% on NDCG@10, indicating the effectiveness of this component.

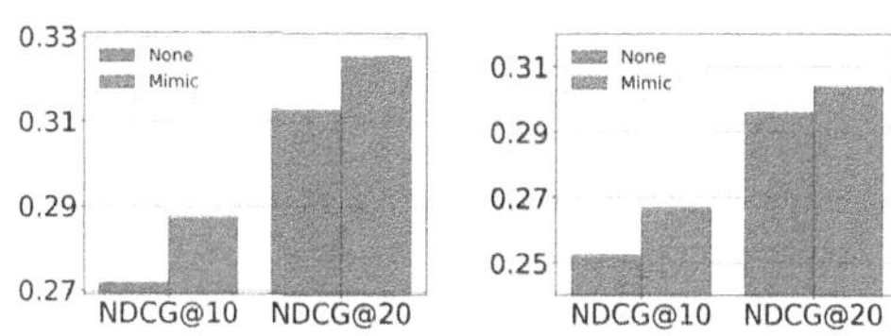

(a) Flickr (b) Yelp

Fig. 3. Test on the Mimic Learning.

5.4 Effects of Hyper-Parameters(RQ3)

Analysis of Ratio of Inactive User. Considering there is not definition of 'inactive' in selected public datasets (i.e., Flickr and Yelp), we also label the last 30% users as 'inactive' in Table 1. To address any concerns regarding ratio selection, we provide a comparison between DiffNet++ (the best baseline) and

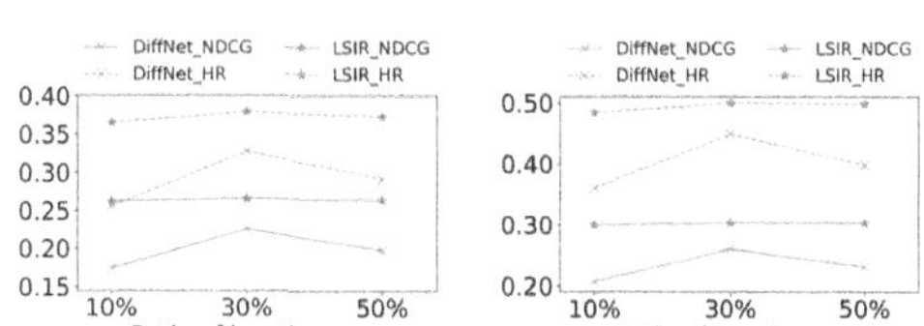

(a) Top-10 (b) Top-20

Fig. 4. Test on the ratio of inactive users.

LSIR on Yelp, considering different ratios of inactive users. As shown in Fig. 4, LSIR outperforms DiffNet++ consistently and exhibits greater stability across various ratios of inactive users. Therefore, the selection of ratio will influence less on the performance of LSIR.

Analysis of Cluster C. In this analysis, we examine the sensitivity of number of cluster C utilized in active interest clusters, plotted in Fig. 5. In this figure, points with lighter colors indicate better performance, while the best point is denoted by a red star. We find that the optimal number of clusters for both the Flickr and Yelp datasets is 30. Using a lower number of clusters would result in an inadequate representation of the diversity of active interests. Conversely, employing a higher number of clusters would make it more challenging for inactive users to find a suitable cluster.

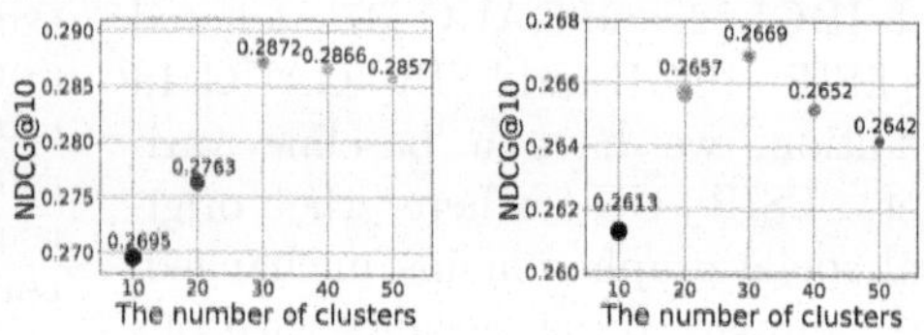

(a) Flickr (b) Yelp

Fig. 5. Test on the number of cluster C.

6 Related Work

Social Recommendation. GNN based social recommenders [24] has emerged as a promising direction in recent years. GraphRec [6] directly utilizes attention strategy to aggregate neighbors for both users and items. DiffNet++ [30] captures the recursive dynamic social diffusion in users and items. DESIGN [25] jointly models user-item and user-user relations, and enhances it with distillation from separately modeling these two relations. GDSRec [1] designs a decentralized interaction graph to vectorize biases from users and items. DH-HGCN [11] constructs hypergraphs separately for each social relation and conducts hypergraph convolution on item-item similar graphs. S^2-MHCN [34] applys multi-channel hypergraph to enhancing mutual information. DcRec [29] disentangles user behaviors into social domain and collaborative domain, and utilizes contrastive learning to transfer knowledge from the former to the latter. SDCRec [5] utilizes contrastive learning to eliminate popularity bias in cold-start recommendation. SEPT [33] combining tri-training strategy and self-supervised learning. The above methods only operate on the raw social graph without any refinement. Inversely, our LSIR carefully modifies social graph to mitigate the mentioned flaws, and specifically benefits inactive users.

Cold-Start Recommendation. Cold-start recommenders [10] mainly rely on meta learning. For example, MetaHIN [21] incorporates heterogeneous information to capture rich semantics of meta-paths. MAMO [4] introduces two feature-specific and task-specific memory matrix based on user profiles. MeLU [17] deploys the framework of MAML [7] in recommendation. TaNP [18] utilizes meta-learning, designing a task-adaptive neural process to enhance cold-start user modeling. Nevertheless, these meta learning based models are not rely on social connections, which immensely limits their performance.

Graph Structure Learning. Graph structure learning aims to estimate a better structure for original graph [39]. LDS [8] jointly optimizes the probability for each node pair and GNN in a bilevel way. Pro-GNN [15] aims to obtain a clear graph by deploying some regularizers, such as low-rank, sparsity and feature smoothness. SUBLIME [19] explores the possibility to learn a good structure in unsupervised settings with the help of graph contrastive learning. Unfortunately, these models only focus on node classification task, while little effort has been made towards improving social recommendation.

7 Conclusion

In this paper, we systematically observe that real social relations are both insufficient quantity and inferior quality for inactive users. We take an initial attempt to incorporate the graph structure learning into refining social graph for inactive users. By deleting noisy connections and establishing useful links towards active interest clusters of social users, our model LSIR can well alleviate the cold start problem for inactive users, showing the superiority of learning from social structure in recommender systems.

Acknowledgement. This work was supported by Alibaba Group through Alibaba Innovative Research Program.

Declarations Interests. The authors have no competing interests to declare that are relevant to the content of this article.

References

1. Chen, J., Xin, X., Liang, X., He, X., Liu, J.: GDSREC: graph-based decentralized collaborative filtering for social recommendation. In: IEEE TKDE, pp. 4813–4824 (2023)
2. Chen, Y., Wu, L., Zaki, M.: Iterative deep graph learning for graph neural networks: better and robust node embeddings. In: NeurIPS (2020)
3. Deshpande, M., Karypis, G.: Item-based top-N recommendation algorithms. ACM Trans. Inf. Syst. 143–177 (2004)
4. Dong, M., Yuan, F., Yao, L., Xu, X., Zhu, L.: MAMO: memory-augmented meta-optimization for cold-start recommendation. In: KDD, pp. 688–697 (2020)
5. Du, J., Ye, Z., Yao, L., Guo, B., Yu, Z.: Socially-aware dual contrastive learning for cold-start recommendation. In: SIGIR, pp. 1927–1932 (2022)
6. Fan, W., et al.: Graph neural networks for social recommendation. In: TheWebConf, pp. 417–426 (2019)
7. Finn, C., Abbeel, P., Levine, S.: Model-agnostic meta-learning for fast adaptation of deep networks. In: ICML, pp. 1126–1135 (2017)

8. Franceschi, L., Niepert, M., Pontil, M., He, X.: Learning discrete structures for graph neural networks. In: ICML, pp. 1972–1982 (2019)
9. Glorot, X., Bengio, Y.: Understanding the difficulty of training deep feedforward neural networks. In: AISTATS, pp. 249–256 (2010)
10. Gope, J., Jain, S.K.: A survey on solving cold start problem in recommender systems. In: ICCCA, pp. 133–138 (2017)
11. Han, J., Tao, Q., Tang, Y., Xia, Y.: DH-HGCN: dual homogeneity hypergraph convolutional network for multiple social recommendations. In: SIGIR, pp. 2190–2194 (2022)
12. He, X., Deng, K., Wang, X., Li, Y., Zhang, Y., Wang, M.: LightGCN: simplifying and powering graph convolution network for recommendation. In: SIGIR, pp. 639–648 (2020)
13. Huang, T., et al.: MixGCF: an improved training method for graph neural network-based recommender systems. In: KDD, pp. 665–674 (2021)
14. Huang, X., et al.: How powerful are interest diffusion on purchasing prediction: A case study of taocode. In: SIGIR, pp. 1308–1317 (2021)
15. Jin, W., Ma, Y., Liu, X., Tang, X., Wang, S., Tang, J.: Graph structure learning for robust graph neural networks. In: SIGKDD, pp. 66–74 (2020)
16. Kingma, D.P., Ba, J.: Adam: a method for stochastic optimization. In: ICLR (2015)
17. Lee, H., Im, J., Jang, S., Cho, H., Chung, S.: MELU: meta-learned user preference estimator for cold-start recommendation. In: KDD, pp. 1073–1082 (2019)
18. Lin, X., Wu, J., Zhou, C., Pan, S., Cao, Y., Wang, B.: Task-adaptive neural process for user cold-start recommendation. In: TheWenConf, pp. 1306–1316 (2021)
19. Liu, Y., Zheng, Y., Zhang, D., Chen, H., Peng, H., Pan, S.: Towards unsupervised deep graph structure learning. In: TheWebConf, pp. 1392–1403 (2022)
20. Liu, Z., et al.: Learning representations of inactive users: a cross domain approach with graph neural networks. In: CIKM, pp. 3278–3282 (2021)
21. Lu, Y., Fang, Y., Shi, C.: Meta-learning on heterogeneous information networks for cold-start recommendation. In: KDD, pp. 1563–1573 (2020)
22. Oord, A.V.D., Li, Y., Vinyals, O.: Representation learning with contrastive predictive coding. arXiv preprint arXiv:1807.03748 (2018)
23. Rendle, S., Freudenthaler, C., Gantner, Z., Schmidt-Thieme, L.: BPR: Bayesian personalized ranking from implicit feedback. In: UAI, pp. 452–461 (2009)
24. Sharma, K., et al.: A survey of graph neural networks for social recommender systems. arXiv preprint arXiv:2212.04481 (2022)
25. Tao, Y., Li, Y., Zhang, S., Hou, Z., Wu, Z.: Revisiting graph based social recommendation: a distillation enhanced social graph network. In: TheWebConf, pp. 2830–2838 (2022)
26. Vaswani, A., et al.: Attention is all you need. In: NIPS, pp. 5998–6008 (2017)
27. Veličković, P., Cucurull, G., Casanova, A., Romero, A., Lio, P., Bengio, Y.: Graph attention networks. arXiv preprint arXiv:1710.10903 (2017)
28. Wang, X., He, X., Wang, M., Feng, F., Chua, T.: Neural graph collaborative filtering. In: SIGIR, pp. 165–174 (2019)
29. Wu, J., Fan, W., Chen, J., Liu, S., Li, Q., Tang, K.: Disentangled contrastive learning for social recommendation. In: CIKM. pp. 4570–4574 (2022)
30. Wu, L., Li, J., Sun, P., Hong, R., Ge, Y., Wang, M.: Diffnet++: a neural influence and interest diffusion network for social recommendation. IEEE Trans. Knowl. Data Eng. pp. 4753–4766 (2022)
31. Wu, L., Sun, P., Fu, Y., Hong, R., Wang, X., Wang, M.: A neural influence diffusion model for social recommendation. In: SIGIR, pp. 235–244 (2019)

32. Wu, Z., Pan, S., Chen, F., Long, G., Zhang, C., Yu, P.S.: A comprehensive survey on graph neural networks. In: IEEE TNNLS, pp. 4–24 (2021)
33. Yu, J., Yin, H., Gao, M., Xia, X., Zhang, X., Hung, N.Q.V.: Socially-aware self-supervised tri-training for recommendation. In: KDD, pp. 2084–2092 (2021)
34. Yu, J., Yin, H., Li, J., Wang, Q., Hung, N.Q.V., Zhang, X.: Self-supervised multi-channel hypergraph convolutional network for social recommendation. In: TheWebConf. pp. 413–424 (2021)
35. Yu, J., Yin, H., Xia, X., Chen, T., Cui, L., Nguyen, Q.V.H.: Are graph augmentations necessary?: Simple graph contrastive learning for recommendation. In: SIGIR. pp. 1294–1303 (2022)
36. Zhang, H., Cissé, M., Dauphin, Y.N., Lopez-Paz, D.: mixup: Beyond empirical risk minimization. In: ICLR (2018)
37. Zhao, T., Liu, Y., Neves, L., Woodford, O.J., Jiang, M., Shah, N.: Data augmentation for graph neural networks. In: AAAI. pp. 11015–11023 (2021)
38. Zhao, T., McAuley, J.J., King, I.: Leveraging social connections to improve personalized ranking for collaborative filtering. In: CIKM. pp. 261–270 (2014)
39. Zhu, Y., et al.: A survey on graph structure learning: progress and opportunities. arXiv e-prints pp. arXiv–2103 (2021)

MANE: A Multi-cascade Adversarial Network Embedding Model for Anchor Link Prediction

Jingya Zhou[1,2](✉)

[1] School of Computer Science and Technology, Engineering Laboratory of Big Date and Intelligence of Jiangsu Province, Soochow University, Suzhou, China
[2] State Key Laboratory for Novel Software Technology, Nanjing University, Nanjing, China
jy_zhou@suda.edu.cn

Abstract. Anchor link prediction plays a crucial role in social network analysis as it is a fundamental task. Current studies mainly rely on feature embedding of user-generated content or network structure to create latent representations. These representations are then used to determine user correspondences by mapping users between different network representation spaces. However, most studies still face some challenges such as independent representation spaces, weak discrimination of embedded representations, etc. In this paper, we propose a Multi-cascade Adversarial Network Embedding (MANE) model to tackle these challenges. MANE learns multiple representations from different cascade networks for each user, making users distinguishable in the representation space. Furthermore, an adversarial network is integrated with a mapping function to output a high-quality collection of possible anchor links for correspondence matching. Extensive experiments on real-world social network datasets demonstrate that our method can achieve the expected performance, especially in improving the top-1 precision and recall.

Keywords: Anchor link prediction · Cascade network · Multi-cascade · Network embedding

1 Introduction

Many people are joining various social networks, and their multiple accounts accordingly form anchor links that bridge different social networks. Anchor link prediction (ALP) [5], aims to identify these unknown anchor links among different social networks. Nevertheless, it is not a trivial task since quite a few people do not explicitly reveal their identities associated with different social accounts [16]. An intuitive way is to make use of the user's profile like username, age, address, etc. However, users' profiles might be incomplete and even inconsistent for the same person. A study [10] on self-description similarity for 15,595 anchor users across 6 social networks shows that only 7% of them have similar profiles.

M. Onizuka et al. (Eds.): DASFAA 2024, LNCS 14855, pp. 168–184, 2024.
https://doi.org/10.1007/978-981-97-5572-1_11

Recent advances in embedding techniques [14] have inspired many embedding driven studies, and the majority of them fall into two categories, i.e., content-embedding methods and structural embedding methods. The first category of methods determines correspondences based on the consistency assumption of the content generated online by users. A trajectory-based method [15] was proposed to capture every user's unique footprints by extracting trajectory features from a series of time-stamped location data. As another useful type of features, network structure can be extracted for correlating anchor users with their correspondences, and it has attracted many efforts recently. The followership and followeeship of users were embedded as their context to identify pairs of anchor users with similar neighborhood relationships [11]. MGCN [1] took into account both local and hypergraph level graph convolutions to capture rich network features. TransLink [27] encoded both users and their interaction edges into a unified vector space with the favor of translating embedding. TALP [9] presented a unified framework of type-aware ALP across heterogeneous networks based on an attention mechanism. There are still several challenges:

First, existing methods only learn one single vector representation for each user and the embedded feature is not sufficient to predict anchor links exactly. Several recent work [20, 23] proposed to mine features from both aspects of content and network structure to improve the prediction results. However, these methods typically embed users via learning separate features with varied latent spaces, which is not entirely beneficial to the prediction. *Second*, network embedding mainly aims to enable structurally similar vertices having similar vector representations, while anchor link prediction aims to connect anchor users and their correspondences between different networks. Therefore, it is difficult to map an anchor user exactly to her true correspondence in another network as considering the existence of multiple 'similar' correspondences. *Third*, the mapping function trained via known anchor links easily makes all users including both anchor and non-anchor users deviated to the distribution of those known anchor users. Through the biased mapping function, the potential anchor users will be incorrectly mapped and depart from their true correspondences.

Noticing that social users are usually involved in multiple information cascade diffusions. Both structural features (e.g., information diffuses through edges in a cascading manner) and content features (e.g., the diffused information has varied topics) are hidden in these information cascade diffusions. It inspires us to take advantage of information cascade diffusions to extract both types of features in a compatible way. To this end, we present a multi-cascade adversarial network embedding (MANE) model. In social networks, each user may participate in various information cascade diffusions and each information has a topic. In our model, each user has multiple representations and each of them corresponds to the user's diffusion behaviors on one topic. On the one hand, user's involvement in information cascade diffusions are closely correlated with her/his structural connections in the network. The multi-representation naturally combines structural and content features. On the other hand, different users often involve in different cascade diffusions and the probability that every representation of one user has a similar correspondence with her/his neighborhood is very low. Hence, the multi-representation helps discriminating anchor users from their 'similar' neighbors in the vector space. To address the third challenge, MANE leverages an adversarial network to learn the real anchor link distribution under the multi-topic view. Our main contributions are summarized as follows:

• We propose MANE model to learn multiple representations for each user corresponding to multiple topics derived from information diffusions. MANE invents a special paradigm to jointly learn latent features from both structure and content for the ALP problem.

• We integrate the idea of generative adversarial nets in MANE model to learn the unbiased mapping function and fusing function that facilitate potential anchor users toward their true correspondences.

• To evaluate the effectiveness and efficiency of MANE, we conduct extensive experiments on real-world datasets. The results indicate that MANE efficiently improve the prediction performance compared to the state-of-the-art methods.

2 Related Work

Network embedding techniques can be leveraged to learn users' latent features for identities linkage. Content features are often represented to capture special characteristics of user identities. The content features that have been explored by current studies includes interest, writing style, trajectory, etc. Zhang et al. [25] leverage user attribute to guide the prediction. Nie et al. [12] present a long-term topic model to extract the user's core interests. However, the features extracted from the model are not sufficient to characterize a user's identity in a fine-grained manner. Zafarani et al. [21] propose a a term-frequency model to extract the patterns of username selection to distinguish users' identities. However, the linkage accuracy is not satisfactory since different people might use the same username. Riederer et al. [15] propose to extract users' footprint features for prediction. But the trajectory-based model only applies to location-based social networks, and its accuracy is greatly affected by the integrity of users' trajectory data.

Network structural features refer to a user's position in a social network and reflect her/his social relationship with other users. Neighborhood-based features have been extensively explored by recent work [11,17,22,28]. However, the neighborhood only reflects the structural features from a local view, and neighborhood-based methods often suffer from potential inconsistency of matching. To learn more useful latent network features, the structural context is extended to multi-hop neighbors and an auto-encoding style convolution network is presented for edge embedding and anchor link prediction [2]. Zhou et al. [26] propose a deep neural network-based algorithm DeepLink to encode local and global network structures to guide the user identification. TransLink [27] takes advantage of user interactions rather than static relationships and feeds them to a deep neural embedding model to identify pairs of anchor users. Though deep neural networks can help the embedding models to enhance the learning results, the problem of multiple 'similar' correspondences still cannot be addressed well. CrossMNA [3] learns three types of embedding vectors, i.e., network vector, inter-vector and intra-vector, where inter-vector is used to reflect the commonness of anchor users across networks. There is a strong assumption that each user in one network has a correspondence in another network, but current social networks are actually partially aligned.

Through capturing both content and structural features, PCT [23] proposes to identify potential anchor users that link multiple kinds of shared entities in latent space. Wang et al. [20] focus on special features captured from both friend-based and interest-based user co-occurrence, and use the learned features to generate comprehensive representations for potential user identity linkage across networks. MEgo2Vec [24] presents a GNN model to capture ego network features for prediction, where each user's representation includes both attributes of user pairs and topologies of the matched ego network. REGAL [7] formulates the prediction problem as a matrix factorization over a similarity matrix that incorporates both structural similarity and attribute agreement. These studies try to extract multiple independent features while these features belong to different representation spaces.

3 The MANE Model

3.1 Model Overview

The overview of MANE model is illustrated in Fig. 1. It consists of three stages. At the first stage, MANE uses the multi-cascade network embedding to learn multiple representations for each user. Each representation of a user reflects her/his structural features inside a diffusion cascade network on a specific topic. In this way, both structural and content features are jointly integrated. At the second stage, MANE uses an adversarial network to learn the real anchor link distribution. Specifically, for generator, we use a mapping function to obtain users' correspondences in the vector space, and fuse representations of both users and their correspondences into possible anchor link representations. Then a discriminator is used to filter out the fake anchor links. Supervised by known anchor links, the two modules are trained to capture the correlations among multiple topics and learn the anchor link distribution. Considering that the trained discriminator is able to distinguish true anchor links from fake ones, the potential anchor users will not depart from their true correspondences. At the anchor link prediction stage, all 'true' anchor links labeled by the discriminator are put into the corresponding user matching module for identification.

3.2 Problem Definition

In this paper, we focus on the solution of anchor link prediction between two social networks, which can be generalized to the multi-network scenario by conducting multiple pairwise link predictions Considering information diffusion,

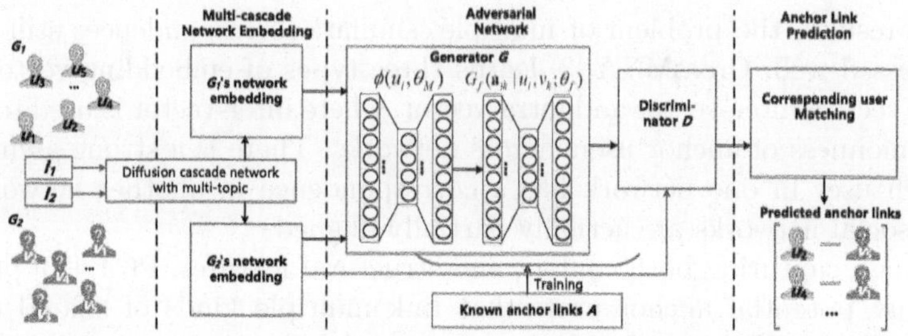

Fig. 1. The overview of MANE model.

users often send information on social networks, and that information may be shared or commented by their followers (friends) and their followers' followers (friends' friends) in a cascading manner, e.g., post sharing on Facebook, retweeting on Twitter. To describe the information diffusion, we define cascade network:

Definition 1 (Cascade Network). *For an information diffusion in social network $\mathcal{G}_1$, its cascade network is denoted by $\mathcal{CN}_{x,1} = (\mathcal{V}_{x,1}, \mathcal{E}_{x,1})$, where x is an index, $\mathcal{V}_{x,1}$ is the set of involved users and $\mathcal{E}_{x,1}$ is the set of involved edges.*

We add cascade network information into the description of social networks. Assume that $\mathcal{G}_1 = (\mathcal{V}_1, \mathcal{E}_1, \mathcal{I}_1)$ and $\mathcal{G}_2 = (\mathcal{V}_2, \mathcal{E}_2, \mathcal{I}_2)$ are two social networks available for anchor link prediction, where $\mathcal{V}_1 = \{u_i | i \in [1, n_1]\}$, $\mathcal{V}_2 = \{v_j | j \in [1, n_2]\}$ are two sets of users, $\mathcal{E}_1 = \{e_{ij} | u_i, u_j \in \mathcal{V}_1\}$, $\mathcal{E}_2 = \{e_{ij} | v_i, v_j \in \mathcal{V}_2\}$ are two sets of (un)directed social edges (e.g., friendship in Facebook, follower/followee in Twitter), and $\mathcal{I}_1 = \{\mathcal{CN}_{x,1} | x \in [1, m_1]\}$, $\mathcal{I}_2 = \{\mathcal{CN}_{y,2} | y \in [1, m_2]\}$ are two sets of cascade networks where $m_1 = |\mathcal{I}_1|$ and $m_2 = |\mathcal{I}_2|$ are set sizes. We define a network mapping function as follows:

Definition 2 (Network Mapping Function). *A function $\phi : \mathcal{V}_1 \rightarrow \mathcal{V}_2$ that maps users' representations from social network $\mathcal{G}_1$ to $\mathcal{G}_2$.*

After having the mapping function $\phi(\cdot)$, we still have to define a distance function $dis(\cdot, \cdot)$ to measure the probability of being an anchor link between arbitrary two users, e.g., $dis(\phi(u_i), v_j)$, $u_i \in \mathcal{V}_1, v_j \in \mathcal{V}_2$. Therefore, the problem can be formally defined as follows:

Definition 3 (Anchor Link Prediction). *Given that $\mathcal{A} = \{a_{kl} | u_k \in \mathcal{V}_1, v_l \in \mathcal{V}_2\}$ is the set of known anchor links between social networks $\mathcal{G}_1$ and $\mathcal{G}_2$, the goal of anchor link prediction is to learn the mapping function $\phi(\cdot)$ to find out every potential anchor link a_{ij} such that*

$$\phi(\cdot) = \underset{\phi, v_j \in \mathcal{V}_2}{\arg\min}\, dis(\phi(u_i), v_j). \tag{1}$$

3.3 Multi-cascade Network Embedding

In social networks, many users are often involved in multiple cascade networks, while the information diffused in each cascade network has one or more topics, so they are also associated with multiple topics. A single representation cannot fully captures a user's features that are hidden in the multi-diffusion activities with multi-topic. In our model, each user is represented by a group of h vectors, e.g., $\{\mathbf{u}_i^{(1)}, ..., \mathbf{u}_i^{(h)}\}$, where h is the number of topics in total, and $\mathbf{u}_i^{(t)}$ is a d-dimensional vector representation of u_i on topic t. For a user u_i, there are probably multiple cascade networks containing diffused information on topic t. To facilitate multi-cascade network embedding, we have to get all topic t-related cascade networks together as a merged one, its formal definition is given below:

Definition 4 (Merged Cascade Network). *For a user u_i in network $\mathcal{G}_1$, its merged cascade network on topic t is denoted by $\mathcal{CN}_{i,1}^{(t)} = \bigcup_{u_i \in \mathcal{V}_{x,1} \wedge t \in \mathcal{T}_{x,1}} \mathcal{CN}_{x,1}$, where $\mathcal{T}_{x,1}$ is the set of topics associated with cascade network $\mathcal{CN}_{x,1}$.*

In a merged cascade network $\mathcal{CN}_{i,1}^{(t)}$ on topic t, u_i and e_{ij} are involved user and diffusion edge, respectively. Then given u_i, the conditional probability that the information passes through edge e_{ij} is defined as follows:

$$p^{(t)}(e_{ij}|u_i) = \frac{\exp(\mathbf{u}_i^{(t)\top} \cdot \mathbf{u}_j^{(t)})}{\sum_{u_w \in \mathcal{V}_{i,1}^{(t)}} \exp(\mathbf{u}_i^{(t)\top} \cdot \mathbf{u}_w^{(t)})}, \tag{2}$$

where $\mathcal{V}_{i,1}^{(t)}$ is the user set of $\mathcal{CN}_{i,1}^{(t)}$, and $\mathbf{u}_i^{(t)\top} \cdot \mathbf{u}_j^{(t)}$ is the inner product of vectors. The above equation defines a conditional distribution $p(\cdot|u_i)$ and its empirical probability is defined as $\hat{p}^{(t)}(e_{ij}|u_i) = num_{ij}^{(t)}/num_{u_i}^{(t)}$, where $num_{ij}^{(t)}$ is the number of diffusions passing through edge e_{ij} on topic t, and $num_{u_i}^{(t)}$ is the number of diffusions involving user u_i on topic t. If u_i has no topic t associated in any cascade network, $\mathbf{u}_i^{(t)}$ will be a zero vector.

Collections $p_i^{(t)} = \{p^{(t)}(e_{ij}|u_i)|u_j \in \mathcal{V}_{i,1}^{(t)}\}$ and $\hat{p}_i^{(t)} = \{\hat{p}^{(t)}(e_{ij}\ |u_i)|u_j \in \mathcal{V}_{i,1}^{(t)}\}$ reflect the distributions of probabilities and empirical probabilities on topic t. To preserve the proximities in both neighborhood and topic for network embedding, the gap between two series of distributions $\hat{p}_i^{(t)}$ and $p_i^{(t)}, t = 1, ..., h$ should be minimized, then the loss of network embedding is defined as follows:

$$L_{embed} = -\sum_{e_{ij} \in \mathcal{E}_1} \sum_{t=1}^{h} \log p^{(t)}(e_{ij}|u_i). \tag{3}$$

To avoid the expensive computation of conditional probability, following the optimization in LINE [18], we adopt negative sampling by sampling k negative samples according to a noisy distribution of edges and use asynchronous stochastic gradient to optimize the objective. The optimization with negative sampling takes $O(d(k+1))$ time. According the conclusion in [18], the number of steps used for optimization is usually proportional to the number of edges used for learning. Let us take network $\mathcal{G}_1$ as an example, for each user $u_i \in \mathcal{V}_1$, she/he has h representations, and each one corresponds to a merged cascade network. Assume that α_1 denotes the average number of cascade networks included in a merged one, ℓ_1 denotes the average overlap factor of edges across different cascade networks, and β_1 denotes the average number of edges included in a cascade network. Then the embedding complexity of network $\mathcal{G}_1$ is $O(h|\mathcal{V}_1|\alpha_1\ell_1\beta_1 dk)$, where $\beta_1 = \frac{\sum_{\mathcal{CN}_{x,1}\in\mathcal{I}_1}|\mathcal{E}_{x,1}|}{|\mathcal{I}_1|}$, α_1 and ℓ_1 are statistics.

3.4 Training with Adversarial Network

We use function $\phi(u_i, \theta_M)$ to realize the mapping of users' representations from $\mathcal{G}_1$ to $\mathcal{G}_2$, and the mapping function is implemented with neural networks where θ_M is the set of parameters. With the help of known anchor links acting as labeled data, the mapping function is trained by minimizing the objective defined in Eq. 1. However, the mapping function learned above easily causes overfitting that all users including anchor users and non-anchor users in $\mathcal{G}_1$ are toward approximating the distribution of known anchor users in $\mathcal{G}_2$. Then the potential anchor users in $\mathcal{G}_1$ will be mapped departing from their true correspondences in $\mathcal{G}_2$. Inspired by the idea of GAN [6], we present an adversarial network-driven mechanism, and one of its goals is to assist the mapping function being trained to make the potential anchor users discriminable from the training ones and meanwhile approaching their true correspondences. The adversarial network consists of two modules: *1)* given the real anchor link $a_{ij} \sim q_{anchor}$, a *generator* G fuses a fake anchor link a_{ik} between u_i and v_k, where q_{anchor} denotes the real anchor link distribution; *2)* a *discriminator* D aims to distinguish the real anchor link a_{ij} from the fake one a_{ik}.

Multi-topic Generator. In our model, each anchor link a_{ij} is represented by a h-dimensional vector, i.e., $\mathbf{a}_{ij} = \{a_{ij}^{(1)}, ..., a_{ij}^{(h)}\}$. To achieve generator's goal in the multi-topic view, we design the generator as a fusing function $f(\cdot,\cdot)$ to generate anchor links given a pair of users. The fusing generator is implemented by a multi-layer perceptron that is able to capture the correlation between anchor users and their correspondences, and this is also another goal of the adversarial network. We use $G(u_i, v_j; \theta_f)$ to denote the fused representation (anchor link representation) where θ_f is the set of parameters. The fused representation is used to calculate the probability of $a_{ij}^{(t)} = 1$ for every topic t. We formalize the multi-topic generator as

$$G(\mathbf{a}_{ij}|u_i, v_j; \theta_G) = \prod_{t=1}^{h} G_f^{(t)}(a_{ij}^{(t)}|\mathbf{u}_i^{(t)}, \mathbf{v}_j^{(t)}; \theta_f), \tag{4}$$

where $G_f^{(t)}$ is the fusing generator on topic t. To fool the discriminator D, for each true anchor link $a_{ij} \sim q_{anchor}$, we use negative sampling to generate fake

Algorithm 1: Adversarial Network Training

Input: known set of anchor links $\mathcal{A}$, topic number h, dimension d, sampling sizes r, s.
Output: θ_D, θ_f, θ_M.

Initialize generator G and discriminator D;
while *not converged* **do**
  for G **do**
    Choose a set of r real anchor users $\{u_i\}$ from $\mathcal{A}$;
    for *each* u_i **do**
      Choose v_k based on $\phi(u_i;\theta_M)$ and $\mathcal{A}$;
      Generate a fake anchor link a_{ik} via $G(\mathbf{a}_{ik}|u_i, v_k;\theta_f)$;
    Update θ_f, θ_M based on Eq. 9;
  for D **do**
    Sample a set of s real anchor links $\{a_{ij}\}$ from $\mathcal{A}$;
    Sample a set of corresponding fake links $\{a_{ik}\}$ generated by G;
    Compute all $D(a_{ik};\theta_D)$ based on Eq. 6;
    Update θ_D based on Eq. 8;
return θ_D, θ_f, θ_M ;

anchor link a_{ik}. Specifically, we use u_i's mapping correspondence to select a negative user v_k who is close to $\phi(u_i;\theta_M)$ and $a_{ik} \notin \mathcal{A}$. The fake link a_{ik} is generated by fusing u_i and v_k together. Fake link a_{ik} should be similar enough to the real one a_{ij}, e.g., $a_{ik} \sim q_G$ with $a_{ik} \approx a_{ij}$ where q_G is the distribution of generated links, so as to be more likely to fool the anchor link discriminator D. Formally, we update parameters θ_M of mapping function to obtain v_k and update parameters θ_f of fusing generator to generate link a_{ik}, such that a_{ik} has the highest probability, i.e.,

$$\arg\max_{a_{ik}} \prod_{t=1}^{h} G_f^{(t)}(a_{ij}^{(t)} = a_{ik}^{(t)}|\mathbf{u}_i^{(t)}, \mathbf{v}_k^{(t)};\theta_G), where\, \theta_G\, refers\, to\, \{\theta_M, \theta_f\}. \quad (5)$$

Anchor Link Discriminator. The discriminator D's objective is to discriminate between true anchor links chosen from real anchor link distribution under multi-topic view and the fake anchor links produced by generator G, so as to enforce G to more accurately fit the distribution of multi-topic anchor links. To achieve this, D is expected to return 1 when a true anchor link is sampled, while it returns 0 once a fake link is sampled. Then G has to update parameters to get a relatively high probability for fake (or negative) samples. The discriminator can be implemented by any differentiable function as long as its domain belongs to the range [0, 1]. In our model, we use a sigmoid function to define D

by considering its stable property, i.e.,

$$D(a_{ij};\theta_D) = \frac{1}{h}\sum_{t=1}^{h} \frac{\exp(\mathbf{u}_i^{(t)\top} \cdot \mathbf{v}_j^{(t)})}{1+\exp(\mathbf{u}_i^{(t)\top} \cdot \mathbf{v}_j^{(t)})}. \tag{6}$$

Algorithm 2: Anchor Link Prediction

Input: $\mathcal{G}_1$, $\mathcal{G}_2$, the input of Algorithm 1.
Output: predicted set $\mathcal{A}_p$ of anchor links.

Learn representations of networks $\mathcal{G}_1, \mathcal{G}_2$ by minimizing Eq. 3;
Train the adversarial network by Algorithm 1;
Use the trained network to obtain all possible anchor links $\mathcal{A}_D$ and anchor users $\mathcal{V}_D \in \mathcal{V}_1$;
for *each user* $u_i \in \mathcal{V}_D$ **do**
  $v_j \leftarrow \arg\min_{v_l \in \mathcal{V}_2} dis(\phi(u_i), v_l)$;
  Append a_{ij} to $\mathcal{A}_p$;
return $\mathcal{A}_p$;

Training. The adversarial network can be considered as a minimax game in which the generator G and discriminator D iteratively playing against each other. Let $L_{G,D}$ denote the loss function, then we formalize the game as follows:

$$\min_{\theta_G}\max_{\theta_D} L_{G,D} = \sum_{u_i \in V_1} (\mathbb{E}_{a_{ij}\sim q_{anchor}}[\log D(a_{ij};\theta_D)] + \mathbb{E}_{a_{ik}\sim q_G}[\log(1 - D(a_{ik};\theta_D))]), \tag{7}$$

where discriminator D updates its parameters θ_D to maximize its capability to discriminate the fake links from the true anchor links, on the contrary, generator G minimizes $\log(1 - D(a_{ik};\theta_D))$ to fool D by tuning parameters θ_M in the mapping function and θ_f in the fusing function.

The specific training process is to alternately update θ_D and θ_G with mini-batch gradient descent, and the training data are chosen from the known set $\mathcal{A}$ of anchor links. This training process obeys the empirical training principle proposed in [6], and the training algorithm is illustrated in Algorithm 1.

For discriminator D, as we have assumed, it is differentiable with respect to the loss defined in Eq. 6, and then the gradient of θ_D is calculated by:

$$\begin{aligned}\nabla_{\theta_D} L_{G,D} = &\sum_{u_i \in V_1 \wedge a_{ij} \in A} (\mathbb{E}_{a_{ij}\sim q_{anchor}}[\nabla_{\theta_D} \log D(a_{ij})] \\ &+ \mathbb{E}_{a_{ik}\sim q_G}[\nabla_{\theta_D} \log(1 - D(a_{ik}))]).\end{aligned} \tag{8}$$

For generator G, we use the policy gradient to update the parameters $\theta_G = \{\theta_f, \theta_M\}$, and its gradient is given by:

$$\begin{aligned}
\nabla_{\theta_G} L_{G,D} &= \nabla_{\theta_G} \sum_{u_i \in V_1 \wedge a_{ij} \in \mathcal{A}} \mathbb{E}_{a_{ik} \sim q_G}[\log(1 - D(a_{ik}; \theta_D))] \\
&= \nabla_{\theta_G} \sum_{u_i \in V_1 \wedge a_{ij} \in \mathcal{A}} \sum_{v_k = \phi(u_i)} G(\mathbf{a}_{ik} | u_i, v_k)[\log(1 - D(a_{ik}; \theta_D))] \\
&= \begin{cases} \sum_{u_i \in V_1 \wedge a_{ij} \in \mathcal{A}} \mathbb{E}_{a_{ik} \sim q_G}[\nabla_{\theta_f} \Theta_M], if\, \theta_G = \theta_f, \\ \sum_{u_i \in V_1 \wedge a_{ij} \in \mathcal{A}} \mathbb{E}_{a_{ik} \sim q_G}[\nabla_{\theta_M} \Psi_f], if\, \theta_G = \theta_M, \end{cases}
\end{aligned} \tag{9}$$

where

$$\Theta_M = \log G_f(\mathbf{a}_{ik} | u_i, v_k; \theta_M) \log(1 - D(a_{ik}; \theta_D))$$
$$\Psi_f = \log G_f(\mathbf{a}_{ik} | u_i, v_k; \theta_f))) \log(1 - D(a_{ik}; \theta_D))$$

The third goal of the adversarial network is to pick out possible anchor links that are very close to the distribution of real anchor links. For the generator, its main operation is generating fake anchor links which takes $O(|\mathcal{A}|hd)$ time. For the discriminator, its main operation is the computation of sigmoid function which takes $O(|\mathcal{A}|khd)$ time. Assume the number of training iterations is denoted by R, then the time complexity of Algorithm 1 is $O(R|\mathcal{A}|hd(k+1))$.

3.5 Anchor Link Prediction

In the MANE model, we embed both networks into vector spaces and obtain the multi-cascade representations for every user. Before making predictions, we utilize the known anchor links to train the adversarial network and obtain the trained parameters θ_D, θ_f and θ_M. Thereafter, we use the mapping function with θ_M to obtain each user u_i's ($u_i \in \mathcal{V}_1/\mathcal{A}$) mapped representation in the latent space, and then use the fusing function with θ_f to generate a potential anchor link a_{ik}, and finally use the trained discriminator to decide whether a_{ik} is a true anchor link. Thus, we obtain all possible anchor links $\mathcal{A}_D$ that follow the distribution of real anchor links under multi-topic view, and the possible anchor users are denoted by set $\mathcal{V}_D = \{u_i | (u_i, \phi(u_i)) \in \mathcal{A}_D\}$. For each user $u_i \in \mathcal{V}_D$, to make a prediction, we identify the most likely corresponding user $v_j \in \mathcal{V}_2$ who has the minimum distance between $\phi(u_i)$ and v_j based on Eq. 1. The prediction algorithm is summarized in Algorithm 2 where the prediction operation takes $O(|\mathcal{V}_D||\mathcal{V}_2|hd)$ time. The overall time complexity of MANE is the sum of complexities at each stage, i.e., $O\left(hdk(|\mathcal{V}_1|\alpha_1\ell_1\beta_1 + |\mathcal{V}_2|\alpha_2\ell_2\beta_2) + R|\mathcal{A}|hd(k+1) + |\mathcal{V}_D||\mathcal{V}_2|hd\right)$, which is linear to both sizes of social network and cascade network.

4 Experiments

4.1 Experimental Setup

Datasets. For evaluation purposes, we need a collection of real-world network datasets that contain users' profiles and activities on different social networks.

But it is difficult to find out the qualified open-access datasets. So we have to build social network datasets by crawling on multiple social networks. How to obtain the ground truth on anchor links is another difficulty for our dataset building. Thanks to the social aggregation website About.me that records multiple accounts of registered users. We tested our method on two pairs of real-world social network datasets. The statistics are summarized in Table 1 and the details are introduced as follows.

- **Twitter-Tumblr:** This dataset is provided by [10]. It collects a set of 180,000 user profiles registered on About.me. Among these users, 30,746 users are found having accounts on both Twitter and Tumblr, so they are selected as the ground truth for this dataset. We used the released user profiles to identify user accounts on both social networks and obtained 127,736 Twitter accounts and 103,427 Tumblr accounts. The dataset also includes the index of users' activities, e.g., post ID. Given the post IDs, we pulled the corresponding actual posts from two networks, respectively, and consequently built 25,635 cascade networks in Twitter and 20,482 cascade networks in Tumblr. The topics are selected from the contents tagged with '#' (if they have) or original posts and classified by leveraging BERT [4].
- **Facebook-Twitter:** We collected an independent group of 10,284 user profiles via About.me APIs, and selected 2,712 users who have account links toward both Facebook and Twitter. Then we let the selected set of users working as seed nodes to crawl more users on two social networks and finally, we obtain 237,754 Facebook accounts and 189,735 Twitter accounts. With the help of About.me, we identified extra 8,836 anchor users from the crawled accounts. Considering that not all users register themselves on About.me, some potential anchor users still have not been identified. The rest accounts were automatically processed by comparing account names, addresses, cross-network sharing URLs, and finally 7,937 users were manually selected to append to the ground truth.

Table 1. Statistics of two pairs of social network datasets.

No.	Dataset	$\lvert\mathcal{V}\rvert$	$\lvert\mathcal{E}\rvert$	$\lvert\mathcal{I}\rvert$	h	anchor links
#1	Twitter	127,736	3,106,396	25,635	73	30,746
	Tumblr	103,427	2,512,583	20,482		
#2	Facebook	237,754	3,924,772	33,752	65	19,485
	Twitter	189,735	5,210,946	32,673		

We define $\frac{2|\text{ground-truth}|}{|V_1|+|V_2|}$ to reflect the overlap between a pair of networks. The overlap is 26.6% for Twitter-Tumblr and 9.1% for Facebook-Twitter. The overlap fraction may be less than the numbers we saw in public reports, i.e., MarketingTech reported that 98% of people have at least one social network account. That is because the overlap between an arbitrary pair of networks is definitely less than 98% on the one hand, on the other hand, the size of user set in our dataset is smaller than that of the actual network and the overlap between a pair of partial networks is inevitably less than the overlap between two entire networks. Two datasets are just used to stand for different overlap levels, i.e., moderate-overlapped networks and low-overlapped networks. We fed the diffused information content to HDP model [19] and obtained the number of topics h and the distribution of topics among those cascade networks.

Settings. We use the cosine similarity function as the distance function $dis(\cdot, \cdot)$ in the implementation of our proposed method. The hyper-parameters are set by default as follows: d =80, the learning rate is set to 0.01 and the training ratio is set to 0.7. We select the following state-of-the-art methods:

1) Structural embedding methods: **IONE** [11], learns a network embedding with the follower-ship/followee-ship of each user and conducts link prediction accordingly; **TransLink** [27], jointly embeds users and their interactions into a unified latent space based on translating embeddings; **CrossMNA** [3], conducts cross-network embedding by combining the network vector and intra-vector.

2) Structural + content embedding methods: **MEgo2Vec** [24], learns ego network representations by jointly capturing user attributes and ego network structures; **REGAL** [7], devises a cross-network matrix factorization scheme that maximally preserves structural and attribute content similarity.

3) GAN-based methods: **SNNA** [8], designs a GAN to minimize the distance between the distributions of user identities; **RANA** [13], proposes a dual GAN model to guarantee the distribution consistency between two networks in terms of both distance and shape.

Table 2. Comparison results under varied k.

	#1 Twitter-Tumblr						#2 Facebook-Twitter					
P@k, R@k	P@1	P@10	P@30	R@1	R@10	R@30	P@1	P@10	P@30	R@1	R@10	R@30
MANE	**0.418**	**0.702**	**0.784**	**0.528**	**0.785**	**0.924**	**0.331**	**0.613**	**0.720**	**0.420**	**0.667**	**0.771**
IONE	0.156	0.345	0.520	0.336	0.504	0.652	0.128	0.362	0.465	0.266	0.451	0.603
TransLink	0.253	0.537	0.702	0.352	0.569	0.654	0.243	0.504	0.634	0.221	0.432	0.586
CrossMNA	0.262	0.568	0.711	0.356	0.583	0.702	0.251	0.521	0.654	0.232	0.457	0.622
MEgo2Vec	0.276	0.607	0.722	0.304	0.487	0.613	0.264	0.552	0.661	0.177	0.382	0.578
REGAL	0.283	0.592	0.716	0.318	0.506	0.622	0.267	0.545	0.660	0.210	0.445	0.604
SNNA	0.260	0.561	0.701	0.363	0.584	0.685	0.245	0.524	0.626	0.256	0.456	0.614
RANA	0.267	0.584	0.710	0.381	0.615	0.713	0.259	0.530	0.648	0.274	0.471	0.620

Evaluation Metrics. We use $P@k$ and $R@k$ to evaluate the performance, where $@k$ refers to the top-k anchor links returned for a user. Therefore, precision $P@k$ refers to the fraction of top-k true positive anchor links among the anchor links returned, and recall rate $R@k$ refers to the fraction of top-k true positive anchor links over the ground-truth.

4.2 Results

Evaluation Under Different Sizes of *k*. In this group of experiments, we evaluate various methods on both metrics over different values of k (from top-1 to top-30). According to the results reported in Table 2, we conclude that:

(1) Our proposed method consistently outperforms the baselines. It demonstrates that the combination of multi-cascade network embedding and the adversarial network is effective for tackling the anchor link prediction problem no matter whether the two networks are moderately overlapped or lowly overlapped.
(2) It is worth mentioning that our proposed methods achieve significantly promotes of $P@1$ and $R@1$ on both datasets, e.g., on average 37%∼65% higher $P@1$ and 58%∼67% higher $R@1$, compared to the state-of-the-art methods. It verifies that the combination of multi-cascade network embedding and the adversarial network training can improve the results by identifying the potential anchor user pairs more accurately and comprehensively.
(3) In comparison, IONE identifies anchor user pairs by means of local structure consistency which does not always hold, and then its precision is relatively low. MEgo2Vec significantly improves the precision by incorporating ego network with user attributes. The candidate user pairs are generated by considering only pairs with similar usernames, which in turn leads to the low recall. TransLink also shows an obvious improvement on precision over IONE, which implies the translating embedding of users and interactive edges is more effective than follower/followee structure embedding for precision enhancement. CrossMNA relies on the commonness of anchor users learned from cross-network embedding and performs slightly better than TransLink. REGAL also conducts cross-network embedding by leveraging the matrix factorization technique as well as incorporating user attribute content. REGAL achieves higher precision but lower recall than CrossMNA. Both SNNA and RANA leverage the idea of GAN to solve the prediction problem. SNNA directly relies on the distribution closeness of users in a unified space to make prediction, while RANA develops a dual GAN to match the distributions of two networks with inconsistent features based on their learned common embedding features. Different from the above methods, GAN in our model is devised to couple with multi-cascade network embedding to approximate the true anchor link distribution, which contributes to the superior performance of MANE over SNNA and RANA.

Evaluation Under Different Training Ratios. We evaluate various methods under different training ratios. The results on the dataset Twitter-Tumblr are shown in Table 3 and the results on Facebook-Twitter have similar trends, so we omit them here. From the results, we conclude that:

(1) The performance of our method tends to be getting saturated as the training ratio reaches 0.7 and no more large upgrades are observed after 0.7. Even for training ratio settings as low as 0.1, our method can still obtain acceptable

Table 3. Comparison results under varied training ratios.

Training ratio	0.1	0.3	0.5	0.7	0.9	0.1	0.3	0.5	0.7	0.9
	P@30					R@30				
MANE	**0.437**	**0.605**	**0.694**	**0.784**	**0.815**	**0.627**	**0.735**	**0.890**	**0.924**	**0.938**
IONE	0.153	0.308	0.447	0.520	0.561	0.347	0.494	0.601	0.652	0.690
TransLink	0.336	0.509	0.623	0.702	0.755	0.337	0.476	0.593	0.654	0.703
CrossMNA	0.348	0.545	0.648	0.711	0.746	0.365	0.508	0.614	0.702	0.732
MEgo2Vec	0.316	0.523	0.654	0.722	0.750	0.288	0.424	0.537	0.613	0.647
SNNA	0.303	0.487	0.636	0.701	0.715	0.342	0.487	0.621	0.685	0.717
RANA	0.311	0.496	0.658	0.710	0.722	0.335	0.501	0.669	0.713	0.730

performance (as high as 43.7% on $P@30$ and 62.7% on $R@30$). This can be empirically considered as the benefit of our method which requires less training data to boost the model.

(2) A similar trend is also observed from the performance of baseline methods, but their performances are far below MANE's level especially when the training ratio is set to 0.1. For MEgo2Vec, the incorporation of user attribute embeddings cannot alleviate the problem of insufficient training due to a low training ratio. In our model, each user has multiple representations corresponding to diverse topics, which contributes to learning more refined anchor user corresponding relationships and compensates for the loss of a low training ratio to some extent.

Running Time Evaluation. We evaluate the efficiency by comparing the running time of different methods. Each method's running time is measured by the overall duration from network input to results output including embedding, training, and prediction time. As shown in Table 4, we conclude that:

(1) The integration of adversarial networks in MANE does not bring a significant increase in running time. MANE requires a relatively small number of iterations to train the mapping and fusing functions. In contrast, TransLink has to spend more time on interaction metapath embedding. IONE requires more iterations to achieve high performance, its embedding complexity is linear to the number of edges, and its anchor user mapping is conducted in an entire searching space $\mathcal{V}_1 \times \mathcal{V}_2$, which results in a high running time. REGAL has a sub-quadratic time complexity and its running time is second only to MANE. MEgo2Vec utilizes graph neural network to represent users' ego networks, which requires a relatively long running time. CrossMNA's running time is second only to REGAL. Because CrossMNA utilizes cross-network embedding to synchronously learn two network representations and the size of its mapping searching space is compressed to $|\mathcal{V}_1| \times \log|\mathcal{V}_2|$. SNNA implements a bidirectional projection model which requires more time than MANE to

make predictions. RANA spends more time than SNNA to perform the dual adversarial autoencoder to guarantee distribution consistency.

(2) The running time performance on both datasets partially verifies that the time complexity is approximately linear to the network size. The running times of all methods are reported based on the execution on a single machine with 2*GTX 2080ti 2*Xeon E5-2640 v4. For large social networks with tens of millions of users, the running time is expected to be dozens of hours by conducting algorithms on more powerful machines, which is acceptable for an offline analysis task.

Table 4. The running time comparison (h: hour, m: minute).

	#1	#2		#1	#2
MANE	**2h17m**	**4h3m**	MEgo2Vec	5h43m	12h22m
IONE	8h32m	19h21m	REGAL	2h23m	4h37m
TransLink	4h56m	11h8m	SNNA	3h47m	7h25m
CrossMNA	3h31m	7h18m	RANA	4h13m	8h34m

Ablation Study. In order to evaluate the effectiveness of key components in our model, i.e., multi-cascade embedding and adversarial network, we conduct an ablation study and simplify MANE by removing these components as follows:

- **Nocas-NoGAN**: It directly conducts network embedding to learn representations of two networks respectively. The representations reflect the structural features without topic included. In the next step, the representations are input to train a mapping function supervised by the set of known anchor links.
- **Nocas-GAN**: It first learns the representations of two networks. The following steps are the same as MANE, i.e., input the representations into an adversarial network to learn both mapping and fusing functions for prediction.
- **Multicas-NoGAN**: It learns multi-cascade network embeddings in the same way as MANE, but no adversarial network is used in training.

The comparison results are shown in Table 5:

(1) Compared to Nocas-NoGAN and Nocas-GAN, Multicas-NoGAN achieves higher precision on both datasets and better recall than that of Nocas-NoGAN. It implies that multi-cascade embedding can help network alignment obtain better performance, especially precision. The multi-involvement in varied diffusions with different topics is captured by user's multiple representations, which helps to discriminate real anchor users from 'similar' users.

(2) Compared to Nocas-NoGAN and Multicas-NoGAN, Nocas-GAN achieves higher recall on both datasets and better precision than that of Nocas-NoGAN. Adversarial network is proven to be able to learn the unbiased distribution of anchor links and to improve alignment performance, especially recall.
(3) MANE integrates both two components and each component corresponds to a stage in the model. The multi-cascade network embedding generated in the first stage provides the diversity of user representations which helps adversarial network in the second stage to accurately learn the distribution of anchor links under the multi-topic view.

Table 5. The evaluation of different components in MANE.

Dataset	Nocas-NoGAN		Nocas-GAN		Multicas-NoGAN		MANE	
	P@30	R@30	P@30	R@30	P@30	R@30	P@30	R@30
#1 Twitter-Tumblr	0.534	0.623	0.653	0.863	0.726	0.796	**0.784**	**0.924**
#2 Facebook-Twitter	0.453	0.582	0.609	0.721	0.672	0.685	**0.720**	**0.771**

5 Conclusion

We investigated a fundamental problem of predicting anchor links across different social networks and presented an effective model MANE. We present an insight into the integration of different types of features, where MANE naturally embeds both content features (e.g., multi-topic) and structural features (e.g., neighboring edges inside cascade networks) from the observed information cascade diffusions. The multi-representation can effectively alleviate the weak discrimination difficulty. In addition, an adversarial network is incorporated to couple with multi-cascade embedding to learn the real anchor link distribution and identify possible anchor links with high confidence. Extensive experiments on real-world datasets verify the effectiveness of our method.

Acknowledgement. This work was partially supported by the National Natural Science Foundation of China under Grant No. 61972272, the Natural Science Foundation of Jiangsu Higher Education Institutions of China under Grant No. 21KJA520008, Qinlan Project of Jiangsu Province of China, and Project Funded by the Priority Academic Program Development of Jiangsu Higher Education Institutions.

References

1. Chen, H., Yin, H., Sun, X., Chen, T.: Multi-level graph convolutional networks for cross-platform anchor link prediction. In: SIGKDD, pp. 1503–1511 (2020)

2. Cheng, A., et al.: Deep active learning for anchor user prediction. In: IJCAI, pp. 2151–2157 (2019)
3. Chu, X., Fan, X., Yao, D., Zhu, Z., Huang, J., Bi, J.: Cross-network embedding for multi-network alignment. In: WWW, pp. 273–284 (2019)
4. Devlin, J., Chang, M., Lee, K., Toutanova, K.: BERT: pre-training of deep bidirectional transformers for language understanding. CoRR abs/1810.04805 (2018)
5. Du, X., Yan, J., Zha, H.: Joint link prediction and network alignment via cross-graph embedding. In: IJCAI, pp. 2251–2257 (2019)
6. Goodfellow, I.J., et al.: Generative adversarial nets. In: NeurIPS (2014)
7. Heimann, M., Shen, H., Safavi, T., Koutra, D.: REGAL: representation learning-based graph alignment. In: CIKM, pp. 117–126 (2018)
8. Li, C., et al.: Adversarial learning for weakly-supervised social network alignment. In: AAAI, pp. 996–1003 (2019)
9. Li, X., Shang, Y., Cao, Y., Li, Y., Tan, J.: Type-aware anchor link prediction across heterogeneous networks based on graph attention network. In: AAAI (2020)
10. Lim, B.H., Lu, D., Chen, T., Kan, M.: #mytweet via instagram: exploring user behaviour across multiple social networks. In: ASONAM, pp. 113–120 (2015)
11. Liu, L., Cheung, W.K., Li, X., Liao, L.: Aligning users across social networks using network embedding. In: IJCAI (2016)
12. Nie, Y., Jia, Y., Li, S., Zhu, X., Li, A., Zhou, B.: Identifying users across social networks based on dynamic core interests. Neurocomputing **210**, 107–115 (2016)
13. Ren, J., Zhou, Y., Jin, R., Zhang, Z., Dou, D., Wang, P.: Dual adversarial learning based network alignment. In: Wang, J., Shim, K., Wu, X. (eds.) ICDM (2019)
14. Ribeiro, L.F.R., Saverese, P.H.P., Figueiredo, D.R.: *struc2vec*: Learning node representations from structural identity. In: SIGKDD, pp. 385–394 (2017)
15. Riederer, C.J., Kim, Y., Chaintreau, A., Korula, N., Lattanzi, S.: Linking users across domains with location data: Theory and validation. In: WWW (2016)
16. Shu, K., Wang, S., Tang, J., Zafarani, R., Liu, H.: User identity linkage across online social networks: A review. SIGKDD Explorations **18**(2), 5–17 (2016)
17. Tan, S., Guan, Z., Cai, D., Qin, X., Bu, J., Chen, C.: Mapping users across networks by manifold alignment on hypergraph. In: AAAI (2014)
18. Tang, J., Qu, M., Wang, M., Zhang, M., Yan, J., Mei, Q.: LINE: large-scale information network embedding. In: WWW, pp. 1067–1077 (2015)
19. Teh, Y.W., Jordan, M.I., Beal, M.J., Blei, D.M.: Sharing clusters among related groups: Hierarchical dirichlet processes. In: NeurIPS, pp. 1385–1392 (2004)
20. Wang, Y., Feng, C., Chen, L.: User identity linkage across social networks via linked heterogeneous network embedding. WWW **22**(6), 2611–2632 (2019)
21. Zafarani, R., Liu, H.: Connecting users across social media sites: a behavioral-modeling approach. In: SIGKDD, pp. 41–49 (2013)
22. Zafarani, R., Tang, L., Liu, H.: User identification across social media. TKDD **10**(2), 16:1–16:30 (2015)
23. Zhang, J., Yu, P.S.: PCT: partial co-alignment of social networks. In: WWW (2016)
24. Zhang, J., Chen, B., Wang, X., Chen, H., and, C.L.: Mego2vec: Embedding matched ego networks for user alignment across social networks. In: CIKM (2018)
25. Zhang, S., Tong, H.: FINAL: fast attributed network alignment. In: SIGKDD, pp. 1345–1354 (2016)
26. Zhou, F., Liu, L., Zhang, K., Trajcevski, G., Wu, J., Zhong, T.: Deeplink: a deep learning approach for user identity linkage. In: INFOCOM, pp. 1313–1321 (2018)
27. Zhou, J., Fan, J.: Translink: user identity linkage across heterogeneous social networks via translating embeddings. In: INFOCOM (2019)
28. Zhou, X., Liang, X., Zhang, H., Ma, Y.: Cross-platform identification of anonymous identical users in multiple social media networks. TKDE **28**(2), 411–424 (2016)

uTransfer: Unified Transferability Metric Incorporating Heterogeneous User Data in Social Network

Nuo Li, Bin Guo(✉), Yao Jing, and Zhiwen Yu

School of Computer Science, Northwestern Polytechnical University, 1 Dongxiang Road, Xi'an 710129, Shaanxi, China
linuo@mail.nwpu.edu.cn,
{guob,zhiwenyu}@nwpu.edu.cn

Abstract. Numerous users in social networks exhibit few-shot behaviors, and identifying appropriate neighbors has emerged as a promising solution. However, traditional similarity metrics often yield redundant neighbor information and fail to adequately consider the scarcity of user behaviors, thereby diminishing their effectiveness. This study endeavors to delve into the transferability between users by analyzing their heterogeneous data, to identify the most suitable users for knowledge exchange and reduce the impact of negative transfers. Existing transferability metrics mainly target homogeneous data, without considering the inherent characteristics and complementarity of heterogeneous data. To solve this, this paper proposes a novel metric, uTransfer, measuring the transferability between users with heterogeneous data in a more fine-grained and accurate way. Specifically, uTransfer first unifies user heterogeneous data into the behavior space to facilitate the fusion of heterogeneous knowledge. Then, uTransfer innovatively considers the specificity of heterogeneous data, proposes static and dynamic transfer modes, and models them separately to obtain finer-grained transferability results. Moreover, uTransfer uniquely models the complementarity between heterogeneous data to obtain more accurate transferability results. Finally, we integrate the complementarity and transferability results to measure the transferability between users. Extensive experiments demonstrate that uTransfer can effectively measure user transferability.

Keywords: User transferability · Heterogeneous data · Social network

1 Introduction

Knowledge can be shared between users to facilitate better understanding and prediction of user behavior, optimize recommendation systems, and improve overall user experience within the social network. Especially for users in social networks who interact sparsely, neighbors play an important role. Identifying appropriate neighbors has emerged as a promising solution [20]. Existing methods

M. Onizuka et al. (Eds.): DASFAA 2024, LNCS 14855, pp. 185–202, 2024.
https://doi.org/10.1007/978-981-97-5572-1_12

explore similarity indicators between different types of data and fuse them at the decision-making level [13] to transfer efficient knowledge among users with high similarity. However, similarity metrics often result in excessive redundancy of neighbor information and fail to adequately address the behavior scarcity. This flaw becomes particularly noticeable in terms of performance when users exhibit few-shot behaviors. In this paper, we investigate the transferability measurement problem for multi-source heterogeneous data. We nonlinearly model the correlation and complementarity in these data, enabling a more accurate measurement of transferability between users. This not only facilitates the prediction of users with few-shot behaviors but also reduces the risk of negative transfer.

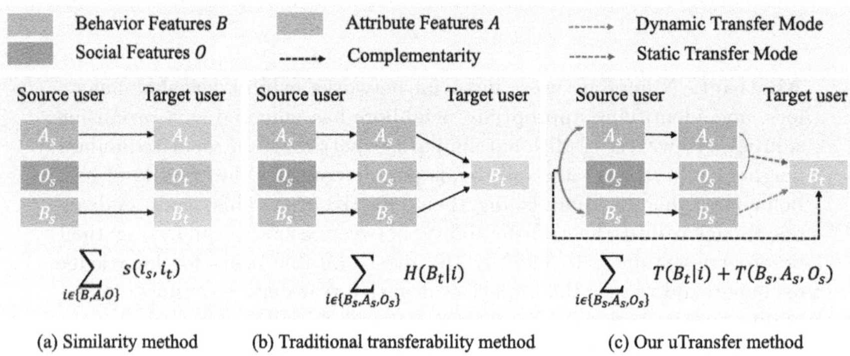

Fig. 1. Comparison of similarity metrics, traditional transferability metrics and our uTransfer metric. Compared to similarity metrics, transferability metrics can quantify the nonlinear statistical correlation between variables from a more intrinsically informative perspective. However, they both ignore the correlation between heterogeneous data. Our uTransfer metric distinguishes the differences in transfer modes at a finer granularity and captures the complementarity between heterogeneous data.

Existing transferability methods mainly target homogeneous data and aim to model the correlation between source features/labels and target features/labels. The difference compared to similarity metrics is shown in Fig. 1. Transferability methods have stronger robustness and stability due to their ability to enable nonlinear modeling of the correlations between variables. For example, the Negative Conditional Entropy (NCE) method [19] utilizes the negative conditional entropy of the source task label and the predicted target task label to measure the transferability between two tasks. MS_LEEP and E_LEEP [1] extends LEEP metric [12], making it applicable to scenarios involving multiple source models. In particular, MS_LEEP emphasizes feature-level fusion. It operates under the assumption that source models are independent and calculates the joint probability distribution predicted by these multiple source models. E_LEEP focuses on decision-level fusion, which calculates the average of prediction results of all source models. The OTCE method [18] proposes an optimal transport method that considers both domain differences and task differences when measuring task transferability. Based on this, OSBORN [3] additionally reflects on the cohesiveness of source models to obtain the ensemble transferability of source models.

In social networks, user data usually comes from different sources and in different types, making it heterogeneous. For example, attribute data is the personal information provided by the user or collected by the system such as age, gender, profession, etc. This structured data gives a basic profile of the user and can be critical in understanding demographics, user classification, and personal preferences. Social relationship refers to the friend lists in social networks. This networked data helps us understand the formation of social networks and the dissemination of social information. Behavior data is generated by the user's actions, including data like browsing history, search queries, click patterns, etc., which directly reflects the user's interests, preferences, and habits on the platform. This sequential data is dynamic and changes over time. However, existing methodologies face challenges in modeling these heterogeneous data. Firstly, these heterogeneous data types exist in separate semantic spaces, making it difficult to perform data fusion and computation directly. Secondly, the transfer modes for heterogeneous data are differentiated. For example, the knowledge transfer between user behavior sequences requires considering the information dynamics [10], while the knowledge transfer between user attributes and social relationships only takes into account the static transfers. Therefore, it is crucial to differentiate and model the transfer modes of heterogeneous data when assessing user transferability. Finally, the information between heterogeneous data may complement each other. For instance, user behavior information captures historical behavior preferences [7], and attribute information captures the behavior preferences of groups with the same attributes [9]. Behavior information and attribute information can complement each other, eliminating the redundancy of attribute information, and enhancing the diversity of behavior information [14,21]. Therefore, how to model the complementarity between heterogeneous data and measure the transferability between users is an important challenge.

To address the above challenges, this paper proposes a novel metric, uTransfer, which can measure the transferability between users with heterogeneous data in a more fine-grained and accurate way, thereby selecting appropriate users for knowledge transfer and reducing the risk of negative transfer. Specifically, the *Unifying Heterogeneous Knowledge Spaces* module unifies user heterogeneous data into the user behavior space, facilitating the fusion of user heterogeneous knowledge. The *Measuring the Transferability of Different Transfer Modes Independently* module then categorizes transfer modes into dynamic and static modes and calculates transferability under each mode separately. Finally, the *Measuring the Complementarity of Heterogeneous Data* module models the complementarity among heterogeneous data to enhance the assessment of user transferability. The overall contribution can be summarized as follows:

- We propose the problem of transferability measurement between heterogeneous data. Our proposal models correlation and complementarity between different types of data, accurately measuring transferability between users.

This approach is particularly beneficial in predicting the behavior of users with few-shot behaviors, while also mitigating the risk of negative transfer.
- We propose the uTransfer metrics to model the consistency of the semantic space, the differences in transfer modes, and the complementarity between heterogeneous data, to measure the transferability between two users more accurately and finely.
- Extensive experiments on three widely used datasets demonstrate that uTransfer can effectively measure the transferability between users.

2 Related Work

2.1 Similarity Measurement

In social networks, user data, such as attribute data, social relationships, and behavior data, comes from different sources and in different types, making it multi-source and heterogeneous. These data are not only distinct but also interrelated and complementary. This interconnectivity can deepen user analysis and understanding, especially for users with few-shot behaviors. To fully harness the available information, several researchers have explored the similarity metrics for heterogeneous data fusion. The primary objective of this exploration is to augment the precision of similarity computation. Concurrently, it also serves to enhance the performance of behavior prediction models, particularly when dealing with users who exhibit a limited range of behaviors, commonly referred to as few-shot behaviors [13]. Among them, the most classic methods are Pearson correlation coefficient (abbreviated as Pearson) [2], Jaccard similarity (abbreviated as Jaccard) [4], and Cosine similarity (abbreviated as Cosine) [2]. For instance, CORE-VAE [20] aims to obtain the similarity between users by utilizing the sum of Jaccard similarity based on user behaviors and social relationships. AGNN [13] aims to achieve similarity between users by utilizing the sum of Cosine similarity based on user behaviors and attributes.

A high similarity between two users indicates that there is a certain level of knowledge or experience sharing between them, which can provide guidance and a foundation for transferability between users. However, in transfer learning, knowledge similarity does not necessarily mean improvement in transfer performance or enhancement of transfer effects after joining the source. Furthermore, similarity metrics can easily lead to redundancy in neighbor information, which may result in negative transfer. Therefore, to avoid this situation, in this paper, we aim to explore the transferability between users with heterogeneous data.

2.2 Transferability Measurement

Transferability refers to identifying the appropriate source task, model, or layer that can effectively facilitate the transfer of knowledge to a target task. This process can assist in predicting or classifying the target task [8]. The existing transferability methods aim to model the correlation between sources and targets at the feature or label level from an information theory perspective. For example, NCE [19] utilizes the negative conditional entropy of the source task label and the predicted target task label to measure the transferability between two tasks. MS_LEEP and E_LEEP [1] extends LEEP metric [12], making it applicable to scenarios with multiple source models. MS_LEEP and E_LEEP are two methods used to calculate transferability. MS_LEEP computes the joint probability distribution predicted by multiple source models. This refers to feature-level fusion, and the calculation is based on the assumption that source models are independent. In contrast, E_LEEP focuses on decision-level fusion, which calculates the average of all source model prediction results. OTCE [18] takes into account both domain and task differences and proposes an optimal transport method to measure task transferability. Based on this, OSBORN [3] goes a step further and considers the cohesiveness of source models. This additional consideration allows for the calculation of the ensemble transferability of source models.

However, the aforementioned transferability metrics are all applicable to homogeneous data. In the context of social networks, user data is derived from multiple heterogeneous sources. It is difficult to directly and effectively fuse user attribute data, social relationships, and behavior data in different feature spaces. Moreover, the transfer modes may be various for different types of data, which have not been considered in the current research. Consequently, our goal is to build a model that maintains the consistency of the semantic space in heterogeneous data and takes into account the correlation and complementarity of different information. This will allow us to gain a more accurate and detailed understanding of transferability between users.

3 Methodology

3.1 Problem Formulation

There is the user set $U = \{u_1, u_2, \dots, u_M\}$ and item set $I = \{i_1, i_2, \dots, i_N\}$ in social network, where M and N denote the number of users and items, respectively. The data of user u is heterogeneous, including user attribute data $A^u = \{a_1^u, a_2^u, \dots, a_{N_A}^u\}$, user social relationships $O^u = \left\{u_1^u, u_2^u, \dots, u_{N_O^u}^u\right\}$, and user behavior data $B^u = \left\{i_1^u, i_2^u, \dots, i_{N_B^u}^u\right\}$, where N_A denotes the number of user attributes, and N_B^u and N_O^u represent the length of behavior and social of user u. Users in our study are divided into two groups based on the disparity in behavior frequency: the source users and the target users. Specifically, the source user's behavior data is rich, indicating high engagement and interaction, while target users exhibit less frequent behaviors, and the data is sparse. Given the heterogeneous data of source and target users, this paper aims to measure the knowledge transferability from source users to target users.

3.2 Our Method: uTransfer

Transferability [19,22] refers to the extent to which the knowledge from one source can be applied to a different target. In this context, the transferability of knowledge between a source user and a target user can be described as the level to which uncertainties concerning the target user's behavior B^t are reduced when given the data from the source user s, that is $H\left(B^t \mid B^s, O^s, A^s\right)$.

Due to the heterogeneity of user data, $H\left(B^t \mid B^s, O^s, A^s\right)$ is difficult to calculate directly. To solve this issue, we propose a novel metric called uTransfer in this paper. This innovative method provides a fine-grained and accurate way to measure the knowledge transferability of heterogeneous user data. The overall framework is shown in Fig. 2, including three modules: *Unifying Heterogeneous Knowledge Spaces*, *Measuring the Transferability of Different Transfer Modes Independently*, and *Measuring the Complementarity of Heterogeneous Data*. Specifically, the *Unifying Heterogeneous Knowledge Spaces* module integrates various types of user data into a unified user behavior space, which facilitates the process of merging and computing the user's heterogeneous knowledge. Then, the *Measuring the Transferability of Different Transfer Modes Independently* module categorizes transfer modes into dynamic and static based on data characteristics and measures the knowledge transferability for each mode individually. Following this, the *Measuring the Complementarity of Heterogeneous Data* module further assesses the complementarity of heterogeneous data, enhancing the calculation of user transferability. Finally, we integrate the transferability results of different transfer modes and the complementarity results of heterogeneous data to measure the transferability between two users.

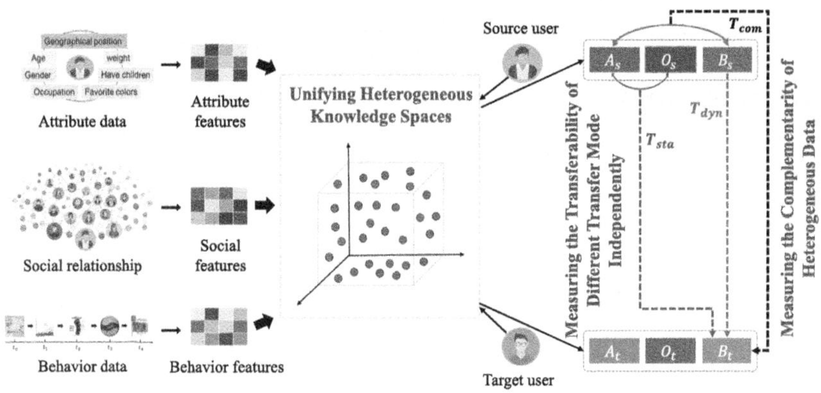

Fig. 2. The framework of our uTransfer metric.

Unifying Heterogeneous Knowledge Spaces. In social networks, user attribute, social relationships, and behavior data are in different semantic spaces, making it difficult to calculate H directly. One solution is to map attribute data, social relationships, and behavior data into the same semantic space to perform the knowledge fusion and calculation. LLAE [9] maps attribute data into the behavior space to facilitate behavior prediction of new users based on their attribute data. Similarly, LoCo [16] maps social relationships into the behavior space for predicting the behaviors of new users based on their social relationships. Inspired by these, we propose a model that unifies heterogeneous knowledge. In particular, it maps both attribute data and social relationships into the behavior space for heterogeneous knowledge fusion.

Usually, user behavior semantics can be comprehensive and detailed, including a wide range of actions and patterns while interacting with the system. However, in practice, user behavior data often proves to be sparse due to factors such as users being new and not very active, or exhibiting inconsistent or infrequent actions. It can be difficult to represent or model these behaviors, especially when there isn't a lot of data to work with. Therefore, we extract the multi-concept semantic information from behavior sequences following [10], to make user behavior representations more robust. Specifically, we first utilize the BPR algorithm [15] to obtain item representations EI, $EI \in R^{N\times d}$, where d denotes the embedding size, and then we apply the K-means method [11] to cluster all items to obtain the concept matrix C, $C \in R^{|C|\times d}$, where $|C|$ is the number of concepts, i.e., the number of clusters. Then the behavior semantic $C\left(B^u\right)$ of user u can be characterized as:

$$C\left(B^u\right)=\delta\left(\operatorname{sim}\left(C, E B^u\right)\right) \tag{1}$$

where $EB^u = B^u \times EI$ indicates the behavior semantic of user u, $\operatorname{sim}(,)$ denotes the Euclidean distance, and $\delta\left(\right)$ represents the Indicator function, denotes as:

$$\delta(x)=\begin{cases}1, & x>\alpha \\ 0, & \text{otherwise}\end{cases} \tag{2}$$

where α is the threshold of the distance.

The behaviors of users with the same attributes are similar [5]. Therefore, for the user attribute data, we can apply the behavior of users with the same attributes to characterize it, i.e., $A^u = \left\{B^j \mid \operatorname{sim}(A^u, A^j)=0\right\}$, where $\operatorname{sim}(A^u, A^j)=0$ indicates that the attributes of user u and user j are the same. And then the attribute semantic $C\left(A^u\right)$ of user u can be achieved by Eq. 1.

The user behavior information can characterize the behavior preferences of that user [23], so for the user social relationships, we can utilize the neighbors' behavior information of that user to characterize it, i.e., $O^u = \left\{B^j \mid j \in O^u\right\}$. And then the social semantic $C\left(O^u\right)$ of user u can be achieved by Eq. 1.

Consequently, the knowledge transferability from source user u_s to target user u_t can be calculated as:

$$\begin{aligned}\operatorname{Trans}\left(u_s \rightarrow u_t\right) & =H\left(B^{u_t} \mid B^{u_s}, A^{u_s}, S^{u_s}\right) \\ & =-p\left(B^{u_t}, B^{u_s}, A^{u_s}, O^{u_s}\right) \log_2 p\left(B^{u_t} \mid B^{u_s}, A^{u_s}, O^{u_s}\right)\end{aligned} \tag{3}$$

where,

$$\begin{aligned}p\left(B^{u_t}, B^{u_s}, A^{u_s}, O^{u_s}\right) & =\frac{C\left(B^{u_t}\right) \cap C\left(B^{u_s}\right) \cap C\left(A^{u_s}\right) \cap C\left(O^{u_s}\right)}{C\left(B^{u_t}\right) \cup C\left(B^{u_s}\right) \cup C\left(A^{u_s}\right) \cup C\left(O^{u_s}\right)} \\ p\left(B^{u_t} \mid B^{u_s}, A^{u_s}, O^{u_s}\right) & =\frac{C\left(B^{u_t}\right) \cap C\left(B^{u_s}\right) \cap C\left(A^{u_s}\right) \cap C\left(O^{u_s}\right)}{C\left(B^{u_s}\right) \cup C\left(A^{u_s}\right) \cup C\left(O^{u_s}\right)}\end{aligned} \tag{4}$$

where $\cap$ denotes the intersection and $\cup$ indicates the union. For instance, $B^{u_s} \cup B^{u_t}$ denotes the concepts in the behavior semantic of source user u_s and target user u_t, while $B^{u_s} \cup B^{u_s}$ represents all concepts in the behavior semantic of source user u_s and target user u_t.

Measuring the Transferability of Different Transfer Modes Independently. Although the aforementioned method can obtain knowledge transferability between users based on their heterogeneous data, it doesn't account for the characteristics and differences within heterogeneous data. For example, it overlooks the dynamic nature of user behaviors that are changing over time, while the user attributes and social relations usually remain static [6]. As such, we propose distinguishing the transfer modes of heterogeneous data into static knowledge transfer and dynamic knowledge transfer. This differentiation would enable a more accurate assessment of the transferability between users.

For the dynamic knowledge transfer model, inspired by [10], transfer entropy is utilized to calculate the knowledge transferability, i.e.,

$$T_{dyn} = H\left(B^{u_t} \mid B^{u_s}\right) = \sum_{c \in \left\{1:N_B^u\right\}} H\left(B_{(c)}^{u_t} \mid B_{(1:c)}^{u_t}\right) - H\left(B_{(c)}^{u_t} \mid B_{(1:c)}^{u_t}, B_{(1:c)}^{u_s}\right) \quad (5)$$

where $B_{(c)}^{u_t}$ indicates the user behavior at time c, and $B_{(1:c)}^{u_t}$ denotes the user behavior sequence before time c, $B_{(1:c)}^{u_t} = \left[B_1^{u_t}, B_2^{u_t}, \ldots, B_{c-1}^{u_t}\right]$, and,

$$\begin{aligned} H\left(B_{(c)}^{u_t} \mid B_{(1:c)}^{u_t}\right) &= -p\left(B_{(c)}^{u_t}, B_{(1:c)}^{u_t}\right) \log_2 p\left(B_{(c)}^{u_t} \mid B_{(1:c)}^{u_t}\right) \\ H\left(B_{(c)}^{u_t} \mid B_{(1:c)}^{u_t}, B_{(1:c)}^{u_s}\right) &= -p\left(B_{(c)}^{u_t}, B_{(1:c)}^{u_t}, B_{(1:c)}^{u_s}\right) \log_2 p\left(B_{(c)}^{u_t} \mid B_{(1:c)}^{u_t}, B_{(1:c)}^{u_s}\right) \end{aligned} \quad (6)$$

For the static knowledge transfer mode, inspired by [19], conditional entropy is applied to calculate the knowledge transferability, i.e.,

$$T_{sta} = H\left(B^{u_t} \mid A^{u_s}, O^{u_s}\right) = -p\left(B^{u_t}, A^{u_s}, O^{u_s}\right) \log_2 p\left(B^{u_t} \mid A^{u_s}, O^{u_s}\right) \quad (7)$$

Measuring the Complementarity of Heterogeneous Data. In social networks, user attributes, social relationships, behaviors, and other data types exhibit a complementary relationship. For example, a user's behavior information captures their historical behavior preferences, while attribute information reflects the behavior preferences of groups sharing the same attributes. The complementarity between behavioral and attribute information can eliminate redundancy and increase the diversity of behavioral information. Therefore, when modeling the transferability of knowledge from source to target users, it's crucial to take into account the complementarity existing within the user's heterogeneous data.

In information theory, conditional mutual information $I(X;\ Y \mid Z)$ represents the correlation between variable X and variable Y given variable Z, that is, it represents the exclusion of components that overlap with Z in the common part of X and Y. Therefore, in this paper, we measure the complementarity between heterogeneous data, which can be expressed in terms of multiple conditional mutual information, namely:

$$T_{com} = I(B^{u_t}; O^{u_s} \mid A^{u_s}, B^{u_s}) + I(B^{u_t}; A^{u_s} \mid O^{u_s}, B^{u_s}) + I(B^{u_t}; B^{u_s} \mid A^{u_s}, O^{u_s}) \quad (8)$$

where,

$$\begin{aligned} I\left(B^{u_t}; O^{u_s} \mid A^{u_s}, B^{u_s}\right) &= p_{all} \times \log_2 \frac{p_{all}}{p\left(B^{u_t} \mid A^{u_s}, B^{u_s}\right) p\left(O^{u_s}, A^{u_s}, B^{u_s}\right)} \\ I\left(B^{u_t}; A^{u_s} \mid O^{u_s}, B^{u_s}\right) &= p_{all} \times \log_2 \frac{p_{all}}{p\left(B^{u_t} \mid O^{u_s}, B^{u_s}\right) p\left(A^{u_s}, O^{u_s}, B^{u_s}\right)} \\ I\left(B^{u_t}; B^{u_s} \mid A^{u_s}, O^{u_s}\right) &= p_{all} \times \log_2 \frac{p_{all}}{p\left(B^{u_t} \mid A^{u_s}, O^{u_s}\right) p\left(B^{u_s}, A^{u_s}, O^{u_s}\right)} \end{aligned} \quad (9)$$

where $p_{all} = p\left(B^{u_t}, B^{u_s}, A^{u_s}, O^{u_s}\right)$, denoting the joint distribution of heterogeneous data between the source user u_s and the target user u_t.

Ultimately, the knowledge transferability from the source user u_s to the target user u_t can be expressed as:

$$\text{Trans}\left(u_s \rightarrow u_t\right) = T_{dyn} + T_{sta} + T_{com} \quad (10)$$

4 Experiments

4.1 Experimental Settings

Datasets. We verify the effectiveness of our uTransfer on three datasets: Filmtrust, MovieLens and Epinions datasets. Both Filmtrust and MovieLens datasets are movie rating datasets. The Filmtrust dataset includes user behavior data and user social relationships, while the MovieLens dataset includes user behavior data and user attribute data. Epinions dataset is a product rating dataset that includes user behavior and social relationships. Following the methodology in [10], we classify interaction-dense users and interaction-sparse users based on their interaction frequency, and utilize these interaction-sparse users as the test set. In addition, to validate the impact of uTransfer on users with different scarcity behaviors and demonstrate the generalization of uTransfer, we classify the test set into two categories: highly sparse users (the length of user behaviors is less than 5) and sparse users (the length of user behaviors is more than 5 and less than 10). The statistical data of three datasets are shown in Table 1.

Table 1. Statistics of Filmtrust, MovieLens, and Epinions datasets. In this table, Sparsity refers to the sparsity of the user-item interaction matrix, 'Few_1 user_count' denotes the number of users with highly sparse behaviors, and 'Few_5 user_count' denotes the number of users with sparse behaviors.

Type	Filmtrust	MovieLens	Epinions
User_count	427	4810	28305
Item_count	1827	3678	133276
Interaction_count	11848	758970	589715
Sparsity	98.48%	95.71%	99.98%
Few_1 user_count	77	86	10820
Few_5 user_count	48	723	4726
User data	behavior/social	behavior/attribute	behavior/social

Baselines. To validate the performance of our uTransfer metric, we compare existing similarity and transferability methods. The similarity methods: Pearson [2], Jaccard [20], and Cosine [13], and the transferability methods: NCE [19], MS_TEEP [1], E_LEEP [1], OTCE [18], and OSBORN [3]. In addition, uTransfer_direct denotes the result calculated by Eq. 3.

Evaluation. Considering that there is no clear ground-truth for the transferability between users, we validate it based on the user-based collaborative filtering method, aligning with similarity evaluation methodologies [17], To analyze the performance of neighbors selected under different measurement metrics more finely and accurately, we set the number of selected neighbors to 1, 3, 5, and 10, i.e., $N_{Adj} \in \{1, 3, 5, 10\}$, and apply HR (Hitting Rate) and NDCG (Normalized Discounted Cumulative Gain) metrics [10] to measure the performance of the model under recommendation top 10.

4.2 Performance Comparison

Properties of the uTransfer Metric. This section mainly demonstrates that the uTransfer metric changes with different numbers of concepts. The experimental results are shown in Fig. 3. It can be seen that the uTransfer and uTransfer_direct metrics first keep increasing and then stabilize as the number of concepts increases. The Filmtrust, MovieLines and Epinions datasets stabilize when the number of concepts is 50,200 and 100, respectively. Therefore, the uTransfer metrics in subsequent experiments are based on these values.

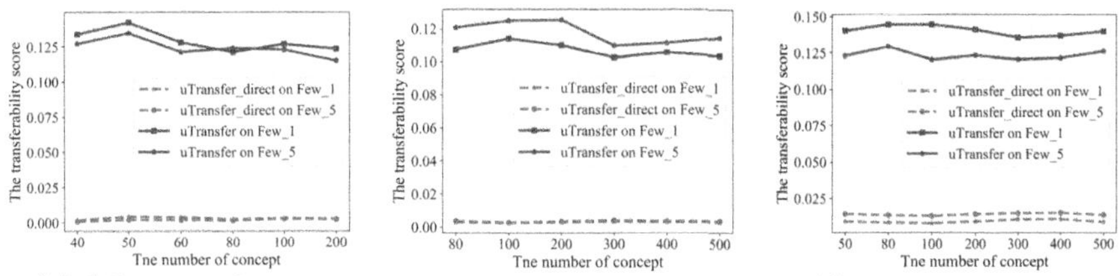

(a) Filmtrust dataset (b) MovieLens dataset (c) Epinions dataset

Fig. 3. The uTransfer scores on Filmtrust, MovieLens and Epinions dataset.

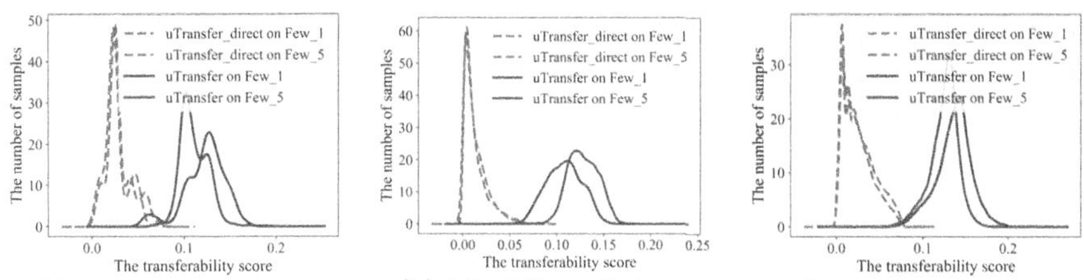

(a) Filmtrust dataset (b) MovieLens dataset (c) Epinions dataset

Fig. 4. The distributions of uTransfer on Filmtrust, MovieLens and Epinions dataset

Distribution of the uTransfer Metric. This section demonstrates the distribution of the uTansfer metric on three datasets. The result is shown in Fig. 4. It can be observed that, under Filmtrust and Epinions datasets, the uTansfer metric on the Few_1 dataset is greater than that on the Few_5 dataset, while it is the opposite under the MovieLens dataset, probably because the MovieLens dataset is denser, the behavior plays a huge role, and the correlation between user attribute and user behavior is higher, so the transferability on the Few_5 dataset is greater than that on the Few_1 dataset.

Performance Comparisons. This section demonstrates the overall performance comparisons with other advanced baselines, as shown in Tables 2, 3 and 4. It can be observed that our uTransfer metric achieves the best performance compared to the existing transferability and similarity metrics.

In these three tables, when N_{Adj}=1, the transferability metrics can achieve better performance compared to the similarity metrics. As the number of neighbors increases, the transferability metrics on the Filmtrust and Epinions datasets remain superior to the similarity metrics, but in the MovieLens dataset, the transferability metrics are inferior to the similarity metrics. This may be because the MovieLens dataset is relatively dense, and the similarity metrics can effectively represent the similarity, especially on the MovieLens Few_5

Table 2. Comparisons of baselines for Few_1 and Few_5 on the Filmtrust dataset. And the best results of our models and baselines are marked in bold and underlined.

		$N_{Adj} = 1$		$N_{Adj} = 3$		$N_{Adj} = 5$		$N_{Adj} = 10$	
		HR	NDCG	HR	NDCG	HR	NDCG	HR	NDCG
Few_1	Jaccard	0.331	0.225	0.409	0.267	0.448	0.296	0.552	0.376
	Pearson	0.377	0.251	0.552	0.351	0.643	0.460	0.682	0.512
	Cosine	0.526	0.310	0.630	0.386	0.636	0.429	0.727	0.512
	NCE	0.657	0.424	0.682	0.482	0.662	0.494	0.701	0.499
	MS_LEEP	0.584	0.414	0.610	0.463	0.623	0.466	0.669	0.527
	E_LEEP	0.584	0.414	0.610	0.463	0.623	0.466	0.662	0.525
	OTCE	0.566	0.377	0.610	0.429	0.614	0.441	0.620	0.464
	OSBORN	0.578	0.388	0.610	0.433	0.617	0.448	0.636	0.490
	uTransfer_direct	0.584	0.340	0.610	0.395	0.623	0.411	0.643	0.411
	uTransfer	**0.672**	**0.448**	**0.714**	**0.478**	**0.740**	**0.504**	**0.734**	**0.519**
Few_5	Jaccard	0.375	0.261	0.584	0.460	0.698	0.500	0.781	0.629
	Pearson	0.459	0.324	0.635	0.450	0.771	0.578	0.844	0.642
	Cosine	0.656	0.371	0.729	0.489	0.771	0.565	0.833	0.650
	NCE	0.750	0.538	0.782	0.574	0.812	0.602	0.844	0.636
	MS_LEEP	0.708	0.528	0.812	0.604	0.823	0.611	0.844	0.652
	E_LEEP	0.708	0.528	0.812	0.604	0.823	0.611	0.844	0.652
	OTCE	0.758	0.512	0.808	0.619	0.818	0.639	0.820	0.654
	OSBORN	**0.761**	0.508	0.812	0.619	0.823	0.640	0.823	0.657
	uTransfer_direct	0.708	0.436	0.740	0.471	0.750	0.506	0.729	0.497
	uTransfer	**0.761**	**0.547**	**0.823**	**0.629**	**0.842**	**0.644**	**0.865**	**0.667**

sub-datasets. For the Filmtrust dataset, the Cosine metric achieves relatively good performance in similarity metrics. Among the transferability indicators, NCE, MS_LEEP, and E_LEEP achieve good performance. For the MovieLens dataset and Epinions dataset, the Pearson metric achieves relatively good performance in similarity metrics. Among the transferability indicators, MS_LEEP and E_LEEP achieve good performance. Therefore, for datasets with dense interactions, Cosine similarity can be applied, while for datasets with sparse interactions, Pearson similarity can be utilized to better characterize transferability. Regardless of sparsity, MS_LEEP and E_LEEP can be chosen as the transferability metric.

Ablation Study. To verify the validity of each module of our uTransfer metric, we conduct an ablation study on three datasets, as shown in Tables 5, 6 and

Table 3. Comparisons of baselines for Few_1 and Few_5 on the Movielines dataset. And the best results of our models and baselines are marked in bold and underlined.

		$N_{Adj} = 1$		$N_{Adj} = 3$		$N_{Adj} = 5$		$N_{Adj} = 10$	
		HR	NDCG	HR	NDCG	HR	NDCG	HR	NDCG
Few_1	Jaccard	0.372	0.208	0.500	0.308	0.587	0.375	0.681	0.475
	Pearson	0.378	0.183	0.524	0.280	0.651	0.394	0.750	0.557
	Cosine	0.355	0.171	0.518	0.278	0.646	0.400	0.721	0.571
	NCE	0.413	0.237	0.471	0.272	0.512	0.309	0.524	0.306
	MS_LEEP	0.425	0.231	0.494	0.265	0.477	0.273	0.582	0.321
	E_LEEP	0.453	0.245	0.494	0.264	0.488	0.276	0.570	0.314
	OTCE	0.441	0.226	0.496	0.260	0.527	0.279	0.530	0.296
	OSBORN	0.436	0.220	0.506	0.265	0.518	0.305	0.559	0.327
	uTransfer_direct	0.401	0.207	0.448	0.246	0.459	0.257	0.448	0.244
	uTransfer	**0.529**	**0.286**	**0.698**	**0.418**	**0.750**	**0.502**	**0.773**	**0.601**
Few_5	Jaccard	0.422	0.252	0.527	0.338	0.588	0.395	0.687	0.489
	Pearson	0.349	0.197	0.505	0.313	0.621	0.424	0.726	0.537
	Cosine	0.348	0.198	0.515	0.324	0.625	0.422	0.728	0.535
	NCE	0.405	0.240	0.443	0.265	0.457	0.281	0.482	0.304
	MS_LEEP	0.404	0.238	0.442	0.266	0.458	0.280	0.508	0.317
	E_LEEP	0.403	0.230	0.437	0.261	0.472	0.287	0.510	0.312
	OTCE	0.371	0.223	0.423	0.253	0.456	0.275	0.500	0.298
	OSBORN	0.392	0.238	0.431	0.266	0.464	0.279	0.497	0.304
	uTransfer_direct	0.382	0.227	0.414	0.247	0.429	0.254	0.443	0.263
	uTransfer	**0.483**	**0.292**	**0.587**	**0.381**	**0.674**	**0.448**	**0.756**	**0.540**

7. In these tables, -w denotes inclusion, and -w/o denotes removal. Thus, -w Behavior means that only dynamic transfer mode is included, -w Side reports that only static transfer mode is included, -w/o interaction represents that the complementarity module is removed, and NCE -w interaction indicates the NCE indicator with the complementarity module.

These three tables show that: 1) Dynamic transferability plays a greater role than static transferability, especially as the number of neighbors increases, demonstrating the unique advantages of user behavior in knowledge transfer in social networks. 2) The NCE metric aims to replace the dynamic knowledge transfer model in uTransfer with the static knowledge transfer mode. Comparing with NCE and -w/o interaction, NCE -w interaction and uTransfer, it can be observed that the advantage of dynamic transferability, especially on the Movie-

Table 4. Comparisons of baselines for Few_1 and Few_5 on the Epinions dataset. And the best results of our models and baselines are marked in bold and underlined.

		$N_{Adj}=1$		$N_{Adj}=3$		$N_{Adj}=5$		$N_{Adj}=10$	
		HR	NDCG	HR	NDCG	HR	NDCG	HR	NDCG
Few_1	Jaccard	0.049	0.043	0.055	0.048	0.062	0.054	0.079	0.070
	Pearson	<u>0.063</u>	<u>0.057</u>	<u>0.086</u>	<u>0.075</u>	<u>0.097</u>	<u>0.090</u>	<u>0.117</u>	<u>0.104</u>
	Cosine	0.049	0.043	0.049	0.043	0.049	0.042	0.098	0.076
	NCE	0.059	0.052	0.075	0.066	0.088	0.077	0.108	0.096
	MS_LEEP	0.061	0.054	0.079	0.070	0.091	0.079	0.111	0.098
	E_LEEP	0.061	0.054	0.078	0.069	0.091	0.079	0.111	0.098
	OTCE	0.057	0.050	0.069	0.061	0.080	0.070	0.098	0.086
	OSBORN	0.056	0.049	0.069	0.061	0.080	0.070	0.099	0.087
	uTransfer_direct	0.054	0.048	0.065	0.058	0.072	0.064	0.087	0.077
	uTransfer	**0.069**	**0.061**	**0.088**	**0.078**	**0.112**	**0.098**	**0.122**	**0.107**
Few_5	Jaccard	0.044	0.039	0.054	0.046	0.060	0.052	0.071	0.063
	Pearson	0.052	0.046	0.071	0.065	0.084	0.076	0.102	0.094
	Cosine	0.044	0.039	0.044	0.039	0.044	0.039	0.090	0.071
	NCE	0.058	0.052	0.088	0.079	0.108	<u>0.096</u>	0.139	0.124
	MS_LEEP	<u>0.064</u>	<u>0.057</u>	0.091	0.080	<u>0.110</u>	<u>0.096</u>	<u>0.142</u>	<u>0.126</u>
	E_LEEP	0.063	<u>0.057</u>	<u>0.092</u>	<u>0.081</u>	<u>0.110</u>	<u>0.096</u>	0.142	0.125
	OTCE	0.060	0.054	0.081	0.071	0.097	0.085	0.127	0.111
	OSBORN	0.058	0.052	0.079	0.070	0.100	0.086	0.137	0.119
	uTransfer_direct	0.058	0.050	0.081	0.072	0.096	0.085	0.120	0.103
	uTransfer	**0.068**	**0.060**	**0.094**	**0.083**	**0.113**	**0.100**	**0.144**	**0.128**

Lens dataset, may be due to the dense nature of the dataset, where the user behavior plays a significant role. Therefore, fine-grained modeling of behavior can achieve better results. 3) Comparing with -w/o interaction and uTransfer, NCE and NCE - w interaction, it can be found that the advantages of the complementarity module, indicating that it is necessary to model the complementarity between heterogeneous data when modeling user transferability. 4) Comparing with uTransfer, uTransfer_direct and NCE, the performance of uTransfer outperforms that of NCE, and the performance of NCE outperforms that of uTransfer_direct, indicating the advantage of modeling static knowledge transfer mode and dynamic knowledge transfer mode independently.

Table 5. Comparisons of different components of our proposed uTransfer on the Filmtrust dataset. In this table, the best results are marked in bold.

		$N_{Adj} = 1$		$N_{Adj} = 3$		$N_{Adj} = 5$		$N_{Adj} = 10$	
		HR	NDCG	HR	NDCG	HR	NDCG	HR	NDCG
Few_1	-w Behavior	0.519	0.292	0.597	0.350	0.682	0.403	0.701	0.463
	-w Side	0.552	0.355	0.591	0.378	0.604	0.409	0.623	0.418
	-w/o interaction	0.669	0.446	0.682	0.477	0.656	0.462	0.688	0.506
	NCE	0.657	0.424	0.682	0.482	0.662	0.494	0.701	0.499
	NCE -w interaction	0.662	0.445	0.708	0.500	0.701	0.507	0.740	0.543
	uTransfer_direct	0.584	0.340	0.610	0.395	0.623	0.411	0.643	0.411
	uTransfer	**0.672**	**0.448**	**0.714**	**0.478**	**0.740**	**0.504**	**0.734**	**0.519**
Few_5	-w Behavior	0.688	0.371	0.729	0.466	0.833	0.558	0.854	0.639
	-w Side	0.708	0.436	0.750	0.516	0.771	0.571	0.782	0.565
	-w/o interaction	0.750	0.527	0.792	0.596	0.802	0.621	0.844	0.643
	NCE	0.750	0.538	0.782	0.574	0.812	0.602	0.844	0.636
	NCE -w interaction	0.761	0.541	0.782	0.593	0.812	0.634	0.854	0.656
	uTransfer_direct	0.708	0.436	0.740	0.471	0.750	0.506	0.729	0.497
	uTransfer	**0.761**	**0.547**	**0.823**	**0.629**	**0.842**	**0.644**	**0.865**	**0.667**

Table 6. Comparisons of different components of our proposed uTransfer on the MovieLens dataset. In this table, the best results are marked in bold.

		$N_{Adj} = 1$		$N_{Adj} = 3$		$N_{Adj} = 5$		$N_{Adj} = 10$	
		HR	NDCG	HR	NDCG	HR	NDCG	HR	NDCG
Few_1	-w Behavior	0.465	0.268	0.558	0.326	0.564	0.349	0.657	0.390
	-w Side	0.471	0.249	0.489	0.272	0.506	0.278	0.506	0.306
	-w/o interaction	0.477	0.276	0.570	0.328	0.663	0.396	0.756	0.510
	NCE	0.413	0.237	0.471	0.272	0.512	0.309	0.524	0.306
	NCE -w interaction	0.477	0.270	0.593	0.357	0.692	0.435	0.762	0.557
	uTransfer_direct	0.401	0.207	0.448	0.246	0.459	0.257	0.448	0.244
	uTransfer	**0.529**	**0.286**	**0.698**	**0.418**	**0.750**	**0.502**	**0.773**	**0.601**
Few_5	-w Behavior	0.464	0.281	0.534	0.328	0.561	0.343	0.604	0.382
	-w Side	0.403	0.232	0.436	0.257	0.462	0.276	0.468	0.280
	-w/o interaction	0.458	0.274	0.536	0.329	0.589	0.375	0.677	0.451
	NCE	0.405	0.240	0.443	0.265	0.457	0.281	0.482	0.304
	NCE -w interaction	0.462	0.284	0.543	0.339	0.607	0.390	0.708	0.492
	uTransfer_direct	0.382	0.227	0.414	0.247	0.429	0.254	0.443	0.263
	uTransfer	**0.483**	**0.292**	**0.587**	**0.381**	**0.674**	**0.448**	**0.756**	**0.540**

Table 7. Comparisons of different components of our proposed uTransfer on the Epinions dataset. In this table, the best results are marked in bold.

		$N_{Adj}=1$		$N_{Adj}=3$		$N_{Adj}=5$		$N_{Adj}=10$	
		HR	NDCG	HR	NDCG	HR	NDCG	HR	NDCG
Few_1	-w Behavior	0.061	0.054	0.080	0.070	0.094	0.083	0.111	0.099
	-w Side	0.057	0.049	0.068	0.060	0.076	0.067	0.091	0.081
	-w/o interaction	0.063	0.055	0.077	0.068	0.096	0.085	0.113	0.100
	NCE	0.059	0.052	0.075	0.066	0.088	0.077	0.108	0.096
	NCE -w interaction	0.059	0.053	0.075	0.067	0.088	0.079	0.110	0.099
	uTransfer_direct	0.054	0.048	0.065	0.058	0.072	0.064	0.087	0.077
	uTransfer	**0.069**	**0.061**	**0.088**	**0.078**	**0.112**	**0.098**	**0.122**	**0.107**
Few_5	-w Behavior	0.063	0.055	0.092	0.082	0.111	0.097	0.141	0.123
	-w Side	0.059	0.052	0.081	0.072	0.096	0.083	0.122	0.105
	-w/o interaction	0.066	0.058	0.093	0.082	0.111	0.099	0.142	0.126
	NCE	0.058	0.052	0.088	0.079	0.108	0.096	0.139	0.124
	NCE -w interaction	0.062	0.054	0.091	0.079	0.111	0.097	0.142	0.128
	uTransfer_direct	0.058	0.050	0.081	0.072	0.096	0.085	0.120	0.103
	uTransfer	**0.068**	**0.060**	**0.094**	**0.083**	**0.113**	**0.100**	**0.144**	**0.128**

5 Conclusion

We propose a novel user transferability metric, uTransfer, designed to accurately characterize user transferability based on heterogeneous data. This facilitates the selection of suitable source users for knowledge transfer, thus mitigating the risk of negative knowledge transfer. Unlike existing transferability metrics, uTransfer considers the heterogeneity of user data and the complementarity between heterogeneous data, which enables a more accurate measurement of the transferability between the source and the target user. Extensive experiments on three widely used datasets demonstrate the effectiveness of uTransfer. It has been shown to accurately select useful source users for knowledge transfer, and enhance behavior prediction performance for users exhibiting few-shot behaviors.

Acknowledgment. This work was partially supported by the National Science Fund for Distinguished Young Scholars(62025205), and the National Natural Science Foundation of China (No.62032020, 61960206008).

References

1. Agostinelli, A., Uijlings, J., Mensink, T., Ferrari, V.: Transferability metrics for selecting source model ensembles. In: Proceedings of the IEEE/CVF Conference on Computer Vision and Pattern Recognition, pp. 7936–7946 (2022)
2. Ahn, H.J.: A new similarity measure for collaborative filtering to alleviate the new user cold-starting problem. Inf. Sci. **178**(1), 37–51 (2008)
3. Bachu, S., Garg, T., Narasimhan, N.L., Konuru, R., Balasubramanian, V.N., et al.: Building a winning team: Selecting source model ensembles using a submodular transferability estimation approach. In: Proceedings of the IEEE/CVF International Conference on Computer Vision, pp. 11609–11620 (2023)
4. Bag, S., Kumar, S.K., Tiwari, M.K.: An efficient recommendation generation using relevant Jaccard similarity. Inf. Sci. **483**, 53–64 (2019)
5. Cohen, D., Aharon, M., Koren, Y., Somekh, O., Nissim, R.: Expediting exploration by attribute-to-feature mapping for cold-start recommendations. In: Proceedings of the Eleventh ACM Conference on Recommender Systems, pp. 184–192 (2017)
6. Dai, X.: Design of hybrid recommendation algorithm based on user dynamic behavior and static attributes. In: Journal of Physics: Conference Series, vol. 1881, p. 022035. IOP Publishing (2021)
7. Hou, Y., He, Z., McAuley, J., Zhao, W.X.: Learning vector-quantized item representation for transferable sequential recommenders. In: Proceedings of the ACM Web Conference 2023, pp. 1162–1171 (2023)
8. Jiang, J., Shu, Y., Wang, J., Long, M.: Transferability in deep learning: a survey. arXiv preprint arXiv:2201.05867 (2022)
9. Li, J., Jing, M., Lu, K., Zhu, L., Yang, Y., Huang, Z.: From zero-shot learning to cold-start recommendation. In: Proceedings of the AAAI Conference on Artificial Intelligence, vol. 33, pp. 4189–4196 (2019)
10. Li, N., et al.: Transfer how much: a fine-grained measure of the knowledge transferability of user behavior sequences in social network. Data Min. Knowl. Disc. **36**(6), 2214–2236 (2022)
11. MacQueen, J., et al.: Some methods for classification and analysis of multivariate observations. In: Proceedings of the Fifth Berkeley Symposium on Mathematical Statistics and Probability, Oakland, CA, USA, vol. 1, pp. 281–297 (1967)
12. Nguyen, C., Hassner, T., Seeger, M., Archambeau, C.: LEEP: a new measure to evaluate transferability of learned representations. In: International Conference on Machine Learning, pp. 7294–7305. PMLR (2020)
13. Qian, T., Liang, Y., Li, Q., Xiong, H.: Attribute graph neural networks for strict cold start recommendation. IEEE Trans. Knowl. Data Eng. **34**(8), 3597–3610 (2020)
14. Ren, Z., et al.: Variational reasoning about user preferences for conversational recommendation. In: Proceedings of the 45th International ACM SIGIR Conference on Research and Development in Information Retrieval, pp. 165–175 (2022)
15. Rendle, S., Freudenthaler, C., Gantner, Z., Schmidt-Thieme, L.: Bayesian personalized ranking from implicit feedback. In: Proceedings of Uncertainty in Artificial Intelligence, pp. 452–461 (2014)
16. Sedhain, S., Menon, A., Sanner, S., Xie, L., Braziunas, D.: Low-rank linear cold-start recommendation from social data. In: Proceedings of the AAAI Conference on Artificial Intelligence, vol. 31 (2017)
17. Singh, P.K., Pramanik, P.K.D., Choudhury, P.: A comparative study of different similarity metrics in highly sparse rating dataset. In: Balas, V.E., Sharma, N.,

Chakrabarti, A. (eds.) Data Management, Analytics and Innovation. AISC, vol. 839, pp. 45–60. Springer, Singapore (2019). https://doi.org/10.1007/978-981-13-1274-8_4
18. Tan, Y., Li, Y., Huang, S.L.: OTCE: a transferability metric for cross-domain cross-task representations. In: Proceedings of the IEEE/CVF Conference on Computer Vision and Pattern Recognition, pp. 15779–15788 (2021)
19. Tran, A.T., Nguyen, C.V., Hassner, T.: Transferability and hardness of supervised classification tasks. In: Proceedings of the IEEE/CVF International Conference on Computer Vision, pp. 1395–1405 (2019)
20. Walker, J., Zhang, F., Zhong, T., Zhou, F., Baagyere, E.Y.: Variational cold-start resistant recommendation. Inf. Sci. **605**, 267–285 (2022)
21. Xie, Y., Zhou, P., Kim, S.: Decoupled side information fusion for sequential recommendation. In: Proceedings of the 45th International ACM SIGIR Conference on Research and Development in Information Retrieval, pp. 1611–1621 (2022)
22. Yosinski, J., Clune, J., Bengio, Y., Lipson, H.: How transferable are features in deep neural networks? In: Advances in Neural Information Processing Systems, vol. 27 (2014)
23. Zhang, P., et al.: Efficiently leveraging multi-level user intent for session-based recommendation via atten-mixer network. In: Proceedings of the Sixteenth ACM International Conference on Web Search and Data Mining, pp. 168–176 (2023)

GPSR: Graph Prompt for Session-Based Recommendation

Cheng Li[1,2,3], Pei-Yuan Lai[3,4](✉), Yi-Hong Lu[2], De-Zhang Liao[3], Xiao-Dong Huang[5], and Chang-Dong Wang[2]

[1] Faculty of Innovation Engineering, Macau University of Science and Technology, Macau, China

[2] School of Computer Science and Engineering, Sun Yat-sen University, Guangzhou, Guangdong, China
luyh66@mail2.sysu.edu.cn

[3] South China Technology Commercialization Center, Guangzhou, Guangdong, China
{lpy,ldz}@sctcc.cn

[4] School of Information Engineering, Guangdong University of Technology, Guangzhou, Guangdong, China

[5] South China Agricultural University, Guangzhou, Guangdong, China
huangxiaodong@stu.scau.edu.cn

Abstract. Prompt learning is an emerging method that effectively bridges the gap between downstream tasks and pretrained models. Currently, graph prompt learning is typically confined to pure graph learning tasks, such as node classification or link prediction. So far, graph prompt learning has been scarcely applied to recommender system, particularly in sequential recommendation. In this paper, we propose a pioneering approach, Graph Prompt for Session-based Recommendation (GPSR), which is the first to apply the "pretraining-prompt-finetuning" paradigm to session graphs in recommender systems. GPSR efficiently utilizes learnable prompt vectors to guide the pretrained Graph Neural Network (GNN) in adapting to new data or tasks. The prompt learning endows GPSR with the ability of working on limited supervision, significantly enhancing the generality and effectiveness of session-based recommendation. Specifically, we first study the item transition pattern by constructing session graphs, based on which the GNN model is pretrained. Then, we introduce a universal prompt-based tuning method called Graph Prompt Feature (GPF), for adapting the pretrained GNN model to the downstream session-based recommendation task. Finally, we use the frozen graph model and trained prompt vectors for the next item recommendation in the downstream dataset. Extensive experiments have been conducted on two public datasets to evaluate the effectiveness of our GPSR method.

Keywords: Session-based recommender system · Graph neural network · Prompt learning · Pretraining · Finetuning

M. Onizuka et al. (Eds.): DASFAA 2024, LNCS 14855, pp. 203–219, 2024.
https://doi.org/10.1007/978-981-97-5572-1_13

1 Introduction

In the quest for enhancing Session-Based Recommender Systems (SBRSs) [5,13], Graph Neural Networks (GNNs) have emerged as a powerful contender, adept at unraveling the complex item transition patterns within sessions. Many recent studies have focused on using GNNs to learn node embedding, which aims to collect information from neighbors for node embedding [11,22], or use contrastive learning for recommender system [25]. These advanced models excel at deciphering the intricate transition patterns that are pivotal for understanding user behavior and preferences in highly dynamic session data [20]. Therefore, many GNN-based methods have been developed for SBRSs in the recent years [2,12,21,28].

Despite the significant success achieved by GNNs, the existing GNN-based SBRSs methods usually suffer from the poor generalization and serious supervision dependency problems. Specifically, like the cases of the other GNN-based applications [4], when the item transition patterns in the target session-based recommendation tasks differ significantly from those in the previous sessions, the graph model trained in the session graphs constructed from the previous sessions fails to be generalized to the target tasks. In addition, the existing GNN-based SBRSs methods rely significantly on the supervision information hidden in the training sessions. If the training sessions are not sufficient, the performance of the GNN-based SBRSs methods may degenerate seriously. Therefore, it is necessary to make full use of the large amount of universal sessions beyond the training pool for the target tasks, which however remains largely unsolved.

In this paper, to address the above issues, we propose a new method called Graph Prompt for Session-based Recommendation (GPSR), which to our best knowledge is the first attempt to leverage the "pretraining-prompt-finetuning" paradigm for SBRSs. Different from the existing GNN-based SBRSs methods consisting of only training stage, the proposed GPSR method consists of two main stages namely pretraining stage and finetuning stage. In the pretraining stage, the graph model is pretrained by using a large amount of session graphs constructed from universal sessions available beyond the training pool. In the finetuning stage, the pretrained graph model is frozen and a set of prompt vectors are trained for adapting the frozen graph model into the target downstream sessions. Finally, the pretrained graph model and trained prompt vectors are used together for predicting the next-item in the testing sessions.

The main contributions of this paper are summarized as follows.

1. To our best knowledge, this work is the first attempt to introduce the "pretraining-prompt-finetuning" paradigm to session-based recommendation. A new method called Graph Prompt for Session-based Recommendation (GPSR) is proposed, which can significantly enhance the generality and effectiveness of the GNN-based methods for SBRSs.
2. A universal prompt-based tuning method called Graph Prompt Feature (GPF), is leveraged for adapting the pretrained GNN model to the downstream session-based recommendation task, which is able to simultaneously

align the item transition patterns between the pretrained and downstream tasks.
3. Extensive experiments are conducted to show the effectiveness of the proposed GPSR method over both the counterpart algorithm and the state-of-the-art methods.

The rest of this paper is organized as follows. The related work will be reviewed in Sect. 2, including session-based recommender systems and graph pretraining methods. In Sect. 3, the proposed GPSR method will be described in detail, including session graph construction, graph model pretraining, prompt and fine-tuning, and next-item prediction. The experimental results will be reported in Sect. 4. Finally, Sect. 5 will conclude this paper.

2 Related Work

2.1 Session-Based Recommender Systems

Early efforts primarily rely on shallow machine learning methods to capture sequential patterns within user sessions for SBRSs. Markov chains (MC) model the transitions over interactions within sessions to predict the probable next interaction. Representative MC-based methods include Factorized Personalized Markov Chains (FPMC) [14] and recommender system based on Markov Decision Processes (MDPs) [15]. However, these MC methods suffer a huge drawback that the assumption on the conditional independence is too strong [3].

One prominent advancement in session-based recommendation involves Recurrent Neural Networks (RNNs), which inherently consider the temporal order of user interactions, allowing for a more nuanced understanding of user behavior over time. GRU4REC [6] applies the Gated Recurrent Unit (GRU) for the sequential behavior modeling. Neural Attentive Recommendation Machine (NARM) [8] capture sequential behavioral features by using an encoder-decoder architecture. Nevertheless, RNNs emphasize on sequential data, potentially overlooking global interactions and intricate structures in sessions.

To take the complex transition relationship between items in the session, Graph Neural Networks (GNNs) have attracted much interest in session-based recommendation. Session-based Recommendation with Graph Neural Networks (SR-GNN) [21] first constructs directed graphs from historical session sequences and employs the GNN for session-based recommendation. Full Graph Neural Network (FGNN) [12] learns the inherent order of the item transition patterns in sessions with a multiple weighted graph attention. Hybrid-order gated GNN (HGNN) [2] uses an attention mechanism to learn different weights of hybrid-order in propagation. Recently, Co-guided Heterogeneous Hypergraph Network (CoHHN) [28] devises a customized heterogeneous hypergraph network to model complex high-order dependencies among various features of items for session-based recommendation. Typically, GNNs work in an end-to-end manner, and rely heavily on abundant task-specific labeled data for performance. This approach poses challenges in obtaining costly task-specific supervision.

2.2 Graph Pretraining

Graph pretraining [7] aims to learn some general knowledge for the graph model with easily accessible information to reduce the annotation costs of new tasks [18]. In recent years, the "pretraining-finetuning" paradigm has garnered significant attention as a way to learn generalized models. Generally, the pretraining models are trained on larger datasets to preserve various intrinsic graph properties. Finetuning is then performed on a smaller dataset to adjust model and reduce the downstream task loss. Some effective pretraining strategies include Deep Graph Infomax (DGI) [19], InfoGraph [16], and Graph Constractive Learning (GraphCL) [27]. They work in the "pretraining-finetuning" paradigm. However, such methods may suffer from the problem of inconsistent objectives between pretraining and downstream tasks, resulting in suboptimal performance [9].

To bridge the gap between pretraining and downstream tasks, prompting [1] has first been proposed for language models. More recently, prompting has also been introduced to graph pretraining. In general, prompting on graph freezes the parameters of the pretrained model, and introduces additional learnable prompt vectors for finetuning the model in downstream tasks, thereby enhancing the compatibility between downstream tasks and pretrained models. However, current research on graph prompts is still limited, due to the inherent complexity of graph structures. Currently, graph prompt learning is typically confined to pure graph learning tasks, such as node classification or link prediction tasks. Some pioneering works like Graph Pre-training and Prompt Tuning (GPPT) [17], GraphPrompt [10] apply graph prompt learning on the GNNs. To our knowledge, graph prompt learning has been scarcely applied to recommender system, particularly in the session-based recommendation.

3 The Proposed GPSR Method

In this section, we will describe in detail the proposed GPSR method. As shown in Fig. 1, it is a "pretraining-prompt-finetuning" pipeline consisting of the following four modules. The first module is the session graph construction, which prepares the data used in GPSR by constructing the session graphs from sessions and converting the next-item into the output label of the graph model. The second module is the graph model pretraining module, which pretrains the graph neural network by using the large-amount of (session graph, output label) pairs. The third module is the prompt and finetuning module, which freezes the graph model pretrained in the second module and trains the prompt vectors for adapting the frozen graph model into the target downstream session-based recommendations. Finally, the fourth module introduces how to use the frozen graph model and trained prompt vectors for predicting the next-item in the testing sessions. Table 1 summarizes the main symbols used in this paper.

3.1 Session Graph Construction

The goal of session-based recommendation is to predict the next-click item with an anonymous session. For this task, we construct session graphs from sessions and model the next-item prediction problem as the (session graph, output label) classification problem where the output label is the next-click item [2]. Let $\mathcal{V} = \{v_1, v_2, ..., v_n\}$ denote the universal set of items in all sessions. And let $s = [v_1^s \rightarrow v_2^s \rightarrow \cdots \rightarrow v_m^s]$ denote session s with length m, which is composed by the clicked items according to the click order. Notice that some items may occur more than one times in session s.

Table 1. Symbol Description.

Symbol	Description
$\mathcal{V} = \{v_1, v_2, \ldots, v_n\}$	Universal set of n items
$s = [v_1^s, v_2^s, \ldots, v_m^s]$	Clicked item sequence in session s with length m
$\mathcal{G}^s = (\mathcal{U}^s, \mathcal{E}^s, \mathbf{U}^s)$	Session graph of session s
$\mathcal{U}^s = \{u_1^s, u_2^s, \ldots, u_l^s\}$	Node set in session s consisting of l nodes
$\mathcal{E}^s$	Set of directed weighted edges in $\mathcal{G}^s$
$\mathbf{U}^s = [\mathbf{u}_1^s, \mathbf{u}_2^s, \ldots, \mathbf{u}_l^s]$	Node feature matrix in $\mathcal{G}^s$
$\mathbf{e}_i$	Embedding vector of item v_i
f	Graph neural network function FGNN
$\mathbf{g}^s$	Graph-level representation vector of session s
Φ	Trainable parameters in f
$\hat{\mathbf{y}}^s$	Probability distribution of the next-click item
$\mathbf{y}^s$	One-hot encoding vector of the ground-truth next-click item
$\mathcal{S}^{\text{Pretrain}}$	Set of all sessions used for pretraining
$\bar{\Phi}$	Pretrained optimal parameters of f
$\mathbf{p}_i$	Learnable prompt vector adding to $\mathbf{e}_i$
$\mathbf{e}_i^*$	Prompted feature vector of item v_i
$\mathcal{G}^{s*}$	Prompted session graph for finetuning
$\mathbf{g}^{s*}$	Prompted graph-level representation vector of session s
$\mathcal{S}^{\text{Finetune}}$	Set of all sessions used for finetuning

An attributed weighted directed session graph is constructed from session s as follows:

$$\mathcal{G}^s = (\mathcal{U}^s, \mathcal{E}^s, \mathbf{U}^s). \tag{1}$$

In the session graph, the node set $\mathcal{U}^s = \{u_1^s, u_2^s, \ldots, u_l^s\}$ consists of l nodes with each node $u_i^s, \forall i = 1, 2, \ldots, l$ representing the unique item in session s, i.e. $\exists t$ s.t. $u_i^s = v_t^s$. And each edge $(u_i^s \rightarrow u_j^s) \in \mathcal{E}^s$ represents the consecutive item-click

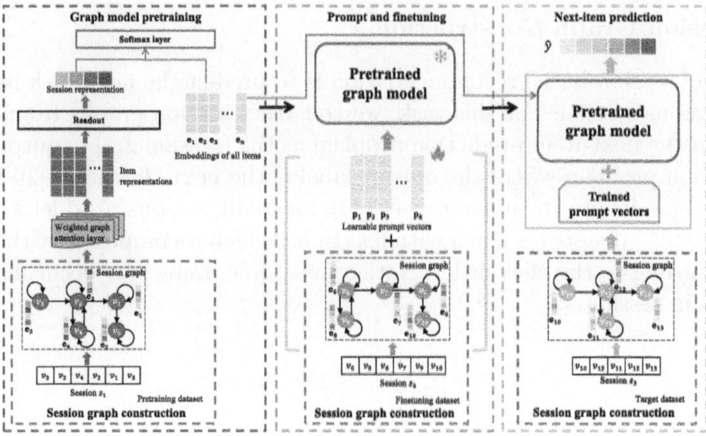

Fig. 1. Overall framework of GPSR, which consists of four components: session graph construction module, graph model pretraining module, prompt and finetuning module and next-item prediction module.

order, i.e., item u_j^s is clicked immediately after item u_i^s. In each edge $(u_i^s \rightarrow u_j^s)$, the weight $w_{i,j}^s$ indicates the number of the corresponding consecutive item-click order in session s. And the node feature matrix $\mathbf{U}^s$ will be introduced later. For input to the graph model, we embed each item $v_i \in \mathcal{V}$ to the d-dimensional embedding vector:

$$\mathbf{e}_i = \text{Embed}(\text{one-hot}(v_i)). \tag{2}$$

The one-hot$(v_i) \in \mathbb{R}^{n \times 1}$ of item $v_i \in \mathcal{V}$ is the vector with the i-th element being 1 and the others being 0. $\mathbf{e}_i \in \mathbb{R}^{d \times 1}$ is the embedding vector of item $v_i \in \mathcal{V}$. Accordingly, the feature vector of each node u_i^s in session graph $\mathcal{G}^s$ can be initialized as the embedding vector of the corresponding item. Therefore, the node feature matrix $\mathbf{U}^s \in \mathbf{R}^{d \times l} = [\mathbf{u}_1^s, \mathbf{u}_2^s, \ldots, \mathbf{u}_l^s]$ can be obtained with the i-th column vector being the embedding vector of the corresponding item.

3.2 Graph Model Pretraining

In this paper, as an example, we take Full Graph Neural Network (FGNN) [12] as the standard graph neural network. However, our technique can be easily extended to the other graph neural networks.

Specifically, let f denote the graph neural network function that maps the input graph $\mathcal{G}^s = \{\mathcal{U}^s, \mathcal{E}^s, \mathbf{U}^s\}$ to the graph-level representation vector $\mathbf{g}^s \in \mathbb{R}^{d \times 1}$ of the entire session graph $\mathcal{G}^s$, i.e.

$$\mathbf{g}^s = f(\mathcal{G}^s, \Phi) \tag{3}$$

where Φ denotes the trainable parameters in the FGNN model. According to [12], the mapping function f contains two main components. The first one is Weighted

Graph Attentional Layer (WGAT), which obtains node-level embedding of items within the session graph by considering both of the click-order (edge direction) and its number (edge weight). The second one is the Readout component, which obtains the graph-level embedding by considering the item-click order information. Please refer to [12] for the detailed description of FGNN.

In the pretraining stage, for each session $s = [v_1^s, v_2^s, \ldots, v_m^s]$ used for pretraining, the constructed session graph $\mathcal{G}^s$ is fed into FGNN to obtain the session representation $\mathbf{g}^s$. And the score $\hat{\mathbf{z}}_j^s$ of item $v_j \in \mathcal{V}$ in the current session s can be obtained by the inner product of the session representation $\mathbf{g}^s$ and the embedding vector $\mathbf{e}_j$ as follows:

$$\hat{\mathbf{z}}_j^s = {\mathbf{g}^s}^\top \mathbf{e}_j. \tag{4}$$

Based on $\hat{\mathbf{z}}_j^s$, the score vector $\hat{\mathbf{z}}^s \in \mathbb{R}^{n\times 1}$ can be obtained with $\hat{\mathbf{z}}_j^s$ being the j-th element of the score vector $\hat{\mathbf{z}}^s$. At last, the prediction vector $\hat{\mathbf{y}}^s \in \mathbb{R}^{n\times 1}$ can be obtained by a softmax function, which can be written as follows:

$$\hat{\mathbf{y}}^s = \mathrm{softmax}(\hat{\mathbf{z}}^s) \tag{5}$$

where the j-th element $\hat{\mathbf{y}}_j^s$ denotes the probability of clicking item v_j next in session s, i.e., the $(m+1)$-th clicked item. For session s, the pretraining loss function is defined as the cross-entropy loss:

$$\mathcal{L}^{\text{Pretrain}}(\Phi) = -\sum_{j=1}^{n} \mathbf{y}_j^s \log(\hat{\mathbf{y}}_j^s) + (1-\mathbf{y}_j^s)\log(1-\hat{\mathbf{y}}_j^s), \quad \forall s \in \mathcal{S}^{\text{Pretrain}} \tag{6}$$

where $\mathbf{y}^s \in \mathbb{R}^{n\times 1}$ denotes the one-hot encoding vector of the ground truth label of the next-click item. That is, if $v_j \in \mathcal{V}$ is the next-clicked item v_{m+1}^s for session s, only the j-th element of $\mathbf{y}_s$ is 1 and the others are 0. And $\mathcal{S}^{\text{Pretrain}}$ denotes the set of all sessions used for pretraining. Following [12], we use the Back-Propagation Through Time (BPTT) algorithm to pretrain the FGNN model. The results of this pretraining stage are the optimal parameters $\bar{\Phi}$ of the graph model and the item embedding vectors $\{\mathbf{e}_1, \mathbf{e}_2, \ldots, \mathbf{e}_n\}$, which will be frozen in the finetuning stage.

3.3 Prompt and Finetuning

Based on the Graph Model Pretraining module, to adapt the pretrained graph model into the target downstream session-based recommendation task, we introduce Graph Prompt Feature (GPF) [4], which freezes the parameters of the pretrained model, i.e., Φ, and assigns an independent learnable prompt vector $\mathbf{p}_i \in \mathbb{R}^{d\times 1}$ to the embedding vector of each item v_i. Specifically, a prompted feature vector of item v_i, denoted as $\mathbf{e}_i^*$, is generated by adding the learnable prompt vector $\mathbf{p}_i$ to the initial embedding vector $\mathbf{e}_i$,

$$\mathbf{e}_i^* = \mathbf{e}_i + \mathbf{p}_i. \tag{7}$$

In order to avoid learning n prompt vectors, only a set of k independent basis vectors are trained, i.e.,

$$\mathbf{p}_1^b, \mathbf{p}_2^b, \ldots, \mathbf{p}_k^b \in \mathbb{R}^{d\times 1} \tag{8}$$

And each prompt vector $\mathbf{p}_i$ can be generated by using the attentive aggregation of these basis vectors

$$\mathbf{p}_i = \sum_j^k \alpha_{i,j}\mathbf{p}_j^b, \quad \alpha_{i,j} = \frac{\exp(\mathbf{a}_j^\top \mathbf{e}_i)}{\sum_h^k \exp(\mathbf{a}_h^\top \mathbf{e}_i)} \tag{9}$$

where $\mathbf{a}_j \in \mathbb{R}^{d\times 1}, \forall j = 1, 2, \ldots, k$ are k learnable linear projections.

For each session s used in the finetuning stage, a prompted session graph is constructed as follows:

$$\mathcal{G}^{s*} = \{\mathcal{U}^s, \mathcal{E}^s, \mathbf{U}^{s*}\} \tag{10}$$

where the node set $\mathcal{U}^s$ and the directed weighted edge set $\mathcal{E}^s$ are completely the same as those in the session graph construction module but the node feature matrix is updated to be the prompted feature vectors of the corresponding items.

After constructing the prompted session graph $\mathcal{G}^{s*}$, it is fed into the frozen graph model with the fixed parameters $\bar{\Phi}$ to obtain the prompted graph-level representation vector $\mathbf{g}^{s*} \in \mathbb{R}^{d\times 1}$, i.e.,

$$\mathbf{g}^{s*} = f(\mathcal{G}^{s*}, \bar{\Phi}). \tag{11}$$

Similar to Eq. (4) and Eq. (5), the score $\hat{\mathbf{z}}_j^s$ of item $v_j \in \mathcal{V}$ and the prediction vector $\hat{y}^s$ can be obtained as follows:

$$\hat{\mathbf{z}}_j^s = {\mathbf{g}^{s*}}^\top \mathbf{e}_j^*, \quad \hat{\mathbf{y}}^s = \text{softmax}(\hat{\mathbf{z}}^s). \tag{12}$$

Similarly, the following cross-entropy loss is used to train the k basis vectors $\{\mathbf{p}_1^b, \mathbf{p}_2^b, \ldots, \mathbf{p}_k^b\}$ and k learnable linear projections $\{\mathbf{a}_1, \mathbf{a}_2, \ldots, \mathbf{a}_k\}$:

$$\begin{aligned} &\mathcal{L}^{\text{Finetune}}(\{\mathbf{p}_1^b, \mathbf{p}_2^b, \ldots, \mathbf{p}_k^b\}, \{\mathbf{a}_1, \mathbf{a}_2, \ldots, \mathbf{a}_k\}) \\ &= -\sum_{j=1}^n \mathbf{y}_j^s \log(\hat{\mathbf{y}}_j^s) + (1 - \mathbf{y}_j^s)\log(1 - \hat{\mathbf{y}}_j^s), \quad \forall s \in \mathcal{S}^{\text{Finetune}} \end{aligned} \tag{13}$$

where $\mathcal{S}^{\text{Finetune}}$ denotes the set of all sessions used for finetuning. The same optimization strategy as [4] is adopted for learning $\{\mathbf{p}_1^b, \mathbf{p}_2^b, \ldots, \mathbf{p}_k^b\}$ and $\{\mathbf{a}_1, \mathbf{a}_2, \ldots, \mathbf{a}_k\}$, based on which the prompt vector $\mathbf{p}_i \in \mathbb{R}^{d\times 1}$ is obtained for each item v_i.

3.4 Next-Item Prediction and Algorithm Summary

Once the frozen pretrained graph model $\bar{\Phi}$ and trained prompt features $\{\mathbf{p}_1, \mathbf{p}_2, \ldots, \mathbf{p}_n\}$ are obtained, we can use them to make a next-item prediction. In particular, for session s, a prompted session graph $\mathcal{G}^{s*} = \{\mathcal{U}^s, \mathcal{E}^s, \mathbf{U}^{s*}\}$ is constructed by using the same way as the prompt and finetuning module, which is then fed into the frozen graph model with the fixed parameters $\bar{\Phi}$ to obtain the prompted graph-level representation vector $\mathbf{g}^{s*} \in \mathbb{R}^{d\times 1}$. And the prediction vector $\hat{y}^s$ can be obtained by Eq. (12), based on which the next-clicked item can be obtained.

For clarity, Algorithm 1 summarizes the main procedure of the proposed GPSR method.

4 Experiments

4.1 Experimental Settings

Datasets and Processing. In our experiments, two widely adopted session-based recommendation datasets are used, namely Yoochoose and Diginetica.

- Yoochoose: The Yoochoose dataset is comprised of click stream data gathered from an e-commerce website over a period of six months.
- Diginetica: The Diginetica dataset, originating from the CIKM Cup 2016, is exclusively composed of transaction data.

However, different from the existing usage of these two datasets, a separate pretraining dataset is constructed from the part of these datasets.

For the Yoochoose dataset, following the conventional setting, the last day's sessions are regarded as the testing sessions, which consist of 55898 sessions. And the remaining sessions (denoted as *Yoochoose_remaining sessions*) are used for constructing various pretraining/training datasets. Specifically all the *Yoochoose_remaining sessions* are partitioned into four equal parts according to the timestamp of sessions. The third 1/4 is regarded as the pretraining dataset for pretraining the GPSR method. And from the fourth 1/4, four different training datasets (for GPSR, they are used for finetuning) are constructed, namely Yoochoose1/64, Yoochoose1/32, Yoochoose1/16, and Yoochoose1/8. The Yoochoose1/64 (*resp.* Yoochoose1/32, Yoochoose1/16, and Yoochoose1/8) dataset is composed of the last 1/64 (*resp.* 1/32, 1/16, and 1/8) sessions of *Yoochoose_remaining sessions.*

Similarly, for the Diginetica dataset, the last week's sessions are regarded as the testing sessions, which consist of 60858 sessions. And the remaining sessions (denoted as *Diginetica_remaining sessions*) are used for constructing pretraining/training datasets. Specifically all the *Diginetica_remaining sessions* are partitioned into five equal parts according to the timestamp of sessions. The first 4/5 (i.e., 80%) is regarded as the pretraining dataset for pretraining the GPSR method. And the last 1/5 is regarded as the training dataset (for GPSR, it is used for finetuning).

For clarity, Table 2 lists the statistic details of the five datasets, including the number of training (finetuning for GPSR) sessions, the number of testing sessions, the number of all items, and the average length of sessions. Notice that, for GPSR, the same pretraining dataset, namely the third 1/4 of *Yoochoose_remaining sessions* is used for pretraining GPSR in Yoochoose, which four different downstream datasets are used for finetuning.

Algorithm 1. The algorithm framework of the proposed GPSR method.

Input: Universal item set $\mathcal{V} = \{v_1, v_2, \ldots, v_n\}$; Pretraining sessions $\mathcal{S}^{\text{Pretrain}}$ and their output labels (i.e. the next-click items); Finetuning sessions $\mathcal{S}^{Finetune}$ and their output labels (i.e. the next-click items); Basis vector number k; Embedding dimension d, n_epochs.
Output: The pretrained graph model parameters $\bar{\Phi}$, the item embedding vectors $\{\mathbf{e}_1, \mathbf{e}_2, \ldots, \mathbf{e}_n\}$ and the prompt vectors $\{\mathbf{p}_1, \mathbf{p}_2, \ldots, \mathbf{p}_n\}$.

Pretraining stage:

Randomly initialize the graph model parameters Φ.
Construct session graph $\mathcal{G}^s$ from each session $s \in \mathcal{S}^{\text{Pretrain}}$ via Eq. (1).
for $epoch = 1$ to n_epochs **do**
    for each $batch \subseteq \mathcal{S}^{\text{Pretrain}}$ **do**
        for each $s \in batch$ **do**
            Obtain the session graph representation $\mathbf{g}^s$ via Eq. (3).
            Obtain the prediction vector $\hat{\mathbf{y}}^s$ via Eq. (5).
            Perform the Back-Propagation Through Time (BPTT) to optimize $\mathcal{L}^{\text{Pretrain}}(\Phi)$.
        end for
    end for
end for
return $\bar{\Phi}$ and $\{\mathbf{e}_1, \mathbf{e}_2, \ldots, \mathbf{e}_n\}$.

Finetuning stage:

Freeze $\bar{\Phi}$ and $\{\mathbf{e}_1, \mathbf{e}_2, \ldots, \mathbf{e}_n\}$.
Randomly initialize the k basis vectors $\{\mathbf{p}_1^b, \mathbf{p}_2^b, \ldots, \mathbf{p}_k^b\}$ and k learnable linear projections $\{\mathbf{a}_1, \mathbf{a}_2, \ldots, \mathbf{a}_k\}$.
Construct prompted session graph $\mathcal{G}^{s*}$ from each session $s \in \mathcal{S}^{\text{Finetune}}$ via Eq. (10).

for $epoch = 1$ to n_epochs **do**
    for each $batch \subseteq \mathcal{S}^{\text{Finetune}}$ **do**
        for each $s \in batch$ **do**
            Obtain the prompted session graph representation $\mathbf{g}^{s*}$ via Eq. (11).
            Obtain the prediction vector $\hat{\mathbf{y}}^s$ via Eq. (12).
            Perform the gradient descent to optimize $\mathcal{L}^{\text{Finetune}}(\{\mathbf{p}_1^b, \mathbf{p}_2^b, \ldots, \mathbf{p}_k^b\}, \{\mathbf{a}_1, \mathbf{a}_2, \ldots, \mathbf{a}_k\})$.
        end for
    end for
end for
Compute the prompt vectors $\{\mathbf{p}_1, \mathbf{p}_2, \ldots, \mathbf{p}_n\}$ via Eq. (9).
return $\{\mathbf{p}_1, \mathbf{p}_2, \ldots, \mathbf{p}_n\}$.

Evaluation Protocols. The following evaluation protocols are adopted.

- **MRR@K**(Mean Reciprocal Rank): MRR@K is the target item's reciprocal rank average. If the target item is not in the top **K** position, the reciprocal

Table 2. Statistic details of the five datasets.

Datasets	#Training sessions	#Testing sessions	#Items	Avg.length
Yoochoose1/64	369859	55898	16373	6.11
Yoochoose1/32	739718	55898	19100	5.86
Yoochoose1/16	1479436	55898	22350	5.76
Yoochoose1/8	2958872	55898	25336	5.76
Diginetica	719470	60858	42973	5.13

is set to 0. Higher rankings of the target items are indicated by a greater MRR@K.

Baselines. In our experiments, the following baselines are adopted:

- **NARM** [8] is able to explicitly emphasise the most important parts of the input by extending an attention layer to integrate encoded states of RNN.
- **GC-SAN** [26] generates session representations using the original GGNN with multiple self-attention layers.
- **COTREC** [23] combines self-supervised learning with co-training by exploiting two views that exhibit the internal and external connectivities of sessions to enhance session-based recommendation.
- **DHCN** [24] proposes a hypergraph convolutional network to capture the beyond-pairwise relations among items through hypergraph modeling so as to improve session-based recommendation.
- **HGNN** [2] proposes a novel hybrid-order GGNN model to avoid insignificant patterns and capture more complex dependencies among items in propagation with utilizing gated recurrent units.
- **FGNN** [12] proposes a weighted attention graph layer and a Readout function to learn the embeddings of items and sessions.

Parameter Setup. In this part, we mainly describe the default hyper-parameter settings across all experiments. For the pre-training phase, we apply a three layer WGAT and each with eight heads as our node representation encoder and a three processing steps of our Readout function. The size of item embedding is set to 100 for every layer including the initial embedding layer. We use the Adam optimizer across 10 epochs, where the initial learning rate is set to 0.001 and the linear schedule decay rate is set to 0.1 for every 3 epochs. The batch size for mini-batch optimization is 100 and the L2 regularization is set to 0.00001 for avoiding overfitting. For the value of the basis vector number k, different values are set for different datasets. Specifically, for the Yoochoose1/64 dataset we set k to 10, the Yoochoose1/32 dataset to 3, the Yoochoose1/16 dataset to 2, the Yoochoose1/16 dataset to 8 and the Diginetica dataset to 7. For the above five baseline methods, the hyper-parameters are tuned according to the default values suggested by the original authors so as to attain the optimal performance.

Table 3. Comparision with baseline methods on the Yoochoose1/64 dataset.

Measure	Methods						
	NARM	GC-SAN	COTREC	DHCN	HGNN	FGNN	GPSR
MRR@5	26.33	27.31	26.87	9.03	27.54	27.27	**27.91**
MRR@10	28.03	29.05	28.69	9.25	29.31	28.99	**29.67**
MRR@15	28.54	29.56	29.22	9.34	29.83	29.50	**30.18**
MRR@20	28.77	29.78	29.46	9.39	30.06	29.74	**30.41**

4.2 Performance Improvement over Non-Pretraining Counterpart

We compare our GPSR model with the non-pretrained FGNN model on five datasets, as shown in the last two columns of Tables 3, 4, 5, 6 and 7. In Table 3 (Yoochoose1/64 dataset), GPSR consistently outperforms FGNN, achieving higher MRR across all cutoff points. Similar trends are observed in Tables 4, 5, and 6 for the Yoochoose1/32, Yoochoose1/16, and Yoochoose1/8 datasets, respectively. Notably, the improvements are consistent and demonstrate the effectiveness of our proposed model on various downstream datasets.

Table 4. Comparision with baseline methods on the Yoochoose1/32 dataset.

Measure	Methods						
	NARM	GC-SAN	COTREC	DHCN	HGNN	FGNN	GPSR
MRR@5	27.14	27.91	26.84	9.01	27.99	27.27	**28.12**
MRR@10	28.82	29.66	28.67	9.25	29.77	29.02	**29.90**
MRR@15	29.28	30.18	29.23	9.34	30.28	29.54	**30.42**
MRR@20	29.50	30.40	29.47	9.39	30.51	29.78	**30.66**

Table 5. Comparision with baseline methods on the Yoochoose1/16 dataset.

Measure	Methods						
	NARM	GC-SAN	COTREC	DHCN	HGNN	FGNN	GPSR
MRR@5	28.29	28.21	26.83	7.99	27.85	26.99	**28.30**
MRR@10	29.81	29.94	28.68	8.23	29.60	28.75	**30.08**
MRR@15	30.22	30.45	29.24	8.32	30.12	29.28	**30.59**
MRR@20	30.41	30.68	29.47	8.37	30.38	29.52	**30.84**

In the case of the Diginetica dataset (Table 7), our GPSR model stands out as a remarkable performer compared to the its counterpart FGNN model. Across

Table 6. Comparision with baseline methods on the Yoochoose1/8 dataset.

Measure	Methods						
	NARM	GC-SAN	COTREC	DHCN	HGNN	FGNN	GPSR
MRR@5	28.23	28.08	26.83	8.18	27.72	26.65	**28.39**
MRR@10	29.89	29.81	28.65	8.39	29.47	28.38	**30.16**
MRR@15	30.34	30.32	29.19	8.47	30.01	28.90	**30.68**
MRR@20	30.54	30.56	29.42	8.52	30.24	29.15	**30.91**

Table 7. Comparision with baseline methods on the Diginetica dataset.

Measure	Methods						
	NARM	GC-SAN	COTREC	DHCN	HGNN	FGNN	GPSR
MRR@5	13.25	14.26	15.95	10.11	13.07	12.79	**21.58**
MRR@10	13.97	15.69	17.58	10.38	14.55	14.32	**22.70**
MRR@15	14.23	16.26	18.20	10.46	15.15	14.90	**23.09**
MRR@20	14.36	16.55	18.52	10.50	15.47	15.22	**23.30**

all evaluation metrics, GPSR consistently outperforms FGNN by a substantial margin. The superior results on the Diginetica dataset further underscore the adaptability and generalizability of GPSR on diverse datasets.

Notice that the pretrained graph model adopted in our GPSR is FGNN, i.e., the same graph neural network is adopted for both GPSR and FGNN. However, the comparison results in this subsection show that GPSR outperforms its non-pretraining counterpart FGNN across the five datasets, demonstrating the effectiveness of GPSR in transferring knowledge from the pretraining datasets to the downstream tasks. It is noteworthy that the "pretraining-prompt-finetuning" pipeline proposed in our method exhibits advantages compared to the traditional training method. Because our proposed GPSR only requires pretraining on a universal dataset, the fixed pretrained model can be applied to various downstream datasets with minimal parameter adjustments. Therefore, we believe that applying the "pretraining-prompt-finetuning" paradigm to other graph models is reasonable and practicable, potentially leading to performance improvements. Our proposed GPSR not only boosts the training efficiency but also enhances model accuracy, demonstrating its strong usability in practical applications.

4.3 Comparison with Baseline Methods

We conduct a comparative analysis between our model and several state-of-the-art baselines across five different datasets. The evaluation metrics, including MRR@5, MRR@10, MRR@15, and MRR@20, are used to assess the model performance. To ensure fair comparison, we rerun the compared baselines using our datasets. The comparison results are listed in Tables 3 to 7, and we highlight

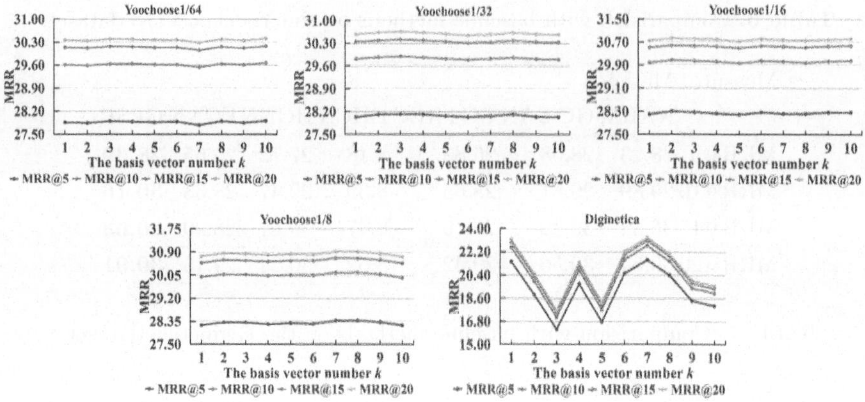

Fig. 2. Parameter analysis: The impact of the basis vector number k on the performance of GPSR.

the best results of each evaluation metric. Based on the obtained results, we have the following observations.

- Across all datasets and evaluation metrics, GPSR consistently outperforms all baseline methods. In top-K recommendation, GPSR improves over the state-of-the-art baselines w.r.t. MRR@5 by 0.6, 0.21, 0.01, 0.16 and 5.63, MRR@10 by 0.62, 0.24, 0.14, 0.27 and 5.12, MRR@15 by 0.62, 0.24, 0.14, 0.34 and 4.89, and MRR@20 by 0.63, 0.26, 0.16, 0.35 and 4.78 on the Yoochoose1/64, Yoochoose1/32, Yoochoose1/16, Yoochoose1/8 and Diginetica datasets, respectively.
- Besides, compared to the GNN-based models, including GC-SAN, COTREC, DHCN, HGNN and FGNN, the RNN-based model namely NARM can sometimes achieve comparable or even better performance. Additionally, FGNN exhibits relatively poor performance among several state-of-the-art baselines. However, with the enhancements introduced in FGNN, i.e., GPSR, it can achieve better performance than all baselines, further demonstrating the effectiveness of the "pretraining-prompt-finetuning" pipeline for GNN-based SBRSs.
- Our proposed GPSR significantly outperforms the state-of-the-art baselines on these five datasets. This again demonstrates that GPSR achieves superior performance across different downstream datasets by finetuning with the learnable prompt vector. Our proposed GPSR demonstrates robust generalization capabilities, achieving competitive results across diverse downstream datasets. Due to GPSR pretraining on a universal dataset, the fixed pretrained model can be seamlessly applied to various downstream datasets with minimal parameter adjustments, while achieving superior performance compared to the state-of-the-art baselines.

4.4 Analysis on the Basis Vector Number

To gain a deeper understanding of the relationship between the model performance and the prompt vector, we conduct a detailed analysis on the effect of the basis vector number k in the performance of GPSR. The results of the parameter tuning analysis are presented in Fig. 2, each subfigure corresponding to a different dataset. The graphs plot the MRR values against the varying values of k in the range from 1 to 10. The performance metric used to assess the impact of these variations is the MRR@K, namely MRR@5, MRR@10, MRR@15, and MRR@20.

The findings indicate that the impact of the parameter k on MRR varies across datasets, which suggests that each dataset may require a unique setting to parameter tuning. For the Yoochoose datasets, it can be observed that the MRR metric does not fluctuate significantly with the adjustment of the parameter k, indicating that we can learn a relatively general representation on the pretrained dataset with only a few prompt vectors. In our experiments, the optimal k value used for Yoochoose1/64, Yoochoose1/32, Yoochoose1/16, and Yoochoose1/8 are set to be 10, 3, 2, and 8 respectively. On the contrary, for the Diginetica dataset, as the parameter k is adjusted, the fluctuation of the MRR metric becomes more pronounced, suggesting the need for a targeted parameter design for k. The optimal k value for this dataset is set to be 7. These results highlight the necessity of customizing the tuning of k to enhance the predictive accuracy of recommendation systems.

5 Conclusion

On account of the poor generalization and serious supervision dependency problems when applying GNN to session-based recommendation, we propose a novel approach, i.e., GPSR, which is the first to apply the "pretraining-prompt-finetuning" paradigm to SBRSs. Specifically, GPSR pretrains the GNN model by using the large-amount of session graphs, and then introduces learnable prompt vectors for finetuning to adapt the pretrained graph model to the downstream session-based recommendation task. By this way, GPSR can effectively transfer knowledge from the pretraining datasets to the downstream tasks. The experiments show that the proposed GPSR method outperforms the representative and state-of-the-art methods. This indicates that our GPSR has robust generalization capabilities, achieving competitive results across diverse downstream datasets.

Acknowledgments. This work was supported by Guangdong Provincial Engineering Research Center of Intelligent Matching for Technology Commercialization (2022A175), Guangdong Basic and Applied Basic Research Foundation (2022B1515120059), and NSFC (62276277).

References

1. Brown, T.B., et al.: Language models are few-shot learners. In: NIPS (2020)
2. Chen, Y.H., Huang, L., Wang, C.D., Lai, J.H.: Hybrid-order gated graph neural network for session-based recommendation. IEEE Trans. Ind. Informatics **18**(3), 1458–1467 (2022)
3. Deng, Z.H., Wang, C.D., Huang, L., Lai, J.H., Yu, P.S.: G^3SR: global graph guided session-based recommendation. IEEE Trans. Neural Networks Learn. Syst. **34**(12), 9671–9684 (2023)
4. Fang, T., Zhang, Y., Yang, Y., Wang, C., Chen, L.: Universal prompt tuning for graph neural networks. In: NIPS (2023)
5. Hidasi, B., Karatzoglou, A.: Recurrent neural networks with top-k gains for session-based recommendations. In: CIKM, pp. 843–852 (2018)
6. Hidasi, B., Karatzoglou, A., Baltrunas, L., Tikk, D.: Session-based recommendations with recurrent neural networks. In: ICLR (2016)
7. Jin, W., et al.: Self-supervised learning on graphs: Deep insights and new direction. CoRR abs/2006.10141 (2020)
8. Li, J., Ren, P., Chen, Z., Ren, Z., Lian, T., Ma, J.: Neural attentive session-based recommendation. In: CIKM, pp. 1419–1428 (2017)
9. Liu, P., Yuan, W., Fu, J., Jiang, Z., Hayashi, H., Neubig, G.: Pre-train, prompt, and predict: a systematic survey of prompting methods in natural language processing. ACM Comput. Surv. **55**, 195:1–195:35 (2023)
10. Liu, Z., Yu, X., Fang, Y., Zhang, X.: Graphprompt: unifying pre-training and downstream tasks for graph neural networks. In: WWW, pp. 417–428 (2023)
11. Lu, Z., Yu, Q., Li, X., Li, X., Yang, Q.: Learning weight signed network embedding with graph neural networks. Data Sci. Eng. **8**(1), 36–46 (2023)
12. Qiu, R., Li, J., Huang, Z., Yin, H.: Rethinking the item order in session-based recommendation with graph neural networks. In: CIKM, pp. 579–588 (2019)
13. Quadrana, M., Cremonesi, P., Jannach, D.: Sequence-aware recommender systems. In: UMAP, pp. 373–374 (2018)
14. Rendle, S., Freudenthaler, C., Schmidt-Thieme, L.: Factorizing personalized markov chains for next-basket recommendation. In: WWW, pp. 811–820 (2010)
15. Shani, G., Brafman, R.I., Heckerman, D.: An MDP-based recommender system. In: UAI, pp. 453–460 (2002)
16. Sun, F.Y., Hoffmann, J., Verma, V., Tang, J.: InfoGraph: unsupervised and semi-supervised graph-level representation learning via mutual information maximization. In: ICLR (2020)
17. Sun, M., Zhou, K., He, X., Wang, Y., Wang, X.: GPPT: graph pre-training and prompt tuning to generalize graph neural networks. In: KDD, pp. 1717–1727 (2022)
18. Sun, X., Cheng, H., Li, J., Liu, B., Guan, J.: All in one: multi-task prompting for graph neural networks. In: SIGKDD, pp. 2120–2131 (2023)
19. Velickovic, P., Fedus, W., Hamilton, W.L., Liò, P., Bengio, Y., Hjelm, R.D.: Deep graph infomax. In: ICLR (2019)
20. Wang, S., Cao, L., Wang, Y., Sheng, Q.Z., Orgun, M.A., Lian, D.: A survey on session-based recommender systems. ACM Comput. Surv. **54**(7), 154:1–154:38 (2022)
21. Wu, S., Tang, Y., Zhu, Y., Wang, L., Xie, X., Tan, T.: Session-based recommendation with graph neural networks. In: AAAI, pp. 346–353 (2019)

22. Xi, W., Huang, L., Wang, C., Zheng, Y., Lai, J.: BPAM: recommendation based on BP neural network with attention mechanism. In: Kraus, S. (ed.) Proceedings of the Twenty-Eighth International Joint Conference on Artificial Intelligence, IJCAI 2019, Macao, China, August 10–16, 2019, pp. 3905–3911. ijcai.org (2019)
23. Xia, X., Yin, H., Yu, J., Shao, Y., Cui, L.: Self-supervised graph co-training for session-based recommendation. In: CIKM, pp. 2180–2190 (2021)
24. Xia, X., Yin, H., Yu, J., Wang, Q., Cui, L., Zhang, X.: Self-supervised hypergraph convolutional networks for session-based recommendation. In: AAAI, pp. 4503–4511 (2021)
25. Xiao, S., Zhu, D., Tang, C., Huang, Z.: Combining graph contrastive embedding and multi-head cross-attention transfer for cross-domain recommendation. Data Sci. Eng. **8**(3), 247–262 (2023)
26. Xu, C., et al.: Graph contextualized self-attention network for session-based recommendation. In: IJCAI, pp. 3940–3946 (2019)
27. You, Y., Chen, T., Sui, Y., Chen, T., Wang, Z., Shen, Y.: Graph contrastive learning with augmentations. In: NIPS (2020)
28. Zhang, X., et al.: Price DOES matter!: Modeling price and interest preferences in session-based recommendation. In: SIGIR, pp. 1684–1693 (2022)

Guiding Graph Learning with Denoised Modality for Multi-modal Recommendation

Yuexian Wang, Wenze Ma, Yanmin Zhu(✉), Chunyang Wang, Zhaobo Wang, Feilong Tang, and Jiadi Yu

Shanghai Jiao Tong University, Shanghai 201109, China
{Wangyx1027,mawenze991226,yzhu,wangchy, w19990112,tang-fl,jiadiyu}@sjtu.edu.cn

Abstract. Multi-modal recommendation improves the recommendation accuracy by leveraging various modalities (*e.g.*, visual, textual, and acoustic) of rich item content. However, most existing studies overlook that modality features can be noisy for recommendation. Recently, several graph-based methods have attempted to alleviate the noise issue via graph structure learning, but they roughly focus on edge denoising while neglecting node denoising. This will lead to the following circumstances: 1) noisy factors at the node level and 2) insufficient edge-level denoising. To address the limitations, we propose a **D**enoised **M**odality-guided **G**raph **L**earning paradigm (DMGL), which could jointly and iteratively eliminate both the node-level and edge-level noise for multi-modal recommendation. Meanwhile, masked feature autoencoder and contrastive learning mechanism are introduced to handle intra- and inter-modality node-level noise, respectively. Extensive experiments on real-world datasets demonstrate the superior performance of our proposed model. *The codes are available here.*

Keywords: Multi-modal Recommendation · Graph Structure Learning · Self-supervised Learning

1 Introduction

Multi-modal recommendation systems, which usually leverage various modalities of items, such as visual, textual, and acoustic to enhance recommendation accuracy, have been widely employed in video-sharing and E-commerce platforms. These systems can help alleviate the issues of data sparsity and cold start commonly encountered in traditional recommender systems, which only consider user-item historical interactions.

Many efforts have been made to benefit recommendations from multi-modal features. Previous studies often integrate modality features into collaborative filtering (CF) framework [7,16]. More recently, with the progress of graph-based recommendation [20,31], many researchers start to use the message-passing

M. Onizuka et al. (Eds.): DASFAA 2024, LNCS 14855, pp. 220–235, 2024.
https://doi.org/10.1007/978-981-97-5572-1_14

mechanism of Graph Neural Networks (GNNs) to capture high-order relations between users and items [23,24]. For instance, MMGCN [24] performs modality feature propagation and aggregation for items, while others [19,22,28,32] tend to use modality features to guide the construction of auxiliary graphs, enhancing the learning of user and item representations.

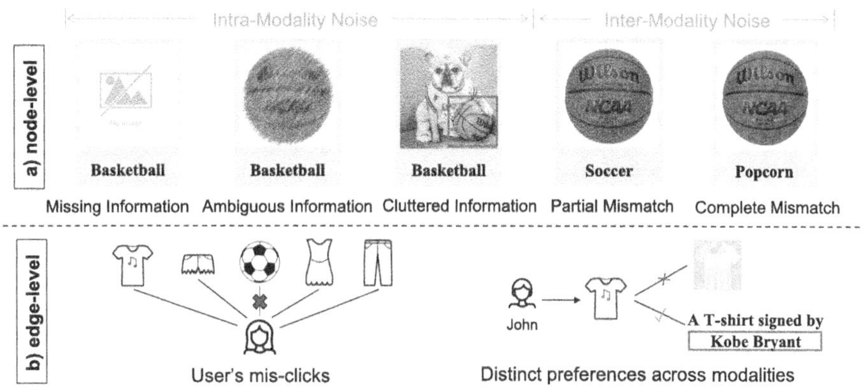

Fig. 1. An illustration of common noise, including a) node-level and b) edge-level. Set "basketball" as the sample item in node-level.

Despite recent graph-based studies have shown effectiveness, they usually overlook the fact that the graph utilized in multi-modal recommendation may be noisy in both node-level and edge-level. *At the node-level*, the noise may commonly exists in both intra- and inter-modality. For intra-modality noise, as illustrated in Fig. 1(a), it can appear in various ways: (1) Missing information: the image's information is missing. (2) Ambiguous information: the image may be damaged or very blurry. (3) Cluttered information: the image contains multiple pieces of information such as pets, pet clothes, basketball, etc., making it hard to distinguish the valuable information it aims to convey. As for inter-modality noise, an error may occur in one or more modalities, leading to (1) partially mismatched information (e.g., basketball and soccer, both being ball sports) or (2) completely mismatched information (e.g., basketball and popcorn) among modalities. *At the edge-level*, taking common user-item interaction graphs as example, the observed user-item interactions may inherently contain noise due to users' mis-clicks. Besides, different users have distinct preferences across various modalities. For instance, as illustrated in Fig. 1(b), John, a fan of Kobe Bryant, may choose to purchase a T-shirt that has textual descriptions with Kobe's signature but is not visually appealing in its image.

Although several studies have attempted to alleviate the noise issues in multi-modal recommendations, most of them focus only on edge denoising while neglecting node denoising. This will lead to the following limitations. *Firstly*, intra- and inter-modality noise factors exist at the node level. Noise in representations can not be efficiently removed after projection (MLP), convolution

(GCN), and fusion, resulting in poor recommendation performance. *Secondly*, current edge-level denoising might be insufficient. The premise for sufficient edge-level denoising is noiseless node representations, which are not satisfied with raw modality features. As shown in Fig. 2, current edge-denoising models usually exploit raw node modality features to rebuild graph structures. For example, GRCN [23] refines the user-item bipartite graph by pruning the noisy edges identified by node modality features, and MMSSL [22] utilizes multimodal node features to reconstruct the modality-aware interaction edges. However, noise in node modality features will be inherited in the subsequent graph structure learning phase, leading to extra dirty edges in denoised graphs.

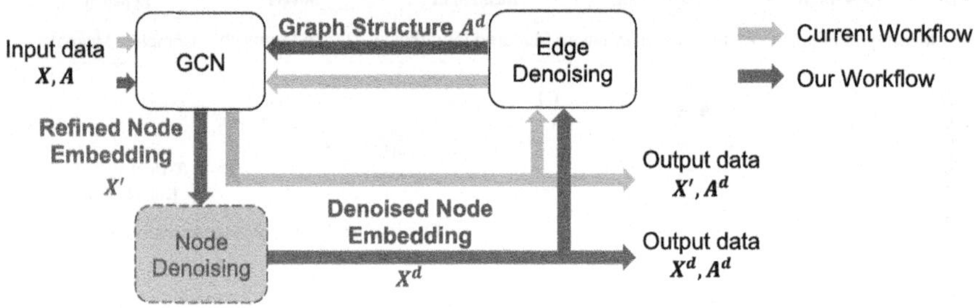

Fig. 2. The workflow of applying denoising framework in graph. Instead of only focusing on edge denoising, DMGL jointly and iteratively denoises both at the node and edge level in multi-modal interaction graphs.

To address the aforementioned limitations, we focus on jointly and iteratively eliminating both node- and edge-level noise in graph representation learning. As shown in Fig. 2, we propose a denoised modality-guided graph learning paradigm (DMGL), aiming at extracting informative and noiseless modality features (node denoising) and constructing modality-guided collaborative graph structure (edge denoising). This paradigm can excavate the latent preference of users among different modalities, leading to better recommendation performance. Specifically, *to address the first limitation of overlooking the intra- and inter-modality noise*, DMGL adopts masked modality feature autoencoder and contrastive learning for self-supervised node representation learning. For intra-modality noise, the autoencoder structure employs a GCN-based graph encoder and a re-mask decoding strategy to reconstruct the masked modality features, so as to minimize the discrepancy between input and reconstructed features on masked nodes. As for inter-modality noise, a contrastive auxiliary task is designed to maximize the shared information among different modalities and discard cross-modality redundancies. *To address the second limitation of insufficient edge-denoising*, instead of using raw modality features, DMGL performs modality-guided structure denoising learning based on denoised modality features. This graph learning mechanism adopts node similarity metric learning [3] to estimate users' preferences and incorporates graph sparsification to formulate an optimized graph for each modality. It is beneficial to perform a two-stage denoising scheme to complement each other at node and edge levels.

In conclusion, the major contributions of our study are as follows:

- We propose to learn informative and noiseless node modality features by combining a masked feature autoencoder to eliminate intra-modality noise and contrastive learning to eliminate inter-modality noise.
- We propose DMGL, which performs node denoising and edge denoising jointly and iteratively for multi-modal recommendation. DMGL further discusses the noise at node-level, in order to obtain more informative and noiseless node representations.
- Extensive experiments have been conducted on three real-world datasets to validate the superior performance of our proposed model, compared with state-of-the-art multi-modal recommendation methods.

2 Related Work

2.1 Multi-modal Recommendation

Many studies have been devoted to enhancing the performance of recommendation systems by integrating multi-modal content. In the early years, multi-modal recommendation models have utilized deep learning techniques to explore users' preferences in combination with Collaborative Filtering (CF) methods. Previous works, such as VBPR [7], extend the BPR [16] method by concatenating visual and style features of items with the item's ID embeddings to make predictions. Besides, attention mechanisms have also been exploited in models such as VECF [2] and MAML [13] to differentiate user's preferences on multimodal information, by focusing on different image regions or textual and visual features.

More recently, Graph Neural Networks (GNNs) have gained attention for their ability to incorporate structural information from user-item interaction graphs and auxiliary graphs, leading to improved user and item representations. MMGCN [24] leverages the message passing mechanism of Graph Convolutional Networks (GCNs) to exploit multi-modal information. It constructs a modality-specific user-item bipartite graph, capturing information from multi-hop neighbors to improve user and item representations. DualGNN [19] considers that user interests may dynamically change over time. It extends MMGCN by introducing a user co-occurrence graph and a preference learning module to capture user preferences for features from different modalities. However, LATTICE [28] argues that the above model implicitly mines semantic information between items, which may result in poor performance. Therefore, it explicitly mines semantic information between items by constructing item-item relation graphs for each modality and fusing them together. FREEDOM [32] proposes freezing the item-item graph for more effective and efficient recommendation. Moreover, inspired by the progress of self-supervised learning, MMSSL [22] proposes a dual-stage self-supervised learning method for modality-aware data augmentation to tackle the label scarcity issue in recommendation. BM3 [33] simplifies the current SSL framework by using a simple latent representation dropout mechanism to generate the target view for contrastive learning without negative samples. In this paper, DMGL utilizes denoised node modality features to guide the reconstruction of modality-aware interaction graphs, mining users' latent preferences among each modality, and resulting in better recommendation results.

2.2 Graph Denoising Network

Research on denoising graph structures can be roughly classified according to whether or not they generate new edges. In edge pruning-only studies, the former work [15,21,25] utilizes attention mechanisms to calculate the attention scores between nodes and their 1-hop neighbors. For edges with lower scores, they are either pruned directly or assigned with a lower weight. The latter work [4,11] samples a sparsified graph following a parameterized distribution, which is learned in a supervised manner. whereas these works only prune edges that are already present in the original graph, the work on graph structure learning [3,10,17,30,34] wants to adaptively learn an optimized graph structure, which not only prunes edges but also adds reliable edges.

In multi-modal recommendation, graph denoising methods have also been widely used. For example, GRCN [23] refines the user-item bipartite graph by pruning the false-positive edges identified by the node modality features. FREEDOM [32] devises a degree-sensitive edge pruning method, to reject the possibly noisy edges. Meanwhile, MMSSL [22] uses the graph structure learning method to generate the modality-aware interaction graph. All of these methods have drawbacks in that they only focus on the noise at the edge-level while overlooking the noise at the node-level, which is important in multi-modal recommendations. In this paper, our proposed model will jointly and iteratively eliminate both the node and edge level noise, to obtain a better recommendation result.

3 Preliminary

3.1 Modality-Aware User-Item Graph

In the multi-modal recommendation scenario, let $\mathcal{U} = \{u_1, u_2, ..., u_{|\mathcal{U}|}\}$, $\mathcal{I} = \{i_1, i_2, ..., i_{|\mathcal{I}|}\}$ denote the set of users and items, respectively, where $|\mathcal{U}|$ is the number of users and $|\mathcal{I}|$ is the number of items. Let $\mathcal{G} = \{\mathcal{V}, \mathcal{A}\}$ denote an undirected user-item interaction graph, where node set $\mathcal{V} = \mathcal{U} \cup \mathcal{I}$, and $\mathcal{A} \in \{0,1\}^{|\mathcal{U}|\times|\mathcal{I}|}$ is the user-item adjacency matrix.

Beyond the interactions, we incorporate the multi-modal information $\bar{F} = \{\bar{F}^1, ..., \bar{F}^{|\mathcal{M}|}\}$, where $|\mathcal{M}|$ is the number of modalities, into the user-item interaction graph $\mathcal{G}$, to generate modality-aware user-item graph $\bar{\mathcal{G}}^m = \{\mathcal{V}, \mathcal{A}, \bar{F}^m\}$. For simplicity, we use $m \in \mathcal{M} = \{v, t, a\}$ as the modality indicator, where v, t, and a correspond to the visual, textual, and acoustic modalities, respectively. The raw feature of item i with modality m is denoted by $\bar{f}_i^m \in \mathbb{R}^{d_m}$.

3.2 Task Formulation

Given the target user u, the generated modality-aware user-item graph $\bar{\mathcal{G}}^m$, our task of multi-modal recommendation is to learn a function recommend a top-k item list that user u will be most likely to interact with.

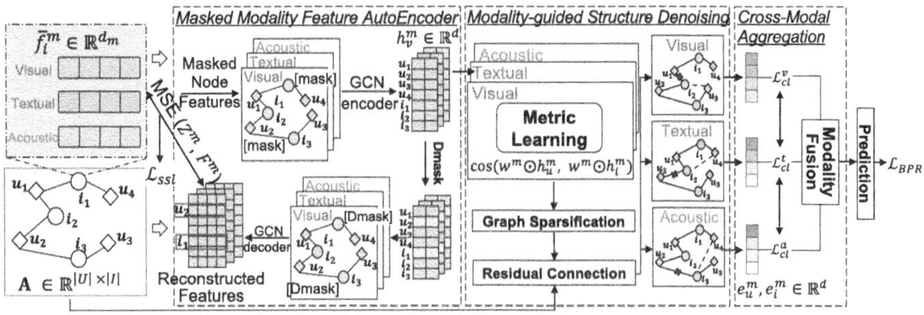

Fig. 3. The overall architecture of DMGL can be divided into three parts. The first part is the node-denoising module to generate the denoised modality features H^m. The second part demonstrates the modality-guided edge-denoising structure, and the third part is the cross-modal aggregation with the contrastive learning method to generate the final representations.

4 Methodology

In this section, we elaborate on our denoised modality-guided graph learning model, which encompasses three components as illustrated in Fig. 3: (a) Masked Modality Feature AutoEncoder, (b) Modality-guided Structure Denoising, and (c) Cross-Modal Aggregation.

4.1 Masked Modality Feature AutoEncoder

In order to address the intra-modality noise problems brought by multi-modal features, we utilize an autoencoder that is composed of an encoder and a decoder to learn modality-specific representation without supervision. The core idea of this method is centered around the reconstruction of masked node features, and in order to enhance this approach, we also introduce a re-mask decoding strategy in the decoder scheme.

Masked Node Features. The autoencoder usually corrupts the input data on purpose to denoise the node features in graphs. Typically, the idea of employing masking as a form of corruption in autoencoders has been widely applied in previous works [6,12]. Inspired by their success, we propose to adopt a masking strategy for modality features reconstruction.

$\bar{\mathcal{G}}^m = \{\mathcal{V}, \mathcal{A}, \bar{F}^m\}$ is the raw modality-aware interaction graph, where $\bar{F}^m \in \mathbb{R}^{|\mathcal{I}|\times d_m}$ is the raw modality features only for items, and $\bar{f}_i^m$ is the i-th row of $\bar{F}^m$. By integrating collaborative effects into $\bar{f}_i^m$, we obtain high-order user and item modality features f_u^m, f_i^m (vectors from F^m):

$$f_u^m = \sum_{i' \in \mathcal{N}_u} \frac{1}{\sqrt{|\mathcal{N}_u|}} \bar{f}_{i'}^m; \qquad f_i^m = \sum_{u' \in \mathcal{N}_i} \frac{1}{\sqrt{|\mathcal{N}_i|}} f_{u'}^m \tag{1}$$

where $\mathcal{N}_u$ and $\mathcal{N}_i$ denote the neighborhood set of user $u \in \mathcal{U}$ and item $i \in \mathcal{I}$ in the user-item interaction graph $\mathcal{G}$. $|\mathcal{N}_u|$ and $|\mathcal{N}_i|$ is the number of them, respectively.

Then, we perform a masking strategy on the input modality-aware user-item graph $\mathcal{G}^m = \{\mathcal{V}, \mathcal{A}, F^m\}$. Formally, we sample a subset of nodes $\tilde{\mathcal{V}}^m \in \mathcal{V}$, and mask each of their modality features f_v^m, $v \in \tilde{\mathcal{V}}^m$ with a pre-defined mask token [MASK], *i.e.*, a learnable vector $x_m \in \mathbb{R}^{d_m}$. Also, in order to mitigate the discrepancy caused by the absence of the mask token during inference [26], we followed the approach described in BERT [12], taking 10% of the masked nodes with another random token $x'_m \in \mathbb{R}^{d_m}$. Thus, the modality-aware masked node feature $\tilde{f}_v^m$ for $v \in \mathcal{V}$ in $\tilde{F}^m$ can be defined as:

$$\tilde{f}_v^m = \begin{cases} x_m \; or \; x'_m, & v \in \tilde{\mathcal{V}}^m \\ f_v^m, & v \notin \tilde{\mathcal{V}}^m \end{cases} \tag{2}$$

Following this strategy, the masked modality-aware user-item graph $\tilde{\mathcal{G}}^m = \{\mathcal{V}, \mathcal{A}, \tilde{F}^m\}$ is constructed, where $\tilde{F}^m$ is the masked feature matrix.

GCN-Based Graph Encoder. Historical interactions can effectively depict the representations of masked nodes. Specifically, historical interactions of users can provide insights into their interests and identify similarities in behavior with other users. The user group associated with an item can also offer supplementary information to its content. Therefore, for a masked node in the masked modality-aware user-item graph $\tilde{\mathcal{G}}^m$, we employ the message-passing mechanism of GCN to encode the input data as follows:

$$H^m = g_{enc}(\mathcal{V}, \mathcal{A}, \tilde{F}^m) \tag{3}$$

where $g_{enc}(\cdot)$ denotes the graph encoder, which is GCN in DMGL. Besides, following the research in [9], the residual connection is incorporated to address the issue of gradient vanishing. Here, the representations from all hidden layers in GCN are combined, resulting in the ultimate embedding $h_v^m \in \mathbb{R}^d$, $d < d_m$, for the m-th modality of node v in H^m. By employing this graph encoder based on GCN, the proposed modal can encode the structural information of the observed interaction graph into node modality-specific features to get the modality-specific latent space representations H^m and utilize it for recommendation as well as reconstruction of the masked nodes.

Decoder for Reconstruction. After encoding the input node features for each modality, the decoder aims to map the latent space representations H^m back to the input modality features F^m. To facilitate the encoder in learning a high-level latent representation, rather than having the latent representation become nearly identical to the input feature, our modal resorts to a more expressive single-layer GCN rather than MLP as its decoder, leveraging information from a set of nodes instead of only considering the node itself.

Furthermore, to promote the encoder's learning of compressed representations, we introduce a re-masking decoding strategy for processing latent space code H^m used for decoding. This technique involves replacing the masked node in H^m with another mask token [DMASK], *i.e.*, a random vector $\ddot{x}_m \in \mathbb{R}^d$. Specifically, the re-mask modality-specific node latent representation $\tilde{h}_v^m$ for $v \in \mathcal{V}$ in re-mask latent space code $\tilde{H}^m$ can be defined as:

$$\tilde{h}_v^m = \begin{cases} \ddot{x}_m, & v \in \tilde{\mathcal{V}}^m \\ h_v^m, & v \notin \tilde{\mathcal{V}}^m \end{cases} \tag{4}$$

By utilizing the re-mask strategy and the aforementioned GCN decoder, a masked node is compelled to reconstruct its input feature by using the adjacent unmasked latent representations. Our goal is to make the reconstructed features as similar as possible to the input features and use the mean square error (MSE) as the criterion. Formally, given the input m-th modality feature F^m and reconstructed output $Z^m = g_{dec}(\mathcal{V}, \mathcal{A}, \tilde{H}^m)$, where g_{dec} is the graph decoder, the loss function can be defined as:

$$\mathcal{L}_{recon} = \sum_{m \in \mathcal{M}} \mathcal{L}_{recon}^m \qquad \mathcal{L}_{recon}^m = \frac{1}{|\tilde{\mathcal{V}}^m|} \sum_{v \in \tilde{\mathcal{V}}^m} ||z_v^m - f_v^m||^2 \tag{5}$$

where z_v^m, f_v^m is the v-th row of Z^m, F^m. Similar to BERT [12], the loss is computed only on masked node, and is calculated independently among modalities.

4.2 Modality-Guided Structure Denoising Learning

After obtaining denoising modality features h_v^m generated by the graph encoder, this section endeavors to facilitate modality-guided graph structure learning over observed real user-item interactions, so as to construct modality-specific collaborative relations for each modality. The objective is to eliminate noise at the edge level stemming not only from users' mis-clicks in the inter-modality context but also from their fine-grained preference at the intra-modality level.

Similarity Metric Learning. To realize such functionalities, we adopt similarity metric learning [3] to estimate the probability that user u likes item i within a given multi-modal context. Specifically, for each modality, the modality-guided collaborative relationships are conducted by learning the preference score matrix $S^m \in \mathbb{R}^{|\mathcal{U}| \times |\mathcal{I}|}$ with multi-modal representations:

$$S^m[u, i] = cos(\mathbf{w}^m \odot h_u^m, \mathbf{w}^m \odot h_i^m) = \frac{(\mathbf{w}^m \odot h_u^m)^{\mathrm{T}}(\mathbf{w}^m \odot h_i^m)}{||\mathbf{w}^m \odot h_u^m||_2 \cdot ||\mathbf{w}^m \odot h_i^m||_2} \tag{6}$$

where $\odot$ denotes the Hadamard product, h_u^m and h_i^m are the noise-free m-th modality features for user $u \in \mathcal{U}$ and item $i \in \mathcal{I}$ generated by Graph Encoder. $\mathbf{w}^m \in \mathbb{R}^d$ is the shared learnable weight matrix for users and items.

Graph Sparsification and Residual Connection. It's obvious that the learned modality-specific user-item similarity score matrix S^m ranges between $[-1, 1]$, which is not only computationally expensive but also might introduce noise [3]. Thus, we conduct top-k sparsification [1] on the dense matrix, where for each user u, we only keep edges with the top-k similarity scores:

$$\tilde{S}^m[u,i] = \begin{cases} S^m[u,i], & S^m[u,i] \in \text{top-}k(S^m[u,:]) \\ 0, & otherwise \end{cases} \tag{7}$$

where $S^m[u,:]$ denotes the score of whole items for user u, and $\tilde{S}^m$ is the resulting sparsified modality-guided collaborative relation matrix.

However, although the initial graph could be noisy, it typically still carries rich and useful information regarding true graph topology [3]. Thus, we use the learned denoised modality-aware relation matrix $\tilde{S}^m$ as supplementary to the original graph topology $\mathcal{A}$ to formulate an optimized graph $\mathcal{A}^m$ for modality m. Formally, set the weight as α, we propose to form a residual connection between the original graph and the learned graph as:

$$\mathcal{A}^m = \alpha\mathcal{A} + (1-\alpha)\tilde{S}^m \tag{8}$$

and we get the denoised modality-guided user-item graphs $\hat{\mathcal{G}}^m = \{\mathcal{V}, \mathcal{A}^m, H^m\}$.

4.3 Cross-Modal Contrastive Aggregation

Cross-Modal Refined Embeddings. To capture the modality-specific semantics of users and items, we applied a light-weight Graph Convolutional Network (lightGCN) [8] to propagate and aggregate information on each denoised modality-aware interaction graph $\hat{\mathcal{G}}^m$:

$$e_u^{m,l+1} = \sum_{i' \in \mathcal{N}_u^m} \frac{1}{\sqrt{|\mathcal{N}_u^m|}} e_{i'}^{m,l}; \qquad e_i^{m,l+1} = \sum_{u' \in \mathcal{N}_i^m} \frac{1}{\sqrt{|\mathcal{N}_i^m|}} e_{u'}^{m,l} \tag{9}$$

where $\mathcal{N}_u^m$ and $\mathcal{N}_i^m$ represent the neighborhood set of user $u \in \mathcal{U}$ and item $i \in \mathcal{I}$ in the denoised modality-aware user-item graph $\hat{\mathcal{G}}^m$. $e_u^{m,l}$ and $e_i^{m,l}$ denote the l-th layer user and item representations of modality m, $e_u^{m,0}$ and $e_i^{m,0}$ are Xavier-initialized corresponding vectors, all modalities share the same initial embeddings $e_u, e_i \in \mathbb{R}^d$. We stack L convolutional layers on each modality-aware interaction graph and set the last layer embeddings $e_u^{m,L}$ and $e_i^{m,L}$ as the final representations $e_u^m, e_i^m \in \mathbb{R}^d$ for the multimodal view.

Then, we aggregate the modality-specific representations by mean-pooling, *i.e.*, $\hat{e}_v = \sum_{m \in \mathcal{M}}^{|\mathcal{M}|} e_v^m / |\mathcal{M}|$ to get both the user's and item's refined representations

$\hat{e}_u, \hat{e}_i \in \mathbb{R}^d$. To better utilize the modality information, we combine the denoised modality features with the refined representations to generate the final user's and items' representations served for recommendation:

$$\ddot{e}_v = \hat{e}_v + \beta \sum_{m \in \mathcal{M}} \frac{h_v^m}{||h_v^m||_2} \tag{10}$$

where β is the aggregation weight, $\ddot{e}_v$ is the final representation for a node v. The value scale difference between two type of views is alleviate by normalization.

Contrastive Auxiliary Task. We design a contrastive auxiliary task which proposes the auxiliary self-supervised signals [18] to address the issue of intra-modality noise at the node level. Specifically, we adopt contrastive loss, *i.e.*, InfoNCE loss [5], to maximize the shared information among user's modality features e_u^m and the fused multi-modal representations $\hat{e}_u$:

$$\mathcal{L}_{CL} = -\sum_{u \in \mathcal{U}} \{ \sum_{m \in \mathcal{M}} log \frac{exp(\phi(e_u^m, \hat{e}_u)/\tau)}{\sum_{u' \in \mathcal{U}} (exp(\phi(e_u^m, \hat{e}_{u'})/\tau) + exp(\phi(e_u^m, e_{u'}^m)/\tau))} \} \tag{11}$$

$$\phi(e_u^m, \hat{e}_u) = \frac{(e_u^m)^{\mathrm{T}} \cdot \hat{e}_u}{||e_u^m||_2 \cdot ||\hat{e}_u||_2} \tag{12}$$

where $\phi(\cdot, \cdot)$ is the critic function implemented by simple cosine similarity, $\tau \in \mathbb{R}$ is the temperature coefficient, which is 0.5 in DMGL.

4.4 Prediction and Optimization

Once we obtain the final user and item representations $\ddot{e}_u, \ddot{e}_i \in \mathbb{R}^d$, we adopt the Bayesian Personalized Ranking (BPR) loss [16] to calculate the pair-wise ranking, aiming to maximize the difference in predicted scores between positive and negative samples:

$$\mathcal{L}_{BPR} = -\sum_{u \in \mathcal{U}} \sum_{i \in \mathcal{A}_u} \sum_{j \notin \mathcal{A}_u} log(\sigma(\hat{y}_{ui} - \hat{y}_{uj})) \tag{13}$$

where $\sigma(\cdot)$ is the sigmoid function, $\mathcal{A}_u$ indicates the observed item set interacted with user u, $i, j \in \mathcal{I}$ denote the sampled positive and negative items. The prediction score between user u and item i is $\hat{y}_{ui} = \ddot{e}_u^{\mathrm{T}} \cdot \ddot{e}_i$.

Overall, to optimize our framework, we leverage a two-stage multi-task learning scheme involving i) the first stage: self-supervised node denoising task $\mathcal{L}_{recon}$, ii) the second stage: cross-modal contrastive augmentation task $\mathcal{L}_{CL}$, and user-item interaction prediction task $\mathcal{L}_{BPR}$. The objectives become:

$$\begin{cases} \textbf{First stage}: \mathcal{L} = \mathcal{L}_{recon} \\ \textbf{Second stage}: \mathcal{L} = \mathcal{L}_{BPR} + \lambda_1 \cdot \mathcal{L}_{CL} + \lambda_2 \cdot ||\Theta||^2 \end{cases} \tag{14}$$

where λ_1, λ_2 are hyper-parameters for controlling the effect of each task in the second stage, and $||\Theta||^2$ is the weight-decay regularization against over-fitting.

5 Experiments

5.1 Experimental Setup

Datasets. Following previous studies [22,28], we evaluate DMGL by conducting extensive experiments on three real-world datasets: Baby[1], Allrecipes[2], and Tiktok[3]. The characteristics of these datasets are summarized in Table 1. All three datasets contain visual and textual modalities, while Tiktok also contains acoustic modality.

Table 1. Dataset statistics with preprocessed multi-modal feature dimensions of visual (V), textual (T), and acoustic (A).

Dataset	#Users	#Items	#Interactions	Sparsity	V	T	A
Baby	19445	7050	139110	99.89%	4096	1048	–
Allrecipes	19805	10067	58922	99.97%	2048	20	–
Tiktok	9319	6710	59541	99.90%	128	768	128

Evaluation Protocols. We randomly split the dataset into training, validation, and testing sets with 8:1:1 ratio, and create the training triples (user, pos_item, neg_item) based on random negative sampling. We adopt three widely used Top-K ranking metrics: Recall (R@K), Precision (P@K), and Normalized Discounted Cumulative Gain (NDCG@K) to measure their performance. The average scores of all users in test sets are reported.

Baselines. To demonstrate the effectiveness of our proposed model, we compare it with several state-of-the-art baselines, including popular graph-based collaborative filtering model and multi-modal recommendation models. More details about baselines are listed below:

i) graph-based collaborative filtering model
- **LightGCN** [8]: This method simplifies the message passing for GCN to make it more suitable for recommendation.

ii) multi-modal recommendation models
- **LightGCN-*M***: This method utilizes LightGCN as backbone and fuses the multi-modal item features during the message passing mechanism.
- **MMGCN** [24]: This method leverages GCN to propagate the modality-specific features and combines them for micro-video recommendation.

1 http://jmcauley.ucsd.edu/data/amazon/links.html.
2 https://www.allrecipes.com/.
3 https://github.com/HKUDS/MMSSL.

- **GRCN** [23]: This method improves MMGCN by refining and denoising the user-item bipartite graph with the removal of false-positive edges.
- **LATTICE** [28]: This method constructs an item-item homogeneous graph for each modality to mine the latent relationships among items. Graph convolution networks are performed on both the item-item graph and user-item graph to learn user and item representations.
- **MIRCO** [29]: This method based on LATTICE, adopts a contrastive auxiliary task to fuse features after performing graph convolution.
- **MMGCL** [27]: This method devises edge dropout and modality masking to enhance the contrastive multi-modal representation learning.
- **BM3** [33]: This method abandons the commonly used negative sampling process, and designs a simple dropout mechanism instead of graph augmentation to generate different views for contrastive learning.
- **MMSSL** [22]: This method leverages adversarial self-supervised learning method to construct modality-aware interactions and uses cross-view contrastive signals to generate better user and item representations.

Hyperparameter Settings. All models are implemented in Pytorch. For fair comparison, we fix the batch size of all models to be 1024, the embedding dimensionality to be 64 and adopting AdamW [14] as the optimizer with the learning rate of $5.5e-3$. We adopt grid search to find the best settings, including the contrastive learning coefficient λ_1 in $\{0.1, 0.01, 0.001\}$, and the decay of regularization term λ_2 in $\{1e\text{-}3, 1e\text{-}5, 1e\text{-}7\}$.

Table 2. Performance comparison of DMGL with different baselines in terms of Recall@20 (R@20), Precision@20 (P@20), and NDCG@20 (N@20).

Models	Baby			Allrecipes			Tiktok		
	R@20	P@20	N@20	R@20	P@20	N@20	R@20	P@20	N@20
LightGCN	0.2257	0.0128	0.1057	0.1465	0.0073	0.0543	0.2232	0.0112	0.0977
LightGCN-*M*	0.2306	0.0131	0.1091	0.1511	0.0076	0.0570	0.2328	0.0116	0.1024
MMGCN	0.2050	0.0116	0.0971	0.1753	0.0087	0.0682	0.2361	0.0118	0.1075
GRCN	0.2407	0.0137	0.1190	0.1811	0.0091	0.0690	0.2383	0.0119	0.1077
LATTICE	0.2499	0.0142	0.1203	0.1804	0.0090	0.0689	0.2472	0.0124	0.1103
MIRCO	0.2632	0.0146	0.1282	0.1778	0.0088	0.0685	0.2397	0.0120	0.1035
MMGCL	0.2376	0.0135	0.1127	0.1799	0.0090	0.0688	0.2354	0.0118	0.1034
BM3	0.2611	0.0143	0.1261	<u>0.1890</u>	<u>0.0095</u>	0.0726	0.2603	0.0130	0.1199
MMSSL	<u>0.2642</u>	<u>0.0147</u>	<u>0.1287</u>	0.1874	0.0093	<u>0.0730</u>	<u>0.2620</u>	<u>0.0131</u>	<u>0.1203</u>
DMGL	**0.2727**	**0.0154**	**0.1337**	**0.2034**	**0.0102**	**0.0797**	**0.2807**	**0.0141**	**0.1272**
improv.	3.21%	3.35%	3.89%	7.62%	7.36%	9.18%	7.14%	6.87%	5.73%

5.2 Overall Performance

The comparison of recommendation accuracy among different methods on all three datasets is reported in Table 2, where the performance of DMGL is marked in **boldface** and the best-performed baseline is underlined. For all of these metrics, the higher the value, the better the performance. From the results, we have the following observations:

- DMGL surpasses all baselines across datasets, demonstrating its effectiveness. These improvements are attributed to the joint and iterative denoising components both on nodes and edges in graphs. As shown in Table 2, compared with MMGCN, GRCN prunes the noisy edges, achieving better performance. On the other hand, MMSSL's graph reconstruction also makes it achieves the best performance, underscoring denoising's value in multimodal recommendation. However, none of the above methods consider node-level noise in the multi-modal features, while DMGL considers dual denoising at the node and edge level, demonstrating competitive results.
- The experimental results show that most of the methods that guide the construction of auxiliary graphs (*e.g.*, LATTICE) tend to yield better recommendation results compared to propagating and fusing multi-modal features on the original graph (*e.g.*, MMGCN). One possible reason is that when simply summing up all modal features, identifying significant contributions of each modality could be a challenge.

5.3 Ablation Study

To verify the effectiveness of each key component of DMGL, *i.e.*, node denoising, edge denoising, and contrastive learning, we conduct ablation studies on each dataset. The results are shown in Table 3 in terms of R@20 and N@20.

DMGL$_{w/o-\mathcal{V}}$: This variant disables the node denoising module, using raw multi-modal features to denoise the interactions and learn representations for recommendation tasks. Compared with this variant, DMGL has relative improvements which demonstrate the effectiveness of node denoising.

DMGL$_{w/o-\mathcal{A}}$: This variant disables the edge denoising module, using the observed user-item interaction graph for each modality. The observed improvements of DMGL demonstrate its effectiveness.

DMGL$_{w/o-CL}$: In this variant, the representations of users and items are learned without considering the inter-modality noise. Experimental results show that contrastive learning plays a critical role in DMGL.

Furthermore, we can see that by jointly performing node-denoising, edge-denoising, and contrastive learning, our proposed model could achieve the best. In conclusion, the joint denoising is beneficial for recommendation.

5.4 Hyper-parameter Analysis

To guide the selection of hyper-parameters of DMGL, we further perform a hyper-parameter study with regard to the recommendation accuracy. Due to

Table 3. Ablation studies on key components of DMGL.

Datasets	Metrics	$\text{DMGL}_{w/o-\mathcal{V}}$	$\text{DMGL}_{w/o-\mathcal{A}}$	$\text{DMGL}_{w/o-CL}$	DMGL
Baby	R@20	0.2704	0.2635	0.2711	**0.2727**
	N@20	0.1330	0.1305	0.1334	**0.1337**
Allrecipes	R@20	0.1971	0.1934	0.2010	**0.2032**
	N@20	0.0771	0.0757	0.0778	**0.0783**
Tiktok	R@20	0.2747	0.2705	0.2778	**0.2807**
	N@20	0.1259	0.1249	0.1265	**0.1272**

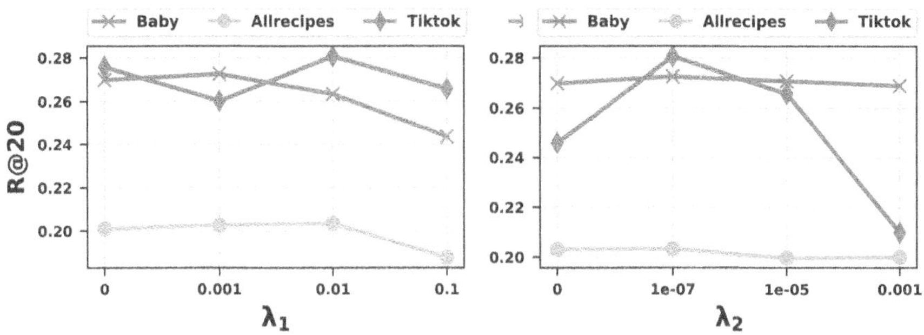

Fig. 4. Effect of hyper-parameters.

the space limit, we only demonstrate the results of two key hyper-parameters, the contrastive learning coefficient λ_1, and the regularization coefficient factor λ_2. Figure 4 displays the results in terms of Recall@20.

We tune λ_1 in $\{0, 0.001, 0.01, 0.1\}$ and vary λ_2 in $\{0, \text{1e-7}, \text{1e-5}, \text{1e-3}\}$. Note that $\lambda_1 = 0$ and $\lambda_2 = 0$ correspond to the version that disables contrastive learning and L_2 regularization, respectively. We find that DMGL performs the best when $\lambda_1 = 0.001$ in dataset Baby, and $\lambda_1 = 0.1$ in Allrecipes and Tiktok. Compared to Baby, Allrecipes and Tiktok have more noise across modalities. Besides, when $\lambda_2 = 1e-7$, DMGL achieves the best under all datasets. The performance of DMGL is more sensitive to the settings of regularization coefficient on small datasets (*i.e.*, Tiktok).

6 Conclusion

In this paper, we propose a denoised modality-guided graph learning module (denoted as DMGL), which jointly and iteratively eliminates the noise in both node and edge levels for multi-modal recommendation. For node denoising, masked feature autoencoder and contrastive learning mechanism are employed to address the node noise issues in intra- and inter-modality, respectively. Extensive experiments are conducted to demonstrate the effectiveness. For future works, we can explore other ways for graph denoising.

Acknowledgments. This research is supported by National Science Foundation of China(No. 62072304), Shanghai Municipal Science and Technology Commission(No. 21511104700), the Shanghai East Talents Program, the Oceanic Interdisciplinary Program of Shanghai Jiao Tong University(No. SL2020MS032), and Zhejiang Aoxin Co. Ltd.

References

1. Chen, J., Fang, H.r., Saad, Y.: Fast approximate knn graph construction for high dimensional data via recursive lanczos bisection. J. Mach. Learn. Res. **10**(9) (2009)
2. Chen, X., et al.: Personalized fashion recommendation with visual explanations based on multimodal attention network: Towards visually explainable recommendation. In: Proceedings of the 42nd International ACM SIGIR Conference on Research and Development in Information Retrieval, pp. 765–774 (2019)
3. Chen, Y., Wu, L., Zaki, M.: Iterative deep graph learning for graph neural networks: Better and robust node embeddings. Adv. Neural. Inf. Process. Syst. **33**, 19314–19326 (2020)
4. Franceschi, L., Niepert, M., Pontil, M., He, X.: Learning discrete structures for graph neural networks. In: International Conference on Machine Learning (2019)
5. Gutmann, M., Hyvärinen, A.: Noise-contrastive estimation: a new estimation principle for unnormalized statistical models. In: Proceedings of the Thirteenth International Conference on Artificial Intelligence and Statistics, pp. 297–304 (2010)
6. He, K., Chen, X., Xie, S., Li, Y., Dollár, P., Girshick, R.: Masked autoencoders are scalable vision learners. In: Proceedings of the IEEE/CVF Conference on Computer Vision and Pattern Recognition, pp. 16000–16009 (2022)
7. He, R., McAuley, J.: VBPR: visual Bayesian personalized ranking from implicit feedback. In: Proceedings of the AAAI Conference on Artificial Intelligence (2016)
8. He, X., Deng, K., Wang, X., Li, Y., Zhang, Y., Wang, M.: Lightgcn: simplifying and powering graph convolution network for recommendation. In: Proceedings of the 43rd International ACM SIGIR Conference on Research and Development in Information Retrieval, pp. 639–648 (2020)
9. Jian, S., Kaiming, H., Shaoqing, R., Xiangyu, Z.: Deep residual learning for image recognition. In: IEEE Conference on Computer Vision and Pattern Recognition, pp. 770–778 (2016)
10. Jin, W., Ma, Y., Liu, X., Tang, X., Wang, S., Tang, J.: Graph structure learning for robust graph neural networks. In: Proceedings of the 26th ACM SIGKDD International Conference on Knowledge Discovery Data Mining, pp. 66–74 (2020)mi
11. Kazi, A., Cosmo, L., Ahmadi, S.A., Navab, N., Bronstein, M.M.: Differentiable graph module (dgm) for graph convolutional networks. IEEE Transactions on Pattern Analysis and Machine Intelligence, pp. 1606–1617 (2022)
12. Kenton, J.D.M.W.C., Toutanova, L.K.: Bert: Pre-training of deep bidirectional transformers for language understanding. In: Proceedings of naacL-HLT (2019)
13. Liu, F., Cheng, Z., Sun, C., Wang, Y., Nie, L., Kankanhalli, M.: User diverse preference modeling by multimodal attentive metric learning. In: Proceedings of the 27th ACM International Conference on Multimedia, pp. 1526–1534 (2019)
14. Loshchilov, I., Hutter, F.: Decoupled weight decay regularization. arXiv preprint arXiv:1711.05101 (2017)
15. Luo, D., et al.: Learning to drop: Robust graph neural network via topological denoising. In: Proceedings of the 14th ACM International Conference on Web Search and Data Mining (2021)

16. Rendle, S., Freudenthaler, C., Gantner, Z., Schmidt-Thieme, L.: BPR: Bayesian personalized ranking from implicit feedback. arXiv preprint arXiv:1205.2618 (2012)
17. Sun, Q., et al.: Graph structure learning with variational information bottleneck. In: Proceedings of the AAAI Conference on Artificial Intelligence, pp. 4165–4174 (2022)
18. Tian, Y., Krishnan, D., Isola, P.: Contrastive multiview coding. In: Computer Vision–ECCV 2020: 16th European Conference, Glasgow, UK, August 23–28, 2020, Proceedings, Part XI 16, pp. 776–794 (2020)
19. Wang, Q., Wei, Y., Yin, J., Wu, J., Song, X., Nie, L.: Dualgnn: dual graph neural network for multimedia recommendation. IEEE Transactions on Multimedia (2021)
20. Wang, S., et al.: Graph learning based recommender systems: a review. arXiv preprint arXiv:2105.06339 (2021)
21. Wang, W., Feng, F., He, X., Nie, L., Chua, T.S.: Denoising implicit feedback for recommendation. In: Proceedings of the 14th ACM International Conference on Web Search and Data Mining, pp. 373–381 (2021)
22. Wei, W., Huang, C., Xia, L., Zhang, C.: Multi-modal self-supervised learning for recommendation. In: Proceedings of the ACM Web Conference 2023 (2023)
23. Wei, Y., Wang, X., Nie, L., He, X., Chua, T.S.: Graph-refined convolutional network for multimedia recommendation with implicit feedback. In: Proceedings of the 28th ACM International Conference on Multimedia, pp. 3541–3549 (2020)
24. Wei, Y., Wang, X., Nie, L., He, X., Hong, R., Chua, T.S.: MMGCN: multi-modal graph convolution network for personalized recommendation of micro-video. In: Proceedings of the 27th ACM International Conference on Multimedia (2019)
25. Xu, Z., Du, B., Tong, H.: Graph sanitation with application to node classification. In: Proceedings of the ACM Web Conference 2022, pp. 1136–1147 (2022)
26. Yang, Z., Dai, Z., Yang, Y., Carbonell, J., Salakhutdinov, R.R., Le, Q.V.: Xlnet: generalized autoregressive pretraining for language understanding. In: Advances in Neural Information Processing Systems vol. 32D (2019)
27. Yi, Z., Wang, X., Ounis, I., Macdonald, C.: Multi-modal graph contrastive learning for micro-video recommendation. In: Proceedings of the 45th International ACM SIGIR Conference on Research and Development in Information Retrieval, pp. 1807–1811 (2022)
28. Zhang, J., Zhu, Y., Liu, Q., Wu, S., Wang, S., Wang, L.: Mining latent structures for multimedia recommendation. In: Proceedings of the 29th ACM International Conference on Multimedia, pp. 3872–3880 (2021)
29. Zhang, J., Zhu, Y., Liu, Q., Zhang, M., Wu, S., Wang, L.: Latent structure mining with contrastive modality fusion for multimedia recommendation. IEEE Transactions on Knowledge and Data Engineering (2022)
30. Zhao, J., Wen, Q., Ju, M., Zhang, C., Ye, Y.: Self-supervised graph structure refinement for graph neural networks. In: Proceedings of the Sixteenth ACM International Conference on Web Search and Data Mining, pp. 159–167 (2023)
31. Zhou, X., Lin, D., Liu, Y., Miao, C.: Layer-refined graph convolutional networks for recommendation. In: 2023 IEEE 39th International Conference on Data Engineering (ICDE), pp. 1247–1259 (2023)
32. Zhou, X., Shen, Z.: A tale of two graphs: Freezing and denoising graph structures for multimodal recommendation. In: Proceedings of the 31st ACM International Conference on Multimedia, pp. 935–943 (2023)
33. Zhou, X., et al.: Bootstrap latent representations for multi-modal recommendation. In: Proceedings of the ACM Web Conference 2023, pp. 845–854 (2023)
34. Zhu, Y., et al.: A survey on graph structure learning: Progress and opportunities. arXiv preprint arXiv:2103.03036 (2021)

Enhancing Multi-view Contrastive Learning for Graph Anomaly Detection

Qingcheng Lu[1], Nannan Wu[1(✉)], Yiming Zhao[1], Wenjun Wang[1], and Quannan Zu[2(✉)]

[1] College of Intelligence and Computing, Tianjin University, Tianjin , China
[2] Department of Management and Economics, Tianjin University, Tianjin , China
{lqc2022,nannan.wu,zhaoym,wjwang,quannan}@tju.edu.cn

Abstract. Graph anomaly detection (GAD) has garnered considerable attention owing to its wide-ranging applications in real-world scenarios, including financial network and cybersecurity. Due to the difficulty associated with learning intricate connectivity and high-dimensional attributes in graphs, GAD task poses significant challenges. Existing deep learning methods extract local information between the target node and its neighbors for GAD. Nevertheless, they neglect the global information in the graph which results in suboptimal performance. In this paper, we propose a self-supervised learning framework integrating contrastive learning module and attribute reconstruction module, which can effectively capture both local and global information for GAD. Specifically, we regard original graph as local view, and global view is generated based on this local view. In contrastive learning module, target node and subgraph are sampled from both views to construct multi-scale contrast combinations. With intra- and intre- view contrastive strategy, we can effectively capture both local and global information of anomaly nodes. In attribute reconstruction module, masked graph auto-encoder is employed to further capture anomaly information within the attribute space. Extensive experiments are conducted on six benchmark datasets and results reveal that our approach outperforms the current state-of-the-art methods.

Keywords: Graph Data Mining · Graph Anomaly Detection · Contrastive Learning

1 Introduction

Recent years, the application of deep learning methods in graph research has attracted great attention [11]. Graph anomaly detection (GAD) has demonstrated powerful capabilities in tasks such as preventing financial fraud [8,24], and detecting malicious users and spam in social networks [13], making it one of the most important tasks in the world. Anomalies in GAD are typically defined as instances, which can be nodes, edges, or subgraphs, that markedly deviate from the majority within the sample [25]. And two typical anomalies are defined based

M. Onizuka et al. (Eds.): DASFAA 2024, LNCS 14855, pp. 236–251, 2024.
https://doi.org/10.1007/978-981-97-5572-1_15

on the mismatch of structure information and attribute information: structure anomaly and attribute anomaly, respectively. The former relates to structure information, manifested in anomalous connection patterns between nodes, while the latter pertains to attribute information that significantly differs from that of other nodes.

To detect these two anomalies, many methods have been proposed. The shallow methods including AMEN [20] and Anomalous [19] detect anomalies with shallow mechanisms, such as residual analysis and CUR decomposition. However, these methods exhibit limitations in addressing non-linear, highly-unstructured network, as well as high-dimensional attributes, making it difficult to further enhance detection effectiveness. Recently, deep learning methods based on graph neural networks have been widely applied to GAD task [16]. The pioneer work, DOMINANT [3], employs a graph auto-encoder (GAE) to reconstruct both structure and attribute information. Anomaly nodes are then identified based on the reconstruction errors. CoLA [15] introduces a graph anomaly detection framework based on contrastive learning, which can capture local information from the node-subgraph contrast pair and detect anomaly nodes based on discriminative scores. ANEMONE [10] constructs a multi-scale contrastive learning framework by introducing node-node contrast pair, and focuses on capturing node-level anomaly information. SLGAD [32] integrates contrastive module and generative module to capture features at both node and subgraph levels, thereby augmenting the efficacy of anomaly detection. However, the aforementioned methods concentrate solely on local information, overlooking global information which leads to the inadequate extraction of information from the graph.

Despite the remarkable achievements of exiting methods, they still exhibit certain shortcomings. The GAE based approaches, represented by DOMINANT, is designed to minimize the differences between original and reconstructed data, while GAD task requires maximizing the discovery of differences between instances to effectively identify anomaly nodes. Discrepancies in task objectives indicates that the GAE framework is not ideally suited for GAD. Prevailing contrastive learning methods, such as CoLA, ANEMONE and SL-GAD are constrained to capturing local information from the one-hop neighbors, and neglect global information beyond one-hop neighbors. Nevertheless, global information encompasses the comprehensive structure information which is essential for detecting structure anomaly and locating anomaly nodes. As shown in Fig. 1, learning local information of target node cannot adequately capture anomaly information in the local view. Nevertheless, this challenge can be addressed by constructing a global view to learn global information. Meanwhile, learning global structure information of graph can also enhances the model's ability to generalize graph structure effectively for GAD. Unfortunately, effectively extracting local and global information simultaneously still faces challenges.

To alleviate the above drawbacks, we introduce a novel framework integrated **Co**ntrastive learning module and attribute reconstruction moduel for **G**raph **A**nomaly **D**etection (termed Co-GAD), which can effectively extract local and global information. Specifically, in contrastive learning module, Co-GAD gen-

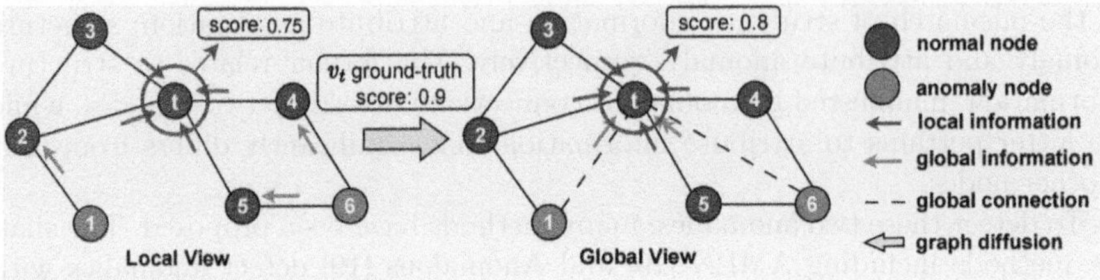

Fig. 1. Example to illustrate local and global information. In the local view, the model learns the representation of target node v_t without incorporating global information from nodes v_1 and v_6. While in the global view, leveraging the global connections formed through graph diffusion benefits the model to capture more comprehensive global information. Obviously, this facilitates the calculation of more precise anomaly scores.

erates global view firstly with graph diffusion, while regard original graph as local view. Then, we sample node and subgraph from two views to construct multi-scale contrast combinations. In the intra-view contrast, contrast combination contains node-node and node-subgraph contrast pair in each view, which focus on anomalies in and around the node itself, respectively. In the intre-view contrast, contrast combination contains subgraph-subgraph and node-node contrast pair between two views. By comparing the node- and subgraph- level embedding representations between global and local views separately, the consistency of node representations in two views is enhanced to allow model to capture local and global information adequately. In reconstruction module, we employ masked auto-encoder to further capture attribute anomaly. Finally, we capture anomaly information from two modules and compute the anomaly score for each node. The main contributions are summarized as follows:

- We present a novel framework Co-GAD for GAD integrated contrastive learning module and reconstruction module. With a multi-contrast strategy combining intra- and intre- view contrasts, Co-GAD can effectively extract both structure and attribute anomaly information.
- We design multi-scale contrast combinations for enhancing the capability of the contrastive learning module to adequately capture both local and global information simultaneously.
- We conduct extensive experiments on six benchmark datasets to evaluate the performance of Co-GAD. Experiment results demonstrate that our model surpasses current state-of-the-art methods across the majority dedasets.

2 Related Work

2.1 Graph Anomaly Detection

Early works in GAD identify anomalous information based on node features and network structure. But these methods have notable shortcomings when confronted with nonlinear, high-dimensional node features. Recently, the introduction of deep learning models has garnered substantial interest for research in GAD [16]. DOMINANT [3] employs GCN as encoder to extract node representations and decoder is formulated to reconstruct both structure and attributes to detect anomaly nodes. Dual-SVDAE [29] aims to learn the hypersphere boundary on the normal nodes' latent space based on the graph structure and attribute. GUIDE [28] leverages the high-order structure in modeling the complex interaction. HCM [9] is a hop-count based model with the PairwiseDistance algorithm. It consists of GCNs for learning graph representation and MLP module for calculating anomaly scores. GATAE [27] embeds the input attributed graph by using multiple graph attention layers and reconstruct structure and attribute, respectively. ASGAE [31] reconstructs subgraph context through GAE model, identifies anomalous subgraphs by minimizing reconstruction error.

2.2 Graph Contrastive Learning

Given the notable success of contrastive learning in Computer Vision (CV) [2] and Natural Language Processing (NLP) [26], its application in graph domains has also attracted great attention. Through training pretext task tailored for anomaly detection task, the contrastive learning model can acquire an enhanced representation to detect anomaly. DGI [22] learns the embedding of graph by maximizing the mutual information between nodes and graphs. MVGRL [6] performs contrastive learning on graphs from one-hop neighbors and graph diffusion on two views. CoLA [15] and ANEMONE [10] capture local information by learning the differences between nodes and their subgraphs to capture anomalies. CG3 [23] exploits contrastive learning based on graph structure and limited labelled information. SLGAD [32] utilizes generative module and contrastive module to extract local information and compute anomaly score for each node.

3 Problem Formulation

In this section, we formalize graph anomaly detection task. For convenience, we use calligraphic fonts (e.g. $\mathcal{G}$), bold uppercase letters (e.g. $\mathbf{X}$), and lowercase letters (e.g. $\mathbf{x}$) to denote sets, matrices and vectors. Accordingly, the definition of graph is given as follows:

Definition 1 ***Graph***. *Graph can be denoted as* $\mathcal{G} = (\mathcal{V}, \mathcal{E}, \mathbf{X})$, *where* $\mathcal{V} = \{v_1, v_2, ..., v_N\}$ *represents the node set and* $n = |\mathcal{V}|$, $\mathcal{E}$ *is the edge set with* m *edges, and* $\mathbf{X} \in \mathbb{R}^{n \times d}$ *is attribute matrix. Specifically,* $\mathbf{x}_i \in \mathbb{R}^d (i = 1, ..., N)$ *represents the attribute of* i^{th} *node.* $\mathbf{A} \in \mathbb{R}^{n \times n}$ *is adjacency matrix, where* $\mathbf{A}_{i,j} = 1$ *if there is an edge between nodes* v_i *and* v_i, *otherwise* $\mathbf{A}_{i,j} = 0$.

Definition 2 ***Graph anomaly detection***. *Given a graph* $\mathcal{G} = (\mathcal{V}, \mathcal{E}, \mathbf{X})$, *the purpose of our model is to learn a anomaly score function* $F(\cdot)$, *which measures the abnormality degree for each node. Finally, anomalous nodes will be identified through a descend sorted list of all nodes anomaly scores.*

4 Method

In this section, we introduce the proposed framework Co-GAD, as shown in Fig. 2. Overall, it consists four components, namely sampling module, contrastive learning, attribute reconstruction and anomaly calculation. In the sampling module, we regard original graph as local view and employ graph diffusion to generate global view. Then, in each view, we sample subgraphs around each node as contrast elements and construct multi-scale contrast pairs with these elements, including node-node, node-subgraph and subgraph-subgraph contrast pair. Based on these pairs, we construct two contrast combinations, intra-view and intre-view contrast combination respectively. In the contrastive learning module, each contrast combination is input to encoders to extract local and global information, and discriminators calculate the similarity score for each pair in the contrast combination. In the attribute reconstruction module, we reconstruct the representation of the node and calculate the reconstruction error. In the anomaly score calculating module, we calculate the anomaly score for each node by integrating contrast similarity and reconstruction error. Eventually, anomaly nodes are identified from the descend list of anomaly score.

4.1 Global View Generation and Contrast Element Sample

Global View Generation. Recent researches have indicated that the introduction of multi-view facilitates the model in extracting richer information. MVGRL [6] states out that converting adjacency matrix to diffusion matrix can effectively learn global structure information in the graph. In our work, we consider the original graph as local view, and generate global view based on local view with graph diffusion. Graph diffusion can aggregate global information more adequately [5]. There are two popular implementations of graph diffusion, Heat Kernel (HK) [12] and Personalized PageRank (PPR) [18] respectively. The process of graph diffusion is formulated as follows:

$$\mathbf{S} = \sum_{k=0}^{\infty} \Theta_k \mathbf{T}^k, \tag{1}$$

where $\mathbf{S}$ is global view of the initial graph, Θ_k is the weighting coefficient to control the proportion of local and global information. $\mathbf{T}^k \in \mathbb{R}^{n \times n}$ is generalized transition matrix. In this paper, we adopt HK to power the graph diffusion. Given the adjacency matrix $\mathbf{A}$ and the degree matrix $\mathbf{D}$, graph diffusion can be reformulated as:

$$\mathbf{S}^{\mathrm{HK}} = \exp\left(t\mathbf{A}\mathbf{D}^{-1} - t\right), \tag{2}$$

where transition matrix $\mathbf{T}$ is formulated as $\mathbf{T} = \mathbf{A}\mathbf{D}^{-1}$, t denotes diffusion time.

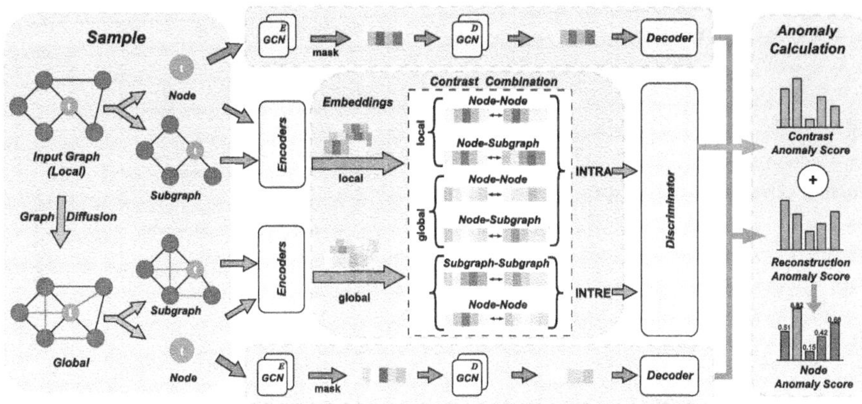

Fig. 2. Overview of the Co-GAD. (1):We employ graph diffusion to generate global view and sample subgraphs around the target node. (2):In the contrastive learning module, we construct contrast combinations with nodes' and subgraphs' embeddings and calculate similarity score for each contrast pair. (3):In attribute reconstruction module, we use masked graph auto-encoder to reconstruct representations and calculate reconstruction error. (4):We calculate anomaly score for each node. Finally, anomaly node can be detected easily, e.g. orange nodes.

Contrast Element Sample. Contrastive learning model can be effectively employed to anomaly detection task with well-designed contrast pair [15]. To adequately capture local and global information, we construct two contrast combinations, intra-view contrast and intre-view contrast combination respectively. Intra-view contrast combination contains node-subgraph and node-node contrast pair, while intre-view contrast combination contains subgraph-subgraph and node-node contrast pair. As elements in contrast pair, node and subgraph are sampled from each view. The sampling strategy is illustrated below:

Target Node Sample. We traverse each node in the graph in a random order as target node.

Subgraph Sample. Subgraph is defined as the neighbor structure around initial node. Thereby, we adopt restarted random walk (RWR) [21] to sample subgraph with any node as initial node in the graph (Fig. 3).

4.2 Contrastive Learning Module

Node anomalies typically manifest as mismatch between node and its adjacent neighbors. To adequately capture the structure information, we propose a contrast combination based contrastive learning framework. Our contrastive learning model consists of intra-view and intre-view contrast. In the intra-view contrast, intra-view contrast combination consisted of node-subgraph and node-node contrast pairs is designed to capture the local and global information

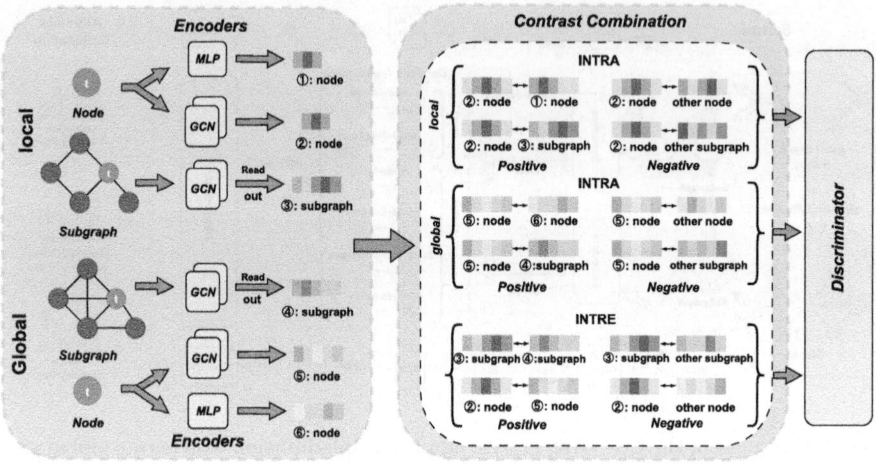

Fig. 3. The sampling process of contrast combination. We first sample target node and subgraph from each view. Then, we use GCN and MLP to learn two types of embeddings for target node and GCN to learn subgraphs' embedding. Then, we construct two contrast combinations with three contrast pairs based on these embeddings, intra-view and intre-view combinations. Note that both local and global combinations are intra-view contrast combination.

simultaneously. The node-subgraph contrast pair extracts the anomaly information between node and its neighbors, while node-node contrast pair extracts anomaly information from the node itself. In the intre-view contrast, we design intre-veiw contrast combination to enhance the consistency of node level and subgraph level representations between two views, which improves the model's ability to capture local and global information simultaneously. In the aforementioned contrast process, we meticulously formulate the loss function to enhance the ability of anomaly detection. Our contrastive learning module consists of three components: GNN encoders, Readout module and Discriminator.

Intra-view Contrast. Intra-view contrast performs in local and global views respectively. We present Intra-view contrast in the local view as a representative. The Intra-view contrast combination contains node-subgraph and node-node contrast pair. And our model performs contrast operation at the contrast pair level. As for node-subrgaph contrast pair, positive sample consists of the target node and its localized subgraph, i.e., the initial node of RWR is set as the target node. It is worth noting that the features of target node are masked in the subgraph. And negative sample consists of the target node and the subgraph localized by a random node except of target node in the local view. The same process also applies to global view.

We learn the information of the local subgrph S_i by GNN encoders. In this paper, we adopt a single GCN layer as GNN encoder, which is represented as follows:

$$\mathbf{H}_i^{(\ell)} = \sigma\left(\widetilde{\mathbf{D}}_i^{-\frac{1}{2}}\widetilde{\mathbf{A}}_i\widetilde{\mathbf{D}}_i^{-\frac{1}{2}}\mathbf{H}_i^{(\ell-1)}\mathbf{W}^{(\ell-1)}\right), \tag{3}$$

where $\widetilde{\mathbf{A}} = \mathbf{A}+\mathbf{I}$ denotes subgraph adjacency matrix with self-loop. $\mathbf{D}$ represents degree matrix, $\widetilde{\mathbf{D}}_i^{-\frac{1}{2}}\widetilde{\mathbf{A}}_i\widetilde{\mathbf{D}}_i^{-\frac{1}{2}}$ denotes adjacency matrix normalization. $\mathbf{H}_i^{(\ell-1)}$ and $\mathbf{H}_i^{(\ell)}$ are hidden representations of the subgraph in $(l-1)^{\text{th}}$ and l^{th} layer. The input of $\mathbf{H}_i^{(0)}$ is the attribute matrix of subgraph, and the output $\mathbf{H}_i^{(l)}$ is the embedding of subgraph, denoted as $\mathbf{Z}_i$. $W^{(0)} \in \mathbb{R}^{d\times d'}$ is learnable weight matrix and $\sigma(\cdot)$ is activation function PReLU.

Then, the final representation $\mathbf{z}_i$ of subgraph S_i is calculated by Readout module, an average pooling function is employed as our Readout module:

$$\mathbf{z}_i = Readout\left(\mathbf{Z}_i\right) = \sum_{j=1}^{n_i} \frac{\left(\mathbf{Z}_i\right)_j}{n_i}. \tag{4}$$

In terms of learning target node embedding, we employ a single GCN layer as encoder, similarly. The difference is that they employ different parameters. Thus, the GCN encoder for target node can be represented as:

$$\mathbf{H}_i^{\prime(\ell)} = \sigma\left(\widetilde{\mathbf{D}'}_i^{-\frac{1}{2}}\widetilde{\mathbf{A}'}_i\widetilde{\mathbf{D}'}_i^{-\frac{1}{2}}\mathbf{H}_i^{\prime(\ell)}\mathbf{W}^{\prime(\ell-1)}\right), \tag{5}$$

where $\mathbf{H}_i^{\prime(\ell)}$ denotes the representation of subgraph around the target node in the lth layer, which consists of node representations $\mathbf{h}_i^{(\ell)}$. The first line of $\mathbf{H}_i'$ is the representation of target node. The final representation of target node is $\mathbf{e}_i$. We use discriminator to generate a similarity score s_i by comparing the representations of node and subgraph. And the abnormality degree of target node is related to this similarity score. We adopt a Bilinear model to evaluate the similarity of node-subgraph contrast pair.

$$s_i = \phi\left(\mathbf{e}_i^T\mathbf{W}\mathbf{z}_i\right), \tag{6}$$

where $\phi(\cdot)$ is sigmoid function, maps anomaly score between 0 and 1. When representations of node and subgraph are similar (typically positive sample), s_i approaches 1, otherwise s_i approaches 0 (typically negative sample). We employ the binary cross-entropy loss to optimize contrast process.

$$\mathcal{L}_{ns}^{\text{local}} = -\sum_{i=1}^{N}\left(y_i \log\left(s_i^{\text{local}}\right) + (1-y_i)\log\left(1-s_i^{\text{local}}\right)\right), \tag{7}$$

where $y_i = 1$ in positive pair, otherwise $y_i = 0$. $\mathcal{L}_{ns}^{\text{local}}$ is the loss of node-subgraph contrast in the local view, and the contrast process in the global view is similar. Thus, the final node-subgraph contrast loss is:

$$\mathcal{L}_{ns} = \alpha\mathcal{L}_{ns}^{\text{local}} + (1-\alpha)\mathcal{L}_{ns}^{\text{global}}, \tag{8}$$

where $\alpha \in (0,1)$ is a weight parameter, which measures the importance between local and global view.

As for node-node contrast pair, positive sample consists two embeddings of the same target node based on GCN and MLP encoders, while negative sample consists the embeddings of target node and a random node except of target node. The first node representation is the same in the node-subgraph contrast pair. And the second node representation $\hat{\mathbf{h}}_i$ is obtained by a single MLP layer module, which can be represented as:

$$\hat{\mathbf{h}}_i^{(\ell)} = \sigma\left(\hat{\mathbf{h}}_i^{(\ell-1)}\mathbf{W}'^{(\ell-1)}\right), \tag{9}$$

where $\mathbf{W}'$ is shared with Eq. (5). Similarly, we adopt Bilinear model to evaluate the similarity $\hat{s}_i$ of node-node contrast pair. And binary cross-entropy loss is employed to calculate the contrastive loss which is formulated as:

$$\mathcal{L}_{nn}^{\text{local}} = -\sum_{i=1}^{N}\left(\hat{y}_i \log\left(\hat{s}_i^{\text{local}}\right) + (1-\hat{y}_i)\log\left(1-\hat{s}_i^{\text{local}}\right)\right). \tag{10}$$

The final node-node contrast loss is expressed as:

$$\mathcal{L}_{nn} = \alpha\mathcal{L}_{nn}^{\text{local}} + (1-\alpha)\mathcal{L}_{nn}^{\text{global}}, \tag{11}$$

where α is shared with Eq. (8). Finally, we can derive the total loss of intra-view contrast combination in the local and global view:

$$\mathcal{L}_{\text{intra}} = \beta\mathcal{L}_{ns} + (1-\beta)\mathcal{L}_{nn}, \tag{12}$$

where $\beta \in (0,1)$ is a weight parameter to balance the importance between node-subgraph and node-node contrast pairs.

Intre-view Contrast. We design a intre-view contrast combination consisted and node-node contrast pair between local and global view. As for subgraph-subgraph contrast pair, positive sample consists subgraphs sampled with the same initial node v_i from local and global view, $\mathbf{z}_i^{\text{local}}$ and $\mathbf{z}_i^{\text{global}}$ respectively. And the negative sample consists subgraphs sampled with different initial node v_j, $\mathbf{z}_j^{\text{local}}$ and $\mathbf{z}_j^{\text{global}}$. Inspired by [17], the subgraph-subgraph contrast loss is formulated as:

$$\mathcal{L}_{ss} = -\sum_{i=1}^{n}\log\frac{\exp\left(\mathbf{z}_i^{\text{local}}\cdot\mathbf{z}_i^{\text{global}}\right)}{\exp\left(\mathbf{z}_i^{\text{local}}\cdot\mathbf{z}_j^{\text{local}}\right)+\exp\left(\mathbf{z}_i^{\text{local}}\cdot\mathbf{z}_j^{\text{global}}\right)}. \tag{13}$$

Similarly, positive and negative samples of node-node contrast pair are same as in subgraph-subgraph contrast pair. Therefore, we can formulate node-node contrast loss as follows:

$$\mathcal{L}_{\text{nn}\prime} = -\sum_{i=1}^{n}\log\frac{\exp\left(\mathbf{e}_i^{\text{local}}\cdot\mathbf{e}_i^{\text{global}}\right)}{\exp\left(\mathbf{e}_i^{\text{local}}\cdot\mathbf{e}_j^{\text{local}}\right)+\exp\left(\mathbf{e}_i^{\text{local}}\cdot\mathbf{e}_j^{\text{global}}\right)}, \tag{14}$$

where $\mathbf{e}_i^{\text{local}}$ and $\mathbf{e}_i^{\text{global}}$ are embeddings of node v_i in two views, and $\mathbf{e}_j^{\text{local}}$ and $\mathbf{e}_j^{\text{global}}$ are embeddings of node v_j in two views.

Therefore, the loss of intre-contrast can be represented as:

$$\mathcal{L}_{\text{intre}} = \mathcal{L}_{ss} + \mathcal{L}_{\text{nn}'}. \tag{15}$$

By integrating both intra- and intre- view contrast loss, the ultimate contrast loss is derived as:

$$\mathcal{L}_{\text{con}} = \mathcal{L}_{intra} + \mathcal{L}_{\text{intre}}. \tag{16}$$

4.3 Reconstruction Module

The above contrastive learning module enhances the capture of global structure information, while attribute information should also be further utilized. GraphMAE [7] points out that masked auto-encoder can effectively learn the attribute information by reconstructing features. Inspired by this, we introduce the attribute reconstruction module to leverage attribute information adequately. We perform reconstruction error as a part of overall loss function to optimize the entire model.

In the reconstruction model, we adopt Guassian distribution based mask strategy [1] to mask node features. Single GCN layer is employed as encoder to transfer high-dimensional attributes to a low-dimensional space. A single GCN layer is adopted as decoder to reconstruct node features. We perform attribute reconstruction in both local and global views. Taking the local view as an example, $\mathbf{x}$ and $\hat{\mathbf{x}}$ denote initial attribute vector and reconstruction attribute vector. We adopt mse loss function to calculate the reconstruction error, which is defined as:

$$\mathcal{L}_{\text{res}}^{\text{local}}(v_i) = \frac{1}{d}\|\mathbf{x}_{local} - \hat{\mathbf{x}}_{local}\|_F^2. \tag{17}$$

The above process is same in global view. Thus, the final reconstruction loss is formulated as:

$$\mathcal{L}_{res} = \frac{1}{2N}\sum_{i=1}^{N}(\mathcal{L}_{res}^{\text{local}}(v_i) + \mathcal{L}_{res}^{\text{global}}(v_i)). \tag{18}$$

The overall framework of Co-GAD consists of contrastive learning and attribute reconstruction modules. To balance the importance of these two modules, $\gamma \in (0, 1)$ is introduced. The total loss function is expressed as:

$$\mathcal{L} = \gamma\mathcal{L}_{\text{con}} + (1-\gamma)\,\mathcal{L}_{\text{res}}. \tag{19}$$

4.4 Anomaly Detection Calculation

According to the loss calculation process, the anomaly scores of nodes consists contrastive learning and attribute reconstruction anomaly scores. The contrastive anomaly score is highly related to the similarity of contrast combinations. As for positive contrast sample, the score $s(p)$ typically approaches to 1, otherwise approaches $s(p)$ to 0. Since intre-view contrast is intended to enhance the consistency of local and global view encodings, the node anomaly score is calculated from intra-view contrast combination. The anomaly score of node-subgraph contrast pair is formulated as follows:

$$s_{ns} = \alpha\left(s_{ns}^{\text{local}}(n) - s_{ns}^{\text{local}}(p)\right) + (1-\alpha)\left(s_{ns}^{\text{global}}(n) - s_{ns}^{\text{global}}(p)\right), \tag{20}$$

where $\mathbf{s}_{ns}$ denotes the anomaly score of node-subgraph contrast pair, α is shared with Eq. (8). The anomaly score calculation of node-node contrast pair is similar to node-subgraph contrast pair, thus, the anomaly score of contrastive learning module is:

$$s_{con} = \beta s_{ns} + (1-\beta)\, s_{nn}, \tag{21}$$

where β is shared with Eq. (12). Since reconstruction anomaly score is similar in local and global views, the anomaly score of reconstruction is:

$$s_{res} = \frac{1}{2d}\left(\|\mathbf{x}_{local} - \hat{\mathbf{x}}_{local}\|_F^2 + \|\mathbf{x}_{global} - \hat{\mathbf{x}}_{global}\|_F^2\right). \tag{22}$$

Share γ with Eq. (19), the final formula of anomaly score is:

$$score(v_i) = \gamma \text{score}_{\text{con}}(v_i) + (1-\gamma)\, \text{score}_{\text{res}}(v_i). \tag{23}$$

5 Experiments

We conduct extensive experiments on six benchmark datasets, and experiment results show that Co-GAD outperforms state-of-the-art anomaly detection methods based on self-supervised learning.

5.1 Datasets

We use six benchmark datasets in our experiments, where Cora [15], ACM [32], CiteSeer [10], Pubmed [10] are anomaly injected datasets, and Books [14], Reddit [14] are real-world datasets contain ground-truth anomaly. The details of these datasets are shown in Table 1.

5.2 Experimental Settings

Baselines and Evaluation Metrics. In the experiments, we compare Co-GAD with six popular baselines. DOMINANT [3] conducts the investigation on the problem of anomaly detection on attributed networks. CoLA [15] is the first contrastive learning based method for GAD. ANEMONE [10] is a recent method which performes patch- and context- level contrastive learning for anomaly detection. SLGAD [32] integrates contrastive and generative module to detect anomaly. Sub-CR [30] constructs multi-view contrastive framework for GAD. GRADATE [4] is a multi-scale instances based contrastive learning method for GAD. In evaluating the aforementioned methods, we employ the widely-utilized anomaly detection metric, Area Under the Curve (AUC).

Table 1. The statistics of the datasets.

dataset	Injected Anomaly				Ground-truth Anomaly	
	Cora	Citeseer	ACM	Pubmed	Books	Reddit
# nodes	2,708	3,327	16,484	19,717	1,418	10,984
# edges	5,429	4,732	71,980	44,338	3,695	168,016
# attributes	1,433	3,703	8,337	500	21	64
# anomalies	150	150	600	600	28	366

Table 2. Performance comparison for AUC. The best and runner-up performing method is in bold and underline, respsctively.

Methods	Cora	CiteSeer	ACM	Pubmed	Books	Reddit
DOMINANT(2019)	0.8155	0.8251	0.7601	0.8081	0.5010	0.5640
ANEMONE(2021)	0.9057	0.9189	0.8353	0.9548	0.5302	0.5745
CoLA(2021)	0.8941	0.8863	0.8218	0.9499	0.4548	0.5891
SLGAD(2021)	0.9110	0.9149	<u>0.8508</u>	**0.9674**	0.5314	0.5689
Sub-CR(2022)	0.8870	<u>0.9350</u>	0.8187	0.9610	0.4753	0.5649
GRADATE(2023)	<u>0.9146</u>	0.8282	0.8478	0.8787	<u>0.5691</u>	<u>0.6202</u>
Co-GAD	**0.9298**	**0.9420**	**0.8941**	<u>0.9633</u>	**0.5999**	**0.6495**

Parameter Settings. To show the optimal performance of Co-GAD, we set the following parameters when conducting experiments. In contrastive learning module, we adopt graph diffusion (HK) to generate global view with diffusion time equals to 1. We fixed the subgraph size(number of nodes) to be 4 during subgraph sampling. In reconstrcution module, we adopt Guassian distribution based masked strategy, setting mean to 0.5 and variance to 0.2.

5.3 Result and Analysis

The results of our experiments are shown in Table 2. We can intuitively find that Co-GAD outperforms baselines on five datasets. Specifically, Co-GAD achieves notable AUC gains of 1.6%, 0.74%, 5.08%, 2.21%, 4.72% on Cora, CiteSeer, ACM, Books amd Reddit, respectively. It is obvious that baseline methods are not as effective on real datasets as injecting anomalous datasets. This is because structure and attribute anomaly are more obvious in the injected anomaly datasets than real-world datasets.

Co-GAD performing better than DOMINANT indicates that the contrastive learning framework is powerful than GAE framework for GAD. CoLA, SLGAD, ANEMONE and GRADATE perform worse shows that global information is crucial for GAD. And Sub-CR does not show the competitive performance because single node-subgraph contrast pair cannot capture structure and attribute information adequately. Thus, Co-GAD can achieve the best performance. Masking strategies based on Gaussian distributions exhibit inferior performance on low-dimensional attribute spaces, leading to suboptimal results for Co-GAD when applied to the Pubmed dataset.

5.4 Ablation Study

Various Contrast Combination Strategy. To verify the effectiveness of contrast combination, we conduct four sets of ablation experiments. As show in Table 3, with the integration of conrtast pairs, the performance of model is better. We can find that, the introduce of subrgaph-subgraph contrast pair in intre-view delivers the greatest performance gains, which is the result of extracting global information. Node-node contrast pair performs better in intra-view because intre-view contrast in used for enhancing the capability of extracting global information.

Table 3. The AUC values of contrast conbination ablation study

	Cora	CiteSeer	ACM	Pubmed	Books	Reddit
(w/) only intra NS	0.8671	0.9079	0.8518	0.9268	0.5387	0.5798
(w/o) intre&intra NN	0.8991	0.9288	0.8854	0.9439	0.5578	0.6159
(w/o) intra NN	0.9046	0.9301	0.8879	0.9505	0.5668	0.6302
(w/o) intre NN	0.9159	0.9356	0.8901	0.9547	0.5712	0.6342
Co-GAD	**0.9298**	**0.9420**	**0.8941**	**0.9633**	**0.5999**	**0.6495**

Various Graph Diffusion. There are two popular approaches to achieve graph diffusion, Personalized PageRank(PPR) and Heat Kernel(HK). Experiment results show that Graph diffusion(HK) can achieve better performance in our model. The result is shown in Table 4.

Table 4. The AUC values of graph diffusion ablation study

	Cora	CiteSeer	ACM	Pubmed	Books	Reddit
Co-GAD (PPR)	0.9166	0.9194	0.8558	0.9413	0.5631	0.5884
Co-GAD (HK)	**0.9298**	**0.9420**	**0.8941**	**0.9633**	**0.5999**	**0.6495**

Reconstruction Module. We study the importance of reconstruction module in our model. We demonstrate the effect of overall framework of Co-GAD through the experiment in the Table 5.

Table 5. The AUC values of reconstruction module ablation study

	Cora	CiteSeer	ACM	Pubmed	Books	Reddit
w/o Res	0.895	0.9074	0.8683	0.9429	0.5698	0.5228
Co-GAD	**0.9298**	**0.9420**	**0.8941**	**0.9633**	**0.5999**	**0.6495**

5.5 Parameter Study

We conduct the model sensitivity analysis on crutical hyper-parameters in Co-GAD, which are α, β, γ, embedding dimension in GCN and training epoch. As shown in Fig 5, when setting α to 0.1, 0.3, 0.1, 0.1, 0.1, 0.1, while β to 0.1, 0.5, 0.5, 0.8, 0.1, 0.1 on Cora, Citeseer, ACM, Pubmed, Books and Reddit, Co-GAD performs best. Figure 4 illustrates the performance variation of Co-GAD with the remaining parameters. As for training epoch, it is set to 400 on Cora, Citeseer, ACM, Pubmed, and 100 on Books and Reddit. Set embedding dimension to 64 on injected anomaly datesets, 128 and 256 on Books and Reddit, respectively.

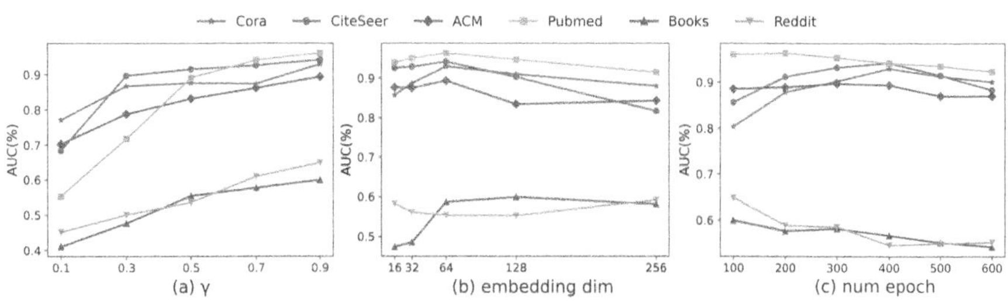

Fig. 4. Performance with different γ, embedding dimensions and training epochs

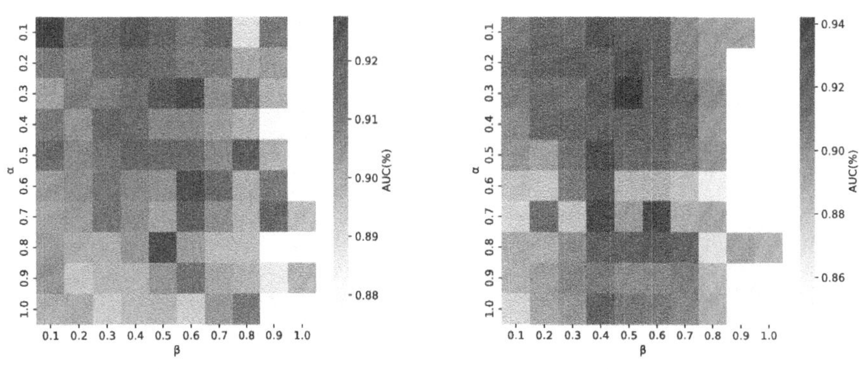

Fig. 5. Sensitivity analysis for the balance parameters α and β w.r.t. AUC on Cora and CiteSeer. Color shades indicate the value of the AUC

6 Conclusion

In this paper, we propose a novel framework integrating multi-view contrastive learning and attribute reconstruction modules, which can effectively extract both local and global information for GAD. Contrast combinations are proposed to enhance the capability of capturing global information. Extensive experiments on six benchmark datasets demonstrate the effectiveness of our Co-GAD.

Acknowledgments. This work was supported by National Key R&D Program of China (2022YFC2602305, 31400), and partly supported by Natural Science Foundation of Inner Mongolia Autonomous Region of China (2022LHMS06008), Key R&D and Transformation Plan of Qinghai Province (2022-QY-218), and University-Industry Collaborative Education Program (202312280037).

References

1. Chen, H., et al.: Masked image training for generalizable deep image denoising. In: Proceedings of the IEEE/CVF Conference on Computer Vision and Pattern Recognition, pp. 1692–1703 (2023)
2. Chen, T., et al.: A simple framework for contrastive learning of visual representations. In: International Conference on Machine Learning, pp. 1597–1607. PMLR (2020)
3. Ding, K., et al.: Deep anomaly detection on attributed networks. In: Proceedings of the 2019 SIAM International Conference on Data Mining, pp. 594–602. SIAM (2019)
4. Duan J, et.al.: Graph anomaly detection via multi-scale contrastive learning networks with augmented view. In: Proceedings of the AAAI Conference on Artificial Intelligence (2023)
5. Gasteiger, J., et al.: Diffusion improves graph learning. In: Advances in Neural Information Processing Systems, vol. 32 (2019)
6. Hassani, K., et al.: Contrastive multi-view representation learning on graphs. In: International Conference on Machine Learning, pp. 4116–4126. PMLR (2020)
7. Hou, Z., et al.: Graphmae: self-supervised masked graph autoencoders. In: Proceedings of the 28th ACM SIGKDD Conference on Knowledge Discovery and Data Mining, pp. 594–604 (2022)
8. Huang, D., et al.: Codetect: financial fraud detection with anomaly feature detection. IEEE Access **6**, 19161–19174 (2018)
9. Huang, T., et al.: Hop-count based self-supervised anomaly detection on attributed networks. In: Joint European Conference on Machine Learning and Knowledge Discovery in Databases. pp. 225–241. Springer (2022)
10. Jin, M., et al.: Anemone: Graph anomaly detection with multi-scale contrastive learning. In: Proceedings of the 30th ACM International Conference on Information and Knowledge Management, pp. 3122–3126 (2021)
11. Kim, H., et al.: Graph anomaly detection with graph neural networks: Current status and challenges. IEEE Access (2022)
12. Kondor, R.I., et al.: Diffusion kernels on graphs and other discrete structures. In: Proceedings of the 19th International Conference on Machine Learning, vol. 2002, pp. 315–322 (2002)

13. Lakha, B., et al.: Anomaly detection in cybersecurity events through graph neural network and transformer based model: A case study with beth dataset. In: 2022 IEEE International Conference on Big Data (Big Data), pp. 5756–5764. IEEE (2022)
14. Liu, K., et al.: Bond: benchmarking unsupervised outlier node detection on static attributed graphs. Adv. Neural. Inf. Process. Syst. **35**, 27021–27035 (2022)
15. Liu, Y., et al.: Anomaly detection on attributed networks via contrastive self-supervised learning. IEEE Trans. Neural Netw. Learn. Syst. **33**(6), 2378–2392 (2021)
16. Ma, X., et al.: A comprehensive survey on graph anomaly detection with deep learning. IEEE Transactions on Knowledge and Data Engineering (2021)
17. Oord, A.v.d., et al.: Representation learning with contrastive predictive coding. arXiv preprint arXiv:1807.03748 (2018)
18. Page, L., et al.: The pagerank citation ranking: Bring order to the web. Tech. rep., Technical report, stanford University (1998)
19. Peng, Z., et al.: Anomalous: a joint modeling approach for anomaly detection on attributed networks. In: IJCAI, pp. 3513–3519 (2018)
20. Perozzi, B., et al.: Scalable anomaly ranking of attributed neighborhoods. In: Proceedings of the 2016 SIAM International Conference on Data Mining, pp. 207–215. SIAM (2016)
21. Tong, H., et al.: Fast random walk with restart and its applications. In: Sixth International Conference on Data Mining (ICDM'06), pp. 613–622. IEEE (2006)
22. Veličković, P., et al.: Deep graph infomax. arXiv preprint arXiv:1809.10341 (2018)
23. Wan, S., et al.: Contrastive and generative graph convolutional networks for graph-based semi-supervised learning. In: Proceedings of the AAAI conference on artificial intelligence, vol. 35, pp. 10049–10057 (2021)
24. Wang, H., et al.: Deep structure learning for fraud detection. In: 2018 IEEE International Conference on Data Mining (ICDM), pp. 567–576. IEEE (2018)
25. Wu, N., Chen, F., Li, J., Huai, J., Zhou, B., Ramakrishnan, N., et al.: A nonparametric approach to uncovering connected anomalies by tree shaped priors. IEEE Trans. Knowl. Data Eng. **31**(10), 1849–1862 (2018)
26. Yan, Y., et al.: Consert: A contrastive framework for self-supervised sentence representation transfer. arXiv preprint arXiv:2105.11741 (2021)
27. You, Z., et al.: Gatae: Graph attention-based anomaly detection on attributed networks. In: 2020 IEEE/CIC International Conference on Communications in China (ICCC), pp. 389–394. IEEE (2020)
28. Yuan, X., et al.: Higher-order structure based anomaly detection on attributed networks. In: 2021 IEEE International Conference on Big Data (Big Data), pp. 2691–2700. IEEE (2021)
29. Zhang, F., et al.: Deep dual support vector data description for anomaly detection on attributed networks. Int. J. Intell. Syst. **37**(2), 1509–1528 (2022)
30. Zhang, J., et al.: Reconstruction enhanced multi-view contrastive learning for anomaly detection on attributed networks. arXiv preprint arXiv:2205.04816 (2022)
31. Zhang, Z., et al.: Unsupervised deep subgraph anomaly detection. In: 2022 IEEE International Conference on Data Mining (ICDM), pp. 753–762. IEEE (2022)
32. Zheng, Y., et al.: Generative and contrastive self-supervised learning for graph anomaly detection. IEEE Transactions on Knowledge and Data Engineering (2021)

Global Route Planning for Large-Scale Requests on Traffic-Aware Road Network

JiaJia Li, Yannan Wang, Rui Zhu, Jing Zhang(✉), and Na Guo

Shenyang Aerospace University, Shenyang, China
{lijiajia,zhurui}@sau.edu.cn, ermengarde2022@163.com,
20230049@email.sau.edu.cn, guona_neu@163.com

Abstract. How to avoid congested roads and spend the least time to reach the destination has become an urgent requirement for citizens. When large-scale origin-destination queries arrive, these queries themselves will affect road congestion. Therefore, it is necessary to plan different routes in advance for these queries to avoid simultaneous occupation of the same roads, thereby reducing congestion and global travel time. Nevertheless, this is not trivial, as calculating which queries will occupy the same road resources is time-consuming. Moreover, re-planning the routes can also affect other queries, and this impact will spread. Existing methods either adopt greedy strategy to consider only the surrounding road condition, or the result quality is poor in the case of dense query. In this paper, we utilize the structural nature of the H2H index, to quickly find potentially conflicting queries and group them together. In this way, only the conflicts within each group need to be resolved in a local area, so the efficiency is improved. Additionally, several alternative paths are calculated and the global optimal routes are found in finite iterations. Extensive experiments conducted on real-world networks with synthetic and real workloads show that our method can provide higher quality results more efficiently.

Keywords: Shortest path query · Large-scale requests · Traffic flow

1 Introduction

With the increase in the number of vehicles in cities, congestion is becoming increasingly serious. Every individual wants to avoid congested roads when traveling in order to reduce their time on the road. There are already electronic map software that can provide users with the route with the least travel time based on historical data and real-time road condition information. However, this is not enough. When large-scale origin-destination queries arrive simultaneously, navigation software may guide them to the same road segments. Therefore, it will bring new and unforeseen congestion. As a platform providing navigation services, it is necessary to guide these vehicles that will occupy the same road and cause congestion to other routes in advance. That is, under limited road resources, we need to complete better scheduling to reduce overall travel time.

M. Onizuka et al. (Eds.): DASFAA 2024, LNCS 14855, pp. 252–268, 2024.
https://doi.org/10.1007/978-981-97-5572-1_16

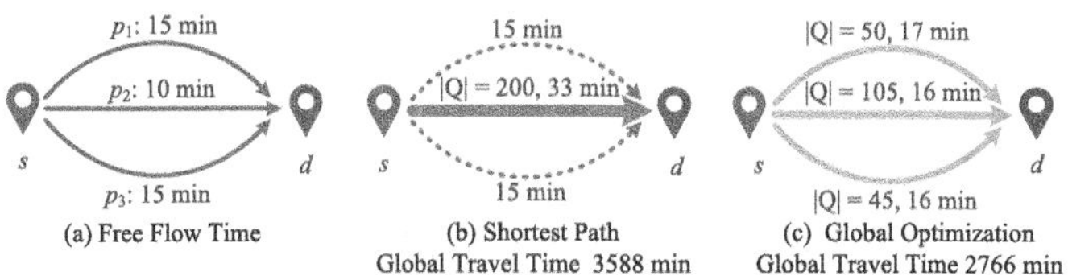

Fig. 1. Comparison of individual vs. global perspectives (road capacity is 50).

For example in Fig. 1-(a), there are three paths from s to d, and their free-flow travel time are 10, 15, 15 respectively. Now, 200 queries arrive, and p_2 is returned to all of them as it is the fastest one as shown in Fig. 1-(b). We say the road network is *traffic-aware*, which means that the travel time is influenced by both the capacity of the road segment and the current traffic flow. The travel time can be calculated using the Bureau of Public Roads (BPR) function, as used in other related works [3,4]. Assuming the road network capacity is 50, obviously, the 200 queries far exceed the capacity, which will undoubtedly cause congestion. After calculating with the BPR function, the slowest vehicle will take 33 min, and the overall time is 3588 min, which is much longer than the expected 2000 min under free-flow. As a comparison, if some vehicles are guided to p_1 and p_3 as shown in Fig. 1-(c), the maximum travel time can be reduced to 16 or 17 min, resulting in an overall cost of 2766 min. Thus, it can improve road traffic efficiency and greatly reduce overall travel time.

Motivated by this, we study the Global Route Planning (*GRP*) problem for large-scale requests. Given a traffic-aware road network, and n original-destination-time queries, *GRP* returns n routes for them with the least total travel cost. The travel cost here is measured in time, and the travel time is calculated by the BPR function, which is related to both the capacity of the edge and the corresponding dynamic traffic flow.

One method to solve the *GRP* problem is to execute the shortest path query one by one, and then update the traffic flow along the route, so that the next query can select a better path based on the newly calculated BPR function value. The algorithm *Next_opt* [12] follows this approach, and adopts a best-first mechanism while expanding queries. It selects the next expanded neighbor edge in a greedy way, using the current traffic flow and estimated time to the destination to heuristic guide the expansion. Nevertheless, it only focuses on the surrounding road conditions and ignores the possible congestion of future road segments. Therefore, the quality of the results is poor and the efficiency is low. Another method *Geo_ITE* [13] first calculates the initial routes for non conflicting queries, and then for those in conflict, reduce the congestion by iteratively "changing route-new conflict-changing route". It groups non conflicting queries first and gradually updates traffic flow while utilizing batch processing models, and then optimizes the routing results by iteration. Because it needs to update traffic information throughout the entire road network, the efficiency is relatively low. In addition, when the query distribution is dense, it may lead to serious conflicts within the same group, which can affect the quality of batch results, and make it difficult for the iterative model to approach accurate solutions.

The idea behind our algorithm is to find queries that occupy the same road resources in the same time period, and then avoid congestion by finding alternative routes for them. However, these two steps pose challenges to both efficiency and quality of results, as calculating which queries will occupy the same road resources is a time-consuming process. Moreover, recalculating the path for some conflicting queries can also affect other queries, and this impact will continue to spread. Firstly, we take advantage of the structural nature of the H2H index, which is the state-of-the-art method for solving shortest distance query, to quickly find potentially conflicting queries by adding auxiliary information and then group them together. In this way, only the conflicts within each group need to be resolved in a local area with the help of a spatial index, and there is no need to search the whole road network, so the efficiency is improved. Secondly, we only conduct re-planning in areas with severe congestion. Intuitively, not all re-planning can reduce overall travel time. For example, when the volume of vehicles during peak hours is much greater than the total capacity, no matter how it is scheduled, the overall travel time will not decrease too much, which means the upper limit of the traffic capacity has been reached. So we cannot expect the number of vehicles on all roads to be less than their capacity. Therefore, we only select those queries that pass through areas with more severe congestion (depending on the threshold set through experiments) for re-planning, in order to improve query efficiency. Finally, we searched for k alternative paths by setting congested areas as restricted areas, thereby completing the re-planning. Here, a grid based merging strategy is designed, which utilizes the quadtree [17] index.

The contributions can be summarized as follows:

- We propose a new idea to solve the *GRP* problem, which involves grouping conflicting queries together and then resolving conflicts through re planning within the group. This approach avoids updating traffic flow across the entire road network and improves query efficiency.
- We use the properties of the H2H index structure to quickly group potentially conflicting queries by storing auxiliary information in the nodes of the tree.
- We design a strategy to select queries worth reprogramming and find alternative paths for them in relatively non congested areas. Furthermore, global path optimization is achieved with only a small number of iterations.
- We conduct extensive evaluation on on real-world networks with synthetic and real workloads and verify the superiority of our approach in terms of quality and efficiency compared with the state-of-the-arts.

2 Related Work

2.1 Shortest Path Planning Algorithm

Shortest path query is a common problem in the field of path planning, which aims to find the path with minimum travel cost for a given origin-destination pair. In general, the travel cost is either static distance or dynamic time. And several methods and index structures have been proposed.

For static ones, some methods include index-free algorithms (*Dijkstra* [6] and *A** [9]) and index-based algorithms (*Contraction Hierarchy (CH), Hub Labeling (HL)*and *H2H* [1,7,15,16]) were also proposed to reduce query time by using some pre-computed auxiliary information. For non-static ones, some work updates the weights of edges in the road network based on real-time traffic flow, and then finds the current optimal path. The batch processing [14,18] can be used to accelerate the search, and index-based methods can be updated using index maintenance [2,8,23] to improve the efficiency of the online path planning. Other works assume that the edge weights are a periodic time-dependent function generated based on historical travel time data. Some classic index structures have been extended to time dependent environments(*TD-Dijkstra, TD-A*, TD-GTree, TD-H2H* [5,10,11,19]) to improve query efficiency.

However, these works only focus on individual travel costs and do not consider their impact on road conditions. Although some path finding methods can be used, they cannot be directly used to solve *GRP* problem.

2.2 Global Route Planning Algorithm

As far as we know, *NextOpt* [12] is the first attempt to solve the *GRP* problem. It assigns labels to each road network edge, recording query arrival and departure times. Then, it adopts a best-first mechanism, and iteratively selects vertices with the minimum departure time for expansion and prioritizes adjacent vertices with the lowest estimated travel time, until all queries reach their destinations. For large-scale queries, as the breadth of expansion increases, the number of neighboring nodes will increase significantly, resulting in low efficiency. In addition, it only focuses on the surrounding road conditions and ignores the possible congestion of future road segments. Therefore, the quality of the results is poor and the efficiency is low. Moreover, in order to quickly obtain the estimated time, it pre-computes and pre-stores the shortest distance between every two vertex, which is space-consuming.

Geo_ITE [21] is a recently proposed approach for solving the *GRP* problem. It groups non-conflicting queries together first, and then it gradually updates traffic while utilizing batch processing models. An iterative model is proposed to reroute some queries to approximate the optimal solution. However, when the query distribution is dense, the quality of the grouping cannot be guaranteed, resulting in conflicting queries being grouped together and causing congestion, increasing global travel time. Moreover, when re-evaluating the routing after each group completes path planning, it takes a lot of time to traverse all nodes that have already planned paths, which increases the response time of query.

Another related work [13] investigated the continuous *GRP* problem, it uses the *A** algorithm for heuristic search and iteratively optimizes the routing results. But it focuses on continuous versions, and the calculation of the initial path is relatively rough. While we mainly focus on snapshot queries, researching how to quickly find a reasonable initial path.

3 Preliminaries

In this section, we introduce some definitions about *GRP* problem and some background knowledge about H2H index, which is utilized by our method.

A road network that considers the impact of traffic flow on travel time is called a traffic-aware road network. That is, the weight of its edge is influenced by the capacity and the current traffic flow. The following is a formal definition.

Definition 1. ***Traffic-aware road network.*** *Given a traffic-aware road network $G(V, E, T)$, where V is the vertex set, E is the edge set, each edge $e(u, v) \in E$ connects two vertices u and v where $u, v \in V$. And $c(u, v)$ denotes the road capacity of the edge and $w(u, v)$ denotes the free-flow travel time of the edge. The function T is the traffic-aware function, where $T(u, v, t)$ returns the actual travel time from u to v by considering the traffic flow of $e(u, v)$ at time t.*

Definition 2. ***Traffic flow.*** *In the road network G, the traffic flow $f(u, v, t)$ on the edge $e(u, v) \in E$ at time t represents the number of vehicles on $e(u, v)$.*

The Bureau of Public Roads(BPR) function is commonly used in the transportation field to calculate the travel time under different traffic flows. We also use BRP here as the traffic-aware function, just like other works [3,20].

$$T(u, v, t) = w(u, v) \times (1 + \alpha(\frac{f(u, v, t)}{c(u, v)})^{\beta}) \tag{1}$$

where α and β are two coefficients with recommended values of 0.15 and 4.

Given a path $p = \{(v_0, t_0), \cdots, (v_j, t_j)\}$, where t_i is the time that arrives at v_i, and the travel time of p is $T(p) = \sum_{i=0}^{j-1} T(v_i, v_{i+1}, t_i)$, where $t_i = t_{i-1} + T(v_{i-1}, v_i, t_{i-1})$. And $T(v_i, v_{i+1}, t_i)$ is calculated by Eq. 1.

Definition 3. ***Global Route Planning Problem (GRP).*** *Given a traffic aware road network G and a set of queries Q, and each query $q_i \in Q$ has tuples of $< s_i, d_i >$ which denotes the source, destination. GRP aims to find a set of paths $P(Q)$ such that $T(P(Q))$ is minimal.*

H2H [15] index uses tree decomposition to organize the graph cut information and distance labels to support efficient distance query on a graph. The *vertex cut* is an important concept of H2H.

Definition 4. ***Vertex Cut*** *[15]. Given a road network $G(V, E)$, a subset of vertices $C \subset V$ is a vertex cut of G if the deletion of C from G splits G into multiple connected components. A vertex set C is called the vertex cut of vertices u and v if u and v are in different connected components by the deletion of C from G.*

Theorem 1. *Given a road network G, let C be a vertex cut for two vertices s and d, we have: $dist(s, d) = min(dist(s, v) + dist(v, d))$, $v \in C$.*

That is, each path from s to t must pass through at least one vertex in C.

A tree decomposition of a graph $G(V, E)$, denoted as TG. We use $X(v)$, $v \in V$, to denote the root node of TG.

Property 1. Given a road network G, its tree decomposition TG , and $\forall X(u) \in TG$, for $\forall X(u) \backslash \{u\}$, is an ancestor of $X(u)$ in TG.

Property 2. Given a tree decomposition TG for a road network G, for any two vertices s and d,$s, d \in V$, suppose $X(s)$ is not an ancestor/decedent of $X(d)$ in TG , let X be the lowest common ancestor(LCA) of $X(s)$ and $X(d)$ in TG , then X is a vertex cut of s and d in G.

4 Global Path Optimization

Figure 2 illustrates the proposed framework. (1) Query grouping. The purpose is to group queries that may enter the same road segment during the same time period. Here, H2H index structure and grid index are used to speed up grouping. (2) Initial path planning. This part parallelly plans paths for queries in each group based on the current traffic flow, and updates traffic flow with new paths to guide subsequent queries to avoid congestion. (3) Local path optimization. Based on pre-set thresholds, some particularly congested sub paths are selected and re-planned. A grid merging strategy is designed to expand the area of local re-planning to avoid new congestion on alternative paths. (4) Iterative optimization. Global iterations are conducted and the optimal paths are obtained.

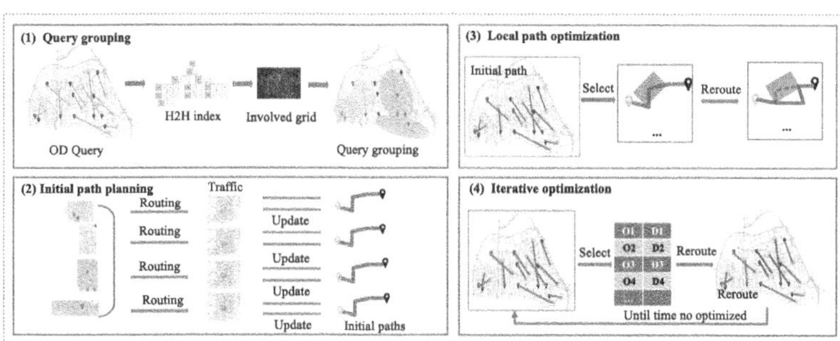

Fig. 2. The proposed framework for solving GRP problem.

4.1 Traffic Evaluation and Weight Update

Traffic Evaluation. Following path planning, the road network state need be evaluated to ensure accurate travel time, so we introduce a traffic evaluation algorithm. Due to the mutual impact of paths, it is not possible to compute

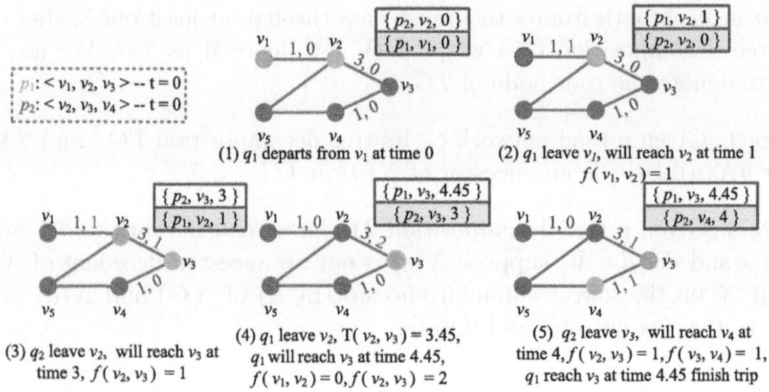

Fig. 3. Example of traffic evaluation. The values on edges denote the travel time of free-flow and the traffic flow respectively. Capacities of all edges are set to 1.

them one by one in order while calculating a set of paths. When the vehicle enters road e at time t, the traffic flow at that time has already determined the actual travel time for the vehicle to pass through e, vehicles entering e after t will not influence the travel time. To find the travel time for the next road on each path, all vertices can be accessed in time-increasing order. We queue the time for each path to the next node, iteratively extract the earliest path to the next node from the queue, calculate its actual travel time, and clear the traffic flow generated on the previous edge while recording the traffic flow generated on the current edge. If the current path reaches destination, add the travel time to $T(P)$. After all paths have reached destination, the global travel time is obtained.

As shown in Fig. 3, there are two paths $p_1\{v_1, v_2, v_3\}$ and $p_2\{v_2, v_3, v_4\}$. We initialize the priority queue as $pq = \{(p_1, v_1, 0), (p_2, v_2, 0)\}$. Because p_1 first removes from pq and q_1 leaves v_1 and reach v_2 at 1, as the traffic flow $f(v_1, v_2) = 0$. And, push $\{(p_1, v_2, 1)\}$ in pq. In the following stages, q_2 leaves v_2 and reach v_3 at 3, pq and edge flows are updated based on each movement. When q_1 moves, due to $f(v_2, v_3)$ being 1, the actual travel time can be calculated based on BPR to be 3.45, so push $\{(p_1, v_3, 4.45)\}$ into pq. Then q_2 reaches its destination at time 4, q_1 reaches its destination at time 4.45. Therefore, we have $T(p1, p2) = 8.45$.

Weight Update. When evaluating traffic, each edge captures entry and exit times, forming a time-dependent function that records traffic flow at different times. The time-dependence function may become too complex in actual applications due to the dispersion of path time. However, we only need to show the congestion situation of the road during each period of time to guide subsequent path planning. Therefore, to simplify the time dependency function, we can record traffic flow in a time period format, that is, calculate the average travel time passing through that section of the road over a period of time, rather than at each specific point in time. The time-dependent function on each edge is denoted as $T_w(u, v, t_i)$, $e(u, v) \in E$. This simplification can retain critical congestion information while reducing the complexity of the function. The time dependent functions T_w of all edges E collectively form a traffic condition of the road network. And the T_w can be calculated during the traffic evaluation process. When the traffic evaluation enters a new time interval, we calculate the average travel time of the previous interval for all pass edges.

4.2 Query Grouping

During the query grouping, our algorithm identifies conflicting queries and only handles local conflicts, reducing search space to improve result quality and computational efficiency. We use the H2H index nature combined with quadtree grid partitioning to identify fast conflict paths, and use temporal and spatial auxiliary information to group them. Path conflicts and congestion only occur when the query reaches the same path at the same time. Hence, it is only after the actual paths are determined that we can accurately identify conflict queries. However, pre-calculating each query's path in a traffic-free state and then grouping through conflict recognition is too time-consuming.

Definition 5. ***Involved vertex.*** *Given a road network G, let C be a vertex cut for two vertices s and d, we have: $dist(s,d) = min(dist(s,v) + dist(v,d))$, and if $dist(s,d) = dist(s,u) + dist(u,d)$, $u, v \in C$, u is the involved vertex of $q(s,t)$.*

The H2H index records the minimum arrival time and involved vertices for any two vertices. As a spatial partitioning structure, quadtree [17] can be used to grid regions based on the number of vertexes. By combining the H2H index with quadtree grid partitioning, we determine the regions and reach time each query needs to pass through. This information serves as a basis for grouping, enabling the assessment of query similarity in both spatial and temporal dimensions.

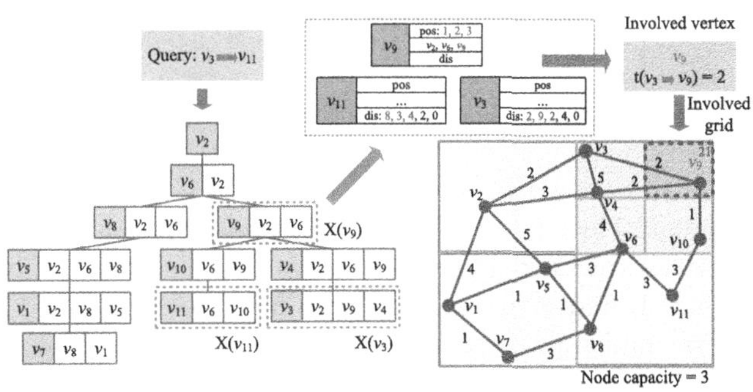

Fig. 4. Example of involved grid calculation for $q = <v_3, v_{11}>$.

The H2H index stores its shortest distance to each ancestor node in each X, $X \in TG$, which means we can calculate the involved vertexes and the shortest time to reach these vertices. When queries overlap in the same area and time, conflicts may arise. So, we extend the involved vertices to involved grids. We use quadtree for grid partitioning to balance vertice numbers in each grid, aiding in the efficient identification of queries that share the same involved grid and may potentially conflict. We divide the recursive map space into four quadrants and ensure that each grid contains fewer vertexes than the capacity. This approach enables the rapid identification of queries reaching the same grid within the same time period. The following example will show the calculation process.

Figure 4 shows an example of involved grid calculation for $q =< v_3, v_{11} >$. The LCA is X_{v_9}, and $dist(v_3, v_{11}) = min\{dist(v_3, v_2) + dist(v_2, v_{11}), dist(v_3, v_6) + dist(v_6, v_{11}), dist(v_3, v_9) + dist(v_9, v_{11})\} = \{2+8, 9+3, 2+4\} = 6$. So the involved vertex of q is v_9, and the time for v_3 to reach v_9 is $dist(v_3, v_9) = 2$. From the quadtree illustrated in the lower right corner, we can see the division of each grid. And v_9 is in grid 21, which will be identified as a involved grid for q.

In recognizing conflict queries, relying only on shared involved grids and similar arrival times may lack accuracy. Conflict queries should have similar search spaces, and judging similarity based solely on reaching the same area may overlook important information such as path direction and angle. If paths have opposite directions or vertical angles, their search spaces are less likely to overlap. Therefore, comparing the path direction and angle of two queries can help determine whether the queries have similar search spaces.

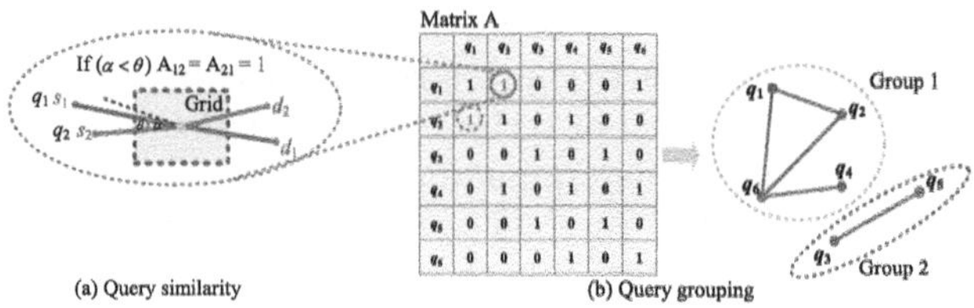

Fig. 5. Example of query grouping.

For queries with the same involved grid and approximate arrival time, we determine similarity by comparing the angle difference between their origin and the grid center line as shown in Fig. 5-(a). If the angle $\alpha < \theta$, we consider two queries to be similar. Through the temporal and spatial pruning techniques, we can filter out queries that are unlikely to cause conflicts. We maintain a query correlation matrix A, and if q_1 and q_2 are similar, we set $A_{12} = 1$. As shown in Fig. 5-(b), q_1 is similar to q_2 and q_4, if q_2 is similar to q_4. This issue can be transformed into a graph connectivity problem. By treating each query as a node, and using the query correlation matrix to denote connectivity, we can determine the number of independent correlated query groups by analyzing the connected components in the graph. We traverse each node by depth-first search, marking and collecting nodes belonging to the same connected graph. Thus, there are as many query groups as the number of connected graphs obtained.

Algorithm 1: GRP_init

Input: Network $G(V, E, T)$; Query group Q_j; Time-dependent function T_w
Output: Path planning results P_j of Q_j
while $q_i \in Q_j$ **do**
  Init: $\forall v \subseteq G.v, v.d = \infty, v.pre = NULL, s_i.d = 0, PQ.push(s_i)$;
  while $PQ \neq \emptyset$ **do**
    $u = PQ.pop()$;
    if $u = d_i$ **then**
      while $PQ \neq \emptyset$ **do**
        $p_i.push(u)$, $u = u.pred$;
    for *each vertex* $v \in G.adj[u]$ **do**
      if $v.d > u.d + t(u, v)$ **then**
        $v.d = u.d + T_w(u, v, u.d)$, $v.pre = u$;
        if $v \notin PQ$ **then**
          $PQ.push(v)$;
  Update $T_w \leftarrow p_i$
return P_j;

4.3 Initial Path Planning

During initial path planning, we alternate routing planning and utilize new paths to update traffic condition within the group to obtain high-quality path results, and batch process each group to improve efficiency. Since conflicts typically arise within each group, we focus on resolving conflicts locally without considering all queries. Query Q is divided into $\{Q_1, \cdots, Q_k\}$, where $Q_i = \{q_1, \cdots, q_m\}, 1 \leq i \leq k$. As the possibility of congestion between query groups where temporal and spatial do not overlap is low, we batch process Q_i. However, there is still a potential for conflicts in the queries in Q_i, so we sequentially plan the path.

We use *Dijkstra* for path planning, where the path planning for first query uses free flow travel time as the weight. For the path planning of all subsequent queries, we use T_w of the new traffic flow generated from all previous paths, to avoid congested areas. To simplify the evaluation process, instead of recalculating the entire traffic evaluation after each path planning, we optimize by only adding the newly planned path to the previous evaluation results.

The details are shown in Algorithm 1. Specifically, in order to solve the above problems, we proposed *GRP_init* algorithm, which improved the *Dijkstra*. During node expansion from the starting point s, the algorithm adjusts the weight value from the edge weight connecting s and the extended point v to the time-dependent function $T_w(s, v, t)$. Then, *GRP_init* updates T_w using p_i to avoid congestion in subsequent path planning. This process is iterated for each query within the query group until path planning for all queries is completed.

4.4 Local Path Optimization

After initial routing, we can identify the congestion areas, because not all reroute can reduce travel time, we only focus on queries that pass through congested areas,and provide traffic guidance to effectively reduce the total travel time. We propose a local rerouting algorithm that uses threshold filtering to reroute some paths, and introduces grid merging technology to expand the local rerouting area to avoid new congestion caused by alternative paths.

Local Re-planning Area. Firstly, we locate grids with severe congestion using traffic evaluation and record the congestion time of each grid to quantify the degree of congestion, we label severe congested grids that exceed the set threshold as "hot grid". Next, we will merge adjacent hot grids in congested paths to determine the area for local re-planning.

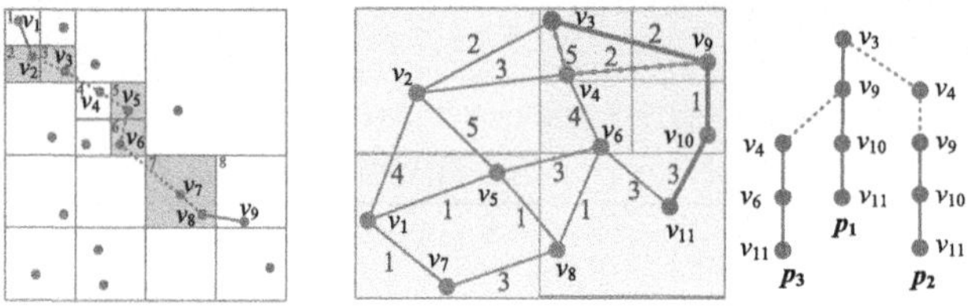

(a) Local re-planning area (b) Yen' s algorithm

Fig. 6. Example of local rerouting.

For each query with severe congestion, we maintain a stack, pushing vertices from the same hot grid onto the stack, until two consecutive non-hot grid vertexes appear in the stack. The search is halted at this moment, and we obtain the region of local re-planning. Because hot grids may continue to generate new congestion, we need to perform additional expansion during grid merging to avoid re-planning paths with severe congestion.

Given a query $q =< v_1, v_9 >$ in Fig. 6-(a), path result $p = \{v_1, v_2, v_3, v_4, v_5, v_6, v_7, v_8, v_9\}$ and hot grids are orange with id $\{3, 5, 6\}$, we aim to find the area for local re-planning. We establish a stack $S = \{\}$, and assess the first vertex v_1. As v_1 is not hot, we push v_1 into S. v_2 is also not hot and being in a different grid with v_1, and ensure that the bottom of S contains one non hot grid, so clear S, and push v_2 into S. Because v_3 is hot, push it into S, $S = \{v_2, v_3\}$. v_4 is not hot (grid 4), and the next v_5 is hot, so $S = \{v_2, v_3, v_4, v_5\}$. The process continues until v_7 and v_8 are not hot (grid 7), and the next v_9 also belongs to a different non-hot grid, the search completes, the area of local re-planning is $\{v_2, v_8\}$.

Local Re-Planning. After obtaining the local re-planning area, we use the k-Shortest Path (kSP) algorithm to find multiple alternative paths, and record the number of re-planning selections for edges. In future queries, while ensuring that the travel time difference is not significant, we prioritize selecting paths with fewer total edge selections to reduce conflicts between edges.

We use *Yen's* [22] to reroute based on the current traffic condition, with the k value set to 3. It first uses *Dijkstra* to obtain the shortest path from origin of re-planning area to destination, adding it to the list of k shortest paths. The *Yen's* involves iteratively removing edges along the shortest path, deviating from it to recalculate the next shortest path. This process is repeated to gradually generate k distinct shortest paths of varying lengths.

As shown in Fig. 6-(b), given a query $q =< v_3, v_{11} >$, $Yen's$ first uses $Dijkstra$ to find the shortest path. Path result $p_1 = \{v_3, v_9, v_{10}, v_{11}\}$ and is added to the list of shortest paths. Next, deviate from v_3 and v_9 in order, the first two iterations result in the paths are $p_2 = \{v_3, v_4, v_9, v_{10}, v_{11}\}$ and $p_3 = \{v_3, v_9, v_4, v_6, v_{11}\}$. Finally, the top 3 shortest paths from v_3 to v_{11} are obtained.

4.5 Iterative Optimization

The initial path planning is constrained by the given query order since it relies on traffic state after traffic estimation. To overcome this limitation, we re-plan some queries. Therefore, to approach the global optimal, those paths that significantly prolong the total time should be re-planned. After traffic evaluation, we arrange the travel time delay in ascending order, which is the actual travel time minus the free flow travel time. Re-plan the path for the top k ranked paths. We iteratively re-plan the paths to approach the optimal solution, but since each re-planning iteration may not necessarily reduce the total travel time, we halt the iteration when the results are no longer improving.

Algorithm 2: *GRP_iter*

Input: An road network $G(V, E, T)$; Quries Q
Output: Path planning results P; Total travel time T

1 **Init:** $T_{min} = INF, P_{min} = \{\}, flag = 0$;
2 $P\_init = GRP\_init(Q)$, $P = GRP\_ksp(P\_init)$;
3 **while** $flag < 2$ **do**
4     $Q_r = Pathselect(Q)$;
5   **for** *each* $q \in Q_r$ **do**
6         $p = Plan(q)$, $P(q) = p$;
7     $T = Traffic\_evaluation(P)$;
8   **if** $T < T_{min}$ **then**
9         $flag = 0$, $T_{min} = T, P_{min} = P$;
10   **else** $flag++$;
11 **return** P_{min};

The iterative optimization algorithm (Algorithm 2) selects the top k queries with the highest travel time delays in each iteration, conducts path planning, and updates path information based on traffic evaluation (lines 7-9). The process iterates until the global travel time shows no improvement for two consecutive iterations (lines 3-10). The algorithm returns the path set P_min.

5 Experimental Study

5.1 Experiment Settings

Comparison Algorithms. The comparison algorithm is as follows:

Dijkstra [6]. The path obtained using the *Dijkstra* algorithm under free flow travel time is labeled as the "Best_t", representing the optimal travel time of ideal conditions without congestion. The actual travel time is labeled as "Real_t".

Geo_BAT [21]. It groups queries to minimize conflicts within the same group, alternately performing batch path planning and traffic evaluation.

Geo_ITE [21]. On the basis of the *Geo_BAT*, this algorithm adds an iterative model to further approach the global optimum.

Next_opt [12]. This algorithm adopts a greedy strategy, iteratively selecting the next road segment with greedy selection and updating routing labels.

GRP_iter. Proposed in this paper, where the initial routing algorithm is denoted as *GRP_init*, the local path planning algorithm is denoted as *GRP_ksp*.

Datasets. We conducted experiments on synthetic and real-world query sets. The synthetic set involved selecting 15 seed OD pairs on the COL ($|V(COL)| = 21,048$, $|E(COL)| = 43,386$) and expanding neighboring vertices to generate 200 pairs of (o,d) for queries [15]. We also added 3000 randomly generated queries. The real-world dataset from DiDi, which included taxi queries in Beijing ($|V(BJ)| = 265,348$, $|E(BJ)| = 688,962$), we performed multiple rounds of random selection, and repeated 5, 3 times, and 2 times to simulate different traffic conditions. And, road capacity is randomly generated based on road length.

All algorithms are implemented in C++ and compiled with GNN GCC full optimization. All experiments are conducted on a machine with an Intel(R) Xeon(R) W-2245 CPU @3.2GHz and 250GB RAM under Linux(Ubuntu, 64bit).

Table 1. Results on synthetic dataset.

Graph	$\|Q\|$	*Best_t*	*Real_t*
Col	6000	614253	1582714
Algorithm	t_{travel}(min)	$\bar{t}_{travel}$(min)	$\bar{t}_{run}$(ms)
GRP_init	645424	107.6	3.83
GRP_ksp	641982	107	0.25
GRP_iter	641666	106.9	5.78
Geo_BAT	742144	120	8.4
Geo_ITE	662728	110	9.8
Next_opt	1582726	263	0.83

Table 2. Parameters.

Params	Values
$\|Q\|$	4000, **6000**, 9000
Traffic	Low, Normal, **High**
k	0.1%, **0.5%**, 1%, 5%
k'	0.05%, 0.1%, **0.5%**, 1%,5%
χ	1%, **5%**, 10%, 15%, 20%

Synthetic Dataset. Table 1 shows the performance of algorithms on the synthetic dataset. The level of conflict among queries in query set can be indicated by the difference between *Real_t* and *Best_t*, where a larger difference indicates more severe congestion. The default parameter settings are listed in Table 2. *Next_opt* is close to *Real_t*, as it greedily chooses the next path. When traffic is high, it cannot accurately predict congestion. The travel time for *GRP_iter* is slightly better than *Geo_ITE*, and both can avoid congestion. However, the dense traffic flow adversely affects the grouping of *Geo_BAT*, increasing travel time and impacting iterative optimization results. Regarding the optimization strategy in *GRP_iter*, despite adding random queries in artificial queries to interfere *GRP_init* grouping, the *GRP_init* avoids congestion. *GRP_ksp* further reduces travel time. *GRP_iter* excels in both efficiency and result quality.

Table 3. Performance on real-world query sets under different numbers of queries.

Avg_travel_time ((Global Travel_time / $\|Q\|$, min)										Avg_run_time (s)		
	Low			*Normal*			*High*			*High*		
Query num	4000	6000	9000	4000	6000	9000	4000	6000	9000	4000	6000	9000
Best_t	14.8	14.9	15	14.8	14.8	14.9	14.5	14.9	15	– –	– –	– –
Real_t	15.7	27.6	6255	25	709	8346	24.8	1077	10339	– –	– –	– –
GRP_iter	**14.9**	**15**	**15.2**	**14.9**	**15**	**15.2**	**14.6**	**15.1**	**15.3**	**0.073**	**0.077**	**0.080**
Geo_BAT	14.9	15	1176	14.9	107	4056	14.7	172	529	0.12	0.12	0.13
Geo_ITE	14.9	14.9	30.6	14.8	15.3	89	14.8	16.1	75	0.12	0.13	0.14
Next_opt	15.7	25.3	5776	24.8	258	7991	25	1155	10463	0.001	0.0009	0.001

Effect of Query Number, $|Q|$. Table 3 shows the global travel time (result quality) and the running time (efficiency) of algorithms under different number of queries. **Result quality**. With a small $|Q|$ and traffic, the travel time of the most algorithms are close to *Best_t* due to a few congestions in the road network. As $|Q|$ and traffic increases, the level of congestion increases. *Next_opt* adopts a greedy strategy, and the quality of the results significantly decreases. The travel time of *Geo_BAT* also increases, as its grouping quality decreases. It also prevents *Geo_ITE* results approaching the optimal solution. However, the result quality of *GRP_iter* hardly fluctuates. Because *GRP_iter* groups queries by using the characteristics of H2H index, the quality of grouping is not affected by the number of queries or congestion. Through iterative re-planning, *GRP_iter* further optimized the results. **Efficiency.** Larger $|Q|$ results in processing more vertices, increasing the running time of queries. The running time of map *BJ* is longer than that of *COL* because map *BJ* has more vertices. During traffic evaluation, it will traverse all the nodes that have passed through, which requires more time. *Next_opt* is the fastest, followed by *GRP_iter*, finally *Geo_BAT* and *Geo_ITE*. Because *Next_opt* only greedily expands the next road rather than iteratively updating, while *Geo_BAT* traverses all passing nodes for traffic evaluation after each group of path planning. *GRP_iter* updates the traffic status using new paths after each path planning within the group, and the time of each update is independent of $|Q|$, so its running time is smaller than *Geo_ITE*.

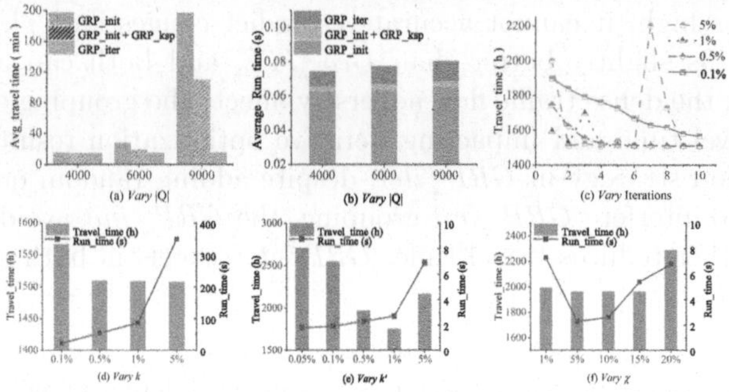

Fig. 7. Performance of algorithms on real query set under different parameters.

Effect of Optimization Strategies. As shown in Fig. 7-(a), *GRP_init* can avoid the areas where congestion may occur, thereby effectively reducing the travel time. *GRP_ksp* alleviates local congestion and optimizes the initial path results. *GRP_iter* further optimizes the results through iteration to reduce the travel time. In Fig. 7-(b), due to *GRP_init* uses *Dijkstra* to calculate the initial route for all queries and update the traffic. Although improving efficiency through batch processing, it is still the main time consumption of the algorithm.

Effect of k. As shown in Fig. 7-(c), as k increases, *GRP_iter* achieves faster improvements during iterations. However, if the value of k becomes too large, the routing quality becomes unstable and deteriorates. In Fig. 7-(d), with k increases, the number of queries that need to be rerouted during each iteration also increases, resulting in a significant increase in runtime.

Effect of k'. As shown in Fig. 7-(e), as the number of local rerouted queries increases, and the total running time also significantly increases. However, if k' is too large, its routing quality becomes unstable and deteriorates.

Effect of χ. As shown in Fig. 7-(f), we only reroute queries which the re-planning area is less than half the length of the path. When there are fewer hot grids, the number of rerouting paths increases, thus prolonging the runtime. As hot grids are not entirely identified, rerouted paths fail to effectively avoid congested grids, increasing travel time. However, as the area requiring rerouting grows larger, some queries are terminated for rerouting, increasing travel time.

The experimental results indicate that *GRP_iter* has significant advantages in terms of efficiency and result quality compared to the comparative algorithms.

6 Conclusion

In this paper, we present a comprehensive study on the *GRP* problem, which aims to return n optimal routes for a batch of n origin-destination queries, such that the overall travel time is minimal on the traffic-aware road network. A new framework is proposed and several optimization strategies have been designed to speed up the planning process while ensuring the quality of the results. Experimental results have shown that our algorithm performs better than others in terms of both efficiency and effectiveness. In future work, we will consider the fairness of the algorithm and make the detour time of each individual as equal as possible.

Acknowledgements. The research work was partially supported by the Key Research and Development Program of Liaoning Province under Grant No. 2023JH26/10300022; and the Shenyang Young and Middle-aged Scientific and Technological Innovation Talent Support Plan under Grant No. RC220504.

References

1. Abraham, I., Delling, D., Goldberg, A.V., Werneck, R.F.: Hierarchical hub labelings for shortest paths. In: Epstein, L., Ferragina, P. (eds.) Algorithms – ESA 2012: 20th Annual European Symposium, Ljubljana, Slovenia, September 10-12, 2012. Proceedings, pp. 24–35. Springer Berlin Heidelberg, Berlin, Heidelberg (2012). https://doi.org/10.1007/978-3-642-33090-2_4
2. Akiba, T., Iwata, Y., Yoshida, Y.: Dynamic and historical shortest-path distance queries on large evolving networks by pruned landmark labeling. In: WWW 2014, pp. 237–248. ACM (2014)
3. Board, T.: Transportation research board special report 309: (2012)
4. Chen, L., Shang, S., Yao, B., Li, J.: Pay your trip for traffic congestion: dynamic pricing in traffic-aware road networks. In: IAAI 2020, pp. 582–589. AAAI (2020)
5. Cooke, K.L., Halsey, E.: The shortest route through a network with time-dependent internodal transit times. JMAA **14**(3), 493–498 (1966)
6. Dijkstra, E.W.: A note on two problems in connexion with graphs. Numer. Math. **1**, 269–271 (1959)
7. Geisberger, R., Sanders, P., Schultes, D., Delling, D.: Contraction hierarchies: faster and simpler hierarchical routing in road networks. In: McGeoch, C.C. (ed.) WEA 2008. LNCS, vol. 5038, pp. 319–333. Springer, Heidelberg (2008). https://doi.org/10.1007/978-3-540-68552-4_24
8. Geisberger, R., Sanders, P., Schultes, D., Vetter, C.: Exact routing in large road networks using contraction hierarchies. Transp. Sci. **46**(3), 388–404 (2012)
9. Hart, P.E., Nilsson, N.J., Raphael, B.: A formal basis for the heuristic determination of minimum cost paths. IEEE Trans. Syst. Sci. Cybern. **4**(2), 100–107 (1968)
10. Kanoulas, E., Du, Y., Xia, T., Zhang, D.: Finding fastest paths on a road network with speed patterns. In: ICDE 2006, p. 10. IEEE Computer Society (2006)
11. Li, J., Ni, C., He, D., Li, L., Xia, X., Zhou, X.: Efficient knn query for moving objects on time-dependent road networks. VLDB J. **32**(3), 575–594 (2023)
12. Li, K., Chen, L., Shang, S.: Towards alleviating traffic congestion: optimal route planning for massive-scale trips. In: IJCAI 2020, pp. 3400–3406 (2020)

13. Li, K., Chen, L., Shang, S., Kalnis, P., Yao, B.: Traffic congestion alleviation over dynamic road networks: continuous optimal route combination for trip query streams. In: IJCAI 2021, pp. 3656–3662 (2021)
14. Li, L., Zhang, M., Hua, W., Zhou, X.: Fast query decomposition for batch shortest path processing in road networks. In: ICDE 2020, pp. 1189–1200. IEEE (2020)
15. Ouyang, D., Qin, L., Chang, L.: When hierarchy meets 2-hop-labeling: efficient shortest distance queries on road networks. In: SIGMOD, pp. 709–724. ACM (2018)
16. Ouyang, D., Wen, D., Qin, L., Chang, L., Lin, X.: When hierarchy meets 2-hop-labeling: efficient shortest distance and path queries on road networks. VLDB J. **32**(6), 1263–1287 (2023)
17. Samet, H.: Foundations of multidimensional and metric data structures. Elsevier/Morgan Kaufmann (2006)
18. Thomsen, J.R., Yiu, M.L., Jensen, C.S.: Effective caching of shortest paths for location-based services. In: SIGMOD 2012, pp. 313–324. ACM (2012)
19. Wang, Y., Li, G., Tang, N.: Querying shortest paths on time dependent road networks. Proc. VLDB Endow. **12**(11), 1249–1261 (2019)
20. Wilkie, D., van den Berg, J.P., Lin, M.C., Manocha, D.: Self-aware traffic route planning. In: AAAI 2011, pp. 1521–1527. AAAI (2011)
21. Xu, Y., Li, L., Zhang, M., Xu, Z., Zhou, X.: Global routing optimization in road networks. In: ICDE 2023, pp. 2524–2537. IEEE (2023)
22. Yen, J.Y.: Finding the k shortest loopless paths in a network. Manage. Sci. **17**(11), 712–716 (1971)
23. Zhang, M., Li, L., Hua, W., Mao, R., Chao, P., Zhou, X.: Dynamic hub labeling for road networks. In: ICDE 2021, pp. 336–347. IEEE (2021)

TransGAD: A Transformer-Based Autoencoder for Graph Anomaly Detection

Zehao Guo, Nannan Wu(✉), Yiming Zhao, and Wenjun Wang

College of Intelligence and Computing, Tianjin University, Tianjin, China
{guozehao,nannan.wu,zhaoym,wjwang}@tju.edu.cn

Abstract. Graph anomaly detection, aimed at identifying anomalous patterns that significantly differ from other nodes, has drawn widespread attention in recent years. Due to the complex topological structures and attribute information inherent in graphs, conventional methods often struggle to effectively identify anomalies. Deep anomaly detection methods based on Graph Neural Networks (GNNs) have achieved significant success. However, they face the challenge of not only obtaining limited neighborhood information but over-smoothing. Over-smoothing is the phenomenon where the representations of nodes gradually become similar and flattened across multiple convolutional layers, thereby limiting the comprehensive learning of neighborhood information. Therefore, we propose a novel anomaly detection framework, TransGAD, to address these challenges. Inspired by the Graph Transformer, we introduce a Transformer-based autoencoder. Treating each node as a sequence and its neighborhood as tokens in the sequence, this autoencoder captures both local and global information. We incorporate cosine positional encoding and masking strategy to obtain more informative node representations and leverage reconstruction error for improved anomaly detection. Experimental results on seven datasets demonstrate that our approach outperforms the state-of-the-art methods.

Keywords: Graph anomaly detection · Transformer · Autoencoder

1 Introduction

Graph anomaly detection, a representative task in graph learning, has garnered increasing attention from numerous researchers [7,15]. The goal of graph anomaly detection is to identify nodes that manifest significantly different behaviors in comparison to other nodes [27]. This undertaking has found widespread applications in areas such as fraud detection [1], network intrusion detection [16,24], and social spam detection [28].

However, anomaly detection on graphs is far from straightforward due to complex structure and high-dimensional attributes. As illustrated in Fig. 1, for a given node, the graph encompasses not only its neighborhood information but

M. Onizuka et al. (Eds.): DASFAA 2024, LNCS 14855, pp. 269–284, 2024.
https://doi.org/10.1007/978-981-97-5572-1_17

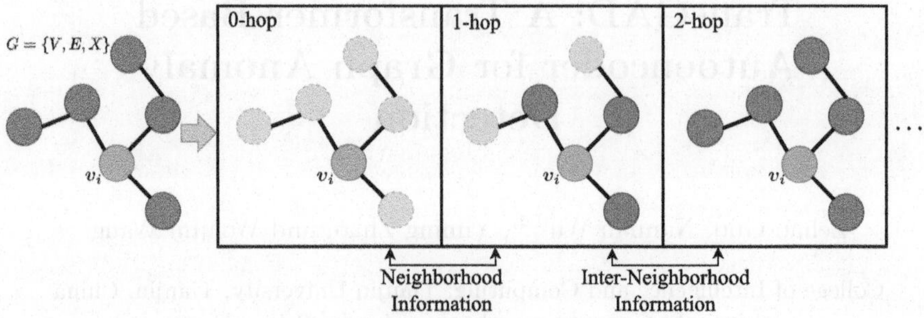

Fig. 1. The neighborhood and inter-neighborhood information available in the graph for a node v_i, where the red node denotes an anomalous node and blue nodes denote normal nodes. The aggregation of information from both the neighborhood and inter-neighborhood is crucial for generating node representations, particularly for anomalous nodes. (Color figure online)

also inter-neighborhood information. We posit that information from nearby neighborhoods represents local information of nodes, whereas information from distant neighborhoods and inter-neighborhood represents global information. Effectively aggregating both local and global information is not only crucial for enhancing the dissimilarity in representations between normal and anomalous nodes but is also a significant challenge.

Several approaches [4,6] employ Graph Neural Networks (GNNs) to aggregate neighborhood information and generate node representations. These approaches assert that anomalous nodes are difficult to reconstruct and utilize the reconstruction error of both structure and attributes to detect anomalous nodes. However, GNNs can solely capture local information by aggregating information of node's immediate neighbors. Merely adding more convolutional layers to aggregate distant neighborhood information gives rise to the over-smoothing, meaning that representations of nodes become progressively indistinguishable [2]. Several contrastive learning methods tackle this challenge by incorporating subgraph views to capture farther neighborhood information [12,32]. Nonetheless, the selection of subgraphs frequently depends on random walk methods, leading to the capture of limited neighborhood information. Furthermore, the obtained node representations overlook contextual information among different hop neighborhoods, suggesting that these methods continue to encounter difficulties in capturing more comprehensive global information.

To overcome the challenges outlined above, we design a Transformer-based autoencoder for Graph Anomaly Detection (TransGAD). In our design, the framework uses a Transformer module as the encoder, enabling the capture of both local and global information. The encoder module treats each node as a sequence, utilizing the representations of multi-hop neighborhoods as input tokens to generate node representations. Additionally, we integrate cosine positional encoding to further incorporate structural information into the node representation, thereby enhancing the aggregation of neighborhood information. Leveraging the node representations derived from the encoder module, the model's decoders autonomously reconstruct the features and structure of the graph. Anomalies are identified based on reconstruction errors. Furthermore, we introduce a masking strategy to more effectively exploit neighborhood information for improved anomaly detection. Our contributions are delineated as follows:

- We propose a Transformer-based autoencoder for Graph Anomaly Detection. The framework employs a Transformer module as the encoder, overcoming the issue of over-smoothing and effectively integrating neighborhood information into node representations. Additionally, the framework reconstructs the graph and utilizes the reconstruction errors to identify anomalous nodes.
- We introduce cosine positional encoding for node features to enhance the acquisition of neighborhood information. Additionally, a masking module is incorporated to optimize the utilization of neighborhood information in generating node representations.
- Experimental results on diverse datasets from various domains demonstrate the superiority of TransGAD over state-of-the-art methods.

2 Related Work

2.1 Graph Neural Networks

Graph Neural Networks (GNN) have proven to be effective tools for modeling graph data, achieving significant success in obtaining node and graph representations. A subtype of GNN, Graph Convolutional Networks (GCN), utilizes spectral convolution as a first-order approximation, enabling information propagation in the spectral domain for aggregating neighborhood information [8]. While GCN excels in node classification tasks, it grapples with the challenge of over-smoothing when integrating neighborhood information across multiple convolutional layers. Graph Attention Networks (GAT), a specific type of Graph Neural Networks (GNN), address the over-smoothing by incorporating an attention mechanism [23]. This mechanism assigns varying attention weights to nodes, fostering more flexible learning of relationships between them and significantly enhancing the performance of Graph Convolutional Networks (GCN).

Although GCN is effective and structurally simple, we avoid using GCN modules for encoding to mitigate over-smoothing. Instead, we integrate GCN as a primary component of the decoder, facilitating the effective reconstruction of node features.

2.2 Graph Anomaly Detection

In recent years, with the rapid development of graph learning, research in graph anomaly detection has attracted increasing attention. Radar [10] and ANOMALOUS [18] detect anomalous nodes through residual analysis. However, these approaches encounter challenges when applied to networks with complex structure and high-dimensional attribute features. With the development of deep learning, approaches leveraging graph neural networks have emerged for graph anomaly detection. A notable method, DOMINANT [4], employs an autoencoder framework with Graph Convolutional Network (GCN) as its backbone. It identifies anomalous nodes by reconstructing the adjacency matrix and feature matrix of the given graph, subsequently calculating reconstruction errors. However, the reliance on GCN for node representations in this method results in a constrained aggregation of neighborhood information. Moreover, with an increasing number of GCN layers, the model faces the challenge of over-smoothing. CoLA [12] and SL-GAD [32] utilize random walk algorithms to sample subgraphs and introduce a subgraph perspective, forming instance pairs for contrastive learning to enhance the utilization of neighborhood information. Nevertheless, subgraphs obtained through sampling still possess limited neighborhood information, and the node representations acquired through contrastive learning overlook contextual information between different hop neighborhoods. In contrast to the aforementioned methods, our framework mitigates over-smoothing by incorporating a Transformer encoder. Attention mechanism in the Transformer can be used to attend to multi-hop neighborhoods of a node, preventing the over-smoothing of the representation.

2.3 Graph Transformer

The Transformer model [22] has garnered considerable success in processing sequence data, especially in the domain of natural language processing.

The key component of Transformer is the self-attention module. Given the input $\mathbf{H} \in \mathbb{R}^{n\times d}$ for the self-attention module, where d is the hidden dimension. The input H is projected by three matrices, $\mathbf{W}_Q \in \mathbb{R}^{d\times d_Q}$, $\mathbf{W}_K \in \mathbb{R}^{d\times d_K}$ and $\mathbf{W}_V \in \mathbb{R}^{d\times d_V}$, to subspaces Q, K and V correspondingly. The output matrix is then calculated as:

$$\mathbf{Q} = \mathbf{H}\mathbf{W}_Q, \mathbf{K} = \mathbf{H}\mathbf{W}_K, \mathbf{V} = \mathbf{H}\mathbf{W}_V \tag{1}$$

$$\mathbf{A} = \frac{\mathbf{Q}\mathbf{K}^\top}{\sqrt{d_K}}, \quad \mathbf{H}' = \text{softmax}\left(\mathbf{A}\right)\mathbf{V}, \tag{2}$$

where the matrix $\text{softmax}\left(\mathbf{A}\right)$ captures the pair-wise similarity of input tokens in the sequence.

Thanks to the multi-head self-attention mechanism, numerous studies seek to extend the Transformer for graph data to address the over-smoothing of node representations [3,9,25,29]. In GraphTrans [25], nodes are treated as independent tokens, and a Graph Neural Network (GNN) serves as an auxiliary module, enabling the Transformer model to capture structure information and learn node representations from graph data. Conversely, NAGphormer [3] regards nodes as sequences, transforming the multi-hop neighborhoods of the node into representations and treating them as tokens to construct sequences for each node.

Inspired by [3], we represent nodes and their neighborhood representations as token sequences inputted to the Transformer for learning node representations.

3 Problem Formulation

In this section, we introduce the definition of graph and describe the research problem addressed in this paper.

Definition 1. ***Graph.*** *Given an attributed graph* $G = (V, E, \mathbf{X})$*, where* $V = \{v_1, v_2, \ldots, v_n\}$ *represents the set of nodes in the graph with* $|V| = n$*,* E *denotes the edge set,* $\mathbf{X} \in \mathbb{R}^{n \times d}$ *represents the features of nodes in the graph, and we use the vector* $\mathbf{x}_i$ *of the* i*-th row to denote the attributes of node* v_i*. We use an adjacency matrix* $\mathbf{A} \in \mathbb{R}^{n \times n}$ *to represent the graph's structure, where* $\mathbf{A}_{ij} = 1$ *if there is an edge between node* v_i *and* v_j*, and 0 otherwise.*

Definition 2. ***Graph Anomaly Detection.*** *Given an attributed graph* $G = (V, E, \mathbf{X})$ *with the node set* $V = \{v_1, v_2, \ldots, v_n\}$*, our objective is to learn an anomaly score function and compute an anomaly score* $k_i = f(v_i)$ *for each node* v_i*. The anomaly score* k_i *indicates the degree of anomaly for each node. Nodes that deviate singularly from the majority of reference nodes should be ranked in high positions by sorting them based on their anomaly scores.*

4 Methodology

In this section, we present the proposed framework of TransGAD in detail. As illustrated in Fig. 2, by treating nodes and their multi-hop neighborhoods as sequences and incorporating both cosine positional encoding and the masking strategy, TransGAD leverages a Transformer-based autoencoder structure for unsupervised anomaly node detection.

4.1 Neighborhood Representation Sequence

We offer a comprehensive description of the transformation process for nodes and their respective neighborhoods into a representation sequence. Following this, we introduce the integration of both cosine positional encoding and masking strategy into the model.

Neighborhood Representation. To facilitate the integration of nodes' neighborhood information into the Transformer framework, it is necessary to transform the neighborhoods of nodes into tokens for Transformer to learn. For a given node v, its k-hop neighborhood is defined as $N_k(v) = \{u \in V \mid d(u, v) \leq k\}$, where $d(u, v)$ represents the shortest distance between node u and node v. We denote the 0-hop neighborhood of a node as the node itself. Following this, We transform the k-hop neighborhood of node v to neighborhood embedding $\mathbf{x}_v^{(k)}$:

$$\mathbf{x}_v^{(k)} = \Phi(N_k(v)), \tag{3}$$

where Φ is an aggregation operation. For a given node v, we can construct its neighborhood sequence $S_v = \{\mathbf{x}_v, \mathbf{x}_v^{(1)}, \ldots, \mathbf{x}_v^{(K)}\}$, where K is fixed as a hyper-parameter representing the propagation step. By constructing neighborhood

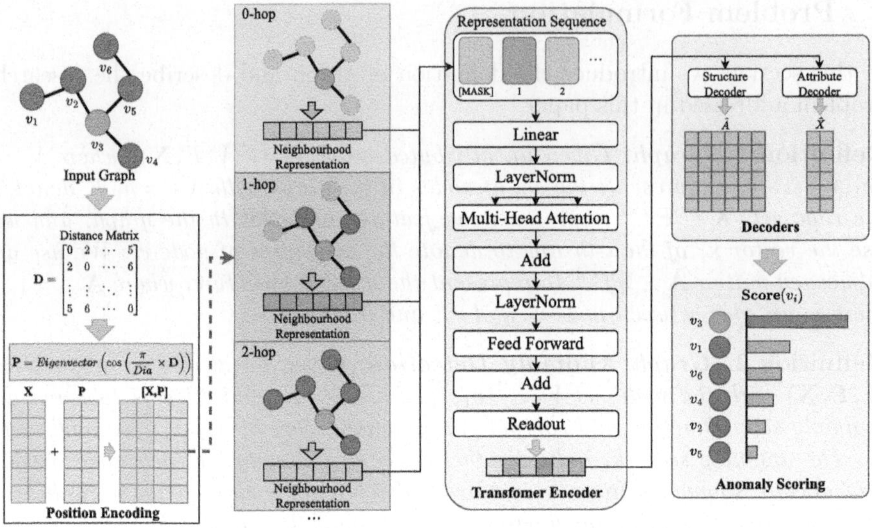

Fig. 2. Framework of TransGAD. A Transformer module is used in the encoder to aggregate local and global information, generating representations of nodes with more comprehensive neighborhood information. Cosine Positional Encoding is introduced to integrate structural information into the Transformer architecture. By reconstructing the structure and features of the graph with decoders, TransGAD detects anomalous nodes based on reconstruction error.

sequences for each node in the graph and concatenating them, we form the tensor $\mathbf{X}_G = \{\mathbf{X}_0, \mathbf{X}_1, \ldots, \mathbf{X}_K\}$, where $\mathbf{X}_G \in \mathbb{R}^{n\times(K+1)\times d}$, $\mathbf{X}_K$ is the K-hop neighborhood matrix, and $\mathbf{X}_0$ is the feature matrix $\mathbf{X}$ of graph G. In practice, multiplying the normalized adjacency matrix $\bar{\mathbf{A}}$ with the feature matrix $\mathbf{X}$ aggregates immediate neighborhood information. Here, $\bar{\mathbf{A}}$ is defined as $\bar{\mathbf{A}} = \tilde{\mathbf{D}}^{-\frac{1}{2}}\tilde{\mathbf{A}}\tilde{\mathbf{D}}^{-\frac{1}{2}}$, where $\tilde{\mathbf{A}}$ denotes the adjacency matrix with self-loops and $\tilde{\mathbf{D}}$ denotes the degree matrix corresponding to $\tilde{\mathbf{A}}$. Repeating this calculation multiple times facilitates the aggregation of information from distant neighborhoods. For example, the 1-hop and 2-hop neighborhood matrix is given by:

$$\mathbf{X}_1 = \bar{\mathbf{A}}\mathbf{X}, \quad \mathbf{X}_2 = \bar{\mathbf{A}}\mathbf{X}_1 = \bar{\mathbf{A}}^2\mathbf{X}. \tag{4}$$

Therefore, the K-hop neighborhood matrix can be represented as:

$$\mathbf{X}_K = \bar{\mathbf{A}}^K\mathbf{X}. \tag{5}$$

Through the iterative computation process described above, we generate a sequence of aggregated neighborhood representations, which can be fed into the Transformer.

It is noteworthy that, to acquire structural information in the graph, it is necessary to construct a feature matrix that integrates structural information as the initial input. The construction will be detailed in the following part.

Cosine Positional Encoding. The key idea of Graph Transformer is the integration of graph structure with the Transformer architecture [17], and the relative positions among nodes inherently contain structural information about the graph. Therefore, Consequently, we employ the minimum distance between nodes to encode the structural information of the graph. We define a symmetric matrix $\mathbf{D} \in \mathbb{R}^{n \times n}$:

$$\mathbf{D}_{ij} = d(i, j). \tag{6}$$

In cases where node i cannot reach node j, $d(i, j) = \infty$. Additionally, for each node i, we define its diameter as:

$$Dia(i) = \max_{j} d(i, j), \quad d(i, j) \neq \infty, \tag{7}$$

where $Dia(i)$ represents the shortest path between node i and its farthest reachable node.

Since nodes may have varying diameters, especially in large graphs where node diameters can be considerable, we normalize the distance matrix D row-wise using the cosine function. We define a matrix $\hat{\mathbf{D}} \in \mathbb{R}^{n \times n}$:

$$\hat{\mathbf{D}}_{ij} = \begin{cases} \cos\left(\frac{\pi}{Dia(i)} \times \mathbf{D}_{ij}\right), & \mathbf{D}_{ij} \neq \infty \\ -1.5, & \mathbf{D}_{ij} = \infty. \end{cases} \tag{8}$$

By capitalizing on the periodic nature of the cosine function, we project distances between reachable nodes in the distance matrix onto the interval $[-1, 1]$, and distances to unreachable nodes uniformly assigned a value of -1.5. Consequently, within the cosine distance matrix $\hat{\mathbf{D}}$, nodes farther apart have smaller values. Specifically, the distance to the node itself is 1, the distance to the furthest reachable node is -1, and the distance to unreachable nodes is set at -1.5, indicating distances surpassing even the farthest reachable node.

Finally, to compress the dimensions of the cosine distance matrix, we compute the eigenvectors of the cosine distance matrix. Specifically, we select eigenvectors corresponding to the s smallest non-trivial eigenvalues to construct the position embedding matrix $\mathbf{P} \in \mathbb{R}^{n \times s}$ and concatenate it with the feature matrix $\mathbf{X}$:

$$\mathbf{X}' = \mathbf{X} || \mathbf{P}, \tag{9}$$

where $\mathbf{X}' \in \mathbb{R}^{n \times d'}$, $d' = d + s$, and $||$ represents the concatenation operation. The combination of the position embedding matrix $\mathbf{P}$ and the feature matrix $\mathbf{X}$ achieves the integration of structural and attribute information.

Masking Strategy. To better utilize neighborhood information for generating node representations, we introduce a masking strategy. Specifically, we randomly sample a node v and replace the first representation in the neighborhood representation sequence S_v, which corresponds to the node v itself, with a mask token [MASK]. We employ a non-replacement uniform random sampling strategy to obtain the masked node, preventing potential biases.

Through the masking strategy, the model can better leverage neighborhood information to generate node representations.

4.2 Transformer-Based Encoder

After acquiring the neighborhood representation sequence S_v with the fused positional encoding for node v, we feed it into the encoder module.

To align the input dimensions with the hidden layer dimensions of the Transformer, we pass all representations in the sequence S_v through a fully connected layer:

$$\mathbf{z}_v^{(i)} = \text{Linear}(\mathbf{x}_v^{(i)}, \mathbf{W}, b), \quad i \in \{0, 1, \ldots, K\}, \tag{10}$$

where $\mathbf{W} \in \mathbb{R}^{d' \times d_m}$. The hidden layer dimension of Transformer is d_m. Finally, we obtain the transformed representation sequence $\mathbf{Z}_v = \{\mathbf{z}_v^{(0)}, \mathbf{z}_v^{(1)}, \ldots, \mathbf{z}_v^{(K)}\}$, where $\mathbf{Z}_v \in \mathbb{R}^{(K+1) \times d_m}$, and feed it into the Transformer.

Following the methodology presented in [22], we implement a single-layer Transformer block consisting of a multi-head self-attention block (MHA) and a position-wise feed-forward network (FFN), with a layer normalization (LN) layer placed before each block [26]:

$$\mathbf{Z}'_v = \text{MHA}\left(\text{LN}\left(\mathbf{Z}_v\right)\right) + \mathbf{Z}_v, \tag{11}$$

$$\mathbf{H}_v = \text{FFN}\left(\text{LN}\left(\mathbf{Z}'_v\right)\right) + \mathbf{Z}'_v, \tag{12}$$

where FFN includes two linear layers with GELU as the non-linear activation function, and the output $\mathbf{H}_v = \{\mathbf{h}_v^{(0)}, \mathbf{h}_v^{(1)}, \ldots, \mathbf{h}_v^{(K)}\}$ contains the embeddings for all neighborhoods of node v, where $\mathbf{H}_v \in \mathbb{R}^{(K+1) \times d_m}$.

Finally, to learn the importance of different neighborhoods in the node representation, we decide to use an attention-based readout function, assigning different weights to different neighborhoods and aggregating them to obtain the representation of node v:

$$\alpha_k = \frac{\exp((\mathbf{h}_v^{(0)} || \mathbf{h}_v^{(k)})\mathbf{W}_a^\top)}{\sum_{i=1}^{K} \exp((\mathbf{h}_v^{(0)} || \mathbf{h}_v^{(i)})\mathbf{W}_a^\top)} \quad , i = 1, 2, \ldots, K, \tag{13}$$

$$\mathbf{h}_v = \mathbf{h}_v^{(0)} + \sum_{k=1}^{K} \alpha_k \mathbf{h}_v^{(k)} \quad , i = 1, 2, \ldots, K, \tag{14}$$

where $\mathbf{W}_a \in \mathbb{R}^{1 \times 2d_m}$ denotes the learnable projection. By concatenating representations of all nodes, we obtain the embedding matrix $\mathbf{H} \in \mathbb{R}^{n \times d_m}$.

4.3 Attribute Decoder and Structure Decoder

The decoder employs the output from the encoder to reconstruct both the structure and attributes of the graph.

Structure Decoder takes the latent representations of two nodes as input and predicts whether there is an edge between them:

$$p(\hat{\mathbf{A}}_{i,j} = 1 | \mathbf{h}_i, \mathbf{h}_j) = \text{sigmoid}(\mathbf{h}_i, \mathbf{h}_j^\top), \tag{15}$$

where $\hat{\mathbf{A}}$ is the reconstructed adjacency matrix. Extending this process to all nodes allows us to obtain the complete reconstructed adjacency matrix:

$$\hat{\mathbf{A}} = \text{sigmoid}(\mathbf{H}\mathbf{H}^\top). \tag{16}$$

Attribute Decoder takes the latent representations of nodes as input and outputs the reconstructed node attributes. We use a 2-layer Graph Convolutional Network (GCN) as the attribute decoder to reconstruct node attributes:

$$\mathbf{H}' = \text{ReLU}(\bar{\mathbf{A}}\mathbf{H}\mathbf{W}^{(0)}), \tag{17}$$

$$\hat{\mathbf{X}} = \text{ReLU}(\bar{\mathbf{A}}\mathbf{H}'\mathbf{W}^{(1)}), \tag{18}$$

where $\bar{\mathbf{A}}$ is normalized adjacency matrix and $\hat{\mathbf{X}}$ is the reconstructed attribute matrix. $\mathbf{W}^{(0)} \in \mathbb{R}^{d_m \times d'_m}$ and $\mathbf{W}^{(1)} \in \mathbb{R}^{d'_m \times d}$ are learnable weight matrix.

4.4 Graph Anomaly Detection

We follow the assumption in [4] that nodes with higher anomalous characteristics present increased difficulty in reconstructing both their structure and attributes. To jointly learn reconstruction errors, the objective function of TransGAD can be formulated as:

$$R_S = ||\mathbf{A} - \hat{\mathbf{A}}||_F^2, \tag{19}$$

$$R_A = ||\mathbf{X} - \hat{\mathbf{X}}||_F^2, \tag{20}$$

$$\mathcal{L} = \alpha R_A + (1 - \alpha) R_S, \tag{21}$$

where $|| \cdot ||_F$ represents the Frobenius norm of a matrix and α is a hyperparameter that balances the importance of attribute reconstruction and structure reconstruction.

By minimizing the above objective function, we formulate the anomaly score function, using the reconstruction errors to measure the anomaly level of each node:

$$\text{Score}(v_i) = \alpha||\mathbf{x_i} - \hat{\mathbf{x}}_\mathbf{i}||_2 + (1 - \alpha)||\mathbf{a_i} - \hat{\mathbf{a}}_\mathbf{i}||_2, \tag{22}$$

where $|| \cdot ||_2$ represents the ℓ_2-norm and nodes with larger reconstruction errors have higher anomaly scores, indicating a higher probability of being anomalous. Following the computation of anomaly scores for each node, we sort them in descending order to better identify anomalous nodes.

5 Experiments

In this section, we validate the effectiveness of the TransGAD framework through experiments on seven datasets. We adopt six state-of-the-art anomaly detection methods as baselines and conduct experiments following their configurations to compare fairly with our approach.

To further investigate the performance of TransGAD, we conduct ablation experiments and parameter sensitivity experiments.

5.1 Dataset Description

The anomaly detection dataset comprises two types: ground-truth graphs and injected anomaly graphs. While many prior studies only use injected anomaly graphs as experimental datasets, we aim for a more comprehensive evaluation of TransGAD. Consequently, the benchmarks of our experiments incorporate both ground-truth graphs and injected anomaly graphs. Specifically, we use five injected anomaly graphs and two ground-truth graphs as benchmarks in our experiments. Details of the seven datasets, including injected anomaly graphs and ground-truth graphs, are shown in Table 1.

Table 1. The statistics of the datasets.

Dataset	Nodes	Edges	Features	Anomalies
Cora [19]	2,708	5,429	1,433	150
Citeseer [19]	3,327	4,732	3,703	150
ACM [20]	16,484	71,980	8,337	600
BlogCatalog [21]	5,196	171,743	8,189	300
DBLP [30]	5,484	8,117	6,775	300
Weibo [11]	8,405	407,963	400	868
Books [11]	1,418	3,695	21	28

Ground-Truth Anomaly Graphs. Books and Weibo are two datasets with ground-truth labels [11]. Books represents a co-purchase network of books from Amazon, and the ground truth labels are derived from the AmazonFail tag information. The Weibo dataset consists of users and related hashtags from the Tencent Weibo platform, forming a graph where nodes represent users connected by edges if they share the same hashtag. Users who made at least five suspicious events are marked as suspicious, while others are labeled as normal. The suspicious users are regarded as anomalies in the graph. The feature vectors for each user consist of the location and bag-of-words features of Weibo posts.

Injected Anomaly Graphs. We use four citation networks (Cora [19], Citeseer [19], DBLP [30], ACM [20]) along with one social network (BlogCatalog [21]) as datasets. The nodes of citation networks represent published papers, with edges denoting citation relationships and textual content serving as node features. The nodes of BlogCatalog represents users of the website, with edges indicating relationships between users and personalized content extracted as node features. Since no ground-truth anomalies are included in these datasets, we follow the approach in [4] to inject an equal number of feature and structure anomalies into each dataset.

5.2 Experimental Setup

Baseline Selection. We compare the TransGAD framework with six powerful anomaly detection methods: **DOMINANT** [4], **CoLA** [12], **SL-GAD** [32], **ComGA** [14], **Sub-CR** [31] and **GRADATE** [5]. Among them, CoLA, SL-GAD, Sub-CR, and GRADATE methods are based on contrastive learning, while DOMINANT and ComGA methods are based on autoencoders.

Evaluation Metrics. To measure the performance of TransGAD and its competitors, we adopt ROC-AUC as the evaluation metric. In the field of anomaly detection, ROC-AUC is widely used to evaluate the performance of various methods. The ROC curve is a plot of the true positive rate against the false positive rate. AUC represents the area under the ROC curve, with values ranging from [0,1], where a higher AUC indicates better detection performance.

Parameter Settings. We perform hyperparameter tuning for each baseline according to the recommended configurations in the original papers. For TransGAD, we use the AdamW [13] optimizer with a learning rate chosen from 5e-3, 3e-3, 1e-3 and weight decay set to 1e-4. Different learning rates are set for different datasets. Specifically, we set the learning rates to 1e-3, 5e-3, 5e-3, 5e-3, 5e-3, 5e-3, 1e-3 for the Cora, Citeseer, BlogCatalog, ACM, DBLP, Weibo, and Books datasets, respectively. For each dataset, we conduct training for 1000 epochs.

5.3 Experimental Result

Comparison with Baselines. The experimental results of TransGAD and six other baselines on seven datasets are shown in Table 2. The results indicate that TransGAD outperforms the competing methods on all benchmark datasets. This is attributed to the Transformer encoder's aggregation of both local and global information through the multi-head self-attention mechanism, enabling the representation of anomalous nodes to be distinguished from normal nodes.

Table 2. AUC values (%) on seven datasets. The values highlighted in bold and underlined represent the best and runner-up results, respectively.

Method	Cora	Citeseer	ACM	BlogCatalog	DBLP	Weibo	Books
DOMINANT(2019)	0.8155	0.8251	0.7601	0.7468	0.8301	0.8501	0.5810
CoLA(2021)	0.8779	0.8968	0.8237	0.7854	0.7824	0.2448	0.4548
SL-GAD(2021)	0.9130	0.9136	<u>0.8538</u>	<u>0.8184</u>	<u>0.8461</u>	0.4290	0.5314
ComGA(2022)	0.8840	0.9167	0.8469	0.8140	0.8287	<u>0.9269</u>	0.5508
Sub-CR(2022)	0.9132	<u>0.9303</u>	0.8428	0.8141	0.8061	0.4867	0.4753
GRADATE(2023)	<u>0.9146</u>	0.8282	0.8478	0.5925	0.7245	0.2112	<u>0.5872</u>
TransGAD	**0.9247**	**0.9354**	**0.8951**	**0.8186**	**0.9554**	**0.9315**	**0.6004**

On the ground-truth datasets, TransGAD achieve the best results. It is noteworthy that contrastive learning methods commonly exhibit poor performance on ground-truth datasets. This might be attributed to the significant disparities between ground-truth anomalies and injected anomalies, such as the infrequent occurrence of fully connected node clusters in the real world. Contrastive learning methods, by selecting subgraphs through random walks, can only aggregate limited neighborhood information, making it challenging to detect anomalies in the real world. In contrast, TransGAD effectively aggregates both local and global information, enabling the successful detection of ground-truth anomalies.

On inject-anomaly datasets, TransGAD achieve the best results, outperforming its competitors, particularly on citation network datasets. TranGAD exhibits a marginal lead on the BlogCatalog dataset, possibly owing to the dataset's higher average degree, where local neighborhood information is already sufficiently rich, and the integration of global information has a limited impact on representations of nodes.

Compared to GNN-based methods DOMINANT and ComGA, TransGAD exhibit superior performance, because TransGAD mitigates over-smoothing of anomalous nodes. While ComGA alleviates over-smoothing by leveraging community structure information, it still encounters limitations in acquiring comprehensive global information, making it challenging to differentiate certain anomalous nodes. TransGAD, through the multi-head self-attention mechanism, captures contextual information between multi-hop neighborhoods, enabling it to avoid over-smoothing and learn more comprehensive global information. This capability enhances its effectiveness in detecting anomalous nodes.

Compared to contrastive learning methods, TransGAD demonstrate better performance, mainly due to the limited subgraph scale introduced by contrastive learning methods, which cannot comprehensively capture neighborhood information and overlook semantic correlations between multi-hop neighborhoods.

Ablation Study. In this section, we investigate the impact of positional encoding and masking on TransGAD. We discuss the performance on the Citeseer, DBLP and Books dataset.

Table 3. Ablation analysis for TransGAD model on three datasets. The values highlighted in bold represent the best results.

	Citeseer		DBLP		Books	
	w/o P.E.	cos P.E.	w/o P.E.	cos P.E.	w/o P.E.	cos P.E.
w/o Mask	0.9316	0.9326	0.9543	0.9552	0.5888	0.5919
Mask	0.9342	**0.9354**	0.9544	**0.9554**	0.5889	**0.6004**

Experiments are conducted both with and without positional encoding, as well as with and without masking. The results are presented in Table 3. The findings demonstrate that both positional encoding and masking contribute to enhancing TransGAD's performance. Cosine positional encoding intuitively integrates neighborhood and structural information, thereby strengthening the Transformer module's ability to learn essential neighborhood features. The masking strategy enables TransGAD to generate representations of nodes covered by the mask, facilitating the model in utilizing neighborhood information to learn critical details and structures.

Parameter Sensitivity Study. In this section, we investigate the influence of the balance factor α and propagation steps K on TransGAD. We discuss the performance on the Cora and Weibo dataset.

Balance factor α: In this experiment, we systematically vary the value of the balance factor (α) to assess its influence on AUC. The performance distinctions are illustrated in Fig. 3. The experimental findings reveal that TransGAD achieves optimal performance when α is set to 0.7 and 0.9 for the Cora and Weibo datasets, respectively. Additionally, when α is set to 0.5, 0.8, 0.8, 0.6 and 0.4, TransGAD achieves optimal performance on Citeseer, BlogCatalog, ACM, DBLP and Books datasets, correspondingly. Notably, it is observed that α greater than 0.5 is optimal on most datasets, indicating that our method, by capturing neighborhood information, excels in distinguishing attribute anomalies.

Propagation steps K: In this experiment, we systematically vary the value of propagation steps K to assess its impact on AUC. The performance distinctions are illustrated in Fig. 4. Notably, when K is small, TransGAD exhibits suboptimal performance across all datasets. As K increases, the model's performance gradually improves. Additionally, when K reaches larger values, the model's performance tends to stabilize without degradation, which substantiates that TransGAD effectively alleviates the over-smoothing of node representations, leading to a better discrimination of anomalous nodes. Specifically, for the Books dataset, TransGAD achieved optimal performance at K=5. This observation is likely attributable to the smaller scale of the Books dataset, where a reduced K proves sufficient for capturing neighborhood information effectively.

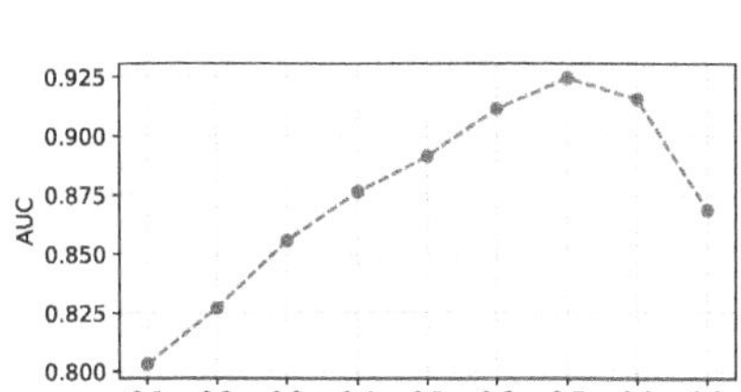

(a) Cora

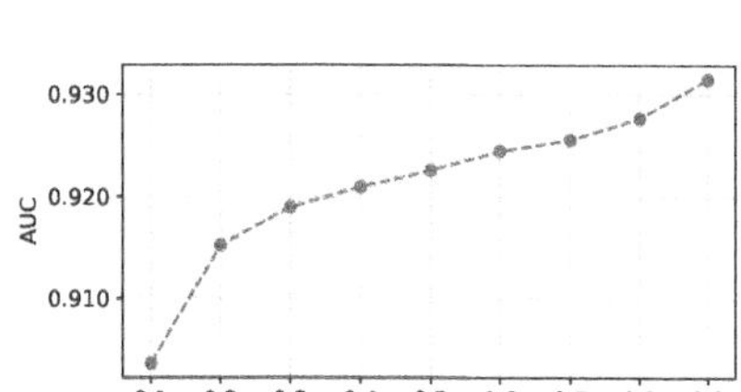

(b) Weibo

Fig. 3. AUC values of TransGAD on Cora and Weibo with different balance factor α.

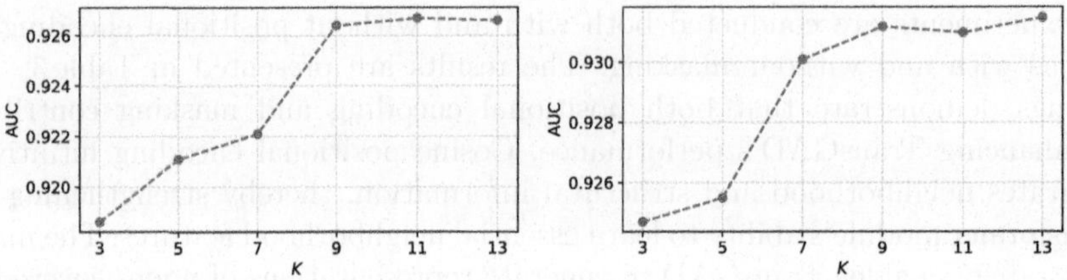

(a) Cora (b) Weibo

Fig. 4. AUC values of TransGAD on Cora and Weibo with different propagation steps K.

6 Conclusion

In this paper, we propose a novel Transformer-based autoencoder, TransGAD, for anomaly detection. By transforming neighborhoods of nodes into representations, we constructed sequences containing representations of multi-hop neighborhoods as input to the Transformer encoder, obtaining node representations that integrate neighborhood information. This approach mitigates over-smoothing of anomalous nodes. Using a decoder for reconstruction, we effectively identified anomalous nodes. Experiments on seven datasets demonstrated the effectiveness of our method. In the future, we will continue exploring graph Transformers to address the over-smoothing more effectively.

Acknowledgements. This work was supported by National Key Research and Development Program of China(31400), and partly supported by Natural Science Foundation of Inner Mongolia Autonomous Region of China (2022LHMS06008), Key R&D and Transformation Plan of Qinghai Province(2022-QY-218), and University-Industry Collaborative Education Program(202312280037).

References

1. Branco, B., Abreu, P., Gomes, A.S., Almeida, M.S., Ascensão, J.T., Bizarro, P.: Interleaved sequence RNNs for fraud detection. In: Proceedings of the 26th ACM SIGKDD International Conference on Knowledge Discovery and Data Mining, pp. 3101–3109 (2020)
2. Chen, D., Lin, Y., Li, W., Li, P., Zhou, J., Sun, X.: Measuring and relieving the over-smoothing problem for graph neural networks from the topological view. In: Proceedings of the AAAI Conference on Artificial Intelligence, vol. 34, pp. 3438–3445 (2020)
3. Chen, J., Gao, K., Li, G., He, K.: Nagphormer: a tokenized graph transformer for node classification in large graphs. In: The Eleventh International Conference on Learning Representations (2022)
4. Ding, K., Li, J., Bhanushali, R., Liu, H.: Deep anomaly detection on attributed networks. In: Proceedings of the 2019 SIAM International Conference on Data Mining, pp. 594–602. SIAM (2019)

5. Duan, J., et al.: Graph anomaly detection via multi-scale contrastive learning networks with augmented view. In: Proceedings of the AAAI Conference on Artificial Intelligence. vol. 37, pp. 7459–7467 (2023)
6. Fan, H., Zhang, F., Li, Z.: Anomalydae: dual autoencoder for anomaly detection on attributed networks. In: ICASSP 2020-2020 IEEE International Conference on Acoustics, Speech and Signal Processing (ICASSP), pp. 5685–5689. IEEE (2020)
7. Kim, H., Lee, B.S., Shin, W.Y., Lim, S.: Graph anomaly detection with graph neural networks: Current status and challenges. IEEE Access (2022)
8. Kipf, T.N., Welling, M.: Semi-supervised classification with graph convolutional networks. arXiv preprint arXiv:1609.02907 (2016)
9. Kreuzer, D., Beaini, D., Hamilton, W., Létourneau, V., Tossou, P.: Rethinking graph transformers with spectral attention. Adv. Neural. Inf. Process. Syst. **34**, 21618–21629 (2021)
10. Li, J., Dani, H., Hu, X., Liu, H.: Radar: residual analysis for anomaly detection in attributed networks. In: IJCAI, vol. 17, pp. 2152–2158 (2017)
11. Liu, K., et al.: Bond: benchmarking unsupervised outlier node detection on static attributed graphs. Adv. Neural. Inf. Process. Syst. **35**, 27021–27035 (2022)
12. Liu, Y., Li, Z., Pan, S., Gong, C., Zhou, C., Karypis, G.: Anomaly detection on attributed networks via contrastive self-supervised learning. IEEE Trans. Neural Netw. Learn. Syst. **33**(6), 2378–2392 (2021)
13. Loshchilov, I., Hutter, F.: Decoupled weight decay regularization. arXiv preprint arXiv:1711.05101 (2017)
14. Luo, X., et al.: Comga: community-aware attributed graph anomaly detection. In: Proceedings of the Fifteenth ACM International Conference on Web Search and Data Mining, pp. 657–665 (2022)
15. Ma, X., et al.: A comprehensive survey on graph anomaly detection with deep learning. IEEE Transactions on Knowledge and Data Engineering (2021)
16. Miao, K., Shi, X., Zhang, W.A.: Attack signal estimation for intrusion detection in industrial control system. Comput. Security **96**, 101926 (2020)
17. Min, E., et al.: Transformer for graphs: an overview from architecture perspective. arXiv preprint arXiv:2202.08455 (2022)
18. Peng, Z., Luo, M., Li, J., Liu, H., Zheng, Q., et al.: Anomalous: a joint modeling approach for anomaly detection on attributed networks. In: IJCAI, pp. 3513–3519 (2018)
19. Sen, P., Namata, G., Bilgic, M., Getoor, L., Galligher, B., Eliassi-Rad, T.: Collective classification in network data. AI Mag. **29**(3), 93–93 (2008)
20. Tang, J., Zhang, J., Yao, L., Li, J., Zhang, L., Su, Z.: Arnetminer: extraction and mining of academic social networks. In: Proceedings of the 14th ACM SIGKDD International Conference on Knowledge Discovery and Data Mining, pp. 990–998 (2008)
21. Tang, L., Liu, H.: Relational learning via latent social dimensions. In: Proceedings of the 15th ACM SIGKDD International Conference on Knowledge Discovery and Data Mining, pp. 817–826 (2009)
22. Vaswani, A., et al.: Attention is all you need. In: Advances in Neural Information Processing Systems, vol. 30 (2017)
23. Veličković, P., Cucurull, G., Casanova, A., Romero, A., Liò, P., Bengio, Y.: Graph attention networks. In: International Conference on Learning Representations (2018)
24. Wu, N., Chen, F., Li, J., Huai, J., Zhou, B., Ramakrishnan, N., et al.: A nonparametric approach to uncovering connected anomalies by tree shaped priors. IEEE Trans. Knowl. Data Eng. **31**(10), 1849–1862 (2018)

25. Wu, Z., Jain, P., Wright, M., Mirhoseini, A., Gonzalez, J.E., Stoica, I.: Representing long-range context for graph neural networks with global attention. Adv. Neural. Inf. Process. Syst. **34**, 13266–13279 (2021)
26. Xiong, R., et al.: On layer normalization in the transformer architecture. In: International Conference on Machine Learning, pp. 10524–10533. PMLR (2020)
27. Xu, H., et al.: Beyond outlier detection: Outlier interpretation by attention-guided triplet deviation network. In: Proceedings of the Web Conference 2021, pp. 1328–1339 (2021)
28. Ye, J., Akoglu, L.: Discovering opinion spammer groups by network footprints. In: Machine Learning and Knowledge Discovery in Databases: European Conference, ECML PKDD 2015, Porto, Portugal, September 7-11, 2015, Proceedings, Part I 15. pp. 267–282. Springer (2015). https://doi.org/10.1007/978-3-319-23528-8_17
29. Ying, C., et al.: Do transformers really perform badly for graph representation? Adv. Neural. Inf. Process. Syst. **34**, 28877–28888 (2021)
30. Yuan, X., Zhou, N., Yu, S., Huang, H., Chen, Z., Xia, F.: Higher-order structure based anomaly detection on attributed networks. In: 2021 IEEE International Conference on Big Data (Big Data), pp. 2691–2700. IEEE (2021)
31. Zhang, J., Wang, S., Chen, S.: Reconstruction enhanced multi-view contrastive learning for anomaly detection on attributed networks. arXiv preprint arXiv:2205.04816 (2022)
32. Zheng, Y., Jin, M., Liu, Y., Chi, L., Phan, K.T., Chen, Y.P.P.: Generative and contrastive self-supervised learning for graph anomaly detection. IEEE Transactions on Knowledge and Data Engineering (2021)

Unsupervised Node Clustering via Contrastive Hard Sampling

Hang Cui(✉) and Tarek Abdelzaher

University of Illinois, Urbana Champaign, Champaign, USA
{hangcui2,zaher}@illinois.edu

Abstract. This paper introduces a fine-grained contrastive learning scheme for unsupervised node clustering. Previous clustering methods only focus on a small feature set (*class-dependent features*), which demonstrates explicit clustering characteristics, ignoring the rest of the feature spaces (*class-invariant features*). This paper exploits *class-invariant features* via graph contrastive learning to discover additional high-quality features for unsupervised clustering. We formulate a novel node-level fine-grained augmentation framework for self-supervised learning, which iteratively identifies competitive contrastive samples from the whole feature spaces, in the form of *positive* and *negative* examples of node relations. While positive examples of node relations are usually expressed as edges in graph homophily, negative examples are implicit without a direct edge. We show, however, that simply sampling nodes beyond the local neighborhood results in less competitive negative pairs, that are less effective for contrastive learning. Inspired by counterfactual augmentation, we instead sample competitive negative node relations by creating virtual nodes that inherit (in a self-supervised fashion) *class-invariant features*, while altering *class-dependent features*, creating contrasting pairs that lie closer to the boundary and offering better contrast. Consequently, our experiments demonstrate significant improvements in supervised node clustering tasks on six baselines and six real-world social network datasets.

Keywords: Contrastive learning · Clustering

1 Introduction

Node clustering, clustering graph nodes into disjoint groups, has been studied extensively in various social network applications such as polarization detection [1], anomaly detection [2–4], and community detection [5]. Due to the sheer volume of data and timely requirement of online applications, the problem is usually considered to be unsupervised, where no labeled nodes are available. Node clustering can also be used as graph pre-training (clustering) before labeled streams arrive. Previous node clustering methods mostly follow (1). multi-stage schemes [1,6–8], which utilize graph encoders, such as variational auto-encoders and self-supervised learning, to encode the graph unsupervised and then

M. Onizuka et al. (Eds.): DASFAA 2024, LNCS 14855, pp. 285–300, 2024.
https://doi.org/10.1007/978-981-97-5572-1_18

filter a small set of feature spaces demonstrating strong clustering characteristics, (2). integrated schemes [9–11], which learn cluster-friendly representations via cluster-related regulators, such as modularity. However, Both schemes only focus on small feature sets, namely **class-dependent features**, while ignoring the vast **class-invariant features**, which can be further exploited to discover additional high-quality features for node clustering.

Contrastive learning has recently shown great performance on many self-supervised tasks, including image classification [12], graph representation learning [13], and sequence recommendation [14]. Its core idea is to generate *hard* positive and negative pairs (hard for a classifier to distinguish) that lie closer to class boundaries [15]. By contrasting pairs that share more similarity, projection into the latent space is forced to focus more on the subset of features that truly matter in distinguishing the classes. For example, by contrasting cats with cat-sized dogs, the mapping of the respective species into the latent space is forced to consider more pertinent features to distinguish the two, than contrasting cats with, say, very large dogs, where less pertinent features (like size) may disproportionately influence the mapping. This intuition calls for fine-grained augmentations that flip the class of a node while *minimally changing* its features.

Unfortunately, previous graph contrastive learning methods [16,17] mostly utilize coarse-grained augmentations, such as node dropping and edge perturbation. Although those methods perform well on self-supervised learning (train the graph via unsupervised contrastive learning, then freeze the trained model for downstream tasks), the obtained self-supervised representations are not clustering-friendly [18], requiring sufficient labeled data to fine-tune downstream tasks.

To further obtain high-quality features for clustering, we present a novel fine-grained contrastive learning scheme that iteratively expands *class-dependent features* by generating high-quality contrastive pairs. Inspired by the popular principle of conductance and modularity [9], which maximizes intra-cluster edges versus inter-cluster edges, we contrast each node's neighbors (as positive examples) against a selected set of negative examples, personalized to the anchor node. However, sampling negative examples is not straightforward. For instance, while buying a product repeatedly constitutes an endorsement, not buying it does not necessarily indicate a 'dislike' [2]. It might have other reasons, like not knowing about the product or missing the opportunity to buy it. Thus, the selection of contrasting edge absences must be done with care.

Efficient augmentation [13] calls for generating the *hardest* positive and negative pairs for the classifier/cluster to distinguish, based on the current learned parameters. Thus, the problem becomes: **what are the hardest negative examples to distinguish from positive ones?** Specifically, to ensure that the selected contrasting pairs are both competitive and negative, we introduce a feature decoupling scheme that captures *class-invariant features* (features invariant across classes) and *class-dependent features* (features variant to classes). We then create virtual nodes via augmenting *class-dependent features* while keeping *class-invariant features* unchanged, in order to create the most similar virtual nodes to help induce contrastive pairs. The above fine-grained augmentation method provides a parameter-free (and yet effective) scheme to generate hard positive/negative pairs.

2 Related Work

2.1 Node Clustering

Early attempts use basic topic models, such as Bayesian models [19] and non-negative matrix factorization [20,21]. Later methods utilize topic modeling with cluster-specific characteristics [5] to maximize modularity (ratio of intra-cluster edges and inter-cluster edges). The recent state of the art utilizes graph neural networks [5–11,17], such as graph auto-encoder and contrastive learning to integrate deep neural models and cluster-specific regulators.

2.2 Contrastive Learning

Contrastive learning has become state-of-the-art in various self-supervised learning tasks, including image recognition [12], sequence recommendation [14], and graph encoding [13]. The fundamental principle is to generate positive and negative pairs for the encoder, to help decouple desired features [15]. The positive pairs are mostly generated as augmentations or similar neighbors. Augmentation methods aim at decoupling the correlations of spurious noise between the representations of augmented positive samples [15]. The key challenge is to not corrupt the desired (sparse) features by augmentations. As a result, most works consult to pre-defined augmentations and distributions.

Previous graph contrastive learning methods resort to graph corruption, such as node masking and edge perturbation, and then contrast representation learning between the original graph and corrupted graphs [17]. Recent works propose to use generative adversarial network (GAN) to automatically generate augmentations by creating *hard* pairs for the encoder in the current epoch [13]. For domains where natural data augmentations are not available, previous literature also derive positive and negative pairs based on inferred (di)similarity to the anchor point. Examples include Mixup interpolation [22], contrastive predictive coding [23], and local neighborhood [24].

3 Problem Formulation and Preliminary

Given an attributed graph $G(V, E, X)$ where V is the vertex set, E is the edge set, and X is the input features of the graph. The graph is assumed to be unsigned, unweighted, and undirected since the edge weights are usually hard to acquire or estimate.

Since recent representation learning models enable easy fusion of various information, we also assume *optional* auxiliary information, in the form of relational graph G_V between each type of vertices. The relational graph can be constructed from (1) auxiliary relations, such as user following/friendship relation, and (2) content similarity derived from any suitable content encoders, such as BERT [25]. Given G and optional $\{G_V\}$, the goal is to cluster V into K disjoint classes C_k, where K is a known hyper-parameter.

Throughout this paper, we denote $\boldsymbol{H}$ as the embedding matrix, $\boldsymbol{h}$ as the embedding vector, $\boldsymbol{A} = \{a_{ij}\}$ as the adjacency matrix, $\boldsymbol{R} = \{r_{ij}\}$ as the class assignment matrix, $\mathcal{N}_u$ as the set of 1-hop neighbours of node u, d_u as the degree of node u. Furthermore, we use subscript d to represent matrices/vectors corresponding to *class-dependent features* and subscript o to *class-invariant features.*

3.1 Graph Contrastive Learning

Our method is based on self-supervised graph contrastive learning [13,17], which is a learning scheme consisting of two steps:
(1) Contrastive self-supervised learning: The graph is encoded unsupervised on a contrastive loss. The most popular graph encoding method is Graph Convolutional Netowrk(GCN) [26]

For an anchor node $\boldsymbol{v}$, the objective of contrastive learning is:

$$\mathcal{L}_{CE} = \mathbb{E}_{u^+ \sim P_p(u|v), u_k^- \sim P_n(u|v)}[-\log \frac{e^{f(v)^\top f(u^+)/\tau}}{\sum_k e^{f(v)^\top f(u_k^-)/\tau}}] \tag{1}$$

where τ is a temperature parameter, P_p and P_n are (often pre-defined) positive and negative distributions, respectively.

4 MeCole

Previous node clustering methods [5–11,17] focus on a small set of features demonstrating strong clustering characteristics, such as modularity and conductivity; while ignoring the rest of the feature spaces. Our method proposes a feature expansion scheme that iteratively **expands** the set of features for clustering:

- Start with a small set of **high-confident features**. The high-confident feature sets include two sub-components (1). **class-dependent feature set**: features of strong clustering characteristics (2). **class-invariant feature set**: features of minimal clustering characteristics.
- Expand the high-confident feature set on both components. In other word, our method iteratively discovers additional *class-dependent features* and *class-invariant features* from the rest of feature spaces.

4.1 Node-Level Fine-Grained Contrastive Learning

We propose a novel node-level fine-grained augmentation, which can efficiently generate *hard* positive/negative examples for any given anchor nodes. Our objective is to contrast between positive and negative examples of node relations, which are in line with the maximization of modularity and conductance (ratio of intra-class edges against inter-class edges). The positive node relations are sampled via edge connection. The negative node relations, however, are considerably more sophisticated to understand, due to the unavailability of associated content and the multitude of potential reasons for not having an edge, such as lack of observations and attention. Therefore, the core contribution of this paper lies in how to **generate a selected set of negative node relations that are competitive for contrastive learning**.

Our method is inspired by **counterfactual augmentations** [27,28]: how would the node interactions change if any feature(s) is augmented? Given an anchor node v, we synthesize a set of virtual nodes $\{\bar{v}\}$ via feature augmentation(s) as counterfactual examples of v. The deviation of node interaction patterns between the anchor node and the virtual nodes represents the counterfactual interactions, which are exploited to sample negative node relations.

To ensure the virtual nodes satisfy both negative and competitive, we propose to synthesize nodes via **minimal augmentation**: highly similar to v (competitiveness), but are clustered into other classes or the boundary (negativity) under the current learned parameters. We summarize the above idea into 4 conditions to generate virtual nodes $\bar{v}$:

- Competitiveness: maximize similarity to the anchor node v. There are two popular options for such similarity metrics: (1). L1 norm: the sum of absolute augmentations (2). L0 norm: the number of augmented features.
- Negativity: minimize the likelihood of being clustered to the same class as v. The reason is to ensure the synthesized node is true negative so that the sampled negative relations are not false negative.
- Randomness: minimize mutual information between pairs of synthesized virtual nodes. This is a contrastive learning condition to improve efficiency.
- Parameter-free: the generator model has few or no additional parameters. This is to reduce over-fitting in the self-supervised setting.

While the above conditions can be satisfied in various ways with different trade-offs, we provide a parameter-free yet effective scheme in this paper as a foundation of our framework.

4.2 Augmentation Scheme

The augmentation scheme is based on the **high-confident feature sets**: *class-invariant features* and *class-dependent features*. Given an anchor node, virtual nodes are created by randomly augmenting *class-dependent features* to randomly sampled nodes clustered into the opposing classes; while keeping *class-invariant features*. The intuition is: (1). *class-dependent features* serve as the short paths towards the opposing classes, which enables *Negativity* while minimizing features augmented to satisfy *Competitiveness*. (2) *class-invariant features* facilitate the learning of high-quality *class-dependent features* in the joint training of contrastive objective. (3). The random sampling scheme ensures *Randomness* condition. (4). The only parameter is the sampling probability, which can be dynamically adjusted on the go.

We then contrast the positive examples of node relations (what the anchor node interacts with) with the negative examples of node relations: nodes that the anchor node does not interact with but the virtual node would interact with.

4.3 Model Overview

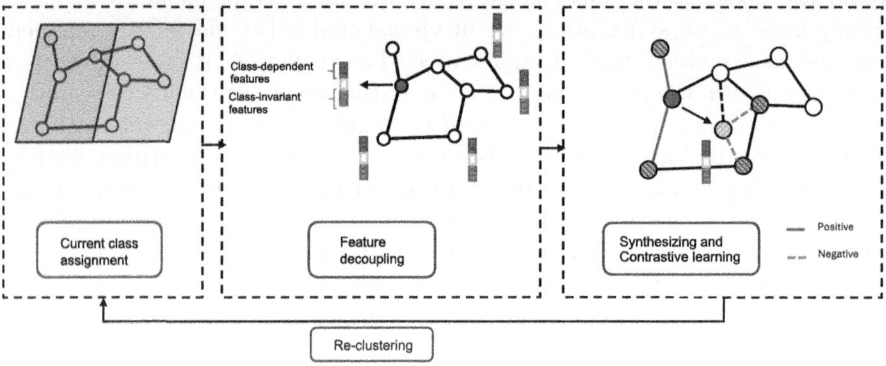

Fig. 1. MeCoLe: (1). Class assignment of last epoch (2). Learn decoupled *class-invariant features* and *class-dependent features* from the underlying graph, contrastive learning loss, and class assignments. (3). Randomly augment (from anchor node red) *class-dependent features* towards nodes in opposing classes to generate synthesized virtual nodes (yellow-shaded) and generate edges for the synthesized nodes. Derive contrastive learning loss between positive pairs (red-shaded) and negative pairs (blue-shaded). (Color figure online)

Our framework consists of three integrated modules: feature decoupling module, synthesizing and contrastive learning module, and decoupled cluster module. Feature decoupling module decouples *class-dependent features* (features correlate with the downstream task) and *class-invariant features* (features that do not correlate). Synthesizing and contrastive learning module generates a set of virtual nodes per anchor node and then samples a set of negative nodes (nodes a virtual node would interact with but the anchor node does not) as negative relations and then contrasts between the sampled negative nodes and sampled positive nodes (nodes the anchor node interacts with). The decoupled cluster module filters out irrelevant nodes and updates node assignments into the K classes (Fig. 1).

In principle, we can use any previous clustering techniques to initialize the clusters. By default, we use GCN with modularity regulator [9]. In each iteration:

- The feature decoupling module updates learned decoupled features based on current class assignments.

- The synthesizing and contrastive learning module synthesizes virtual nodes, and then samples fine-grained positive/negative pairs for contrastive learning.
- The decoupled cluster module updates class assignments.

4.4 Feature Decoupling

Given the graph inputs $\{G\}$, our graph encoding module deploys a decoupling framework to output *class-dependent features* and *class-invariant features* according to the current class assignments.

We design the following generalized objective to decouple learned features:

$$\boldsymbol{H}_d, \boldsymbol{H}_o \sim enc(G) \tag{2a}$$

$$min_{v_1,v_2 \sim V} \mathcal{L}_1(G, Z(\boldsymbol{H}_d, \boldsymbol{H}_o)) \tag{2b}$$

$$min\mathcal{L}_2 = min_{v_1 \sim C_1, v_2 \sim C_2} \frac{d(\boldsymbol{h}_{o,1}, \boldsymbol{h}_{o,2})}{d(\boldsymbol{h}_{d,1}, \boldsymbol{h}_{d,2})} \tag{2c}$$

where $\boldsymbol{H}_d$ and $\boldsymbol{H}_o$ are the matrix of *class-dependent features* and *class-invariant features*, respectively; enc is the graph encoder; C_k are the vertex set of class k; Z is a link prediction model that reconstructs G from $\boldsymbol{H}_d$ and $\boldsymbol{H}_o$; $\mathcal{L}_1$ is the reconstruction loss, such as cross entropy; d is a distance metric that measures discrepancy between nodes in two classes. The above formulation decouples the self-supervised models on class-invariant and class-dependent signals, enabling easy adaptation of existing state-of-the-art graph models.

The formulation is solved as a joint training framework between $\boldsymbol{H}_d, \boldsymbol{H}_o$. Equation(2b) minimizes the reconstruction loss of G and $Eq.$(2c) disentangles *class-dependent features* and *class-invariant features* by minimizing the relative discrepancies. We use two separate GCNs as our default graph encoder on both feature sets.

Link Prediction Model. Graph reconstruction is commonly used in clustering techniques to transfer knowledge from connection patterns [29]. The link prediction model Z reconstructs graph G by estimating the probability of edges between node pairs. The main question is how to combine the *class-dependent features* $\boldsymbol{H}_d$ and *class-invariant features* $\boldsymbol{H}_o$ in the prediction. Our experiments show that a simple black box classifier does not perform well due to the lack of labeled data in an unsupervised setting.

We observe that: in online scenarios, two nodes interact when both *class-dependent features* and *class-invariant features* are satisfied. For example, a Republican supporter supports online items when the post satisfies both *class-dependent features* (i.e., the post favors Republican) and *class-invariant features* (i.e., the post engages him; the post is within the topic of interest, etc.). Therefore, we use a simple yet effective prediction model Z:

$$Z_{uv}(\boldsymbol{H}_d, \boldsymbol{H}_o) \sim Z^d_{uv}(\boldsymbol{H}_d) \times Z^o_{uv}(\boldsymbol{H}_o) \tag{3}$$

The above formulation further disentangles *class-dependent features* and *class-invariant features* in probability relations between both sets of features, while also enabling easy adaptation of state-of-the-art predictors. We use dot product for Z^d_{uv}, Z^o_{uv}, because of its state-of-the-art performance on self-supervised tasks.

Discrepancy Metric. The most popular discrepancy metrics are l-norm and cosine similarity. Examples are: 1-norm (commonly used by domain adaptation); euclidean distance 2-norm; and max-pooling ∞-norm. Our experiment shows our method is resilient to the above choices. Nevertheless, the best performance comes from 1-norm.

4.5 Joint Learning Framework

Our decoupled representation learning framework jointly optimizes the objectives Eq.(2) in a self-supervised manner. Following the discussion in section. 4.4, we follow a graph rewiring procedure [30]. In short, given the input graph G and the current estimation of $\boldsymbol{H}_o$. We rewire G by $G^d = f(G|G^o)$, where G^o is the reconstructed graph from $\boldsymbol{H}_o$; and then use G^d to update $\boldsymbol{H}_d$ in the iteration. From Eq.(3), the back-propagation is:

$$\frac{\delta Z}{\delta \boldsymbol{H}_d} = Z^o(\boldsymbol{H}_o)\frac{\delta Z^d}{\delta \boldsymbol{H}_d} \tag{4}$$

Therefore, the augmentation function f is:

$$e^d_{ij} = f(e_{ij}|e^o_{ij}) = min(\eta, e_{ij}/e^o_{ij}) \qquad \forall e_{ij} \in G \tag{5}$$

where $e^o_{ij} = Z^o(\boldsymbol{h}^o_{v_i}, \boldsymbol{h}^o_{v_j})$ and η is the maximum weight of an edge in the rewired graph.

4.6 Integrate Content Representations

Content latent representations can facilitate the decoupling of *class-dependent features* and *class-invariant features.* We assume the contents are encoded into: (1). latent representations X using suitable content encoders, such as BERT (2). a sequence of key attributes M.

Due to the limited fine-tuning inputs in self-supervised settings, we only focus on extracting high-quality features. For latent representation X, we first construct the content similarity graph to enrich the input graphs:

$$\boldsymbol{A}^{sim}_{ij} \sim sim(\boldsymbol{X}_i, \boldsymbol{X}_j) \tag{6}$$

To filter out low-quality features, we first construct the k-nn graph, such that we only keep the top-k edges of a node. The graph is further filtered by a threshold η_1 by removing any edges of similarity lower than η_{sim}. The graph G_X is then included into the heterogeneous graph encoder in section. 4.4.

The key attributes M are separated according to its distribution. We use *tf-idf* to separate key attributes. In details, we compute *tf-idf* scores $\boldsymbol{S}_k$ of attributes for each classes k and then derive the class-dependent representation of node i as:

$$\boldsymbol{M}_{d,i} = \sum_k r_{ik} \frac{\sum_{m \in M} S_{k,m} \boldsymbol{x}_m}{\sum_{m \in M} S_{k,m}} \tag{7}$$

where x_m is the word embedding of attribute m and r_{ik} is the class assignment of node i to class k. The key attribute *class-dependent features* are then constructed into G_M in the same fashion as G_X.

4.7 Synthesizing Nodes and Contrastive Learning

Anchor node selection favors the *boundary* nodes. In detail, for each class, we perform Gaussian sampling from nodes within the class according to the inverse assignment to the class. Given the anchor node v, we generate positive samples from its 1-hop neighbors. The reason is to demonstrate that the most simple and generalizable positive sampling scheme is sufficient to produce good results. For negative samples, we first create virtual nodes $\tilde{v}$ by randomly augmenting *class-dependent features* with probability P_{CE} to uniformly sampled nodes in the other classes then, sample negative nodes proportional to the edge prediction probability $Z(\tilde{v}, u)$.

$$\mathcal{L}_{CE} = \mathbb{E}_{u^+ \sim P_p(u|v), u_k^- \sim P_n(u|v)} [-\log \frac{e^{f(v)^\top f(u^+)/\tau}}{\sum_k e^{f(v)^\top f(u_k^-)/\tau}}] \tag{8}$$

$$P_p(u^+|v) = \begin{cases} 1/d_v & u^+ \in \mathcal{N}_v \\ 0 & u^+ \notin \mathcal{N}_v \end{cases} \tag{9}$$

$$P_n(u^-|v) \sim Z(\tilde{v}, u) \quad u^- \in \mathcal{N}_{\tilde{v}} - \mathcal{N}_v \tag{10}$$

where $f(v) = \boldsymbol{h}_{d,v}$ is the class-dependent feature of node v, $\mathcal{N}_v$ is the set of neighboring nodes of v, P_p and P_n are the positive and negative distributions respectively. The implication is to distinguish the nodes that v interact with and the nodes the virtual nodes $\tilde{v}$ interact with.

4.8 Decoupled Cluster Module

The cluster module is based on the previous contrastive learning framework [17]: logistic regression for each class.

Compared to previous works [9–11], we do not include any regulators, such as modularity and conductance, to encourage clustering-friendly representations. The reason is two-folded:

1. Our contrastive objective already includes the modularity principle by contrasting positive and negative node relations.
2. Static-form regulators often interfere the self-supervised objectives.

4.9 Put Everything Together

To sum up, the loss function consists of three components:

1. Graph loss: $\mathcal{L}_G = \mathcal{L}_1 + \mathcal{L}_2$
2. Contrastive learning loss: $\mathcal{L}_{CE}$

Our model minimizes:

$$\mathcal{L} = \mathcal{L}_G + \alpha_{CE}\mathcal{L}_{CE} \tag{11}$$

5 Experiments

This section evaluates our framework on four public social network datasets **Wikipedia**[1], **IMDB**[1], **Covid** [2], **Protest** [2], two citation datasets **Cora**, **Citeseer**, and two newly collected datasets. The two new datasets differ from the previous ones, by including a substantial number of irrelevant nodes. The purpose is to simulate the collected raw datasets when the understanding of classes is minimal.

- **2020 presidential election**: is a newly collected data set from Twitter API using keywords: 'presidential election', 'Trump', and 'Biden'. Nodes are users and tweets, and an edge denotes that the user originally posts or positively interacts with the tweets.
- **Police shooting**: is a newly collected data set from Twitter API on the shooting of Daunte Wright, using keywords: 'police shooting', 'Daunte Wright'. Nodes are users and tweets, and an edge denotes that the user originally posts or positively interacts with the tweets.

Table 1. % Dataset statistics

	#nodes	#edges	#classes	Ave. deg	Inputs
Wiki	1124	6205	3	11.04	G, G_X
IMDB	12875	40711	6	6.81	G, G_X
Covid	1175	4429	3	7.53	G, G_V, G_X
Protest	1025	2245	3	4.38	G, G_V, G_X
President	2063	11883	3	11.52	G, G_V, G_X
Police	2011	12438	4	12.37	G, G_V, G_X
Cora	2485	5069	7	4.07	G
Citeseer	2110	3668	6	3.48	G

[1] http://konect.cc/networks/.

In addition to graph G, we also include auxiliary graphs constructed from node relations (G_V) and contents (G_X) using the methods described in section. 4.6. The detailed statistics of datasets are shown in Table 1.

We compare our method with unsupervised clustering methods: **DMoN** [9], **DCRN** [8], **DGCN** [29], **CommDGI** [5], **vGAE** [1], **cvGAE** [7]; and self-supervised methods: **CCA-SSG** [31], **GRACE** [17], **InfoGCL** [32].

5.1 Experiment Results

Table 2. Clustering accuracy

	Wiki	IMDB	Covid	Protest	President	Police	Cora	Citeseer
DMoN	69.5	75.6	70.1	69.6	75.9	75.0	46.3	69.5
DCRN	71.9	76.5	73.0	70.5	75.6	74.9	48.3	70.9
DGCN	74.2	77.6	76.2	70.5	79.1	75.2	72.9	71.6
CommDGI	78.5	80.3	80.1	72.1	82.9	76.2	70.1	69.4
vGAE	74.0	78.1	76.5	70.6	78.6	75.1	73.1	68.2
cvGAE	76.3	80.1	77.4	72.1	80.6	76.2	79.0	71.8
CCA-SSG	79.5	80.3	76.2	71.0	77.6	79.1	70.5	68.7
GRACE	74.2	76.1	73.4	69.5	77.4	75.4	65.8	67.5
InfoGCL	78.9	80.1	75.5	70.4	77.5	78.0	69.6	69.1
MeCoLe	**86.8**	**89.7**	**84.3**	**80.1**	**90.7**	**83.7**	**79.9**	**74.6**

The results are presented in Table 2. Our approach outperforms 9 baselines in all 8 datasets. The contrastive methods, CCA-SSG, GRACE, and InfoGCL, outperform unsupervised clustering methods in the 6 social network datasets, while underperform in the 2 citation datasets. The observation demonstrates their strong capability of extracting robust features from noisy real-world datasets. Our method (1). integrates both contrastive objectives and traditional clustering objectives (2). utilizes the vast *invariant features* to facilitate the learning of *dependent features*; and thus outperform other baselines.

Out of the 6 social network datasets, *Protest* has the lowest clustering accuracy because of the lowest average degree of the underlying graph. *Police* also shows relatively low accuracy despite the high average degree. The reason is the intertwined opinions on related topics across classes.

5.2 Ablation Study

This section performs the ablation analysis of each module, including:

Table 3. Clustering accuracy - ablation study

	Wiki	IMDB	Covid	Protest	Pres.	Police	Cora	Citeseer
Baseline	86.8	89.7	84.3	80.1	90.7	83.7	79.9	74.6
Feature decouple	74.1	77.1	77.2	72.1	83.3	75.2	70.2	66.0
Negative sample	73.2	77.2	77.6	76.2	82.1	74.4	72.5	66.4
MLP	82.2	87.1	80.9	83.7	85.5	78.6	76.3	72.6
- C.L.	75.0	76.9	76.8	71.3	79.2	74.3	70.1	69.0
Graph augment	81.9	83.1	78.5	72.4	80.5	80.0	75.6	71.2
-G_V	-	86.4	81.1	74.7	86.2	78.1	-	-
-G_X	84.2	88.1	81.5	77.4	87.9	80.1	-	-

- **Feature decoupling** is a core module of our framework. We remove the decoupled representation learning and (1) augment all learned features in contrastive learning (2). modifies negative sampling approaches to sample any nodes outside of neighborhood (3) replace the prediction model with MLP.
- **Contrastive Learning**. We first (1) completely remove contrastive learning (the clustering uses the output of the graph encoder), and then (2) replace our node-level augmentations with graph augmentation approaches.
- **Auxiliary graph**. We remove (1) user-user relations G_s (2) item-item relations G_t (3) content representation X (4) key attributes W.

The results are presented in Table 3. In general, our approach can perform relatively well without any of the above components. Out of all modules, the feature decoupling module shows the most significant performance drops, which supports our claim that *hard* positive/negative pairs are essential for self-supervised contrastive learning. Compared to graph-level augmentations [17], our methods are superior in: (1) generating effective contrastive learning pairs of any anchor nodes, (2) utilizing the downstream clustering results to improve pairs generation. Our methods are also resilient in removing input types. The experiment shows G_V, the user relation graph is relatively more important to the performance. The reason behind this is the lack of content information on users (i.e., user profiles) in the datasets.

5.3 Discrepancy Functions

This section evaluates the discrepancy function d used in our method. We focus on l-norm and *cosine* in this paper because of their simple yet state-of-the-art performance in machine learning models.

Our results in Table 4 demonstrates l = 1-norm being the best discrepancy for feature decoupling. The result is in line with previous literature on feature-based domain adaptations [33].

Table 4. Clustering accuracy - discrepancy function

	Wiki	IMDB	Covid	Protest	Pres.	Police	Cora	Citeseer
Baseline ($l = 1$)	86.8	89.7	84.3	80.1	90.7	83.7	79.9	74.6
$l = 2$	84.3	86.8	80.7	76.9	84.3	79.2	79.8	74.8
cosine	83.5	86.1	79.5	74.9	82.2	81.4	79.5	74.2
$l = \infty$	81.6	82.2	71.1	70.0	79.8	77.2	78.2	74.2

5.4 Integrate Contrastive Learning

Our contrastive learning framework can be integrated into most graph encoders. In detail, we replace *enc* in Eq. (2) with the baseline encoders and use the joint learning method described in Sect. 4.5 to train the encoders jointly. For GNN-based methods (COMMDGI, GCN, and vGAE), this is a straightforward adaptation.

Table 5. Clustering accuracy - integrating contrastive learning

		Wiki	IMDB	Covid	Protest	Pres.	Police	Cora	Citeseer
C.DGI		78.5	80.3	80.1	72.1	82.9	76.2	70.1	69.4
	+CL	84.4	85.3	83.2	79.1	88.0	80.2	71.5	70.9
GCN		66.0	68.9	70.8	64.3	71.2	65.3	44.2	51.3
	+CL	75.2	78.6	78.9	74.2	83.9	74.3	60.2	67.2
vGAE		74.0	78.1	76.5	70.6	78.6	75.1	73.1	68.2
	+CL	79.3	84.2	80.7	76.7	85.1	78.3	76.6	70.9
MeCoLe		**86.8**	**89.7**	**84.3**	**80.1**	**90.7**	**83.7**	**79.9**	**74.6**

The results are presented in Table 5. In general, our node-level contrastive learning method demonstrates performance boosts on all 3 baselines on 8 datasets. GNN-based methods benefit from contrastive learning because of their robust representation models on integrated inputs. The sparest dataset, *Protest*, shows the highest performance boosts on all baselines, demonstrating the benefits of contrastive learning on sparse graphs. Another observation is that GCN enjoys a greater performance boost compared to other baselines. Without contrastive loss or cluster-specific regulations, GCN fails to learn robust cluster-friendly representations.

5.5 Sparse Graph

This section evaluates the clustering performance on the sparse section of the datasets. In detail, we remove the top 30% degreed nodes from the datasets and self-training, then evaluate the clustering accuracy of the rest of the datasets. We exclude citation datasets, Cora and Citeseer, due to the numerous disconnected components.

Table 6. Clustering accuracy - Sparse

	Wiki	IMDB	Covid	Protest	President	Police
DMoN	55.4	62.3	65.5	68.8	70.8	72.2
DCRN	58.2	64.1	68.0	67.8	70.9	72.5
DGCN	67.2	72.5	70.5	69.5	71.2	71.9
CommDGI	70.9	76.3	76.5	69.5	78.6	72.8
vGAE	68.0	74.2	73.5	67.4	75.9	70.0
cvGAE	70.1	73.6	74.2	0.702	78.5	70.8
CCA-SSG	74.5	80.5	76.2	70.7	78.7	76.7
GRACE	72.2	77.9	74.8	68.9	77.1	76.3
InfoGCL	71.6	75.3	71.2	66.7	76.7	74.7
MeCoLe	**81.8**	**87.6**	**82.9**	**79.2**	**86.7**	**80.3**

The results are presented in Table 6. Our method shows greater resilience on sparse graphs than other baselines. The unsupervised clustering methods, DMoN, DCRN, and DGCN, suffer greater performance loss in sparse graphs due to their heavy reliance on message-passing models to extract cluster-specific features.

6 Conclusion

This paper presented MeCoLe, a novel fine-grained contrastive learning framework that integrates the commonly ignored *class-invariant features* with *class-dependent features* in unsupervised node clustering tasks. MeCoLe innovates as the first attempt to formulate a finer-grained contrastive learning scheme on node relations: contrast between edges (as positive node relations) and absent edges (no edge as negative relations) as a source of contrastive learning to improve node clustering tasks. To generate competitive contrastive pairs, MeCoLe first decouples *class-dependent features* and *class-invariant features*, then generates virtual nodes by augmenting *class-dependent features* while retaining *class-invariant features*, and finally sample the negative node relations as counterfactual relations.

Acknowledgements. Research reported in this paper was sponsored in part by DARPA award HR001121C0165, DARPA award HR00112290105, and DoD Basic Research Office award HQ00342110002. It was also supported in part by ACE, one of the seven centers in JUMP 2.0, a Semiconductor Research Corporation (SRC) program sponsored by DARPA

References

1. Li, J., et al.: Unsupervised belief representation learning with information-theoretic variational graph auto-encoders (2022)
2. Cui, H., Abdelzaher, T.: The voice of silence: interpreting silence in truth discovery on social media. In: Proceedings of the 2021 IEEE/ACM International Conference on Advances in Social Networks Analysis and Mining, pp. 82–89 (2021)
3. Cui, H., Abdelzaher, T., Kaplan, L.,: A semi-supervised active-learning truth estimator for social networks. In: The World Wide Web Conference, pp. 296–306 (2019)
4. Cui, H., Abdelzaher, T., Kaplan, L.: Recursive truth estimation of time-varying sensing data from online open sources. In: *2018 14th International Conference on Distributed Computing in Sensor Systems (DCOSS)*, IEEE, pp. 25–34 (2018)
5. Zhang, T., Xiong, Y., Zhang, J., Zhang, Y., Jiao, Y., Zhu, Y.: Commdgi: community detection oriented deep graph infomax. In: Proceedings of the 29th ACM International Conference on Information and Knowledge Management, pp. 1843–1852 (2020)
6. Mrabah, N., Bouguessa, M., Ksantini, R.: Beyond the evidence lower bound: dual variational graph auto-encoders for node clustering. In: Proceedings of the 2023 SIAM International Conference on Data Mining (SDM). SIAM, 2023, pp. 100–108 (2023)
7. Mrabah, N., Bouguessa, M., Ksantini, R.: A contrastive variational graph auto-encoder for node clustering. In: Pattern Recognition (2023)
8. Liu, Y., et al.: Deep graph clustering via dual correlation reduction. In: Proceedings of the AAAI conference on artificial intelligence, vol. 36(7), 7603–7611 (2022)
9. Tsitsulin, A., Palowitch, J., Perozzi, B., Müller, E.: Graph clustering with graph neural networks. J. Mach. Learn. Res. **24**(127), 1–21 (2023)
10. Fan, S., Wang, X., Shi, C., Lu, E., Lin, K., Wang, B.: One2multi graph autoencoder for multi-view graph clustering. In: Proceedings of the Web Conference, pp. 3070–3076 (2020)
11. Peng, Z., Liu, H., Jia, Y., Hou, J.: Egrc-net: embedding-induced graph refinement clustering network. IEEE Trans. Image Process. **32**, 6457–6468 (2023)
12. Chen, T., Kornblith, S., Norouzi, M., Hinton, G.: A simple framework for contrastive learning of visual representations. In: International Conference on Machine Learning, PMLR, 2020, pp. 1597–1607 (2020)
13. You, Y., Chen, T., Shen, Y., Wang, Z.: Graph contrastive learning automated. In: International Conference on Machine Learning, PMLR, 2021, pp. 12 121–12 132 (2021)
14. Qiu, R., Huang, Z., Yin, H., Wang, Z.: Contrastive learning for representation degeneration problem in sequential recommendation. In: Proceedings of the Fifteenth ACM International Conference on Web Search and Data Mining, 2022, pp. 813–823 (2022)
15. Wen, Z., Li, Y.: Toward understanding the feature learning process of self-supervised contrastive learning. In: International Conference on Machine Learning. PMLR, 2021, pp. 11 112–11 122 (2021)
16. Velickovic, P., Fedus, W., Hamilton, W.L., Liò, P., Bengio, Y., Hjelm, R.D.: Deep graph infomax. ICLR (Poster) **2**(3), 4 (2019)
17. Zhu, Y., Xu, Y., Yu, F., Liu, Q., Wu, S., Wang, L.: Graph contrastive learning with adaptive augmentation. In: Proceedings of the Web Conference, pp. 2069–2080 (2021)

18. Zhou, S., et al.: A comprehensive survey on deep clustering: Taxonomy, challenges, and future directions. arXiv preprint arXiv:2206.07579 (2022)
19. Akoglu, L.: Quantifying political polarity based on bipartite opinion networks. In: Proceedings of the International AAAI Conference on Web and Social Media, vol. 8, no. 1 (2014)
20. Yang, C., et al.: Hierarchical overlapping belief estimation by structured matrix factorization. In: 2020 IEEE/ACM International Conference on Advances in Social Networks Analysis and Mining (ASONAM), IEEE, 2020, pp. 81–88 (2020)
21. Darwish, K., Stefanov, P., Aupetit, M., Nakov, P.: Unsupervised user stance detection on twitter. In: Proceedings of the International AAAI Conference on Web and Social Media, vol. 14, 2020, pp. 141–152 (2020)
22. Verma, V., Luong, T., Kawaguchi, K., Pham, H, Le, Q.: Towards domain-agnostic contrastive learning. In: International Conference on Machine Learning, PMLR, 2021, pp. 10 530–10 541 (2021)
23. Veličković, P., Fedus, W., Hamilton, W.L., Liò, P., Bengio, Y., Hjelm, R.D.: Deep graph infomax. arXiv preprint arXiv:1809.10341 (2018)
24. Yèche, H., Dresdner, G., Locatello, F., Hüser, M., Rätsch, G.:' Neighborhood contrastive learning applied to online patient monitoring. In: International Conference on Machine Learning. PMLR, 2021, pp. 11 964–11 974 (2021)
25. Peinelt, N., Nguyen, D., Liakata, M.: tbert: topic models and bert joining forces for semantic similarity detection. In: Proceedings of the 58th Annual Meeting of the Association for Computational Linguistics, 2020, pp. 7047–7055 (2020)
26. Hamilton, W., Ying, Z., Leskovec, J.: Inductive representation learning on large graphs. In: Advances in Neural Information Processing Systems, vol. 30 (2017)
27. Ma, J., Guo, R., Wan, M., Yang, L., Zhang, A., Li, J.: Learning fair node representations with graph counterfactual fairness. In: Proceedings of the Fifteenth ACM International Conference on Web Search and Data Mining, 2022, pp. 695–703 (2022)
28. Mesnard, T., et al.: Counterfactual credit assignment in model-free reinforcement learning. arXiv preprint arXiv:2011.09464 (2020)
29. Pan, E., Kang, Z.: Beyond homophily: reconstructing structure for graph-agnostic clustering. In: International Conference on Machine Learning. PMLR, 2023, pp. 26 868–26 877 (2023)
30. Brüel-Gabrielsson, R., Yurochkin, M., Solomon, J.: Rewiring with positional encodings for graph neural networks. arXiv preprint arXiv:2201.12674 (2022)
31. Zhang, H., Wu, Q., Yan, J., Wipf, D., Yu, P.S.: From canonical correlation analysis to self-supervised graph neural networks. In: Advance Neural Information Processing System **34**, 76–89 (2021)
32. Xu, D., Cheng, W., Luo, D., Chen, H., Zhang, X.: Infogcl: information-aware graph contrastive learning. In: Advances in Neural Information Processing Systems, vol. 34, pp. 30 414–30 425 (2021)
33. Ganin, Y., Lempitsky, V.: Unsupervised domain adaptation by backpropagation. In: International conference on machine learning, PMLR, 2015, pp. 1180–1189 (2015)

DySDGNN: Representation Learning in Dynamic Signed Directed Networks

Zihao Liu and Yitong Wang(✉)

School of Computer Science, Fudan University, Shanghai, China
{zihaoliu22,yitongw}@m.fudan.edu.cn

Abstract. Dynamic signed directed networks are common in our daily lives so that representation learning in such networks is critical for many downstream tasks. However, most existing works only focus on either temporal or signed information in the graph. Moreover, taking account of both temporal and signed information into representation learning still faces some challenges. Firstly, edge sign between nodes could change (i.e. from positive to negative) over some time and this will bring the conflict in learning process. Secondly, balance theory and status theory that prior models based on are not suitable to learn signed semantics for dynamic signed network anymore. To tackle these challenges, we propose a novel clip-based Dynamic Signed Directed Graph Neural Networks model (DySDGNN). In DySDGNN, we design three components: the Attention-based TLSTM layer (A-TLSTM) to capture the different kinds of information within each clip, the Masked Temporal Self-Attention layer (MT-SA) to aggregate node embeddings across clips and the loss function related to the Signed Triad Transition Matrix (STTM) to measure the impact of different triads. We conduct extensive experiments and comparisons on three real-world datasets. The experimental results prove the superiority of DySDGNN compared with SOTA baselines on the downstream task of link sign prediction.

Keywords: Graph neural networks · Graph representation learning · Dynamic signed directed networks

1 Introduction

With the development of social medias, dynamic signed directed networks become more and more common in our daily lives. Social relationships in such networks could be both dynamic and with emotion tendencies. They could change over time and also be positive to express approval/trust/support, or negative to express disapproval/distrust/hostility. Thus, how to learn both temporal and signed semantics in node representation effectively is important for many downstream problems.

However, most graph representation learning methods just focus on one aspect of network features: temporal information such as [4,15,16] or signed information such as [3,8,9]. Moreover, simply combining the temporal and signed semantics [2,17] usually bring some conflicts.

M. Onizuka et al. (Eds.): DASFAA 2024, LNCS 14855, pp. 301–310, 2024.
https://doi.org/10.1007/978-981-97-5572-1_19

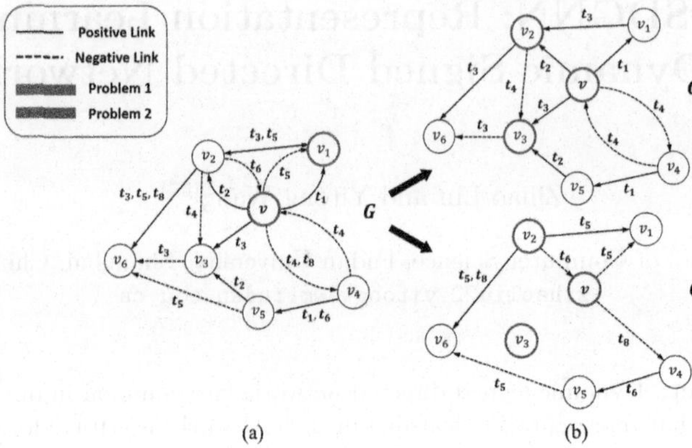

Fig. 1. (a) The continuous-time network. (b) Clips split from (a).

Problem 1. How to model the sign conflict during the evolution of network?

Sign conflict refers to the signed relationship between nodes could be reversed at different times (e.g., as shown in Fig. 1a, the link (v, v_1) is positive at time t_1 and it changes to negative at time t_5). To analyse *sign conflict* in real-world dynamic signed directed networks, we investigate the ratios and frequencies that the sign conflicts appeared on threet datasets. Table 1 shows the ratios and time intervals that sign conflicts appeared of active node (node degree>10).

Table 1. The ratios of reversed nodes and the reverse intervals.

Dataset	Wikirfa	ePINIONS	Weibo
reversed nodes	27.26%	23.89%	35.82%
reversed interval >14 days	95.43%	83.12%	81.52%
reversed interval >30 days	92.30%	73.19%	73.45%
reversed interval >60 days	86.26%	64.21%	62.29%

We have found that *sign reverse between active nodes usually happens after a time interval.* So we propose a new clip-based approach that splits the continuous-time network by the same time window (e.g., 61 days for Wikirfa) into a sequence of clips, which differ from the snapshots because each clip can preserve the temporal information. As shown in Fig. 1, due to continuous clip partition, we could limit that most edge sign reverse only appear between different clips and the edge sign preserves to be more stable within the same clip.

Problem 2. How to analyze the atheoretical triads in dynamic signed networks?

The *atheoretical triads* (such as the triad formed by nodes (v, v_2, v_3) in Fig. 1b) refer to those signed directed triads that do not strictly follow balance theory and status theory (see Sect. 2 for details). Some works ?, [14] find that there are some special triads more likely to form specific transition and new links. We are inspired to study the transition of *atheoretical triads* in Weibo dataset, and find that 54% of the *atheoretical triads* within each clip will cause the formation of new links in future clips, especially including the sign reverse. Thus, we propose the Signed Triad Transition Matrix (STTM) to recognize the network evolution in dynamic signed networks and understand the underlying mechanism of link formation (see details in Sect. 3.3).

In this paper, we propose a novel clip-based embedding model named DySDGNN, **Dy**namic **S**igned **D**irected **G**raph **N**eural **N**etworks. In DySDGNN, we design three components to tackle the problems of sign conflict and atheoretical triads. Firstly, we design the Attention-based TLSTM (A-TLSTM) layer to learn both temporal and signed semantics within each clip. Secondly, we present the Masked Temporal Self-Attention (MT-SA) module to accurately aggregate node embeddings across different clips. Finally, we propose the Signed Triad Transition Matrix (STTM) in the loss function to optimize our model.

The main contributions of our work are summarized as follows:

- We summarize two unsolved problems in modeling dynamic signed directed networks and present possible solutions;
- We propose a novel clip-based model named DySDGNN, which consists of three components: the A-TLSTM, the MT-SA, the loss function related to the STTM.
- We conduct extensive experiments on three benchmark datasets. Our experimental results show the superiority of DySDGNN, compared with the stateof-the-art methods.

2 Related Work

Dynamic unsigned representation learning models can be classified in two kinds: the snapshot-based [4,5] and the continuous-time [12,15] models. DTINE [4] applies node2vec and LSTM to embed the temporal information. However, the snapshot-based methods may cause the loss of temporal information [13]. Nowadays, many researchers prefer to use the continuous-time network. TGN [15] presents Temporal Graph Networks (TGNs) to model timed events.

Static signed embedding methods allow us to model the signed semantics between nodes. Balance theory and status theory are the most widely used theories in signed networks. Balance theory [7] states that balanced triads are more plausible than unbalanced triads and should be more prevalent in real-world social networks. And status theory [11] proposes a new organizational principle for directed signed networks and assumes that stable triads should have consistent status differences among nodes. SGCN [3] introduces balance theory into GCN for message passing. SiGAT [8] successfully generalizes Graph Attention Networks to signed networks. SDGNN [9] considers 4 motifs in signed directed networks and introduce balance theory and status theory into its loss function.

There are only a few works which have attempted to model dynamic signed directed networks. For instance, RW-LSTM [2] considers the temporal information of edges in signed networks during the link prediction process. The latest continuous-time model SEMBA [17], based on the dynamic network TGN [15], introduces balance theory to aggregate signed information and achieves excellent performances in different downstream tasks.

However, the above signed network embedding models are based on the balance theory and the status theory. But in real-world dynamic signed directed networks, the node interaction patterns do not strictly follow these two theories, which requires us to explore new aggregation method.

3 Proposed Method

In this section, we propose the DySDGNN, a novel model that can learn node embeddings in dynamic signed directed networks. As illustrated in Fig. 2, there are three key components in DySDGNN: the Attention-based TLSTM layer (A-TLSTM), the Masked Temporal Self-Attention layer (MT-SA) and the loss function related to the Signed Triad Transition Matrix (STTM). Next, we will elaborate on each component.

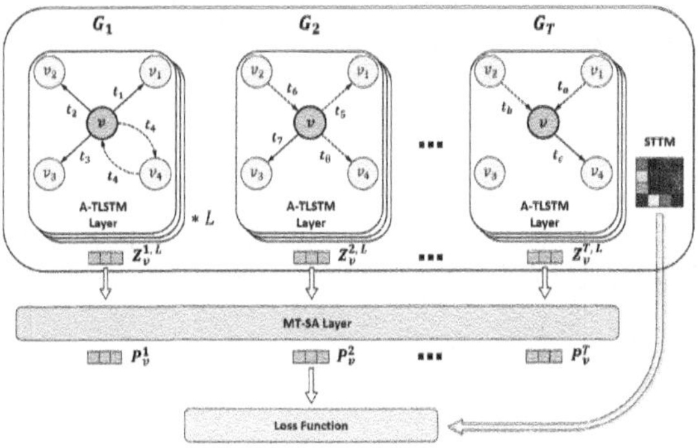

Fig. 2. The model architecture of DySDGNN.

3.1 Attention-Based TLSTM Layer

For dynamic signed directed networks, the temporal, signed and directed information contain different node behavior semantics. Firstly, we define the dynamic signed directed network as follows:

Definition 1. ***Dynamic Signed Directed Network.*** *Given a dynamic signed directed network* $\mathcal{G} =< \mathcal{V}, \mathcal{E}, \mathcal{T}, \mathcal{S} >$ *where* $\mathcal{V}$ *is the set of nodes,* $\mathcal{E}$ *is the set of directed edges in the graph,* $\mathcal{T}$ *and* $\mathcal{S}$ *are functions that map each edge to a timestamp and a sign, respectively. Each edge sign can only take one of two values* $\{+1, -1\}$*. The dynamic signed directed network* $\mathcal{G}$ *can be divided into several clips* $\left\{G^1, G^2, ..., G^T\right\}$ *with the equal time interval* θ*, where* T *is the number of time steps, and* $G^k =< \mathcal{V}, \mathcal{E}^k, \mathcal{T}^k, \mathcal{S}^k >$ *represents the k-th clip. It is unnoticeable that the edges in* $\mathcal{E}^k$ *have both temporal and signed information.*

To capture the different information, we design the Attention-based TLSTM layer. Similar to SDGNN [9], we also define 4 signed directed motifs (i.e., $v \xrightarrow{+} u$, $v \xrightarrow{-} u$, $v \xleftarrow{+} u$, $v \xleftarrow{-} u$). As shown in Fig. 3, based on these 4 motifs $m_i, i \in \{1, 2, 3, 4\}$ we use a multi-layer A-TLSTM to aggregate 4 types of neighbors and use a MLP to encode the different messages into node embeddings, where $\mathcal{E}_{m_i}^{v,k}$ means the m_i edge set of node v in clip G^k and $Z_v^{k,l}$ is the embedding of node v at layer l in clip G^k. When $l = 0$, $Z_v^{k,l} \in R^d$ is the initial feature of the node v.

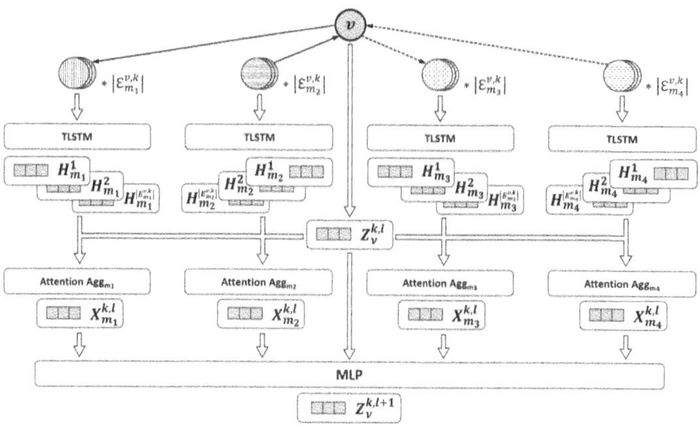

Fig. 3. An illustration of A-TLSTM layer.

Time-Aware LSTM. We transform the Time-Aware LSTM (TLSTM) [1] into an aggregator to distinguish the different impact of each edge. TLSTM can decay the prior cell state $C_v^{k,l}(t-)$ according to the time interval $\triangle_t$. And the TLSTM unit can be formulated as follows:

$$\begin{aligned}
\dot{C}_v^{k,l}(t-) &= tanh(W_s \cdot C_v^{k,l}(t-) + b_s) \\
\ddot{C}_v^{k,l}(t-) &= C_v^{k,l}(t-) - \dot{C}_v^{k,l}(t-) \\
\widetilde{C}_v^{k,l}(t-) &= \ddot{C}_v^{k,l}(t-) + \dot{C}_v^{k,l}(t-) \cdot decay(\triangle_t)
\end{aligned} \tag{1}$$

We choose the logarithmic decay function $decay(\triangle_t) = \frac{1}{1+log(\triangle_t)}$, which means the larger interaction interval is, the less short term memory is kept [12].

Attention-Based Aggregator. To further distinguish the contribution of different neighbors, we propose the Attention-based Aggregator to aggregate each edge features. We calculate the weights of edges between the focal node v and its neighbors with the scaled dot-product attention [18] as follows:

$$Attention(\mathbf{Quary}, \mathbf{Key}, \mathbf{Value}) = softmax(\frac{\mathbf{QK}^T}{\sqrt{d}} + \mathbf{M})\,\mathbf{V} \tag{2}$$

The neighbor hidden states under the motif m_i will be activated and concatenated into a matrix $H_{m_i} = \left\{ tanh(H_{m_i}^1), tanh(H_{m_i}^2), ..., tanh(H_{m_i}^{|\mathcal{E}_{m_i}^{v,k}|_{sample}}) \right\}$. Since multi-head self-attention allows model to jointly consider information from different embedding sub-spaces [18], the aggregation process in each motif m_i can be formulated as follows:

$$\begin{aligned} X_{m_i}^{k,l} &= CONCAT(head_{m_i}^1, head_{m_i}^2, ..., head_{m_i}^h) \cdot W_{m_i}^O \\ head_{m_i}^i &= Attention(Z_v^{k,l} W_{m_i}^Q,\ H_{m_i} W_{m_i}^K,\ H_{m_i} W_{m_i}^V) \end{aligned} \tag{3}$$

where $W_{m_i}^Q, W_{m_i}^K, W_{m_i}^V \in R^{D \times D}$ are learnable parameters. The Query is only relevant to the focal node embedding. Finally, all of the 4 aggregation results and the l-layer focal node embedding are concatenated and inputted into the MLP to calculate the next-layer embedding.

$$Z_v^{k,l+1} = MLP(CONCAT(X_{m_1}^{k,l}, X_{m_2}^{k,l}, X_{m_3}^{k,l}, X_{m_4}^{k,l}, Z_v^{k,l})) \tag{4}$$

3.2 Masked Temporal Self-attention Layer

To capture the relationship across different clips, we design the Masked Temporal Self-Attention layer (MT-SA). For the node v, MT-SA only considers the historical (i.e., prior k-1 clips) and active (i.e., not isolated) clips when calculating the node embeddings in the clip G^k. The MT-SA can be formulated as:

$$\begin{aligned} P_v &= CONCAT(head_v^1, head_v^2, ..., head_v^h) \cdot W_v^O \\ head_v^i &= Attention(Z_v^S W_k^Q,\ Z_v^S W_k^K,\ Z_v^S W_k^V) \end{aligned} \tag{5}$$

where $W_k^Q, W_{m_i}^K, W_{m_i}^V \in R^{D \times F}$ are learnable parameters, and the mask matrix $M_k^{k'}$ is defined as follows:

$$M_k^{k'} = \begin{cases} 0, & k' \le k \quad and \quad |\mathcal{E}_v^k| > 0 \\ -\infty, & otherwise \end{cases} \tag{6}$$

where $|\mathcal{E}_v^k|$ means the number of node v's interactions in the clip G^k.

3.3 Loss Function of STTM

To model the signed relationships between nodes, we firstly design a cross entropy loss function of edge signs between two nodes:

$$\mathcal{L}_{sign} = \sum_{k=0}^{T} \sum_{e \in \xi^k} -s_{ij} log(\sigma(< P_i^k, P_j^k >)) - (1 - s_{ij}) log(1 - \sigma(< P_i^k, P_j^k >)) \tag{7}$$

where ξ^k is the training batch of edges in clip G^k, and each edge $(v_i, v_j, s_{ij}, t_{ij})$ means the edge with sign s_{ij} formed by any two nodes v_i, v_j at time t_{ij} in G^k.

As mentioned earlier, to analyze the transition between triads, we design an additional loss function:

$$\mathcal{L}_{triad} = \sum_{k=0}^{T} \sum_{e \in \xi^k} weight_e \cdot (-s_{ij} log(\sigma(\widehat{y}_{ij})) - (1 - s_{ij}) log(1 - \sigma(\widehat{y}_{ij}))) \quad (8)$$

We use Hadamard Operator $(P_i^k \odot P_j^k)$ [6,16] to compute a feature vector for each node pair and then pass it through a MLP to predict the edge sign $\widehat{y}_{ij}$.

$$\widehat{y}_{ij} = MLP(P_i^k \odot P_j^k) \quad (9)$$

Inspired by the Triad Transition Matrix (TTM) in prior works [14], to estimate the probability of the edge sign reverse between the two nodes, we propose the Signed Triad Transition Matrix (STTM) to calculate the weight of edge in loss function. The formal definition of STTM is as follows:

Definition 2. ***Signed Triad Transition Matrix.*** *In dynamic signed networks, given two adjacent clips* G_k *and* G_{k+1}*, the signed triad transition matrix* $STTM_k \in R^{d_M \times d_M}$ *records the transition probabilities between* G_k *and* G_{k+1}*.*

where $d_M = 9^3 = 729$ and each triad is assigned an ID, such as Tr_x and Tr_y. According to the definition, the values of STTM entries are defined as follows:

$$STTM_k(x, y) = P(Tr_x \rightarrow Tr_y) \quad (10)$$

$STTM_k(x, y)$ is the probability of the triad Tr_x in G^k transiting to Tr_y in next clip G^{k+1}. By studying the transition of triads in networks, the estimated probability of forming a new link can be quantified as the values in STTM. We use the mean value of T-1 STTMs to estimate the STTM of the whole network.

In addition, balance theory [7] also shows that nodes can perceive imbalance in the *atheoretical triads* and will be motivated to correct the imbalance in the future (Problem 2). Since we believe that the model should learn relatively stable node relationships in each clip, the minimum probability of each sign that does not reverse in next clip among all of *atheoretical triads* is its weight in the loss function. So the $weight_e$ can be formulated as follows:

$$weight_e = \underset{Tr_x \in \triangle_{ij}^{atheo}}{MIN} \sum_{Tr_y \in \triangle_{ij}^{stable}} \overline{STTM}(x, y) \quad (11)$$

where $\triangle_{ij}^{atheo}$ is the set of the *atheoretical triads* in current clip, formed by the focal edge (v_i, v_j) and another node, and $\triangle_{ij}^{stable}$ is the set of the stable triads in next clip, which contain the same node pair (v_i, v_j) and the edge sign s_{ij} does not reverse.

Finally, the overall objective loss function is written as:

$$\mathcal{L} = \mathcal{L}_{sign} + \lambda \cdot \mathcal{L}_{triad} \quad (12)$$

where λ is the weight of triad loss.

4 Experiments

4.1 Datasets and Settings

Ten state-of-the-art graph embedding methods are selected as baselines, including two Static Unsigned (S/U) models: Node2Vec [6] and GCN [10], three Dynamic Unsigned (D/U) approaches: DySAT [16], DTINE [4], TGN [15], three Static Signed (S/S) methods: SGCN [3], SiGAT [8], SDGNN [9] and two existing Dynamic Signed (D/S) works: SGC-LSTM [2], SEMBA [17]. We conduct link sign prediction experiments on three real-world dynamic signed directed datasets. The statistics of these datasets are shown in Table 2.

Wikirfa[1]: In the Wikirfa dataset, nodes represent Wikipedia users, and each edge represents the results of "Request for Adminship" votes such as support, neutral, or oppose. We only choose the positive and negative edges.

Epinions[2]: Epinions is a dynamic signed network formed in an online e-commerce website. Users can build trust or distrust relationships with others.

Weibo[3]: Weibo is a social media platform which is similar to Twitter. Users can post positive or negative comments while retweeting other tweets.

Table 2. Summary statistics of three datasets.

Dataset	#nodes	#pos links	#neg links	#days
Wikirfa	1956	35560	10500	365
Epinions	2328	70035	15833	425
Weibo	14021	146789	167757	460

4.2 Experiment Results in Link Sign Prediction

We evaluate the performances of all baselines and our model DySDGNN in link sign prediction experiments. As shown in Table 3, compared with the SOTA baseline model, the improvement of DySDGNN is 2.54% and 3.19% for AUC and Macro-F1 on the Wikirfa dataset, 2.70% and 7.37% for Accuracy and Macro-F1 on the Epinions dataset, 9.14% and 20.77% for Macro-F1 and Binary-F1 on the Weibo dataset. Notice that the improvement on Weibo dataset is the best, which demonstrates the effectiveness of DySDGNN for handling a larger number of negative links. In addition, SiGAT achieves the second best performance on Epinions dataset. This is because signed triads play important roles on Epinions dataset and the 38 motifs in SiGAT can capture the triadic impact.

[1] https://snap.stanford.edu/data/wiki-RfA.html.

[2] https://snap.stanford.edu/data/soc-sign-epinions.html.

[3] https://www.aminer.cn/influencelocality.

Table 3. The Results of Link Sign Prediction on Three Datasets

Datasets		Wikirfa				Epinions				Weibo			
Metrics		AUC	Acc	Mac-F1	Bin-F1	AUC	Acc	Mac-F1	Bin-F1	AUC	Acc	Mac-F1	Bin-F1
S/U	Node2vec	0.5719	0.6768	0.5272	0.7932	0.6717	0.7541	0.5890	0.8495	0.5303	0.5211	0.4525	0.2487
	GCN	0.6188	0.7922	0.5203	0.8815	0.6332	0.8350	0.5244	0.9087	0.5540	0.4241	0.5357	0.5626
D/U	DySAT	0.5575	0.8091	0.4479	0.8945	0.6129	0.8400	0.4565	0.9130	0.5304	0.5428	0.5214	0.4520
	DTINE	0.4696	0.8090	0.4472	0.8944	0.5276	0.8420	0.4902	0.9137	0.5419	0.5537	0.5294	0.4225
	TGN	0.5117	<u>0.8099</u>	0.4722	<u>0.8944</u>	0.7037	0.8498	0.6127	0.9157	0.5496	0.5867	0.5226	0.3477
S/S	SGCN	0.6064	0.8064	0.4595	0.8925	0.6116	0.8446	0.5140	0.9148	0.5720	0.5818	0.5346	0.3578
	SiGAT	0.6350	0.7631	0.5732	0.8579	<u>0.8489</u>	0.8697	0.6842	0.9262	0.5415	0.5330	0.5228	0.4530
	SDGNN	0.6691	0.7820	0.6302	0.8671	0.8172	0.8680	0.6531	0.9261	0.5748	0.5604	0.5489	<u>0.4772</u>
D/S	SGC-LSTM	0.5000	0.8090	0.4472	0.8944	0.5441	0.8325	0.4543	0.9086	0.5483	0.5558	0.5293	0.4178
	SEMBA	<u>0.6820</u>	0.8073	<u>0.6079</u>	0.8875	0.8355	<u>0.8699</u>	<u>0.6919</u>	<u>0.9264</u>	<u>0.5819</u>	<u>0.5932</u>	<u>0.5504</u>	0.4117
DySDGNN		**0.6993**	**0.8264**	**0.6273**	**0.8997**	**0.8499**	**0.8934**	**0.7429**	**0.9396**	**0.6332**	**0.6022**	**0.6007**	**0.5763**
Rela, Improv.		2.54%	2.04%	3.19%	0.59%	0.12%	2.70%	7.37%	1.42%	8.81%	1.52%	9.14%	20.77%

5 Conclusions

In this paper, we propose DySDGNN, which is a novel clip-based dynamic signed directed graph embedding method. In DySDGNN, we design three modules: (1)The A-TLSTM layer to capture the different information within each clip; (2)The MT-SA layer to aggregate node embeddings across clips; (3)The loss function related to STTM to overcome the shortcomings of aggregation strategy based on balance theory and status theory in dynamic signed directed networks. Our experimental results on three real-world datasets also prove the superiority and effectiveness of DySDGNN. In future work, we will continue to explore the relationship between nodes and further improve the performance of our model.

References

1. Baytas, I.M., Xiao, C., Zhang, X., Wang, F., Jain, A.K., Zhou, J.: Patient subtyping via time-aware lstm networks. In: Proceedings of the 23rd ACM SIGKDD International Conference on Knowledge Discovery and Data Mining, pp. 65–74 (2017)
2. Dang, Q.V., Ignat, C.L.: Link-sign prediction in dynamic signed directed networks. In: 2018 IEEE 4th International Conference on Collaboration and Internet Computing (CIC), pp. 36–45. IEEE (2018)
3. Derr, T., Ma, Y., Tang, J.: Signed graph convolutional networks. In: 2018 IEEE International Conference on Data Mining (ICDM), pp. 929–934. IEEE (2018)
4. Gong, M., Ji, S., Xie, Y., Gao, Y., Qin, A.: Exploring temporal information for dynamic network embedding. IEEE Trans. Knowl. Data Eng. **34**(8), 3754–3764 (2020)
5. Goyal, P., Kamra, N., He, X., Liu, Y.: Dyngem: Deep embedding method for dynamic graphs. arXiv preprint arXiv:1805.11273 (2018)

6. Grover, A., Leskovec, J.: node2vec: scalable feature learning for networks. In: Proceedings of the 22nd ACM SIGKDD International Conference on Knowledge Discovery and Data Mining, pp. 855–864 (2016)
7. Heider, F.: Attitudes and cognitive organization. J. Psychol. **21**(1), 107–112 (1946)
8. Huang, J., Shen, H., Hou, L., Cheng, X.: Signed graph attention networks. In: Tetko, I.V., Kůrková, V., Karpov, P., Theis, F. (eds.) ICANN 2019. LNCS, vol. 11731, pp. 566–577. Springer, Cham (2019). https://doi.org/10.1007/978-3-030-30493-5_53
9. Huang, J., Shen, H., Hou, L., Cheng, X.: Sdgnn: learning node representation for signed directed networks. In: Proceedings of the AAAI Conference on Artificial Intelligence, vol. 35, pp. 196–203 (2021)
10. Kipf, T.N., Welling, M.: Semi-supervised classification with graph convolutional networks. arXiv preprint arXiv:1609.02907 (2016)
11. Leskovec, J., Huttenlocher, D., Kleinberg, J.: Signed networks in social media. In: Proceedings of the SIGCHI Conference on Human Factors in Computing Systems, pp. 1361–1370 (2010)
12. Ma, Y., Guo, Z., Ren, Z., Tang, J., Yin, D.: Streaming graph neural networks. In: Proceedings of the 43rd International ACM SIGIR Conference on Research and Development in Information Retrieval, pp. 719–728 (2020)
13. Nguyen, G.H., Lee, J.B., Rossi, R.A., Ahmed, N.K., Koh, E., Kim, S.: Continuous-time dynamic network embeddings. In: Companion Proceedings of the the Web Conference 2018, pp. 969–976 (2018)
14. Qiu, Z., Wu, J., Hu, W., Du, B., Yuan, G., Yu, P.: Temporal link prediction with motifs for social networks. IEEE Trans. Knowl. Data Eng. (2021)
15. Rossi, E., Chamberlain, B., Frasca, F., Eynard, D., Monti, F., Bronstein, M.: Temporal graph networks for deep learning on dynamic graphs. arXiv preprint arXiv:2006.10637 (2020)
16. Sankar, A., Wu, Y., Gou, L., Zhang, W., Yang, H.: Dysat: deep neural representation learning on dynamic graphs via self-attention networks. In: Proceedings of the 13th International Conference on Web search and Data Mining, pp. 519–527 (2020)
17. Sharma, K., Raghavendra, M., Lee, Y.C., Kumar M, A., Kumar, S.: Representation learning in continuous-time dynamic signed networks. In: 32nd ACM International Conference on Information and Knowledge Management (2023)
18. Vaswani, A., et al.: Attention is all you need. Advances in neural information processing systems **30** (2017)

Multi-objective Graph Neural Network Explanatory Model with Local and Global Information Preservation

Yibowen Zhao, Yonghui Xu(✉), Yixin Zhang, Wei He, and Lizhen Cui(✉)

School of Software & Joint SDU-NTU Centre for Artificial Intelligence Research (C-FAIR), Shandong University, Shandong, China
{ybw.zhao,yixinzhang}@mail.sdu.edu.cn, xu.yonghui@hotmail.com,
{hewei,clz}@sdu.edu.cn

Abstract. Graph Neural Network (GNN) has demonstrated significant application potential in fields such as bioinformatics, social network analysis, and financial risk management. However, explainability issues have limited their widespread adoption. Although numerous explanatory models have been developed, as the range of application scenarios expands, there is still an increasing need for diverse explanations. However, current explanatory models struggle to meet the varying requirements of different scenarios. They are often tailored for specific objectives, compromising their adaptability to diverse explanatory needs. Additionally, current models commonly use instance-level explanations that mainly analyze single graph samples for structural features, leading to a focus on local information and neglect of broader global patterns and trends. To address these issues, this paper proposes the Multi-Objective Graph Neural Network Explanatory model (MOE). MOE utilizes a Pareto frontier-based strategy for exploratory subgraphs to harmonize and support multiple explanatory objectives. Additionally, MOE integrates both local and global information from the original data into each explanation, enhancing the explanatory depth. Empirical studies demonstrate that MOE excels in various datasets and explanatory tasks, significantly advancing GNN explainability.

Keywords: Explainable AI · Graph Neural Network · Multi-Objective · Global Information

1 Introduction

Graph Neural Networks (GNNs) exhibit significant potential across diverse domains such as bioinformatics, social network analysis, and financial risk management. Harnessing the capabilities of deep learning, GNNs adeptly capture intricate relationships among graph nodes, yielding valuable insights. Nonetheless, the lack of interpretability poses a notable impediment to their broad adoption [5]. A primary challenge inherent in GNNs lies in their opaque nature. Given

M. Onizuka et al. (Eds.): DASFAA 2024, LNCS 14855, pp. 311–320, 2024.
https://doi.org/10.1007/978-981-97-5572-1_20

their reliance on intricate neural architectures, comprehending the rationale behind model predictions becomes arduous. Consequently, engendering trust in model outputs and justifying their utilization in pivotal decision-making processes becomes challenging. Particularly within fields like bioinformatics and financial risk management, where stakes are high, elucidating model reasoning is imperative for ensuring the reliability and accuracy of outcomes.

To address this challenge, numerous explanatory models are developed to elucidate the inner mechanisms of GNNs, encompassing gradient-based, sensitivity-based, feature-based, and graph structure-based approaches [17]. These models aim to explicate the predictions generated by GNNs in an understandable manner. Despite progress in developing explanatory models for GNNs, an ongoing challenge persists in accommodating the diverse requirements of various application scenarios [3]. Existing models are tailored to specific objectives, thereby limiting their adaptability to diverse explanatory needs. As the spectrum of application scenarios broadens, there arises an escalating demand for diverse explanations capable of catering to distinct user cohorts and their varying expertise levels. For instance, when elucidating the prediction of a molecule's classification within a dataset, researchers may seek to comprehend the intricate atomic interactions and structural specifics contributing to the molecule's categorization. Conversely, lay users may simply seek rudimentary explanations regarding why the molecule falls into a mutagenic category. Thus, the development of more flexible and versatile explanatory models capable of accommodating multi-objective scenarios while striking a balance between explanation clarity and accuracy is paramount.

Furthermore, current GNN explanatory methods encounter challenges in handling both global and local information. While these methods offer a certain level of explainability, they often concentrate solely on local nodes or edges, neglecting the broader global structure and patterns [17]. This limitation can result in incomplete and misleading explanations, as pivotal global features might exert significant influence on the model's predictive outcomes. Additionally, existing explanatory methods grapple with the task of striking a balance between explanation clarity and complexity. Enhancing explanation clarity may necessitate simplifying the structure of explanatory subgraphs or compromising the model's explanation accuracy, presenting a challenging trade-off in practical scenarios. Hence, comprehensively integrating global and local information to furnish more precise and thorough explanatory methods for GNNs emerges as an urgent concern.

To tackle the aforementioned challenges, this paper introduces a pioneering approach named the **M**ulti-**O**bjective Graph Neural Network **E**xplanatory (MOE). MOE reframes the explainability objective of GNNs as a multi-objective optimization problem, drawing upon Pareto frontier theory to yield a set of explanations that mutually dominate, thus catering to diverse user requirements. Addressing the deficiency of global information in instance-level explanations, MOE extracts frequent subgraphs with the highest confidence predictions from a pre-trained GNN within a sample graph set. This serves as the Global View

Reference and is seamlessly integrated as one of the objectives during optimization, ensuring the incorporation of global context into each instance-level explanation. The primary contributions of this study are as follows:

- An explanatory approach for multi-objective scenarios is proposed for the graph neural network method, which breaks through the limitations of the traditional single-objective explanatory method and improves the applicability of graph neural networks in complex scenarios such as multi-objective explanatory demands.
- MOE infuses global context into instance-level explanations, bridging the existing gap in explainable models concerning global insights.
- This paper empirically validates the efficacy of MOE, demonstrating its potential to augment the explainability of GNNs.

The structure of this paper is organized as follows: Sect. 2 delves into the related models and applications of instance-level explanations. Section 3 elucidates the proposed MOE model. Section 4 discusses the experimental setup and results. Lastly, Sect. 5 provides a summary of the paper.

2 Related Works

In recent years, researchers aim to enhance model explainability through global explanations, offering a comprehensive understanding of the model's behavior. Models like XGNN [16] use representative samples to elucidate GNN behavior, optimizing predictive accuracy through reinforcement learning. However, their focus on graph classification raises questions about applicability to node classification tasks. In contrast, the Gem [7] model treats GNN explanation as causal learning, employing Granger causality in its loss function for broader generalization. While global models provide broad insights, the importance of instance-level explanation models persists, offering detailed insights into specific GNN predictions.

To consider the explanation with instances in the dataset, GNNExplainer [15] pioneers the field by offering explanations through subgraphs, which it identifies via perturbation methods. The subsequent introduction of PGExplainer [8] in 2020 addresses some of the limitations of GNNExplainer, particularly the issue of introduced evidence, by learning an approximated discrete mask through the maximization of mutual information. This is followed by SubgraphX [18], which leverages the Monte Carlo Tree Search algorithm and Shapley values to efficiently explore and refine subgraph explanations. GNN-LRP [9] takes a different approach by exploring the significance of graph walks, aligning closely with deep GNNs and offering a novel perspective through higher-order Taylor decomposition. RelEx [19] then combines proxy methods with perturbation-based techniques in a hybrid framework, tailored for explaining node classification models. Despite these advancements, the current instance-level explanation models for GNNs have yet to fully address the breadth of explainability needs, pointing to the ongoing quest for a unified framework that can accommodate diverse objectives and perspectives.

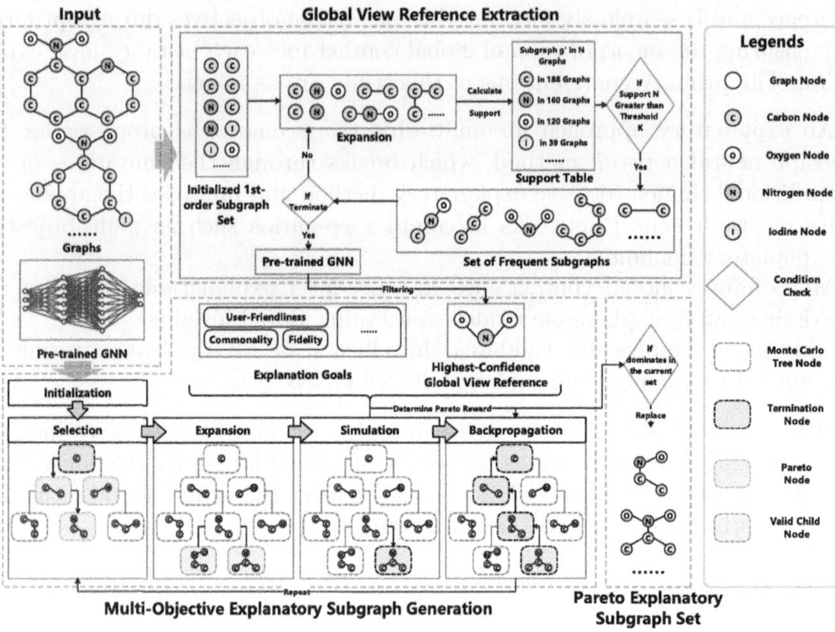

Fig. 1. The working progress of the MOE. From raw graph inputs to Global View Reference Extraction with gSpan and GNN refinement, culminating in Multi-Objective Explanation Generation follows MCTS subgraph searching and Pareto optimization rules.

3 Purposed Method

To address the challenges mentioned above, a novel GNN explanation model named **M**ulti-**O**bjective Graph Neural Network **E**xplanatory model (MOE) is proposed. As depicted in Fig. 1, the model is primarily divided into two modules: global reference view extraction and multi-objective explanatory subgraph generation. The global reference view extraction module is dedicated to extracting global reference information from various data types within the dataset, ensuring a robust foundation for global information embedding. The multi-objective explanatory subgraph generation module, based on the Pareto front theory, is designed to embed this global information and generate a collection of Pareto-optimal explanatory subgraphs that align with multiple explanatory objectives. The details of the modules are illustrated as follows.

3.1 Global View Reference Extraction

To enhance instance-level explainability and incorporate global insights for greater commonality, MOE requires the identification of global categorical information across various class samples in the dataset. To achieve this, MOE develops a strategy for obtaining a global view reference, grounded in the assumption that recurrent subgraphs in a category's graph dataset encapsulate essential category-specific characteristics. Thus, MOE utilizes the gSpan [14] algorithm to detect frequent subgraphs among similar samples and employs a pre-trained Graph Neural Network $f(\cdot)$ to identify high-confidence subgraphs as the global view reference for each category.

In the MOE model, the exploration of frequent subgraphs for a specific category represents a specialized adaptation of the gSpan algorithm. Commencing with a initialized 1st-order subgraph set $\mathcal{G}_c$ comprising single nodes or edges, this set represents potential frequent motifs within the graph set S_i of category i. Through a recursive expansion process, MOE systematically adds new nodes and edges to each candidate subgraph $g \in \mathcal{G}_c$ to form extended subgraphs g', retaining only those that appear frequently across more than the threshold k graphs in S_i. These subgraphs are stored in the set of frequent subgraphs $\mathcal{G}_f$. The iterative process continues to construct more complex subgraph structures until termination conditions are met. Following the recursive expansion, MOE utilizes a pre-trained Graph Neural Network $f(\cdot)$ to compute predictions $f(g_f)$ for each frequent subgraph, selecting the unique subgraph with the highest confidence prediction score as the global view reference R^i_{exp} for the category.

This methodology proficiently pinpoints key and recurrent subgraph structures, furnishing an expansive global perspective for each class within the dataset. Subsequent to this identification phase, the focus shifts to generating explanations for the Graph Neural Network models, in which embedds the discerned information within these subgraph structures.

3.2 Multi-objective Explanatory Subgraph Generation

To generate sets of explanatory subgraphs tailored to diverse explanatory objectives, MOE integrates Pareto frontier theory with the Monte Carlo Tree Search (MCTS) methodology. MCTS is strategically employed within MOE to navigate through explanatory subgraphs. During the search for optimal subgraphs, Pareto frontier theory comes into play, enabling the selection of Pareto optimal explanatory subgraphs that align with various objectives in the predefined explanatory objective set $\mathcal{O}_{\text{exp}}$. Furthermore, embedding global view information, denoted as R^i_{exp} for each category i, is incorporated as a crucial explanatory objective to enhance the commonality across explanations.

In MOE, the exploration process using Monte Carlo Tree Search (MCTS) is meticulously divided into several key modules to ensure a systematic and orderly progression. The exploration commences with the **initialization** and root node selection module, where multiple explorations are conducted for each input graph G, beginning from a randomly selected, non-repeated graph node G_{root} as the root MCTS node. Additionally, this root MCTS node acts as one of the termination nodes, halting the backpropagation process. This MCTS node, representing a part of G, provides a pivotal starting perspective for exploration.

The **selection** module then takes over, recursively selecting the best MCTS Pareto node for further exploration. At this stage, Pareto nodes among all MCTS nodes are identified and compared based on the Pareto Upper Confidence Bound

(UCB1) strategy [12] expanded from the regular UCB1 [1,6], as formulated in Eq. 1. A particular Pareto node is chosen from these for the subsequent expansion phase.

$$UCB1(s,a) = \bar{X}_{s,a} + \sqrt{\frac{2\ln n_s}{n_{s,a}}} \tag{1}$$

Here, $\bar{X}_{s,a}$ is the average reward for action a in state s, n_s is the visit count for state s, and $n_{s,a}$ is the count for action a in state s. This guides the selection of the best Pareto node in MCTS.

The **expansion** module then extends the selected Pareto MCTS node by randomly incorporating an edge or node from the original graph, thereby generating a new set of valid child MCTS nodes which contains potentially larger subgraphs. The **simulation** module assesses the explanatory quality of subgraphs and calculates potential Pareto rewards in these newly expanded MCTS nodes to select a start termination MCTS node for backpropagation. Subsequently, the **backpropagation** module updates the statistical information of all traversed MCTS nodes, including termination nodes on the backpropagation path, based on the Pareto frontier comparison, as depicted in Eq. 2.

$$O_i(x) \leq O_i(y) \ \forall i \text{ and } O_j(x) < O_j(y) \text{ for } \exists\, j \tag{2}$$

It compares solutions x and y across multiple objectives $\mathcal{O}_{\text{exp}}$, updating the MCTS nodes based on their Pareto optimality during backpropagation.

During each simulation and backpropagation, Pareto reward calculations are conducted, and the existing set of Pareto explanatory subgraphs is compared and replaced if a superior set is identified. This crucial updating process ensures that each exploration optimally refines the set of explanatory subgraphs under the multi-objective Pareto definition. The exploration persists until a predefined termination condition, such as reaching the maximum depth of MCTS, is met. The culminating set of explanatory subgraphs is termed the Pareto explanatory subgraph set, marking the end of a comprehensive and nuanced exploration.

Building upon the systematic exploration enabled by its modular structure, MOE then concentrates on three pivotal explanatory objectives: aligning with the Global View Reference, ensuring explanation accuracy, and maintaining sparsity for user-friendliness. These objectives are encapsulated in $\mathcal{O}_{\text{exp}}$ and address different aspects of the explanation to provide a thorough understanding.

Alignment with the Global View Reference is quantified using Cosine Similarity (CS) between the embeddings of R^i_{exp} and G^s obtained from a Graph Neural Network $f(\cdot)$, as shown in Eq. 3.

$$O_{\text{CS}} = \max_{R^i_{\text{exp}}} \frac{f(R^i_{\text{exp}}) \cdot f(G^s)}{\|f(R^i_{\text{exp}})\|_2 \|f(G^s)\|_2} \tag{3}$$

In this expression, $f(R^i_{\text{exp}})$ and $f(G^s)$ symbolize the embeddings of the Global View Reference and the explanatory subgraph, respectively. This objective integrates global information into the explanation process, enhancing commonality.

Explanation accuracy is paramount, ensuring that the prediction from the explanatory subgraph $f(G^s)$ closely aligns with the original graph's prediction $f(G)$, as measured by Eq. 4.

$$O_{\text{Acc}} = \min \|f(G) - f(G^s)\|_2 \tag{4}$$

This objective serves as a quantitative measure of how well the explanatory subgraph replicates the original graph's prediction.

To ensure user-friendliness, the size of the explanatory subgraph G^s is minimized, as indicated by Eq. 5.

$$O_{\text{Size}} = \min |G^s| \tag{5}$$

where the size of G^s is represented as $|G^s|$. This objective promotes a more concise explanation for user to understand.

By iteratively constructing a collection of Pareto optimal explanatory subgraphs, MOE offers detailed and comprehensive explanations that emphasize accuracy, conciseness, and global context integration, enhancing their relatability and applicability.

4 Experiment

4.1 Datasets and Experimental Settings

The experiment is designed to evaluate the explanation performance of the proposed MOE in comparison to representative explanation models, including Deeplift [11], Grad-CAM [10], GNNExplainer [15], GNN-LRP [9], and SubgraphX [18], on two graph datasets: MUTAG [4] and BA-2Motifs [18]. A 3-layer Graph Convolutional Network (GCN) is implemented as the base predictive model, with consecutive layer dimensions of 16, 32, and 64 hidden units, and is trained using the Adam optimization algorithm with a learning rate of 0.001. The base model achieves prediction accuracies of 0.8684 on MUTAG and 0.9190 on BA-2Motifs.

Since both MOE and SubgraphX provide a collection of explanatory subgraphs, the average score of these subgraphs is used as the evaluation criterion. The gSpan threshold ratio for MOE is set to 1.0 for both MUTAG and BA-2Motifs, and the Monte Carlo Tree Search (MCTS) implemented in MOE is configured with a maximum depth of 8 to regulate the search space. A maximum ratio of 0.5 is applied to constrain the size of the largest explainable subgraph across all methods under consideration.

To evaluate the explainability of the models, a comprehensive set of evaluation metrics is employed. These include well-established metrics such as Explanation Accuracy [2], Fidelity [17], Sparsity [17], and Unfaithfulness [13].

Table 1. Explanation Performance Evaluation

		DeepLift	GNNExplainer	SubgraphX	GNN-LRP	Grad-CAM	MOE
Explanation Accuracy	MUTAG	0.5519	0.6324	0.6616	0.6092	0.5442	**0.8280**
	BA-2Motifs	0.5365	0.6173	0.5263	0.5636	0.5368	**0.7672**
Fidelity-	MUTAG	0.3245	0.2979	0.2289	0.3457	**0.0851**	0.1189
	BA-2Motifs	0.5120	0.3900	0.3443	0.4120	0.2440	**0.2179**
Sparsity	MUTAG	0.5519	0.6324	0.4109	0.6092	0.5442	**0.1306**
	BA-2Motifs	0.5365	0.6173	0.4202	0.5636	0.5368	**0.1254**
Unfaithfuless	MUTAG	0.7833	0.7585	1.0000	0.8446	0.5473	**0.3159**
	BA-2Motifs	0.9147	0.9753	0.8186	0.8671	**0.3591**	0.9764

4.2 Evaluation of Explanations

This study conducts experiments to assess the performance of different explanation methods on two common datasets: MUTAG and BA-2Motifs. We employ various performance metrics to evaluate the effectiveness of each method, and here presents the results with a focus on MOE's performance compared to other methods, as shown in Table 1.

Evaluation w.r.t. Explanation Accuracy. In the comprehensive evaluation of Explanation Accuracy, MOE showcases its superior performance. Particularly in the MUTAG dataset, MOE achieves an impressive score of 0.8280, significantly outshining GNNExplainer's 0.6324 and SubgraphX's 0.6616. This remarkable accuracy indicates MOE's profound ability to replicate the target prediction accurately, ensuring reliable and trustworthy explanations. Even in the more challenging BA-2Motifs dataset, MOE's accuracy of 0.7672 far exceeds that of GNNExplainer's 0.6173, demonstrating its robustness and reliability across different contexts.

Evaluation w.r.t. Fidelity-. The Fidelity- metric indicates MOE's general effectiveness, with scores reflecting minimal alteration to the original prediction. In the MUTAG dataset, although MOE's score of 0.1189 is relatively higher than some models, it is significantly better than GNN-LRP's 0.3457, suggesting that while MOE is generally faithful, there's potential for further optimization to ensure that essential features are retained without distorting the original prediction.

Evaluation w.r.t. Sparsity. The Sparsity score results reveal that MOE excels in producing sparse yet informative explanations. Notably, in the MUTAG dataset, MOE's score of 0.1306 is much lower than those of GNNExplainer (0.6324) and GNN-LRP (0.6092), indicating a more streamlined and focused explanation. While MOE's sparsity is not the lowest across all datasets, it strikes a commendable balance between simplification and retaining necessary explanatory details, making it user-friendly and accessible.

Evaluation w.r.t. Unfaithfulness. The Unfaithfulness score offers insights into the reliability of the explanations. In the MUTAG dataset, MOE registers a low score of 0.3159, signifying a high degree of faithfulness to the original model's predictions. This contrasts with its performance in the BA-2Motifs dataset, where it scores 0.9764, indicative of a higher unfaithfulness level, and suggesting that MOE's performance might vary with the complexity and nature of the dataset. This variability points to potential areas for further refinement in MOE's approach to ensure consistency and reliability across various types of data.

In conclusion, the evaluations across various metrics and datasets clearly demonstrate MOE's ability to provide precise, faithful, concise, and commonality-rich explanations.

5 Conclusion

This paper proposes an explanatory method using graph neural networks with multi-objective scenarios. The method solves the problem that traditional explanatory methods are only applicable to specific target scenarios and cannot satisfy multi-target requirements. In addition, the method avoids the limitations due to explanation by fusing and maintaining global and local information in the graph during the explanation process. Experimental results in the real world validate the effectiveness of this paper. While the method proposed in this paper yields improved results, there remains room for further enhancement. Future endeavors will delve deeper into exploring the causal relationships among various explanations in multi-target scenarios.

Acknowledgements. This research is partially supported by the National Key R&D Program of China 2021YFF0900800, Natural Science Foundation of China (No. 92367202, No. 62202279, No. 72293581), the Shandong Provincial Key Research and Development Program (Major Scientific and Technological Innovation Project) (No.2021CXGC010108), Excellent Youth Science Fund Project (Overseas) of Shandong (No. 2023HWYQ-039), Natural Science Foundation of Shandong Province (No. ZR2022QF018, No. ZR2023LZH006), the Fundamental Research Funds of Shandong University.

References

1. Auer, P., Cesa-Bianchi, N., Fischer, P.: Finite-time analysis of the multiarmed bandit problem. Mach. Learn. **47**, 235–256 (2002)
2. Chen, J., Song, L., Wainwright, M., Jordan, M.: Learning to explain: An information-theoretic perspective on model interpretation. In: International Conference on Machine Learning, pp. 883–892. PMLR (2018)
3. Dai, E., et al.: A comprehensive survey on trustworthy graph neural networks: Privacy. Robustness, Fairness, and Explainability. arXiv 2204 (2022)

4. Debnath, A.K., Lopez de Compadre, R.L., Debnath, G., Shusterman, A.J., Hansch, C.: Structure-activity relationship of mutagenic aromatic and heteroaromatic nitro compounds. correlation with molecular orbital energies and hydrophobicity. J. Medicinal Chem. **34**(2), 786–797 (1991)
5. Kakkad, J., Jannu, J., Sharma, K., Aggarwal, C., Medya, S.: A survey on explainability of graph neural networks. arXiv preprint arXiv:2306.01958 (2023)
6. Kocsis, L., Szepesvári, C.: Bandit based Monte-Carlo planning. In: European Conference on Machine Learning, pp. 282–293. Springer (2006). https://doi.org/10.1007/11871842_29
7. Lin, W., Lan, H., Li, B.: Generative causal explanations for graph neural networks. In: International Conference on Machine Learning, pp. 6666–6679. PMLR (2021)
8. Luo, D., Cheng, W., Xu, D., Yu, W., Zong, B., Chen, H., Zhang, X.: Parameterized explainer for graph neural network. Adv. Neural. Inf. Process. Syst. **33**, 19620–19631 (2020)
9. Schnake, T., Eberle, O., Lederer, J., Nakajima, S., Schütt, K.T., Müller, K.R., Montavon, G.: Higher-order explanations of graph neural networks via relevant walks. IEEE Trans. Pattern Anal. Mach. Intell. **44**(11), 7581–7596 (2021)
10. Selvaraju, R.R., Cogswell, M., Das, A., Vedantam, R., Parikh, D., Batra, D.: Grad-cam: visual explanations from deep networks via gradient-based localization. In: Proceedings of the IEEE International Conference on Computer Vision, pp. 618–626 (2017)
11. Shrikumar, A., Greenside, P., Kundaje, A.: Learning important features through propagating activation differences. In: International Conference on Machine Learning, pp. 3145–3153. PMLR (2017)
12. Sun, M., Xing, J., Meng, H., Wang, H., Chen, B., Zhou, J.: Molsearch: search-based multi-objective molecular generation and property optimization. In: Proceedings of the 28th ACM SIGKDD Conference on Knowledge Discovery and Data Mining, pp. 4724–4732 (2022)
13. Wang, Z., et al.: Faithful explanations for deep graph models. arXiv preprint arXiv:2205.11850 (2022)
14. Yan, X., Han, J.: GSPAN: graph-based substructure pattern mining. In: 2002 IEEE International Conference on Data Mining, 2002. Proceedings, pp. 721–724. IEEE (2002)
15. Ying, Z., Bourgeois, D., You, J., Zitnik, M., Leskovec, J.: Gnnexplainer: Generating explanations for graph neural networks. Advances in neural information processing systems **32** (2019)
16. Yuan, H., Tang, J., Hu, X., Ji, S.: Xgnn: Towards model-level explanations of graph neural networks. In: Proceedings of the 26th ACM SIGKDD International Conference on Knowledge Discovery & Data Mining, pp. 430–438 (2020)
17. Yuan, H., Yu, H., Gui, S., Ji, S.: Explainability in graph neural networks: a taxonomic survey. IEEE Trans. Pattern Anal. Mach. Intell. **45**(5), 5782–5799 (2022)
18. Yuan, H., Yu, H., Wang, J., Li, K., Ji, S.: On explainability of graph neural networks via subgraph explorations. In: International Conference on Machine Learning, pp. 12241–12252. PMLR (2021)
19. Zhang, Y., Defazio, D., Ramesh, A.: Relex: a model-agnostic relational model explainer. In: Proceedings of the 2021 AAAI/ACM Conference on AI, Ethics, and Society, pp. 1042–1049 (2021)

EGNN-AD: An Effective Graph Neural Network-Based Approach for Anomaly Detection on Edge-Attributed Graphs

Hewen Wang[1(✉)], Bryan Hooi[1], Dan He[2], Juncheng Liu[1], and Xiaokui Xiao[1]

[1] National University of Singapore, Singapore, Singapore
{wanghewen,juncheng.liu}@u.nus.edu, {dcsbhk,xkxiao}@nus.edu.sg
[2] University of Queensland, Brisbane, Australia
d.he@uq.edu.au

Abstract. The emergence of Graph Neural Networks (GNNs) has led to the development of several GNN-based anomaly detection models that detect anomalies in attributed graphs using graph structural and attribute information. However, most existing solutions focus on detecting anomalous nodes, while some applications require detecting anomalous edges, such as fraudulent financial transactions or product reviews. In this work, we present a method for detecting anomalous edges from a node classification perspective in a transformed node-attributed graph, where each edge in the original graph is converted into an attributed node. We propose an effective solution, `EGNN-AD`, which leverages two GNN models, `EGNN` and `NGNN`, to incorporate information from both edges and nodes for detecting edge anomalies. `EGNN` uses top-k Personalized PageRank (PPR) to establish different edge-to-edge relations in the new graph, while `NGNN` employs trainable node embeddings to enhance performance. Our experiments on real edge-attributed datasets demonstrate that `EGNN-AD` consistently outperforms several baselines in terms of anomaly detection accuracy.

Keywords: Graph Neural Network · Anomaly Detection · Attributed Graph

1 Introduction

Anomaly detection in graph-structured data is essential for identifying unusual patterns in nodes, edges, or sub-graphs, with significant implications for fraud detection and social network analysis. Graph Neural Networks (GNNs) are pivotal in understanding these complex structures, aiding in tasks like node classification and link prediction by utilizing topology and node attributes. This paper investigates GNNs for anomaly detection in edge-attributed graphs, crucial for spotting fraud in financial transactions and sifting genuine from fake reviews.

Supplementary Information The online version contains supplementary material available at https://doi.org/10.1007/978-981-97-5572-1_21.

M. Onizuka et al. (Eds.): DASFAA 2024, LNCS 14855, pp. 321–331, 2024.
https://doi.org/10.1007/978-981-97-5572-1_21

It also includes an analysis of limitations, addressing potential challenges and areas for improvement in applying GNNs to anomaly detection.

In our work, we propose an effective approach that is capable of not only distinguishing between different edge relationships, considering that they convey diverse types of information, but also modeling the behaviour of N-Nodes by leveraging information from neighbouring E-Nodes. Specifically, we introduce `EGNN-AD`, an Anomaly Detection framework with an Edge-centric Graph Nneural Networks comprising two major components: `EGNN` and `NGNN`. `EGNN` focuses on learning correlations among different edges based on edge attributions, considering various groups of edges, and utilizing top-k Personalized Page Rank (PPR) to construct a reduced-size edge-centric graph. This step allows for a more efficient representation of edge relationships. On the other hand, `NGNN` is designed to capture the original topological correlations among nodes while aggregating edge attributes as features for nodes. By combining `EGNN` and `NGNN`, `EGNN-AD` effectively captures the nuances of edge relationships and node behaviours, resulting in improved anomaly detection performance in complex real-life scenarios.

Our contributions are summarized as follows:

- We propose `EGNN-AD`, an edge-centric graph embedding algorithm with `EGNN` and `NGNN` components. `EGNN` constructs a reduced-size edge-centric graph using top-k PPR to capture edge correlations, while `NGNN` aggregates edge attributes to model node behavior effectively.
- Unlike existing approaches, `EGNN-AD` effectively distinguishes diverse edge relationships by considering the behavior of neighboring E-Nodes, resulting in a more nuanced representation of complex graph structures.
- We conduct extensive experiments on real-life edge-attributed datasets, demonstrating `EGNN-AD`'s superior performance in detecting anomalous edges, and its potential impact in real-world applications.

2 Related Work

Anomaly detection, or outlier detection, identifies data instances that significantly differ from most. For in-depth reviews, see [19]. Traditionally, this relied on expert-designed features or statistical models, combined with machine learning. Recent trends favor deep learning for more effective anomaly identification in graphs. Graph Neural Networks (GNNs), borrowing from convolutional networks, are effective for graph data, with variants like GCN focusing on node neighborhoods and others adding layers for enhanced capacity. Recent models also integrate both node and edge features for richer graph representations.

GNNs, inspired by convolutional neural networks, have advanced graph-structured data modeling, with GCN focusing on 1-step node neighborhoods. Recent GNN variants enhance capacity with features like attention [14] and transformer layers [21]. Several GNN-based approaches have been developed for anomaly detection. GEM [10] uses node embeddings from GNNs for detecting anomalies in account-device graphs. FDGars [18] and Player2Vec [22] apply GCN for fraud detection in reviews and key player identification in networks, respectively. SemiGNN [16] offers a semi-supervised model for fraud detection in multi-view graphs, while GraphConsis [9] addresses node inconsistencies in anomaly detection. CARE-GNN [4] enhances detection with a label-aware similarity measure and reinforcement learning. These methods, however, mainly address node anomalies, with potential limitations in edge anomaly detection.

To enhance edge anomaly detection, methods often use node embeddings from edge endpoints. AANE [5] employs GCN with anomaly-aware loss functions for static graphs, while AddGraph [23] integrates GCNs with temporal models for dynamic graphs. TADDY [8] applies transformers to unattributed dynamic graphs, focusing on edge embeddings. More recently, [17] proposes an edge-centric edge representation learning approach on edge classification tasks.

Our approach differs by directly detecting edge anomalies through GNNs, emphasizing edge-to-edge relations and attributes. We distinctively use separate GNN models for nodes and edges, applying top-k Personalized Page Rank for edge relations and trainable embeddings for nodes. This approach effectively captures edge-specific intricacies, improving anomaly detection in graph-structured data.

3 Problem Formalization

In this section, we formalize edge anomaly detection as an edge classification problem on an edge-attributed graph. Given a set of attributed records, such as financial transactions between accounts, each record contains the IDs of two accounts and various attributes describing the transaction. We represent this information using a directed multigraph $G = (V, E, \mathbf{x}_i)$, where V is the node set with a total of n nodes, E is the directed edge set with a total of m edges and $\mathbf{x}_i$ denotes the attribute vector of length d for edge $e_i \in E$. It is important to note that multiple edges can exist between two nodes, representing multiple actions or interactions between the two endpoints. Thus the edges connecting two same nodes with different attributes are considered as different edges in the graph.

An edge e_i in G is considered *anomalous* if it is associated with a label indicating that the transaction involves an anomalous action, such as financial loss, unhelpful reviews, or suspicious purchases. Given a training set E_{tr} of edges ($E_{tr} \subseteq E$) with associated labels $\mathbf{y}$, where $\mathbf{y}_i = 1$ if edge $e_i \in E_{tr}$ is an anomaly, and $\mathbf{y}_i = 0$ otherwise, the task of *graph-based edge anomaly detection* is to predict the labels for edges in $E_{ts} = E \backslash E_{tr}$, which constitutes the testing set.

We formulate edge anomaly detection as a supervised edge classification task and aim to minimize the following binary cross-entropy loss function:

$$\mathcal{L} = -\sum_{e_i \in E_{tr}} \mathbf{y}_i \cdot \log(g(\mathbf{z}_{e_i})) + (1 - \mathbf{y}_i) \cdot \log(1 - g(\mathbf{z}_{e_i})), \tag{1}$$

where $\mathbf{z}_{e_i}$ is a vector that represents edge e_i and is obtained through some feature transformation functions, such as a neural network that maps each edge e_i to an edge feature vector $\mathbf{z}_{e_i}$. $g(\cdot)$ represents a label prediction function, which could be a neural network classifier, mapping the edge feature vector to the corresponding

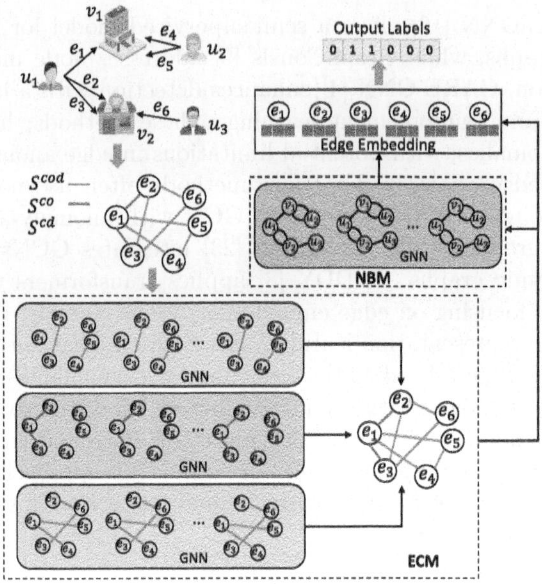

Fig. 1. Framework Architecture

labels. The loss function quantifies the difference between the predicted labels and the ground truth, guiding the learning process to achieve accurate edge anomaly detection.

4 ADEC Model

Our proposed framework, `EGNN-AD`, comprises two main parts, as illustrated in Fig. 1. (1) Edge Correlation Modeling (`EGNN`): This component employs a GNN called `EGNN` to model relationships among different edges. By utilizing top-k Personalized PageRank, it establishes meaningful connections between edges, enhancing the understanding of their inter-dependencies. (2) Node Behavior Modeling (`NGNN`): The second component, `NGNN`, is another GNN tailored to capture correlations originating from nodes. It aggregates edge attributes to form features for nodes, enabling the exploration of node behaviours within the graph. Then, the features from nodes are aggregated to form the edge embedding, concatenating the edge features from `EGNN`, to generate the final output.

4.1 `EGNN`: Edge Correlation Modeling

In edge attribute-based anomaly detection, we transform the original graph into an edge-to-node graph, allowing for multiple distinct-attribute edges between nodes. We use GNN models for information aggregation via message passing.

Algorithm 1: `EGNN`

Input: Original graph G, Derived graphs G^{cod}, G^{co}, G^{cd}, Number of GNN layers L.
Output: *E-Node* embedding vector set Z^E.

// Build edge centrality vector set for all edges, at first call
Obtain top-k PPR values $\mathcal{T} \leftarrow$ `ForwardPush`$(G^{ETN}, \alpha, r_{max}, k)$;
for $e_i \in E$ **do**
  for $(e_i, \hat{\pi}_{e_i}(w)) \in \mathcal{T}(e_i)$ **do**
    if $w = e_j \in E$ **then**
      $\pi_{e_i}(e_j) \leftarrow 1$
    else
      $\pi_{e_i}(e_j) \leftarrow 0$
// Build three graphs G^{ϕ}, $\phi \in \{cod, co, cd\}$, at first call
if G^{cod}, G^{co}, G^{cd} *are empty graphs* **then**
  for $e_i = (u_i, v_i) \in E$ **do**
    for $e_j = (u_i, v_j) \in E$ **do**
      if $v_i = v_j$ **then** Add undirected edge (e_i, e_j) into G^{cod}
      else if $\pi_{e_i}(e_j) = 1$ **then** Add undirected edge (e_i, e_j) into G^{co}
    for $e_j = (u_j, v_i) \in E$ **do**
      if $u_i = u_j$ **then** Add undirected edge (e_i, e_j) into G^{cod}
      else if $\pi_{e_i}(e_j) = 1$ **then** Add undirected edge (e_i, e_j) into G^{cd}
// Use GNN models to transform node features in G^{ϕ}
for $\phi \in \{cod, co, cd\}$ **do**
  for $e_i \in E$ **do**
    $\mathbf{h}_{e_i}^{(0)} \leftarrow \mathbf{x}_{e_i}$;
    for $l \leftarrow 1$ *to* L **do**
      $\mathbf{h}_{e_i}^{(l)} \leftarrow$ `AGGREGATE` $\left(\mathbf{h}_{e_i}^{(l-1)}, \left\{\mathbf{h}_{e_j}^{(l-1)} \mid e_j \in N_{\phi}(e_i)\right\}, \mathbf{W}_{\phi}^{(l)}\right)$
    $\mathbf{z}_{\phi,e_i}^E = \mathbf{h}_{e_i}^{(L)}$;
return $\{\mathbf{z}_{e_i}^E = \mathbf{z}_{cod,e_i}^E \parallel \mathbf{z}_{co,e_i}^E \parallel \mathbf{z}_{cd,e_i}^E | e_i \in E\}$

Instead of a single GNN, edges are categorized by connectivity, with three distinct edge sets modeling their individual correlations. Given a pair of nodes u_i, v_i on the original graph G, we have: $S_i^{cod} = \{(u,v) : u = u_i, v = v_i\}$, $S_i^{co} = \{(u,v) : u = u_i, v \neq v_i\}$, and $S_i^{cd} = \{(u,v) : u \neq u_i, v = v_i\}$ where S_i^{cod} represents the set of edges whose origins and destinations are the same, S_i^{co} indicates the set of edges that share the same origins and different destinations, and S_i^{cd} denotes the ones with the same destinations but different origins.

In the user-transaction-user example from Fig. 1, consider nodes u_1 and v_2. The set $S_{(u_1,v_2)}^{cod} = \{e_2, e_3\}$ represents transactions from a common sender to a common receiver, $S_{(u_1,v_2)}^{co} = \{e_1\}$ denotes transactions from the same sender to different receivers, and $S_{(u_1,v_2)}^{cd} = \{e_6\}$ includes transactions from different

Algorithm 2: NGNN

Input: Edge-attributed graph G, *E-Node* embedding vector set Z^E.
Output: *N-Node* embedding vector set $Z^N = \{\mathbf{z}^N_{u_i} | u_i \in V\} \leftarrow$NGNN (G, Z^E).

for $n_i \in V$ **do**
  if *N-Node embedding vector* $\mathbf{x}_{n_i}$ *is not initialized* **then**
    $\mathbf{x}_{n_i} \leftarrow \texttt{AGGREGATE}\left(\mathbf{0}, \left\{\mathbf{z}^E_{(n_i,n_j)} \mid (n_i, n_j) \in E\right\}, \mathbf{W}^{(0)}\right)$
  $\mathbf{h}^{(0)}_{n_i} \leftarrow \mathbf{x}_{n_i}$;
  for $l \leftarrow 1$ *to* L **do**
    $\mathbf{h}^{(l)}_{n_i} \leftarrow \texttt{AGGREGATE}\left(\mathbf{h}^{(l-1)}_{n_i}, \left\{\mathbf{h}^{(l-1)}_{n_j} \mid n_j \in N(n_i)\right\}, \mathbf{W}^{(l)}\right)$
  $\mathbf{z}^N_{n_i} = \mathbf{h}^{(L)}_{n_i}$;
return $\{\mathbf{z}^E_{n_i} | n_i \in V\}$

senders to the same receiver. To create an "edge-to-node" graph, we focus on three distinct edge sets: S^{cod}_i, S^{co}_i, and S^{cd}_i. Instead of a naive transformation that could overwhelm the graph with edges, we selectively manage edge sets to minimize edge count. For S^{cod}_i edges, which link the same node pairs, we keep all due to their low count. For S^{co}_i and S^{cd}_i, to avoid a bloated graph, we apply edge centrality and top-k Personalized PageRank (PPR) for edge selection. This approach constructs a heterogeneous graph, acknowledging the unique correlations of each edge category. Algorithm 1 presents the pseudo-code for EGNN, which is designed to model *E-Node* relations based on top-k Personalized PageRank (PPR) in the edge-to-node graph G^{ETN}.

4.2 NGNN: Node Behavior Modeling

In edge-attributed graphs, *N-Nodes* represent crucial entities like users or products, encapsulating valuable structural and attribute information. For effective edge anomaly detection, updating and learning *N-Node* embeddings during training is vital. These embeddings summarize entity behaviors and interactions, enhancing understanding of graph patterns and improving anomaly detection accuracy. We detail a method to refine *N-Node* embeddings via GNNs, highlighting their role in capturing essential graph structures for improved anomaly detection. Addressing graphs lacking node attributes, traditional methods like identity matrices, random walks, or matrix factorization have limitations in modeling relationships in edge-only attribute graphs, presenting a unique challenge.

In this section, we introduce NGNN, a novel approach that utilizes GNNs to construct *N-Node* embeddings, thereby effectively modeling *N-Nodes* using edge attributes. The core concept behind NGNN is to employ a GNN model to update *N-Node* embeddings based on the information provided by attributed edges in the training data. To achieve this, we first aggregate *E-Node* embeddings generated by `EGNN` as *N-Node* attributes. Subsequently, these *N-Node* attributes undergo a transformation process through another GNN, enabling the propagation of information and incorporating graph structural knowledge. As a result, we obtain updated *N-Node* embeddings, which provide a comprehensive representation of each node, integrating both edge attributes and graph structural information. By concatenating these new *N-Node* embeddings with *E-Node* embeddings obtained from Sect. 4.1, we derive the final embeddings for *E-Nodes*, which serve as the basis for calculating the classification loss. Crucially, during the training phase, *N-Node* embeddings can be updated through the process of backpropagation, ensuring the model effectively captures relevant information from the edge attributes and enhances its anomaly detection capabilities. Algorithm 2 outlines the procedure of `NGNN`.

Table 1. Data statistics ($K = 10^3, M = 10^6$)

Dataset	#nodes	#edges	#attributes	%anomalies
YelpChi	38K	67K	25	13.23%
-Edge-to-node	106K	375K		
-Line graph	67K	48M		
PayPal Sanction	77K	90K	512	4.77%
-Edge-to-node	167K	529K		
-Line graph	90K	158M		
PayPal Transaction	24K	141K	296	10.44%
-Edge-to-node	165K	730K		
-Line graph	141K	39M		

5 Experiments

We evaluated our framework on three edge-attributed graph datasets against GNN baselines. The YelpChi [11] dataset involves Chicago restaurant and hotel reviews, represented in a user-review-restaurant/hotel graph with 25-dimensional review text vectors as edge features. The PayPal Sanction dataset [12] includes buyer and seller transactions, captured in a buyer-item-seller graph with item details encoded by Sentence-Bert for text representation. The PayPal Transaction dataset contains financial transactions represented in a sender-transaction-receiver graph, with transaction attributes forming 296-dimensional vectors. For YelpChi and PayPal Sanction, edges were split into train, validation, and test sets by an $8:1:1$ ratio. For PayPal Transaction, we used a temporal split for realistic evaluation, ensuring cross-validation and test transactions occurred after those in the training set. For reproducibility, our codes and datasets are made publicly available at https://github.com/wanghewen/ENGNN.

5.1 Baselines

In our experiments, we utilize several widely adopted GNN models as the backbones for constructing node and edge embeddings for anomaly detection. The GNN models we consider include Graph Convolutional Networks (GCN), GraphSAGE [6], Graph Attention Networks (GAT) [14], Graph Isomorphism Networks (GIN) [20], ChebNet [3], Attention-based Graph Neural Networks (AGNN) [13], Graph Transformer [7] and GATv2 [2]. To evaluate the effectiveness of our proposed framework, we compare it against three other methods, namely, Original, Line graph, and Edge-to-node, which are used as baselines for constructing graphs. Additionally, we include TADDY [8], AANE [5] AttrE2Vec [1] and Edge2vec [15] as baselines with the original graphs.

5.2 Performance Analysis

The experimental results in Table 2 reveal distinct performance patterns among the methods. The unsupervised methods, TADDY and AANE, consistently exhibit inferior performance across various settings, except for specific instances on the PayPal dataset. On the original graph, MLP, which solely utilizes edge attribute information shows better performance. GNN-based models generally demonstrate superior performance on the edge-to-node graph compared to the original and line graphs. Their ability to leverage graph structural information allows for improved edge representations. However, the memory limitations of GNNs become apparent on line graphs, particularly on the PayPal dataset. Despite this, `EGNN-AD` consistently outperforms other methods, leveraging both edge-to-edge and edge-to-node relations to enhance edge representations and achieve more effective anomaly detection. Let's now proceed to discuss the individual performance of each dataset in detail.

In summary, our experiments show that edge-to-node graphs are more effective than original graphs in representing edge attributes, implying limited propagation of these attributes in original graphs. Line graphs, focusing solely on neighborhood information, yield the least favorable results for edge anomaly detection. In contrast, `EGNN-AD`'s ability to leverage both edge and node relationships allows it to outperform existing methods consistently, demonstrating its potential for addressing real-world anomaly detection challenges.

5.3 Ablation Study

In this ablation study, we investigate the effectiveness of `EGNN` and `NGNN` within our proposed framework. The results are presented in Table 3. The GNNs applied on `EGNN` and `NGNN` demonstrate improved anomaly detection metrics, especially F1-scores, compared to the GNNs applied on the original graph. This highlights the effectiveness of our proposed methods in enhancing the representation of edges within the graph. Moreover, we observe that leveraging `EGNN` and `NGNN` together achieves the best performance, which further validates the importance of our proposed framework in generating effective edge representations.

Table 2. Experiment results (in %) for YelpChi, PayPal Sanction and PayPal Transaction dataset for the baselines and EGNN-AD. '–' represents the model runs out of the GPU memory (24GB)

		YelpChi				PayPal Sanction				PayPal Transaction			
Method	Backbone	AUC	AP	$F1_A$	$F1_M$	AUC	AP	$F1_A$	$F1_M$	AUC	AP	$F1_A$	$F1_M$
Original	GAT	77.71	32.01	40.56	62.57	79.70	61.75	59.61	65.50	42.48	12.92	20.29	19.76
	GCN	76.52	30.96	39.31	60.66	73.31	57.09	55.26	68.57	55.03	17.53	23.43	29.38
	GraphSAGE	77.90	32.22	40.09	62.86	83.22	59.64	64.71	72.06	56.14	15.36	24.84	35.45
	GIN	77.67	33.64	39.97	62.25	77.61	58.27	63.93	71.09	44.89	11.38	21.27	20.96
	Cheb	78.93	34.21	41.57	62.61	78.32	54.95	61.14	68.47	54.38	15.51	24.00	31.22
	AGNN	78.25	32.96	41.17	62.12	76.60	58.94	55.55	68.47	52.83	13.86	22.15	30.37
	GATv2	77.53	32.51	40.24	62.02	82.62	59.09	64.03	71.46	47.98	13.59	21.37	23.96
	Transformer	77.36	30.82	39.37	50.32	82.50	63.60	62.30	42.05	57.59	15.21	26.97	47.16
	MLP	77.84	32.85	40.25	61.89	83.97	64.94	65.37	72.79	51.09	14.04	20.88	23.37
	TADDY	23.42	13.24	23.40	49.92	–	–	–	–	73.28	7.00	19.42	9.56
	AANE	59.18	16.23	26.35	42.30	–	–	–	–	63.35	14.36	24.18	24.05
	Edge2Vec	–	–	–	–	–	–	–	–	50.04	10.79	19.43	9.98
	AttrE2Vec	80.13	34.01	42.32	64.07	74.79	44.15	57.21	67.56	61.08	15.71	25.01	46.20
Line graph	GAT	52.46	14.04	24.11	19.50	87.02	66.73	68.49	75.49	31.76	11.36	19.51	10.15
	GCN	58.40	19.19	25.28	44.82	83.10	66.70	63.06	70.33	54.30	21.81	24.38	57.77
	GraphSAGE	79.20	33.25	41.14	62.36	81.82	66.82	60.89	68.97	62.63	18.11	25.52	39.61
	GIN	55.72	15.58	24.25	54.09	59.70	42.05	44.61	65.14	53.81	12.73	23.83	30.54
	Cheb	81.00	36.82	43.64	64.49	79.90	62.03	60.02	69.21	75.00	25.20	33.02	59.95
	AGNN	58.96	18.80	25.64	38.59	79.76	63.72	63.95	75.06	35.93	7.79	20.91	22.25
Edge-to-node	GAT	78.42	29.84	40.08	61.74	76.30	55.54	57.89	68.74	36.20	14.63	19.54	11.49
	GCN	80.08	33.53	41.60	63.11	81.30	54.24	62.75	69.47	48.46	19.36	24.57	58.37
	GraphSAGE	79.00	32.37	41.05	62.05	85.80	59.91	68.57	75.42	69.49	22.74	28.97	49.05
	GIN	69.41	27.86	33.55	56.93	63.27	50.90	49.24	50.01	35.78	10.23	19.68	11.86
	Cheb	80.73	35.07	42.99	63.00	84.29	57.00	65.80	75.75	85.81	37.90	53.25	72.06
	AGNN	77.08	31.38	39.16	60.48	75.75	58.20	53.91	67.93	30.42	7.13	19.62	12.72
	GATv2	78.42	29.78	40.45	61.51	76.70	59.45	56.43	66.12	67.51	24.15	27.10	59.21
	Transformer	81.30	36.48	43.25	61.75	87.79	60.01	53.35	70.75	95.37	77.79	72.60	85.51
EGNN-AD	GAT	81.18	36.66	43.53	63.84	77.30	57.82	57.11	61.25	78.36	27.49	35.42	59.27
	GCN	**82.41**	38.20	44.27	**64.65**	83.90	**72.04**	63.18	75.07	94.24	83.07	81.08	89.44
	GraphSAGE	81.32	36.06	42.85	64.41	**88.18**	69.30	69.57	**77.84**	75.87	24.62	32.70	53.35
	GIN	81.23	36.80	42.78	63.23	76.08	50.46	56.34	67.34	**97.22**	83.07	81.94	89.82
	Cheb	80.00	35.12	41.41	61.82	79.12	62.54	62.45	74.45	95.36	**84.20**	**83.36**	**90.71**
	AGNN	77.56	30.91	40.30	62.35	79.17	53.43	59.15	67.08	84.58	36.23	51.12	70.98
	GATv2	81.01	35.96	43.32	64.43	86.91	65.52	**69.78**	77.56	68.33	20.78	28.81	47.45
	Transformer	82.29	**39.12**	**44.35**	58.80	85.60	68.60	67.72	65.36	92.87	81.03	80.41	88.97

The combination of EGNN and NGNN allows for a comprehensive capture of both edge-to-edge and node-to-edge relations, leading to superior edge embeddings for anomaly detection.

Table 3. Ablation results (in %) for YelpChi, PayPal Sanction and PayPal Transaction dataset for different components

Method	Backbone	YelpChi				PayPal Sanction				PayPal Transaction			
		AUC	AP	$F1_A$	$F1_M$	AUC	AP	$F1_A$	$F1_M$	AUC	AP	$F1_A$	$F1_M$
Original	GCN	76.52	30.96	39.31	60.66	73.31	57.09	55.26	68.57	55.03	17.53	23.43	29.38
	GraphSAGE	77.90	32.22	40.09	62.86	83.22	59.64	64.71	72.06	56.14	15.36	24.84	35.45
	GATv2	77.53	32.51	40.24	62.02	82.62	59.09	64.03	71.46	47.98	13.59	21.37	23.96
	Transformer	77.36	30.82	39.37	50.32	82.50	63.60	62.30	42.05	57.59	15.21	26.97	47.16
+EGNN	GCN	82.03	38.07	44.31	64.19	83.73	71.55	64.38	74.33	94.07	82.42	80.20	88.94
	GraphSAGE	80.94	34.97	43.02	63.94	72.65	44.98	54.96	66.16	90.12	76.65	72.44	84.61
	GATv2	80.81	37.06	42.70	63.88	80.47	59.74	62.71	70.16	66.04	20.77	27.82	59.73
	Transformer	82.65	40.16	45.20	56.23	81.73	71.29	62.34	57.32	92.50	81.93	80.65	86.59
+NGNN	GCN	82.41	38.20	44.27	64.65	83.90	72.04	63.18	75.07	94.24	83.07	81.08	89.44
	GraphSAGE	81.32	36.06	42.85	64.41	88.18	69.30	69.57	77.84	75.87	24.62	32.70	53.35
	GATv2	81.01	35.96	43.32	64.43	86.91	65.52	69.78	77.56	68.33	20.78	28.81	47.45
	Transformer	82.29	39.12	44.35	58.80	85.60	68.60	67.72	65.36	92.87	81.03	80.41	88.97

6 Conclusion

In our paper, we introduced a framework for detecting anomalous edges in edge-attributed networks, combining attribute and topological data. This framework categorizes edges, utilizes top-k Personalized PageRank for edge relations, and employs trainable node embeddings. Future work will extend this framework to more complex situations like multi-hop edge relations and dynamic networks.

Acknowledgements. This work is supported by Singapore Ministry of Education Academic Research Fund Tier 3 under MOE's official grant number MOE2017-T3-1-007. The findings herein reflect the work, and are solely the responsibility, of the authors.

References

1. Bielak, P., Kajdanowicz, T., Chawla, N.V.: Attre2vec: unsupervised attributed edge representation learning. Inf. Sci. **592**, 82–96 (2022)
2. Brody, S., Alon, U., Yahav, E.: How Attentive are Graph Attention Networks? In: ICLR (May 2022)
3. Defferrard, M., Bresson, X., Vandergheynst, P.: Convolutional neural networks on graphs with fast localized spectral filtering. NIPS **29** (2016)
4. Dou, Y., Liu, Z., Sun, L., Deng, Y., Peng, H., Yu, P.S.: Enhancing graph neural network-based fraud detectors against camouflaged fraudsters. In: CIKM, pp. 315–324 (2020)
5. Duan, D., Tong, L., Li, Y., Lu, J., Shi, L., Zhang, C.: AANE: anomaly aware network embedding for anomalous link detection. In: ICDM, pp. 1002–1007 (2020)

6. Hamilton, W.L., Ying, R., Leskovec, J.: Inductive representation learning on large graphs. NIPS **2017-Decem**, 1025–1035 (2017)
7. Hu, Z., Dong, Y., Wang, K., Sun, Y.: Heterogeneous graph transformer. In: WWW, pp. 2704–2710 (2020)
8. Liu, Y., et al.: Anomaly detection in dynamic graphs via transformer. In: TKDE, vol. 14 (2021)
9. Liu, Z., Dou, Y., Yu, P.S., Deng, Y., Peng, H.: Alleviating the inconsistency problem of applying graph neural network to fraud detection. In: SIGIR, vol. 4, pp. 1569–1572 (2020)
10. Liu, Z., Zhou, J., Chen, C., Li, X., Yang, X., Song, L.: Heterogeneous graph neural networks for malicious account detection. In: CIKM, pp. 2077–2086 (2018)
11. Rayana, S., Akoglu, L.: Collective opinion spam detection: bridging review networks and metadata. KDD **2015-Augus**, 985–994 (2015)
12. Reimers, N., Gurevych, I.: Sentence-BERT: sentence embeddings using siamese BERT-networks. In: EMNLP-IJCNLP, pp. 3982–3992 (2020)
13. Thekumparampil, K.K., Wang, C., Oh, S., Li, L.J.: Attention-based graph neural network for semi-supervised learning. arXiv preprint arXiv:1803.03735 (2018)
14. Veličković, P., Cucurull, G., Casanova, A., Romero, A., Liò, P., Bengio, Y.: Graph Attention Networks. In: ICLR (2017)
15. Wang, C., Wang, C., Wang, Z., Ye, X., Yu, P.S.: Edge2vec: edge-based social network embedding. TKDD **14**(4), 1–24 (2020)
16. Wang, D., et al.: A semi-supervised graph attentive network for financial fraud detection. In: ICDM, vol. 2019, pp. 598–607 (2019)
17. Wang, H., Yang, R., Huang, K., Xiao, X.: Efficient and effective edge-wise graph representation learning. In: KDD, pp. 2326–2336 (2023)
18. Wang, J., Wen, R., Wu, C., Huang, Y., Xiong, J.: FDGars: fraudster detection via graph convolutional networks in online app review system. In: WWW, pp. 310–316 (2019)
19. Wang, R., et al.: Deep learning for anomaly detection. In: KDD, pp. 3569–3570 (2020)
20. Xu, K., Hu, W., Leskovec, J., Jegelka, S.: How powerful are graph neural networks? arXiv preprint arXiv:1810.00826 (2018)
21. Yun, S., Jeong, M., Kim, R., Kang, J., Kim, H.J.: Graph transformer networks. In: Advances in Neural Information Processing Systems, vol. 32 (2019)
22. Zhang, Y., Fan, Y., Ye, Y., Zhao, L., Shi, C.: Key player identification in underground forums over attributed heterogeneous information network embedding framework. In: CIKM, pp. 549–558 (2019)
23. Zheng, L., Li, Z.Z., Li, J., Li, Z.Z., Gao, J.: Addgraph: anomaly detection in dynamic graph using attention-based temporal GCN. In: IJCAI, vol. 2019-Augus, pp. 4419–4425 (2019)

Diffusion Model-Enhanced Contrastive Learning for Graph Representation

Qi Dai, Yumeng Song, Yu Gu(✉), Fangfang Li, and Xiaohua Li

School of Computer Science and Engineering, Northeastern University, Shenyang, Liaoning, China
2210701@stu.neu.edu.cn, {guyu,lifangfang,lixiaohua}@mail.neu.edu.cn

Abstract. Unsupervised Graph Contrastive Learning (GCL) aims to derive graph representations for downstream tasks without labeled data. While GCL methods have made significant progress, they suffer from limitations including noise amplification and neglecting global structural and semantic information. In this paper, we propose Diffusion Model-Enhanced Graph Contrastive Learning (DiffGCL) to overcome these limitations and enhance graph representation learning for the first time. Specifically, a graph-specific diffusion module is designed to explicitly capture global structural and semantic patterns by controlled Gaussian noise injection and an attention-based graph denoising network. A GCL module focuses on capturing local discriminative information. Through integrating the diffusion model with GCL, a shared graph encoder can acquire both global and local structures and semantic information within the graph while efficiently removing noise, leading to significant performance enhancement. Experimental results on real-world datasets demonstrate the effectiveness of DiffGCL, showing that it outperforms state-of-the-art competitors in graph classification accuracy.

Keywords: Contrastive Learning · Graph Representation Learning · Unsupervised Learning · Diffusion Model

1 Introduction

Graph neural networks (GNNs) have achieved state-of-the-art performances in various graph-based tasks [6,17], such as node classification, graph classification, and recommendation systems. However, their performance heavily relies on labeled data, which is both time-consuming and costly to obtain. To address this issue, recent research has focused on self-supervised graph representation learning, where labels are not required [10].

Among self-supervised learning methods, Graph Contrastive Learning (GCL) has emerged as one of the most popular paradigms [2]. Typically, GCL methods construct multiple views of the input graph through various data augmentation techniques and then train a GNN encoder to learn representations by keeping similar graph embeddings close and dissimilar ones far apart. Existing GCL

M. Onizuka et al. (Eds.): DASFAA 2024, LNCS 14855, pp. 332–341, 2024.
https://doi.org/10.1007/978-981-97-5572-1_22

models improve the performance by designing novel data augmentation strategies [20], or adjusting contrastive objectives [10,13]. However, these methods suffer from some limitations: Firstly, they struggle to effectively handle noise in graph data. The iterative message propagation mechanism of GNN encoders tends to amplify noise in graph data, propagating it throughout the entire graph. Secondly, GCL models have limitations in capturing global structural and semantic information within graphs. GNNs typically treat all nodes uniformly, neglecting node similarity and are often constrained by a limited number of layers (e.g., 2 or 3) [6]. Additionally, the contrastive loss in GCL emphasizes inter-data contrasts between positive and negative pairs across graphs, overlooking valuable intra-data global information within individual graphs.

To tackle the aforementioned issues, we incorporate Denoising Diffusion Probabilistic Models (DDPMs) [5] into GCL and propose Diffusion Model-Enhanced Graph Contrastive Learning (DiffGCL) for graph representation learning. To the best of our knowledge, this is the first GCL model to incorporate the denoising diffusion model to enhance GCL by solving the problems of noise handling and global structure and semantic information preservation.

Applying DDPMs to graph data involves two challenges: (1) Arbitrary noise injection may harm the semantic information of the graph. So how to inject noise into graphs in the forward diffusion process? (2) Learning from randomly initialized Gaussian noise is difficult due to the complex relationships within graphs. Besides, the reverse denoising process in DDPMs does not consider the structure and semantic information of graphs which is important for graph representation learning. So how to learn from initialized noise and capture meaningful semantic information effectively in the reverse denoising process?

To tackle the above challenges, we design two graph-specific modules. Firstly, we design a novel graph-specific diffusion module that applies partial controlled linear Gaussian noise injection to maintain the semantic information of the graph. Secondly, we develop a denoising model with an attention mechanism, which incorporates overall topological structural information and global graph characteristics to guide the denoising process. A shared encoder facilitates input mapping and information transfer between both modules. The encoder effectively capture both intricate local and global structures and semantic information within the graph. In summary, the main contributions of our work are as follows:

- We propose DiffGCL, the first GCL model with a diffusion module, can effectively handle noise and preserve global structural and semantic information in GCL.
- We apply a novel diffusion module specifically tailored for graphs. It can learn from randomly initialized nosiy data, while preserving meaningful semantic information effectively.
- Experimental results on six real-world datasets demonstrate that DiffGCL outperforms several state-of-the-art baselines.

2 Related Work

The primary objective of GCL is to maximize the mutual information between the input and its representations by contrasting positive pairs with negative pairs. InfoGraph [13] obtains expressive graph representations by maximizing the mutual information between graph-level representations. GRACE [21] and GCA [22] employ subgraph sampling and graph perturbation to create augmented pairs, adopting a node-level contrastive strategy. GraphCL [20] learns graph representations by applying predefined types of graph augmentations, while JOAO [19] introduces a unified bi-level optimization framework for selecting the best combination of data augmentations. However, these GCL methods often rely on data augmentation, which require careful selection to generate high-quality views. Moreover, they focus on perturbing local structures or node features, neglecting the rich global semantic structure inherent in graphs. MVGRL [4] addresses this limitation by constructing multiple graph views and then conducting contrastive learning at both the node and graph levels. Although MVGRL explores global information, it relies on graph pooling, which may not adequately capture sufficient global information, particularly in larger or more complex input graphs. Recently, some methods [7,16] exploit augmentation-free contrasting mode, however, they fail to effectively address the issue of noise.

3 Method

In this section, we introduce the proposed DiffGCL model, designed for graph representation learning. Firstly, we introduce the notations and problem definition graph representation learning.

Notations. In this work, we focus on attributed graphs denoted as $G = (V, E)$, where V represents the vertex set and E represents the edge set, $x \in \mathbb{R}^{n \times f}$ represents attribute information with f as the attribute dimension. The adjacency matrix is represented as $A \in \mathbb{R}^{f \times f}$, where A_{ij} indicates the relation between nodes v_i and v_j.

Problem Definition. Given a graph $\mathcal{G} = (V, E)$, the goal of graph representation learning is to generate a low-dimensional embedding representation matrix $G \in \mathbb{R}^{n \times d}$ using the topological structure A and feature matrix X. Formally, this can be expressed as $f : (A, X) \rightarrow G$, where d is the dimension of the embeddings. The learned representation G can be further employed in various downstream tasks.

3.1 Overview

As shown in Fig. 1, the DiffGCL model comprises two modules : the Graph Diffusion Module and the Contrastive Learning Module.

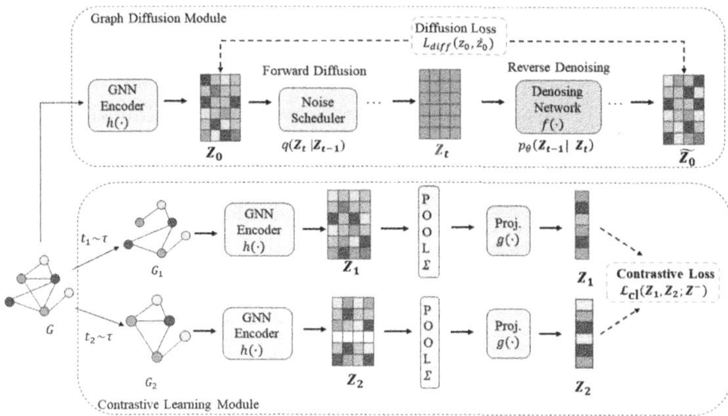

Fig. 1. The framework of the proposed model.

In the graph diffusion module, we first utilize the shared GNN encoder to map the input graph G into lower-dimensional latent embeddings z_0. Then, we introduce controlled Gaussian noise injection to pertubing latent embeddings during the forward diffusion process. Finally, an attention-based graph denoising network reconstructs the latent embeddings incorporating both structural and global information.

In the contrastive learning module, we first conduct two graph views G_1 and G_2 from G by node dropping and feature masking. Then we embed two views by the shared GNN and pool the outputs as the representation of views.

The model of the shared GNN model is flexible to various GNN-based architectures such as GCN, GAT, or GIN. In this paper, we use a GIN [17] as the shared encoder. We train DiffGCL by a joint loss function including a diffusion loss and a contrastive loss in an end-to-end manner.

3.2 Graph Diffusion Model

Existing diffusion models [5,15] on graphs perform denoising diffusion directly in the original graph feature space. In contrast, we operate the diffusion in a low-dimensional latent representation space. This choice provides several advantages: (1) Abstraction and generalization. The latent space provides a more abstract and high-level representation of graph data, facilitating the capture of latent structures and patterns. (2) Bias reduction. We found that denoising in the feature space introduces biases. In contrast, denoising in the latent space results in unbiased embeddings closely resembling the original samples, avoiding biases that may limit contrastive learning's ability to learn discriminative representations for downstream tasks. (3) Control and efficiency. Operating in the latent space allows precise control over the denoising process, reducing data dimensionality, minimizing computational costs, and enhancing efficiency.

Forward Diffusion Process. Based on the embeddings from the encoder, we iteratively introduce Gaussian noise to the embeddings $\boldsymbol{z}$, generating $\boldsymbol{z}_t$ at step t by $\boldsymbol{z}_t = \sqrt{\bar{\alpha}_t}\boldsymbol{z}_0 + \sqrt{1-\bar{\alpha}_t}\epsilon_t$, where $\alpha_t = 1-\beta_t$, $\bar{\alpha}_t = \prod_{t'=1}^{t}\alpha_{t'}$, and $\epsilon \sim \mathcal{N}(0, I)$. When T is sufficiently large, it approximates a standard Gaussian distribution. The noise schedule $\bar{\alpha}_t$ controls noise injection at each step.

The graph structure exhibits intricate relationships and diverse complexities in graph structures may require different levels of noise for optimal learning. To address this issue, we apply partially controlled Gaussian Noise injection on latent representations. The noise schedule is designed as a controllable linear noise β_t :

$$\beta_t = s \cdot [\beta_{\min} + (t-1)/(T-1)(\beta_{\max} - \beta_{\min})], \quad t \in 1, \ldots, T \tag{1}$$

Here, $s \in [0, 1]$ is a hyperparameter controlling the noise scale, and $\beta_{\max}$ and $\beta_{\min} \in (0, 1)$ set the upper and lower bounds for β.

Reverse Denoising Process. The reverse diffusion process aims to recover the original embedding $\boldsymbol{z}_0$ from randomly initialized Gaussian noise $\boldsymbol{z}_T$ through iterative denoising steps. To achieve this, we train a denoising network f_θ, which takes noisy latent representations $\boldsymbol{z}_t$ as input and predicts $\boldsymbol{z}_0$.

However, generating graph representations from uncontrollable, randomly initialized Gaussian noise is challenging. It is difficult to capture and preserve key structural and semantic information effectively, and cannot ensure the quality of generation representation. Therefore, we design a denoising model incorporates overall topological structural information and global graph information for guidance.

To extract structural information, we use the shortest path distance as initial edge features and global graph representations are obtained from the encoder. To incorporate time information, sinusoidal position embedding for the timestep t [14] is added to the initial node representations. Then we feed the embeddings into the graph denoising network, incorporating not only node embeddings $\boldsymbol{z}$ but also edge $\boldsymbol{e}$ and graph embeddings $\boldsymbol{g}$. We use Transformer structures as our graph graph-denoising network, consisting of Multi-Head Attention (MHA) and Feed-Forward Network (FFN) layers, iteratively enhancing representations through these series of stacked layers. We denote the embedding representations updated for the node i at the l-th message-passing layer as follows:

$$\hat{\alpha}_{ij}^{k,l} = \frac{Q_i^{k,l}(K_j^{k,l})^T}{\sqrt{d_k}} \cdot E^{k,l} e_{ij}^l, \tag{2}$$

$$\alpha_{ij}^{k,l} = \mathrm{softmax}(\hat{\alpha}_{ij}^{k,l}), \tag{3}$$

$$\hat{e}_{ij}^{l+1} = ||_{k=1}^{H} \hat{\alpha}_{ij}^{k,l}, \tag{4}$$

$$\hat{h}_i^{l+1} = \left(||_{k=1}^{H} \sum_{j=1}^{N} \alpha_{ij}^{k,l} \cdot V_j^{k,l} \right) \cdot W_g^l G^l, \tag{5}$$

$$\hat{g}^{l+1} = \text{POOLED}(\hat{h}_i^{l+1}), \tag{6}$$

where $Q_i^{k,l} = h_i^l W_Q^{k,l}$, $K_i^{k,l} = h_i^l W_K^{k,l}$, $V_i^{k,l} = h_i^l W_V^{k,l}$, $W_Q^{k,l}$, $W_K^{k,l}$, $W_V^{k,l}$, and $E^{k,l} \in \mathbb{R}^{d_k \times d}$ are Linear projection matrices for the k-th head at layer l, and d_k is the dimension of the key vectors. N_i represents the neighbors of node i, and H is the number of attention heads, and $||$ denotes concatenation of multi-head messages. W_g^l is a learnable weight matrix.

The outputs $\hat{h}_i^{l+1}$, $\hat{e}_{ij}^{l+1}$, and $\hat{g}^{l+1}$ are then further processed through two-layer FFN layers with ReLU activation, layer normalization, dropout strategy, and residual connection layers, contributing to the generation of the final representations.

Diffusion Loss. The diffusion module learns the distribution of latent embeddings to refine meaningful embedding representations. Given the input embeddings $\boldsymbol{Z}_0$ and setting the sampling step t in $t \in [0, T]$ from a uniform distribution, the loss function $\mathcal{L}_{\text{diff}}$ is defined as minimizing the Kullback-Leibler (KL) divergence between the predicted representation $f_\theta(\boldsymbol{Z}_t, E, G, t)$ and the original representation $\boldsymbol{Z}_0$:

$$\mathcal{L}_{\text{diff}} = \mathbb{E}_{(\boldsymbol{Z}_0, t)} \left[\| f_\theta(\boldsymbol{Z}_t, E, G, t) - \boldsymbol{Z}_0 \|_2^2 \right] \tag{7}$$

3.3 Graph Contrastive Learning

For the given input graph $\{G_i\}_{i=1}^N$, we apply two random graph augmentations t_1, t_2 and $t \sim T$, here T is the set of all augmentation functions to obtain two views for $\{G_i\}$, $G_1 = t_1(G_i)$ and $G_2 = t_2(G_i)$. We can then obtain embeddings $z_i = g(h_\phi(G_1))$ and $z_j = g(h\phi(G_2))$, through the encoder h_ϕ and further into a projection head $g(\cdot)$ [2]. Here, the pair (z_i, z_j) is considered as a positive pair, and (z_i, z_v) as a negative pair for all $v \neq i$. We adopt InfoNCE as the contrastive loss function $\mathcal{L}_{\text{cl}}$:

$$\mathcal{L}_{\text{cl}} = -\log \frac{\exp(\text{sim}(z_i, z_j)/\tau)}{\sum_{k=1, k\neq i}^{m} \exp(\text{sim}(z_i, z_k)/\tau)}, \tag{8}$$

where z_i, z_j represent the embedding for positive samples, and z_k represents the embeddings for all negative samples. $\text{sim}(z_i, z_j) = z_i^\top z_j / \|z_i\|_2 \|z_j\|_2$ is the cosine similarity, τ is a temperature hyperparameter. Note that, after training is completed, the projection head $g(\cdot)$ is discarded, and the representations are directly used for downstream tasks.

3.4 Training

We integrate the graph diffusion module and the contrastive learning module, by leveraging their distinct strengths and complementary nature to synergistically enhance the quality of learned graph representations.

To sum up, we jointly optimize the diffusion module and the contrastive learning module. The encoder h_ϕ and the denoiser model f_θ are trained in an end-to-end manner. The overall optimization objective is defined by combining the loss terms from equations Eq. 7 and Eq. 8 as follows:

$$\min \mathcal{L} = \alpha \cdot \mathcal{L}_{\text{diff}} + \beta \cdot \mathcal{L}_{\text{cl}} \tag{9}$$

Here, α and β are hyperparameters, providing flexibility in balancing the influence of diffusion loss and contrastive loss during training. Furthermore, this joint optimization enhances the stability and efficiency of the diffusion model's convergence.

Table 1. Statistics of the evaluation datasets.

Datasets	Graph Num.	Avg.Nodes	Avg.Edges	Classes
PROTEINS	1,113	39.06	72.82	2
DD	1,178	284.32	715.66	2
NCI1	4,110	29.87	32.30	2
IMDB-B	1,000	19.77	96.53	2
REDDIT-B	2,000	429.63	497.75	2
COLLAB	5,000	74.49	2457.78	3

4 Experimental Study

4.1 Experimental Setup

Datasets. We conduct experiments using six real-world graph datasets : PROTEINS, NCI1, DD, IMDB-B, COLLAB, and REDDIT-B [8]. The details of datasets are described in Table 1.

Baselines. We compare the proposed method with several unsupervised learning methods, including graph kernels methods, graph embedding methods, and graph contrastive learning methods. The graph kernel methods include Graphlet Kernel (GK) [12], Weisfeiler-Lehman Kernel (WL) [11], and Deep Graph Kernel (DGK) [18]. For graph embedding methods, we select node2vec [3], sub2vec [1], and graph2vec [9]. For contrastive models, we choose infoGraph [13], MVGRL [4], GraphCL [20], GCA [22], JOAO [19], SimGRACE [16] and DualGCL [7]. For all baselines, we report the results as presented in their respective original papers.

Experimental Settings. We employ a GIN as the encoder. The batch size is set to 128, with a hidden dimensionality of 256, reduced to 128 for larger graphs such as COLLAB due to limited GPU memory. We utilize the pooling operation to obtain the graph representation for all baseline methods. For the denoising network, we employ a Transformer with 2 layers and 4 attention heads. For a fair comparison, We employ 10-fold cross-validation on each dataset. For each fold, we use 80% of the data as the unlabeled data, and 20% as labeled testing data. We feed the encoded graph-level representation into a downstream SVM classifier and the graph classification results represent the average accuracy score across the 10 folds, along with their standard deviations.

Table 2. Summary of the graph classification results (accuracy % ± standard deviation %) on six benchmark datasets. The **bold** numbers denote the best performance.

Methods	PROTEINS	DD	NCI1	COLLAB	IMDB-B	REDDIT-B
GK	-	-	-	-	65.87 ± 0.98	77.34 ± 0.18
WL	72.92 ± 0.56	-	**80.01 ± 0.50**	-	72.30 ± 3.44	68.82 ± 0.41
DGK	73.30 ± 0.82	-	78.31 ± 0.46	-	66.96 ± 0.56	78.04 ± 0.39
node2vec	57.49 ± 3.57	-	54.89 ± 1.61	-	-	-
sub2vec	53.03 ± 5.55	-	52.84 ± 1.47	-	55.26 ± 1.54	71.48 ± 0.41
graph2vec	73.30 ± 2.05	-	73.22 ± 1.81	-	71.10 ± 0.54	75.78 ± 1.03
InfoGraph	74.44 ± 0.31	72.85 ± 1.78	76.20 ± 1.06	70.65 ± 1.13	73.03 ± 0.87	82.50 ± 1.42
GraphCL	74.39 ± 0.45	78.62 ± 0.40	77.87 ± 0.41	71.36 ± 1.15	71.14 ± 0.44	**89.53 ± 0.84**
JOAOv2	71.25 ± 0.85	66.91 ± 1.75	72.99 ± 0.75	70.40 ± 2.21	71.60 ± 0.86	78.35 ± 1.38
SimGRACE	75.35 ± 0.09	77.44 ± 1.11	79.02 ± 0.44	71.72 ± 0.82	71.30 ± 0.77	89.01 ± 1.31
DualGCL	75.60 ± 0.50	-	79.80 ± 0.30	75.30 ± 0.40	-	-
DiffGCL	**75.74 ± 0.33**	**79.09 ± 0.14**	78.94 ± 0.42	**81.74 ± 0.69**	**75.63 ± 0.52**	87.75 ± 0.58

4.2 Comparison with State-of-the-Art Methods

The comparison experiment results for all baselines and our proposed method are shown in Table 2, we also have the following observations: First, the CL-based methods perform better than the non-CL-based methods in general, which indicates the effectiveness of GCL for unsupervised graph learning; Second, compared to various CL methods, our method achieves the highest accuracy across most datasets. This demonstrates the ability of our method to address noise issues and effectively capture global structural and semantic information within the graph. And the result illustrates the effectiveness of incorporating DDPMs into GCL to enhance the performance of GCL.

4.3 Ablation Study

To assess the effectiveness of each component in our proposed model, we conduct a ablation study on DiffGCL and its various variants across four datasets. The details are outlined as follows:

Table 3. Ablation results.

Model	IMDB-B	PROTEINS	DD	NCI1
w/o EG	74.67 ± 0.87	74.86 ± 0.69	76.80 ± 0.78	76.91 ± 0.83
DiffGCL-D	72.64 ± 0.72	72.16 ± 0.88	75.00 ± 0.70	74.39 ± 2.91
DiffGCL-G	73.16 ± 0.88	74.16 ± 0.68	77.56 ± 0.51	75.42 ± 0.49
DiffGCL	75.63 ± 0.65	75.73 ± 0.79	78.64 ± 0.49	78.71 ± 0.48

- w/o EG: denotes removing the component of information edge and the global context in the denoising network.
- DiffGCL-D: denotes the graph diffusion model.
- DiffGCL-C: denotes the graph contrastive learning model.

The results are presented in Table 3, it can be observed that: (1) The integration of structural and global information to guide the denoising process proves to be effective; (2) The combination of both contrastive learning and denoising diffusion significantly enhances performance, outperforming the individual contributions of each module. This indicates the effectiveness of integrating both modules.

5 Conclusion

In this paper, we propose a Diffusion Model-Enhanced Graph Contrastive Learning (DiffGCL) model for graph representation to successfully tackle the challenges of noise handling and the capture of global structural and semantic information in graph representation learning. By integrating Denoising Diffusion Probabilistic Models (DDPMs) in a low-dimensional latent representation space and introducing a graph-specific diffusion module, DiffGCL outperforms existing baselines in unsupervised graph contrastive learning. The shared encoder facilitates effective information transfer between the denoising model and contrastive learning, demonstrating the potential of DiffGCL to enhance graph representation without the reliance on labeled data.

Acknowledgements. This work is supported by the National Natural Science Foundation of China (62072083) and the Fundamental Research Funds of the Central Universities (N2216017).

References

1. Adhikari, B., Zhang, Y., Ramakrishnan, N., Prakash, B.A.: Sub2Vec: feature learning for subgraphs. In: Phung, D., Tseng, V.S., Webb, G.I., Ho, B., Ganji, M., Rashidi, L. (eds.) Advances in Knowledge Discovery and Data Mining: 22nd Pacific-Asia Conference, PAKDD 2018, Melbourne, VIC, Australia, June 3-6, 2018, Proceedings, Part II, pp. 170–182. Springer International Publishing, Cham (2018). https://doi.org/10.1007/978-3-319-93037-4_14

2. Chen, T., Kornblith, S., Norouzi, M., Hinton, G.: A simple framework for contrastive learning of visual representations. In: PMLR, pp. 1597–1607 (2020)
3. Grover, A., Leskovec, J.: node2vec: Scalable feature learning for networks. In: SIGKDD, pp. 855–864 (2016)
4. Hassani, K., Khasahmadi, A.H.: Contrastive multi-view representation learning on graphs. In: PMLR, pp. 4116–4126 (2020)
5. Hoogeboom, E., Satorras, V.G., Vignac, C., Welling, M.: Equivariant diffusion for molecule generation in 3D. In: PMLR, pp. 8867–8887 (2022)
6. Kipf, T.N., Welling, M.: Semi-supervised classification with graph convolutional networks. In: ICLR (2017)
7. Luo, Z., Dong, Y., Zheng, Q., Liu, H., Luo, M.: Dual-channel graph contrastive learning for self-supervised graph-level representation learning. Pattern Recogn. **139**, 109448 (2023)
8. Morris, C., Kriege, N.M., Bause, F., Kersting, K., Mutzel, P., Neumann, M.: Tudataset: A collection of benchmark datasets for learning with graphs. arXiv preprint arXiv:2007.08663 (2020)
9. Narayanan, A., Chandramohan, M., Venkatesan, R., Chen, L., Liu, Y., Jaiswal, S.: graph2vec: Learning distributed representations of graphs. arXiv preprint arXiv:1707.05005 (2017)
10. Peng, Z., et al.: Graph representation learning via graphical mutual information maximization. In: WWW, pp. 259–270 (2020)
11. Shervashidze, N., Schweitzer, P., Van Leeuwen, E.J., Mehlhorn, K., Borgwardt, K.M.: Weisfeiler-lehman graph kernels. J. Mach. Learn. Res. **12**(9) (2011)
12. Shervashidze, N., Vishwanathan, S., Petri, T., Mehlhorn, K., Borgwardt, K.: Efficient graphlet kernels for large graph comparison. In: AISTATS, pp. 488–495 (2009)
13. Sun, F.Y., Hoffmann, J., Verma, V., Tang, J.: Infograph: unsupervised and semi-supervised graph-level representation learning via mutual information maximization. In: ICLR, pp. 2629–2639 (2020)
14. Vaswani, A., et al.: Attention is all you need. In: Advances in Neural Information Processing Systems, vol. 30 (2017)
15. Vignac, C., Krawczuk, I., Siraudin, A., Wang, B., Cevher, V., Frossard, P.: Digress: discrete denoising diffusion for graph generation. arXiv preprint arXiv:2209.14734 (2022)
16. Xia, J., Wu, L., Chen, J., Hu, B., Li, S.Z.: Simgrace: a simple framework for graph contrastive learning without data augmentation. In: WWW, pp. 1070–1079 (2022)
17. Xu, K., Hu, W., Leskovec, J., Jegelka, S.: How powerful are graph neural networks? arXiv preprint arXiv:1810.00826 (2018)
18. Yanardag, P., Vishwanathan, S.: Deep graph kernels. In: KDD, pp. 1365–1374 (2015)
19. You, Y., Chen, T., Shen, Y., Wang, Z.: Graph contrastive learning automated. In: ICML, vol. 139, pp. 12121–12132 (2021)
20. You, Y., Chen, T., Sui, Y., Chen, T., Wang, Z., Shen, Y.: Graph contrastive learning with augmentations. NIPS **33**, 5812–5823 (2020)
21. Zhu, Y., Xu, Y., Yu, F., Liu, Q., Wu, S., Wang, L.: Deep graph contrastive representation learning. arXiv preprint arXiv:2006.04131 (2020)
22. Zhu, Y., Xu, Y., Yu, F., Liu, Q., Wu, S., Wang, L.: Graph contrastive learning with adaptive augmentation. In: WWW, pp. 2069–2080 (2021)

H²GNN: Graph Neural Networks with Homophilic and Heterophilic Feature Aggregations

Shixiong Jing[1], Lingwei Chen[2], Quan Li[1], and Dinghao Wu[1(✉)]

[1] The Pennsylvania State University, State College, University Park, PA 16802, USA
{jing,qbl5082,dinghao}@psu.edu
[2] Wright State University, Dayton, OH 45435, USA
lingwei.chen@wright.edu

Abstract. Graph neural networks (GNNs) rely on the assumption of graph homophily, which, however, does not hold in some real-world scenarios. Graph heterophily compromises them by smoothing node representations and degrading their discrimination capabilities. To address this limitation, we propose H^{2}GNN, which implements **H**omophilic and **H**eterophilic feature aggregations to advance **GNN**s in graphs with homophily or heterophily. H^{2}GNN proceeds by combining *local feature separation* and *adaptive message aggregation*, where each node separates local features into similar and dissimilar feature vectors, and aggregates similarities and dissimilarities from neighbors based on connection property. This allows both similar and dissimilar features for each node to be effectively preserved and propagated, and thus mitigates the impact of heterophily on graph learning process. As dual feature aggregations introduce extra model complexity, we also offer a simplified implementation of H^{2}GNN to reduce training time. Extensive experiments on seven benchmark datasets have demonstrated that H^{2}GNN can significantly improve node classification performance in graphs with different homophily ratios, which outperforms state-of-the-art GNN models.

Keywords: Graph Neural Networks · Heterophily · Node Classification

1 Introduction

Graph neural networks (GNNs) have emerged as prevalent models to address diverse graph-based tasks [10–12,16]. Their impressive learning capabilities can be attributed to the utilization of the message-passing mechanism, which allows for neighborhood aggregation through graph structure. In other words, GNNs rely on the assumption of graph *homophily* suggesting that nodes tend to connect with others sharing similar features [19,27,28]. However, this assumption does not hold in some real-world graph learning problems, such as fraud and bot detection [3,7], where attackers tend to build the relationships between malicious

M. Onizuka et al. (Eds.): DASFAA 2024, LNCS 14855, pp. 342–352, 2024.
https://doi.org/10.1007/978-981-97-5572-1_23

and benign nodes for stealth, leading to graphs with *heterophily*. When neighbors in a graph significantly differ, conventional aggregation mechanisms used by GNNs that directly blend neighborhood features may result in smoothed or indistinct representations, and degrade the discrimination capabilities of GNNs.

To address this limitation, recent methods have been proposed to handle graph heterophily, which broadly falls into three categories: neighbor extension [1,15,21], inter-layer connections [5,17,28], and adaptive message aggregation [4,8,18,26]. The neighbor-extension approach captures long-range dependencies to complement node representations, while the inter-layer connection approach capitalizes residuals to alleviate the disruptions caused by low-level abnormal connections. However, their efficacy is limited when tackling graphs with high levels of heterophily. In contrast, the adaptive message aggregation method trains separate filters to manipulate homophily and heterophily, and aggregates the outputs from these filters to derive node representations. This formulation is more adept at managing local homophily and heterophily but may compromise long-range information propagation in heterophilic graphs [2].

This naturally leads to a pivotal insight to address the heterophily issue: both similar and dissimilar features from neighbors contribute to node characterization, which need to be locally extracted, separately preserved, and further propagated to higher orders to prevent aggregations from blending distinguishable information associated with different labels. As such, we propose H^{2}GNN, which implements **H**omophilic and **H**eterophilic feature aggregations to advance **GNN**s in graphs with homophily or heterophily. Specifically, instead of mixing features from neighbors as a single feature vector, H^{2}GNN proceeds by combining the ideas of *local feature separation* and *adaptive message aggregation*. Each node separates local features into similar and dissimilar feature vectors, and aggregates corresponding features from neighbors based on connection property: when the connection is homophilic, the neighbor's similar and dissimilar features would respectively flow into the target node's similar and dissimilar feature vectors; conversely, in heterophilic connections, the neighbor's two sets of features would be amalgamated into the target node's opposite feature vectors.

By maintaining two distinct feature vectors side by side, both similar and dissimilar features for each node can be effectively preserved and propagated, and thus mitigate the impact of heterophily on the graph learning process. For each aggregation operation, we further elaborate an edge identifier that determines whether a given node pair is homophilic or heterophilic to selectively extract neighborhood features, promoting adaptive message aggregation. Moreover, considering that dual feature aggregations in H^{2}GNN introduce extra model complexity, we offer a simplified implementation to significantly reduce training time.

2 Problem Statement

We denote a given graph as $G = (V, E, \mathbf{X})$, where $V(n = |V|)$ is the set of nodes, E is the set of edges, and $\mathbf{X} \in \mathbb{R}^{n\times d}$ is the feature matrix. Edges E can be encoded as an adjacency matrix $\mathbf{A} \in \mathbb{R}^{n\times n}$ and $\mathbf{A}_{ij} = \{0,1\}$. The

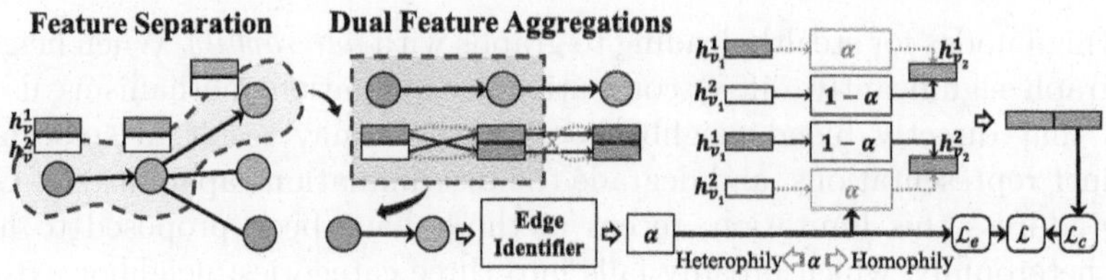

Fig. 1. Overview of our proposed model H^{2}GNN.

neighbors for v_i is represented as $\mathcal{N}(v_i) = \{v_j|(v_i, v_j) \in E)\}$. Each labeled node is associated with a ground truth $y \in Y = \{0, 1, \cdots, k-1\}$. GNN models enforce each node to aggregate information from its neighbors and generate higher-level embedding for downstream node classification. These graphs exhibit different degrees of homophily (or heterophily). Homophily is a concept that describes the tendency of nodes to associate with similar others [27]. In this paper, we focus on homophily in class labels [28], where a graph with high homophily indicates that connected nodes share the same label with a high probability. Heterophily is the opposite of homophily to describe the status of connected nodes belonging to different labels.

3 Proposed Model: H^{2}GNN

In this section, we present our proposed GNN model H^{2}GNN to deal with graph heterophily, the overview of which is illustrated in Fig. 1.

3.1 Feature Separation

To restore the discrimination capability in heterophilic graphs, GNN models need to harness the information of neighborhood differences [18]. Following this idea, various adaptive message aggregation methods [4,8,18,21] have been deployed that train separate filters to extract similar and dissimilar signals from neighbors, enriching node representations while mitigating the impact of heterophily. Nevertheless, these approaches have one limitation: though dissimilarities from neighbors are extracted, they are still mixed with similarities to be formulated as a single feature vector; such straightforward preservation may not be helpful for propagating dissimilarities to higher-order nodes [2].

To address this limitation, we propose to preserve these similarities and dissimilarities separately. Specifically, we devise two distinct feature vectors for each node $v \in V$ to facilitate feature aggregation: similar feature vector $\mathbf{h}_v^1$, and dissimilar feature vector $\mathbf{h}_v^2$. $\mathbf{h}_v^1$ is used to aggregate and preserve similar signals from neighborhood features, while $\mathbf{h}_v^2$ is used to preserve dissimilar signals extracted from neighbors. In this respect, the representation for each node $v \in V$ during aggregation can be formalized as $\mathbf{h}_v = \{\mathbf{h}_v^1, \mathbf{h}_v^2\}$, which is initialized as $\mathbf{h}_v = \{\mathbf{X}_v, \mathbf{0}\}$. Such a bi-vector feature setting can collaboratively preserve local neighborhood similarities and dissimilarities, and promote long-range information propagation, which, in turn, boosts node representation expressiveness and model discrimination power. The details about the feature aggregations on $\mathbf{h}_v^1$ and $\mathbf{h}_v^2$ will be elaborated in the subsequent section.

3.2 Dual Feature Aggregations

Dual feature aggregations significantly rely on edge property specified by the labels of connected nodes, which is not always available in real-world graphs, as most of the nodes are not pre-annotated. To tackle this issue, we introduce an edge identifier to discriminate if a node pair is homophilic or heterophilic.

Edge Identifier. Similar to node classification using node representations, here we devise an edge identifier using the representations of its connected nodes to determine the property of the edge. Considering that node representations vary at each aggregation layer, edge identification is performed in a layer-wise manner. Specifically, at aggregation layer l, the edge identifier can be formalized as:

$$\alpha_{ij} = \sigma((\mathbf{h}_{v_i}^{(l-1)} || \mathbf{h}_{v_j}^{(l-1)})\mathbf{W}_e^{(l)}) \tag{1}$$

where $\sigma(\cdot)$ is a non-linear activation function and $\mathbf{W}_e^{(l)}$ is a learnable weight matrix at layer l. α_{ij} $(0 \leq \alpha_{ij} \leq 1)$ is a prediction score, indicating how possible an edge $e_{v_i v_j}$ is homophilic.

Feature Aggregation. Generally, the normalized adjacency matrix $\tilde{\mathbf{A}}$ is considered a low-lass filter in GNNs to retain the commonality of neighboring features [4,20], while $\mathbf{I} - \tilde{\mathbf{A}}$ provides diversification operation, which is considered as high-pass filter to extract neighborhood differences [18]. This inspires us to directly employ the soft prediction of edge identifier, i.e., prediction score α_{ij} $(0 \leq \alpha_{ij} \leq 1)$ instead of hard prediction to perform dual feature aggregation, which enables α_{ij} and $1 - \alpha_{ij}$ to act like advanced dual filters that flexibly control the number of similarities and dissimilarities from each neighbor to be aggregated and thus further refine the learned node representations. Formally, for a given node v_i, the dual feature aggregations are implemented as follows:

$$\mathbf{h}_{v_i}^{1,(l)} = \sigma(\mathbf{h}_{v_i}^{1,(l-1)}\mathbf{W}_0 + \frac{1}{|\mathcal{N}(v_i)|}\sum_{v_j \in \mathcal{N}(v_i)} \alpha \mathbf{h}_{v_j}^{1,(l-1)}\mathbf{W}_1 + (1-\alpha)\mathbf{h}_{v_j}^{2,(l-1)}\mathbf{W}_2) \tag{2}$$

$$\mathbf{h}_{v_i}^{2,(l)} = \sigma(\mathbf{h}_{v_i}^{2,(l-1)}\mathbf{W}_0 + \frac{1}{|\mathcal{N}(v_i)|}\sum_{v_j \in \mathcal{N}(v_i)} (1-\alpha)\mathbf{h}_{v_j}^{1,(l-1)}\mathbf{W}_1 + \alpha \mathbf{h}_{v_j}^{2,(l-1)}\mathbf{W}_2) \tag{3}$$

where $\sigma(\cdot)$ is specified as ReLU, $\mathbf{W}_0$, $\mathbf{W}_1$, and $\mathbf{W}_2$ are learnable weight matrices to promote feature aggregations. For node v_i at layer l, $\mathbf{h}_{v_i}^{1,(l)}$ aggregates neighboring features and produces its new similarity feature vector, while $\mathbf{h}_{v_i}^{2,(l)}$ produces its new dissimilarity feature vector, both of which contribute to the new hidden representation. By leveraging dual feature aggregations, features contributing to the node's ground truth can be effectively preserved in $\mathbf{h}_v^1$, while features correlated to other classes can be explicitly saved in $\mathbf{h}_v^2$ to support higher-order enhancement, where heterophily can barely impact on the outputs from different classes.

3.3 Loss Optimization

After multi-layer dual feature aggregations, $\mathbf{h}_v^1$ and $\mathbf{h}_v^2$ are concatenated and fed to a fully-connected layer for node classification, which results in a cross-entropy loss $\mathcal{L}_c$ between predictions and nodes' true labels. In addition, for each feature aggregation layer l, edge identifier needs to be optimized as well. We select those edges whose connected nodes are labeled to construct homophilic and heterophilic edge samples for facilitating edge identifier training, which leads to another cross-entropy loss $\mathcal{L}_e^{(l)}$. Accordingly, the final training objective can be formulated as follows:

$$\mathcal{L} = \mathcal{L}_c + \gamma \sum_{l=1}^{L} \mathcal{L}_e^{(l)} \tag{4}$$

where γ is a balance parameter to trade off two losses, and L is the number of feature aggregation layers. All the weights can be updated by minimizing $\mathcal{L}$.

3.4 Simplified Model

We refer to our proposed GNN model as H^{2}GNN-W, since it resorts to different weight matrices for dual feature aggregations. This naturally introduces extra model complexity that increases the training time cost. To keep a better trade-off between model effectiveness and efficiency, we offer a simplified implementation of our framework called H^{2}GNN-S. H^{2}GNN-S reduces training time using a similarity metric $\alpha_{ij} = \text{sim}(\mathbf{h}_{v_i}^{(l-1)}, \mathbf{h}_{v_j}^{(l-1)})$ as the edge identifier. To exclude the weight matrix updates and multiplications during feature aggregations, we assemble all weights into a single linear layer that maps the original feature matrix $\mathbf{X}$ to $\mathbf{X}^{(0)} = f_\theta(\mathbf{X}) \in \mathrm{R}^{d \times m}$ $(m \ll d)$, such that $\mathbf{h}_v (v \in V)$ is initialized as $\mathbf{h}_v = \{\mathbf{X}_v^{(0)}, \mathbf{0}\}$, where $\mathbf{h}_v^1 = \mathbf{X}_v^{(0)}$, $\mathbf{h}_v^2 = \mathbf{0}$. Afterward, feature aggregations are performed without any weighted operations:

$$\mathbf{h}_{v_i}^{1,(l)} = \sigma(\mathbf{h}_{v_i}^{1,(l-1)} + \frac{1}{|\mathcal{N}(v_i)|} \sum_{v_j \in \mathcal{N}(v_i)} \alpha \mathbf{h}_{v_j}^{1,(l-1)} + (1-\alpha)\mathbf{h}_{v_j}^{2,(l-1)}) \tag{5}$$

$$\mathbf{h}_{v_i}^{2,(l)} = \sigma(\mathbf{h}_{v_i}^{2,(l-1)} + \frac{1}{|\mathcal{N}(v_i)|} \sum_{v_j \in \mathcal{N}(v_i)} (1-\alpha)\mathbf{h}_{v_j}^{1,(l-1)} + \alpha \mathbf{h}_{v_j}^{2,(l-1)}) \tag{6}$$

The outputs are also concatenated and fed to a fully connected layer for node classification. Since there's no necessity to train edge identifiers, the loss function $\mathcal{L}$ can be reduced to $\mathcal{L} = \mathcal{L}_c$. This simplified model sustains potent learning capability while also yielding the additional benefit of reduced time consumption.

4 Experimental Results and Analysis

4.1 Experimental Setup

Datasets. We evaluate H^2GNN on seven public datasets with different homophily ratios. Cora and Citeseer [25] are benchmark citation networks; Texas and Wisconsin [21] are subdatasets of WebKB collected from different universities; Actor [22] is a subgraph induced from the film-director-actor-writer network; Penn94 and Genius [14] are two large-scale online social networks. For Penn94 and Genius, we use 50%-25%-25% random split to match the experimental setting in related works [13,14]. For all other datasets, we use the 48%-32%-20% data-split as provided by [8,21]. Table 1 summarizes the data statistics.

Table 1. Statistics of the datasets

Dataset	#Nodes	#Edges	#Features	#Classes
Cora	2,078	5,429	1,433	7
Citeseer	3,327	9,228	3,703	6
Texas	183	309	1,703	5
Wisconsin	251	499	1,703	5
Actor	7,600	30,019	932	5
Penn94	41,554	1,362,229	5	2
Genius	421,961	984,979	12	2

Baselines. We use 3 GNN variants and 9 advanced GNNs for graph heterophily as baselines. Conventional GNNs includes GCN [9], GAT [23], and GIN [24]; advanced GNNs includes GCNII [5], Geom-GCN [21], CPGNN [27], MixHop [1], H2GCN [28], GPRGNN [6], GBK-GNN [8], LinkX [14], and GloGNN [13].

Implementation Details. The number of aggregation layers L is set to 2. Models are trained for 2,000 epochs with patience of 200 using Adam optimizer with learning rate $lr = 0.01$ and $5e-4$ L2 regularization. For H^2GNN-S, $\text{sim}(\cdot)$ is specified as cosine similarity, and the feature dimension is set to $m = 16$. We further introduce consistency loss to provide an advantage of smoothing decision boundaries, which is adjusted by a balance parameter λ. The impacts of λ and γ on the performance of H^2GNN are evaluated in Sect. 4.3.

4.2 Comparison with Baselines

We compare H^2GNN with 12 selected baselines over seven different datasets. Table 2 presents these comparative results. We observe that: (1) Traditional

GNN models perform relatively well in homophilic graphs, but suffer from drastic drops in heterophilic graphs. (2) By contrast, enhanced models tailored for heterophilic graphs manage to bring up the accuracy to a higher level in graphs with extremely low homophily ratios, due to their enhancement in neighbor extension, inter-layer connections, or adaptive message aggregation, but some of these models sacrifice the performance in homophilic graphs. (3) Our proposed model delivers the best or second-best results, outperforming most of the baselines across all graph datasets. Among enhanced GNN models, H^2GNN-W achieves the best performance on most of the datasets. Even the simplified version of our model, H^2GNN-S, offers comparable results on all datasets. In summary, H^2GNN achieves state-of-the-art performance on all seven benchmarks (both homophily and heterophily) and outperforms the leading GNN models in most cases.

Table 2. Comparison of baselines (classification accuracy (%)) on both homophily and heterophily graph datasets. The best result for each dataset is highlighted in blue color.

	Cora	**Citeseer**	**Texas**	**Wisconsin**	**Actor**	**Penn94**	**Genius**
H. ratio r	**0.82**	**0.70**	**0.11**	**0.14**	**0.20**	**0.46**	**0.61**
GCN	86.81±0.79	73.96±1.02	49.52±5.29	46.78±5.36	26.96±1.25	82.47±0.27	87.42±0.37
GAT	84.31±1.21	71.89±1.40	41.94±5.33	63.04±5.33	26.56±0.98	81.53±0.55	55.80±0.87
GIN	81.93±0.57	68.08±0.62	29.03±4.42	45.65±4.71	23.61±0.76	81.78±0.39	87.16±0.51
GCNII	87.22±0.98	75.11±1.18	51.61±5.47	58.70±3.25	28.72±0.60	82.92±0.59	89.82±0.49
GeomGCN	84.10±1.12	76.28±2.06	67.57±5.35	68.63±4.92	30.00±1.26	83.01±0.38	89.75±0.41
CPGNN	79.40±1.39	69.41±0.45	75.68±5.12	76.47±6.16	35.59±0.86	79.76±0.39	85.33±0.92
H2GCN	82.70±0.87	43.70±0.24	72.97±4.53	70.59±5.76	35.39±1.34	81.31±0.60	OOM
MixHop	87.61±0.85	76.26±0.33	77.84±7.73	75.88±4.90	32.22±2.34	83.47±0.71	90.58±0.16
GBK-GNN	88.69±0.42	79.18±0.96	81.08±4.87	84.21±4.33	38.97±0.97	81.99±0.47	90.13±0.50
GPRGNN	87.95±1.18	77.13±1.67	78.38±4.36	82.94±4.21	34.63±1.22	81.38±0.16	90.05±0.31
LinkX	84.64±1.13	73.19±0.99	74.60±8.37	75.49±5.72	36.10±5.55	84.71±0.52	90.77±0.27
GloGNN	88.31±1.13	77.41±1.65	84.32±4.15	87.06±3.53	37.35±1.30	85.57±0.35	90.66±0.11
H^2GNN-S	89.64±1.20	80.59±1.37	83.15±5.67	84.43±5.32	37.41±1.86	85.03±0.41	90.25±0.46
H^2GNN-W	89.97±1.23	81.60±1.95	85.01±5.98	85.11±6.10	38.53±1.46	85.61±0.56	90.80±0.42

4.3 Parameter Evaluation

The performance of H^2GNN can be potentially affected by λ, which is the coefficient controlling the weight of consistency loss, and γ, which is the coefficient controlling the weight of edge identification loss. In this section, we evaluate the impacts of these parameter settings. The experimental results are reported in Fig. 2(a) and (b). It can be observed that different parameters contribute to slightly different results. However, these fluctuations are insignificant, implying that our model's performance is stable across various parameter settings.

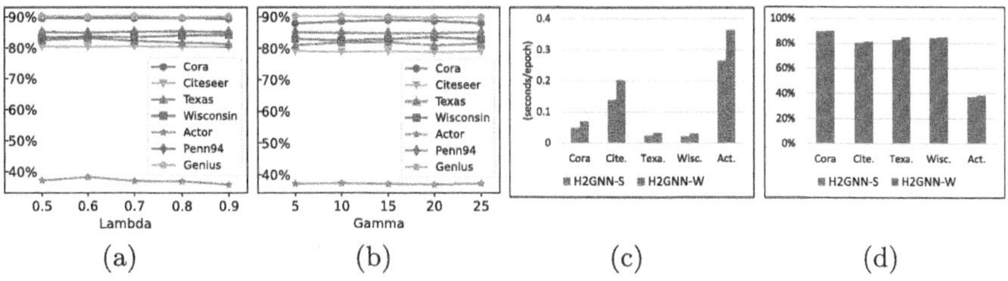

Fig. 2. Impact of parameters and model variants on model performance: (a) accuracy of H^{2}GNN-W with different λ; (b) accuracy of H^{2}GNN-W with different γ; (c) time cost of H^{2}GNN-W and H^{2}GNN-S; (d) accuracy of H^{2}GNN-W and H^{2}GNN-S.

4.4 Comparison Between H^{2}GNN-W and H^{2}GNN-S

In this section, we analyze the advantages and disadvantages between weighted H^{2}GNN (H^{2}GNN-W) and simplified H^{2}GNN (H^{2}GNN-S). We illustrate the comparative results in Fig. 2(c) and (d). Compared to H^{2}GNN-W, H^{2}GNN-S significantly reduces the training time per epoch on all tested datasets by over 20%, while not compromising the node classification performance. Specifically, the accuracy of node classification using H^{2}GNN-S only drops by less than 2% across all tested datasets. The reduced time consumption comes from the removal of weights in the aggregation layers and the exclusion of the edge identifier training. The preserved dual feature aggregations enable H^{2}GNN-S to maintain its potent learning capability. Overall, H^{2}GNN-S provides a better trade-off between effectiveness and training efficiency than H^{2}GNN-W.

4.5 Ablation Study

In this section, we set up an ablation study to investigate how different components contribute to the performance of our model. We investigate three components in our model design: (1) Feature separation, which refers to the combination of local feature separation and dual feature aggregation; (2) Edge identification, which refers to the utilization of α to identify homophilic and heterophilic edges; (3) Weighted operation, which refers to the inclusion of all weight matrices in feature aggregation operation. As illustrated in Table 3, all three components contribute to the performance of H^{2}GNN. Among these three components, the impact of weighted operation is the smallest. However, removing either feature separation or edge identification leads to significant performance drops on heterophilic graphs. This outcome aligns with our expectations. The core functionality of aggregation layers in H^{2}GNN relies on the collaboration of feature separation and edge identification. Therefore, the exclusions of either feature separation or edge identification may enforce a drastic performance drop.

Table 3. Ablation study: Feature Separation (FS), Edge Identification(EI), and Weighted Operation (WO)

FS	EI	WO	Cora	Cite.	Texa.	Wisc.
			86.59	73.04	48.03	48.20
		✓	88.04	70.16	48.35	65.41
	✓	✓	87.77	72.83	69.08	66.37
✓		✓	88.06	72.13	49.22	66.18
✓	✓		89.64	80.59	83.15	84.43
✓	✓	✓	89.97	81.60	85.01	85.11

4.6 Case Study

To validate our claim that H^{2}GNN provides more distinguishable node embeddings, we present a brief case study on two datasets, Cora (homophilic) and Texas (heterophilic), to showcase the difference between embeddings generated by GCN and H^{2}GNN. We map node embeddings into a two-dimensional space using t-SNE, and the resulting embeddings are visualized in Fig. 3. In both cases, H^{2}GNN either generates clusters with better-distinguished boundaries or redistributes nodes more cohesively. These observations reaffirm the effectiveness of H^{2}GNN in learning meaningful and distinguishable node representations.

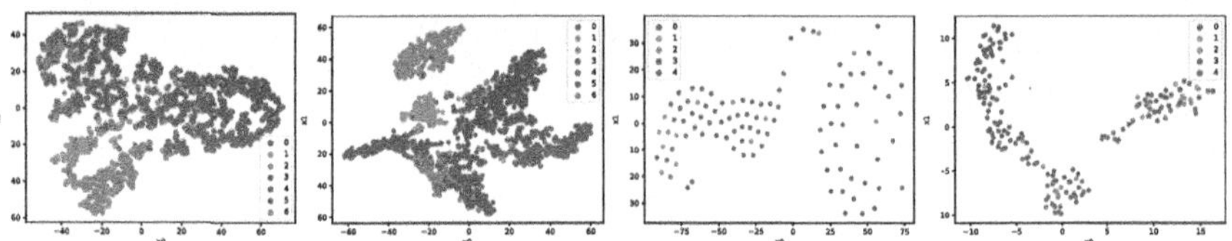

(a) (b) (c) (d)

Fig. 3. Visualization of node embeddings: (a) Cora processed by GCN, (b) Cora processed by H^{2}GNN, (c) Texas processed by GCN, (d) Texas processed by H^{2}GNN.

5 Conclusion

In this paper, we introduce a new model H^{2}GNN for node classification in graphs with homophily or heterophily. H^{2}GNN employs feature separation and dual feature aggregation guided by edge identification to better preserve and propagate both similarities and dissimilarities for each node, which not only boosts the expressiveness of node representations but also mitigates the impact of heterophily on the graph learning process. Evaluation through extensive experiments demonstrates that our model achieves state-of-the-art performance, which affirms its effectiveness in node classification, superiority over baselines, and practical significance in handling both heterophilic and homophilic graphs.

Acknowledgments. L. Chen's work is partially supported by the NSF under grant CNS-2245968.

References

1. Abu-El-Haija, S., et al.: Mixhop: Higher-order graph convolutional architectures via sparsified neighborhood mixing. In: International Conference on Machine Learning, pp. 21–29. PMLR (2019)
2. Alon, U., Yahav, E.: On the bottleneck of graph neural networks and its practical implications. In: International Conference on Learning Representations (ICLR) (2021)
3. Ashmore, B., Chen, L.: Hover: Homophilic oversampling via edge removal for class-imbalanced bot detection on graphs. In: CIKM, pp. 3728–3732 (2023)
4. Bo, D., Wang, X., Shi, C., Shen, H.: Beyond low-frequency information in graph convolutional networks. In: AAAI, pp. 3950–3957 (2021)
5. Chen, M., Wei, Z., Huang, Z., Ding, B., Li, Y.: Simple and deep graph convolutional networks. In: ICML (2020)
6. Chien, E., Peng, J., Li, P., Milenkovic, O.: Adaptive universal generalized pagerank graph neural network. In: ICLR (2021)
7. Dou, Y., Liu, Z., Sun, L., Deng, Y., Peng, H., Yu, P.S.: Enhancing graph neural network-based fraud detectors against camouflaged fraudsters. In: CIKM (2020)
8. Du, L., et al.: Gbk-gnn: gated bi-kernel graph neural networks for modeling both homophily and heterophily. In: Proceedings of the ACM Web Conference 2022, pp. 1550–1558 (2022)
9. Kipf, T.N., Welling, M.: Semi-supervised classification with graph convolutional networks. arXiv preprint arXiv:1609.02907 (2016)
10. Li, Q., Chen, L., Cai, Y., Wu, D.: Hierarchical graph neural network for patient treatment preference prediction with external knowledge. In: PAKDD, pp. 204–215 (2023)
11. Li, Q., Chen, L., Jing, S., Wu, D.: Pseudo-labeling with graph active learning for few-shot node classification. In: ICDM, pp. 1115–1120 (2023)
12. Li, Q., Li, X., Chen, L., Wu, D.: Distilling knowledge on text graph for social media attribute inference. In: SIGIR, pp. 2024–2028 (2022)
13. Li, X., et al.: Finding global homophily in graph neural networks when meeting heterophily. In: International Conference on Machine Learning, pp. 13242–13256. PMLR (2022)
14. Lim, D., et al.: Large scale learning on non-homophilous graphs: New benchmarks and strong simple methods. NeurIPS **34**, 20887–20902 (2021)
15. Liu, S., Chen, L., Dong, H., Wang, Z., Wu, D., Huang, Z.: Higher-order weighted graph convolutional networks. arXiv preprint arXiv:1911.04129 (2019)
16. Liu, S., et al.: Local augmentation for graph neural networks. In: International Conference on Machine Learning (2022)
17. Liu, S., et al.: Graph neural networks with adaptive residual. NeurIPS **34**, 9720–9733 (2021)
18. Luan, S., Hua, C., Lu, Q., et al.: Is heterophily a real nightmare for graph neural networks to do node classification? arXiv preprint arXiv:2109.05641 (2021)

19. Maurya, S.K., Liu, X., Murata, T.: Improving graph neural networks with simple architecture design. arXiv preprint arXiv:2105.07634 (2021)
20. Nt, H., Maehara, T.: Revisiting graph neural networks: All we have is low-pass filters. arXiv preprint arXiv:1905.09550 (2019)
21. Pei, H., Wei, B., Chang, K.C.C., Lei, Y., Yang, B.: Geom-gcn: geometric graph convolutional networks. arXiv preprint arXiv:2002.05287 (2020)
22. Tang, J., Sun, J., Wang, C., Yang, Z.: Social influence analysis in large-scale networks. In: KDD, pp. 807–816 (2009)
23. Veličković, P., Cucurull, G., Casanova, A., Romero, A., Lio, P., Bengio, Y.: Graph attention networks. arXiv preprint arXiv:1710.10903 (2017)
24. Xu, K., Hu, W., Leskovec, J., Jegelka, S.: How powerful are graph neural networks? arXiv preprint arXiv:1810.00826 (2018)
25. Yang, Z., Cohen, W., Salakhudinov, R.: Revisiting semi-supervised learning with graph embeddings. In: ICML, pp. 40–48 (2016)
26. Zheng, X., Liu, Y., Pan, S., Zhang, M., Jin, D., Yu, P.S.: Graph neural networks for graphs with heterophily: a survey. arXiv preprint arXiv:2202.07082 (2022)
27. Zhu, J., et al.: Graph neural networks with heterophily. In: AAAI, pp. 11168–11176 (2021)
28. Zhu, J., Yan, Y., Zhao, L., et al.: Beyond homophily in graph neural networks: current limitations and effective designs. NeurIPS **33**, 7793–7804 (2020)

Super-Node Generation for GNN-Based Recommender Systems: Enhancing Distant Node Integration via Graph Coarsening

Shasha Hu[1,2], Chao Wang[1,3(✉)], Chuan Qin[2,4], Hengshu Zhu[2], and Hui Xiong[3(✉)]

[1] School of Computer Science and Technology, University of Science and Technology of China, HeFei, China
shashahu@mail.ustc.edu.cn

[2] Career Science Lab, BOSS Zhipin, Beijing, China

[3] Guangzhou HKUST Fok Ying Tung Research Institute, The Hong Kong University of Science and Technology (Guangzhou), Guangzhou, China
chadwang2012@gmail.com, xionghui@ust.hk

[4] PBC School of Finance, Tsinghua University, Beijing, China

Abstract. In the realm of recommender systems, Graph Neural Network based collaborative filtering has emerged as a leading approach. This method represents user-item interactions as a bipartite graph, facilitating the extraction of user and item embeddings by aggregating neighborhood data and capturing potential user preferences. However, a glaring limitation persists: GNNs, by design, primarily concentrate on closely located nodes, often sidelining those situated farther apart. This issue becomes more pronounced due to the "long-tail" effect prevalent in recommendation systems. Attempts to enhance GNNs, either by adding more layers or integrating contrastive learning, have only seen limited success. To overcome these hurdles, we present the General Graph Coarsening Recommendation Framework (GGCRF). This novel methodology utilizes graph coarsening algorithms to forge crucial connections between user/item nodes and newly created super-nodes. A key aspect of GGCRF is the division of user-item graphs into smaller segments using community detection, leading to the formation of super-nodes that encapsulate groups of similar nodes. GGCRF also introduces a unique indirect connectivity loss tailored for collaborative filtering, enhancing its adaptability to a range of GNN-based applications. Our extensive evaluations of GGCRF across five real-world datasets confirm its effectiveness and adaptability, underscoring its capability for generalization in various GNN-based models.

Keywords: Graph Neural Network · Recommender System · Graph Coarsening

This work was accomplished when Shasha Hu was an intern in Career Science Lab, BOSS Zhipin, supervised by Chuan Qin and Hengshu Zhu.

M. Onizuka et al. (Eds.): DASFAA 2024, LNCS 14855, pp. 353–363, 2024.
https://doi.org/10.1007/978-981-97-5572-1_24

1 Introduction

Recommender systems are essential in contemporary online platforms to offer personalized content for users [1,19,20,29]. Graph Neural Networks (GNNs) based Collaborative filtering (CF) approaches, representing user-item interactions as a bipartite graph and learning latent embeddings by aggregating neighborhood information, have been a prevalent recommendation approach [4,17,31,32]. However, GNN-based systems tend to prioritize closely interconnected nodes, overlooking distant ones. This bias is problematic in the widely-used 3-layer GNN-based CF models, restricting users to insights from items within a 3-hop distance, limiting exposure to distant nodes. Increasing GNN-based CF model layers will also exacerbate "over-smoothing" problem diminishing embedding personalization and effectiveness [8]. This limitation, along with the long-tail distribution in recommender systems, risks undervaluing insights from sparsely connected users and items, missing significant recommendation value [16].

In the literature, efforts have been made to deepen the understanding of user-item relationships. Several studies tackled the over-smoothing problem in GNNs by adding layers, but this doesn't always improve connectivity to distant nodes [3]. Besides, contrastive learning techniques combined with GNNs enrich user-item interactions but still primarily involve neighboring nodes, leaving distant nodes unexplored [27,28]. Innovations like UltraGCN break away from the conventional layered graph convolution, introducing unique aggregation frameworks [14]. However, these specialized methods are not universally adaptable to standard GNN-based models, indicating a gap in the field's ability to effectively incorporate long-range node information.

Incorporating long-distance node information into GNN-based recommendations is fraught with challenges. Primarily, the absence of direct connections between distant nodes complicates the standard message-passing mechanisms of GNNs, posing challenges in enabling message transmission across these distances. Additionally, conventional loss functions in GNN-based CF methods often treat distant nodes as negative samples, neglecting potential similarities. Thus, it is crucial to integrate these distant yet potentially analogous nodes into the loss function design. Finally, the diversity of GNN variants in recommendation systems adds another layer of complexity, making a universal and model-independent solution for the long-distance node issue challenging.

To address the aforementioned challenges, this paper introduces the General Graph Coarsening Recommendation Framework (GGCRF), which enhances GNN-based recommendation models by incorporating long-distance node information. Specifically, GGCRF employs graph coarsening algorithms to create super-nodes, establishing high-level connections between user/item nodes and them. GGCRF also introduces a novel indirect connectivity loss leveraging high-level connections to identify nodes with shared attributes and enhance learning. This model-agnostic framework seamlessly integrates into various GNN-based applications. Empirical testing of GGCRF across three different GNN-based models, using five real-world datasets, has confirmed its efficiency, especially in scenarios featuring nodes with sparse connections.

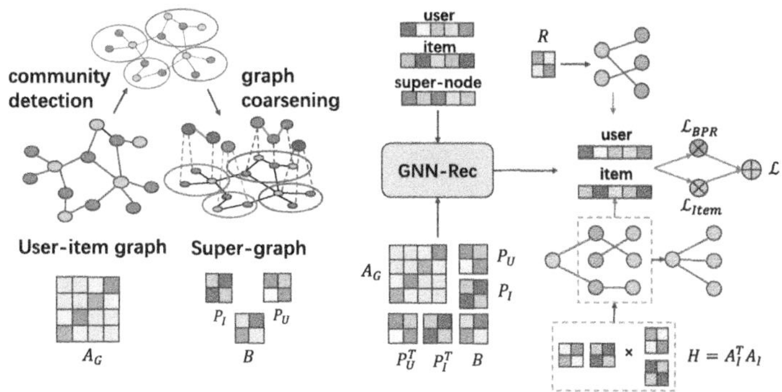

Fig. 1. An illustration of GGCRF-Framework: the user-item graph enhanced by super-nodes is fed into the GNN-based recommender system, and the indirect connectivity loss is calculated by the item-item matrix.

2 Method

This section introduces our framework for enhancing GNN-based recommenders by addressing propagation distance, illustrated in Fig. 1. We first elucidate the process of super-graph construction. Then, we use LightGCN as a case to illustrate the integration of super-nodes into the training process. Finally, we detail the collaborative filtering loss incurred by distant yet similar nodes.

2.1 Super-Graph Construction for Recommendation

To construct the super-graph, we firstly employ community detection algorithms to cluster users and items based on structural similarity. Subsequently, a graph coarsening algorithm is applied to each sub-graph to generate super-nodes.

Community Detection for User-Item Graph. Following collaborative filtering methods [23,24], users select similar items share preferences, connecting closely related nodes. Thus, we use a community detection method to categorize users and items into distinct subgroups for constructing meaningful super-nodes.

Given a user-item graph $G = \langle U, I, E \rangle$, where U and I denote sets of user and item nodes respectively, and E represents their interaction edges, we apply a non-overlapping community detection algorithm [18] to segment G into a collection $\mathcal{S}_G = \{G_1, G_2, \ldots, G_{n_G}\}$ of n_G sub-groups. Each $G_k =< U_k, I_k, E_k >$ is a sub-graph of G, where U_k and I_k are the subsets of U and I, respectively, and E_k denotes the corresponding interaction edges.

Super-Node Generation for User-Item Graph. Each sub-graph G_k from $\mathcal{S}_G$ is coarsened into a higher-level graph by the graph coarsening algorithm [13]. The coarsening results in disjoint node partitions $\mathcal{D}_k = \{D_k^1, D_k^2, \ldots, D_k^{n_{S_k}}\}$.

Each partition $D_k^i =< U_k^i, I_k^i >$ forms a super-node s_k^i, leading to a set of super-nodes $S_k = \{s_k^1, s_k^2, \ldots, s_k^{n_{S_k}}\}$, where n_{S_k} denotes the number of node partitions.

We denote P_U and P_I as matrices of assigning user and item nodes to super-nodes respectively, and the matrix of super-nodes B is then defined as:

$$P = \begin{bmatrix} P_U \\ P_I \end{bmatrix}, \ A_G = \begin{bmatrix} \mathbf{0} & R \\ R^T & \mathbf{0} \end{bmatrix}, \tag{1}$$

$$B = P^T A_G P, \tag{2}$$

where $R \in \mathbb{R}^{|U| \times |I|}$ denotes the user-item matrix of graph G. The node clipping ratio $r = 1 - \frac{\sum_{k=1}^{n_G} n_{S_k}}{|U|+|I|}$ in graph coarsening algorithm is adjusted to regulate the number of super-nodes for balancing the granularity and efficacy.

2.2 Super-Graph Collaborative Filtering Learning

The super-graph comprises three node types: users, items, and super-nodes. And its adjacency matrix A includes: 1) the user-item interaction matrix R; 2) the allocation matrices P_U and P_I mapping original nodes to super-nodes; and 3) the inter-super-node relationship matrix B. Setting n, m, and q as the number of users, items, and super-nodes respectively, the matrix A is defined as:

$$A = \begin{bmatrix} \mathbf{0} & R & P_U \\ R^T & \mathbf{0} & P_I \\ P_U^T & P_I^T & B \end{bmatrix} \in \mathbb{R}^{(n+m+q) \times (n+m+q)}, \tag{3}$$

Taking LightGCN as a case, we capture the l-th layer node embeddings by $E^{l+1} = D^{-\frac{1}{2}} A D^{-\frac{1}{2}} E^l$, where D is the diagonal degree matrix. We finally aggregate the convolved node embeddings by $e = \frac{1}{K+1} \sum_{k=0}^{K} e^k$ for all k layers. The user-item inner product $\widehat{y}_{ui} = e_u^T e_i$ is calculated as the preference score.

Consistent with prior studies [10,28], we adopt the Bayesian Personalized Ranking (BPR) loss and L_2 loss for parameters' optimization:

$$\mathcal{L}_{BPR} = \sum_{(u,i) \in N^+,\ (u,j) \in N^-} - \ln \sigma(\widehat{y}_{ui} - \widehat{y}_{uj}) + \gamma \|E_0\|^2, \tag{4}$$

where N^+ denotes historical pairs and N^- denotes sampled negative pairs. $\sigma(\cdot)$ is the sigmoid function, and γ manages L_2 regularization's impact.

2.3 Indirect Connectivity Loss

Traditional negative sampling treats all non-neighbor items as negatives, increasing the distance from users. In this section, we introduce the indirect connectivity loss by using the super-graph to identify similar distant items for users.

We construct an item-item graph by $H = A_I^T A_I$, where $A_I = \begin{bmatrix} R^T & P_I \end{bmatrix} \in \mathbb{R}^{m \times (n+q)}$ is the association matrix between item nodes and other types in the super-graph. Following UltraGCN [14], we select the top-K similar items $S(i)$

Table 1. The statistics of datasets

Dataset	Users	Items	Interactions	Sparsity
Gowalla [5]	29,858	40,981	1,027,370	0.00084
Amazon [15]	52,643	91,599	2,984,108	0.00268
Yelp2018 [30]	31,668	38,048	1,561,406	0.00130
Ml-1M [9]	6,022	3,043	895,699	0.04888
News [21]	29,785	21,549	766,874	0.00119

for item i based on the item similarity coefficient $\omega_{i,j} = \frac{H_{i,j}}{g_i - H_{i,i}} \sqrt{\frac{\sum_k H_{i,k}}{\sum_k H_{j,k}}}$ to maintain training efficiency. The indirect connectivity loss is then defined as:

$$\mathcal{L}_{Item} = \sum_{(u,i) \in N^+,\ j \in S(i)} -\omega_{i,j} \log(\sigma(e_u^T e_j)), \tag{5}$$

Therefore, our training's overall optimization goal becomes $\mathcal{L} = \mathcal{L}_{BPR} + \lambda \mathcal{L}_{Item}$, where λ is the weight coefficient to control the influence of $\mathcal{L}_{Item}$.

3 Experiments

3.1 Experimental Settings

Dataset Description. We conducted comparative experiments on five widely used datasets as shown in Table 1, including Gowalla [5], Amazon-Book(referred as Amazon) [15], Yelp2018 [30], Movielens-1M [9], and News [21]. Following [14,25], we used the same 10-core setting for all datasets to ensure data quality, and split each dataset into training, validation, and testing set using a ratio of 7:1:2.

Baseline Approaches. We selected several representative recommendation models as baselines to compare with, including MF [22], GC-MC [2], NGCF [25], LR-GCCF [3], Mult-VAE [12] and DGCF [26]. Furthermore, to show our framework's versatility, we also applied it to three competitive GNN-based recommendations: LightGCN [10], SGL [28] and GTN [6].

Evaluation Metrics. We evaluated model performance using the top-K evaluation protocol with two commonly used metrics: Recall and Normalized Discounted Cumulative Gain (NDCG), setting K as 20 by default.

Parameter Settings. All models were implemented using PyTorch, standardized with an embedding size of 64 and a mini-batch size of 1024. Parameters were initialized following the Xavier method [7] and optimized by Adam optimizer [11]. The learning rate was explored within $\{0.0001, 0.001, 0.01, 0.1\}$. We searched L_2 regularization coefficient γ and indirect connectivity loss coefficient λ within $\{1e^{-5}, 1e^{-4}, 1e^{-3}, 1e^{-2}, 1e^{-1}, 1\}$ and chose $1e^{-4}$ as the default value. We explored the graph coarsening rate r within $\{0.05, 0.1, 0.3, 0.5, 0.7, 0.9, 0.95\}$ and investigated the number of communities c across $\{16, 32, 48, 60, 84, 96, 112\}$.

Table 2. Overall performance comparison between the baselines and our framework as applied in LightGCN, SGL, and GTN, respectively, labeled as LightGCN-GC, SGL-GC, and GTN-GC. R@20 and N@20 are used to denote Recall@20 and NDCG@20, respectively. The best results are bold, and the second-best are underlined.

Dataset	Gowalla		Amazon		Yelp2018		Ml-1M		News	
Metric	R@20	N@20	R@20	N@20	R@20	N@20	R@20	N@20	R@20	N@20
MF	0.1348	0.1057	0.0332	0.0251	0.0549	0.0445	0.2604	0.2697	0.1927	0.1222
GC-MC	0.1395	0.1204	0.0288	0.0224	0.0548	0.0445	0.2458	0.2299	0.1889	0.1197
NGCF	0.1561	0.1288	0.0335	0.0261	0.0579	0.0477	0.2619	0.2729	0.1992	0.1289
LR-GCCF	0.1558	0.1313	0.0353	0.0278	0.0561	0.0343	0.2231	0.2124	0.1979	0.1263
Mult-VAE	0.1641	0.1335	0.0407	0.0315	0.0615	0.0495	0.2625	0.2397	0.2075	0.1308
DGCF	0.1842	0.1561	0.0422	0.0322	0.0654	0.0534	0.2640	0.2504	0.2077	0.1358
LightGCN	0.1830	0.1548	0.0417	0.0322	0.0636	0.0520	0.2574	0.2437	0.2058	0.1300
LightGCN-GC	0.1860	0.1554	0.0454	0.0348	0.0660	0.0535	0.2620	0.2489	0.2102	0.1331
SGL	0.1786	0.1505	0.0480	0.0381	0.0673	0.0553	0.2511	0.2373	0.2135	0.1394
SGL-GC	0.1820	0.1531	<u>0.0495</u>	<u>0.0391</u>	0.0684	0.0564	0.2536	0.2393	0.2251	0.1475
GTN	<u>0.1894</u>	<u>0.1593</u>	0.0475	0.0364	<u>0.0697</u>	<u>0.0569</u>	<u>0.2772</u>	<u>0.2622</u>	<u>0.2257</u>	**0.1478**
GTN-GC	**0.1927**	**0.1620**	**0.0538**	**0.0415**	**0.0706**	**0.0579**	**0.2817**	**0.2675**	**0.2263**	<u>0.1475</u>

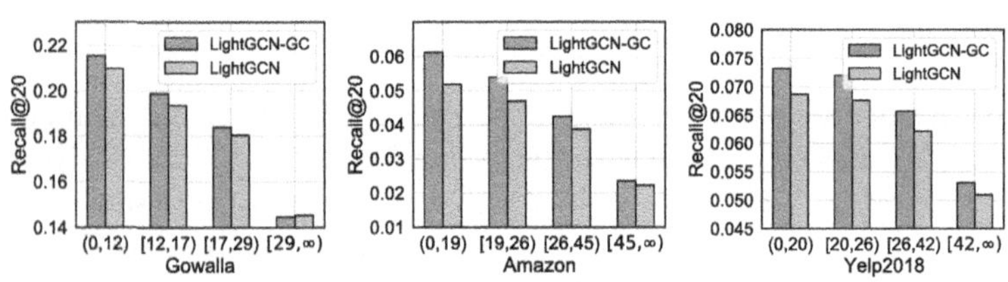

Fig. 2. Performance across user groups categorized by testing data interaction counts.

3.2 Performance Comparison

Overall Comparison. Table 2 demonstrates that GGCRF significantly enhance three representative GNN-based CF models and shows its superiority over baseline methods. For instance, when applied to the base model GTN on Amazon, GGCRF shows an improvement of 13.3% in Recall@20 and 14.0% in NDCG@20. These improvements of GGCRF can be attributed to the following reasons: 1) The generated super-nodes effectively capture user intents or item characteristics, resulting in more representative node embeddings; 2) The super-node interconnections surpassing k-hop neighbors, accessing broader and more valuable information from distant sources beyond historical interactions.

Performance with Sparse User Interaction. We divided users into four intervals ensuring equal testing data interaction counts. And we tested LightGCN and our LightGCN-GC on varying sparsity levels within Gowalla, Amazon, and Yelp2018 datasets, shown as Fig. 2. We note that performance improvements are more pronounced in sparser groups than denser ones. For instance, increasing the sparsity of users in Amazon, LightGCN-GC respectively achieves improvements of 5.9%, 9.6%, 14.7%, and 17.7% in Recall@20 compared to LightGCN.

Table 3. Performance comparison between LightGCN-GC and its variants.

Dataset	Gowalla		Amazon		Yelp2018		Ml-1M		News	
Metric	R@20	N@20	R@20	N@20	R@20	N@20	R@20	N@20	R@20	N@20
LightGCN	0.1830	0.1548	0.0417	0.0322	0.0636	0.0520	0.2574	0.2437	0.2058	0.1300
LightGCN-GC	**0.1860**	0.1554	**0.0454**	**0.0348**	**0.0652**	**0.0532**	**0.2607**	**0.2463**	**0.2102**	**0.1331**
w/o $\mathcal{L}_{Item}$	0.1854	**0.1556**	0.0447	0.0342	0.0642	0.0523	0.2581	0.2439	0.2080	0.1317
w/o s2s	0.1833	0.1540	0.0419	0.0322	0.0650	0.0530	0.2577	0.2444	0.2093	0.1325
w/o s2s &$\mathcal{L}_{Item}$	0.1824	0.1548	0.0430	0.0331	0.0635	0.0519	0.2551	0.2426	0.2069	0.1317

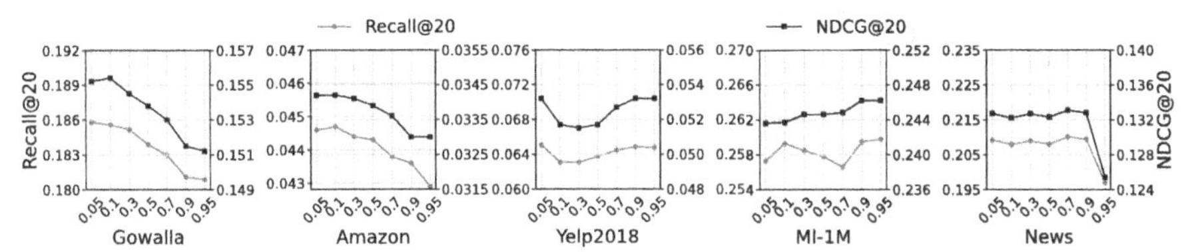

Fig. 3. The impact of graph coarsening rate r. The left y-axis represents Recall@20, and the right y-axis represents NDCG@20.

3.3 Ablation Study

To verify the importance of GGCRF's different components, we constructed three variants of LightGCN-GC: 1) LightGCN-GC$_{w/o\,\mathcal{L}_{Item}}$ indicates training conducted without the indirect connectivity loss $\mathcal{L}_{Item}$; 2) LightGCN-GC$_{w/o\,s2s}$ refers to the super-graph without edges between super-nodes; 3) LightGCN-GC$_{w/o\,s2s\&\mathcal{L}_{Item}}$ denotes removal of both ablative components. Table 3 demonstrates that all components significantly enhance model's performance. The performance of LightGCN-GC$_{w/o\,s2s\&\mathcal{L}_{Item}}$ even below that of LightGCN. It is

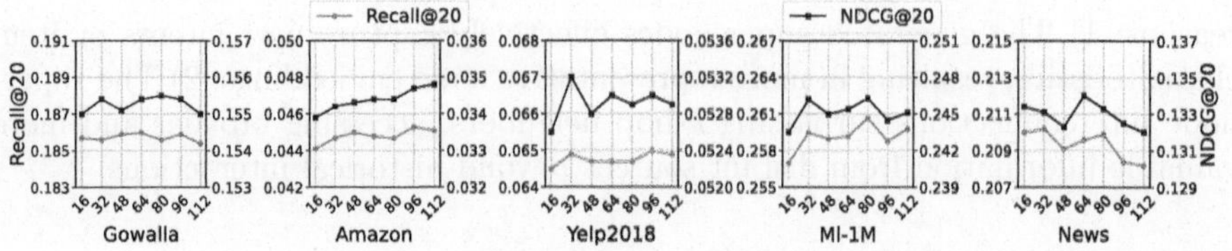

Fig. 4. The impact of number of communities c. The left y-axis represents Recall@20, and the right y-axis represents NDCG@20.

because retaining only edges between original nodes and super-nodes potentially results in overly smooth representations and diminished discriminatory power.

3.4 Parameter Sensitivity Analysis

In this section, we adjust the hyperparameter values to assess their impact on GGCRF's performance. 1) **Impact of graph coarsening rate** r. We searched for r within $\{0.05, 0.1, 0.3, 0.5, 0.7, 0.9, 0.95\}$ to control the super-nodes counts as depicted in Fig. 3. It's evident that a suitable r notably impacts our framework's effectiveness. 2) **Impact of number of communities** c. we explored c within $\{16, 32, 48, 64, 80, 96, 112, 128\}$ as shown in Fig. 4. The sensitivity of different community numbers c is lower than the coarsening rate r. For most datasets, selecting a value from $\{64, 80\}$ ensures stable and satisfactory results.

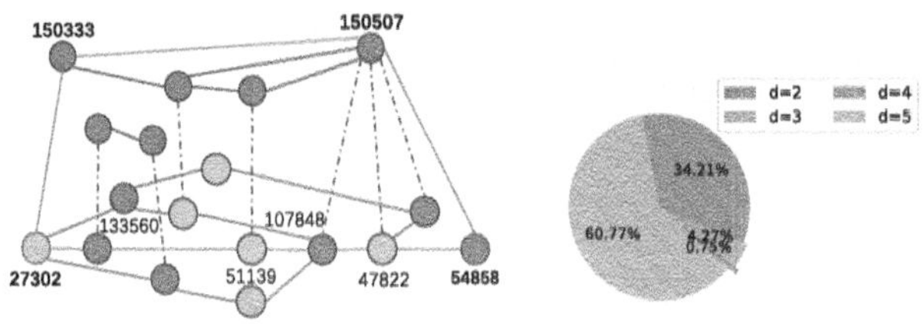

Fig. 5. Case Study: (a) a user-item pair in the Gowalla dataset whose distance is realistically shortened on the super-graph; (b) distance distribution on the super-graph of testing user-item pairs in the Yelp2018 dataset.

3.5 Case Study

To investigate how GGCRF shorten the distance between users and target items, we selected a testing pair (u_{27302}, i_{54858}) from Gowalla and presented the shortest paths on the training graph of LightGCN and LightGCN-GC in Fig. 5a. It is worth noting that i_{54858} is successfully recommended to u_{27302} only by LightGCN-GC. We can find that there exits a shorter propagation path through super-nodes with a length of 3, so that u_{27302} can receive i_{54858}'s information in one epoch. In Fig. 5b, we also calculated the distance distribution of testing pairs from Yelp2018 with a distance ≥ 5. There are nearly 40% of the pairs with a reduced distance, which helps users directly obtain target items' messages.

4 Conclusion

In this study, we addressed the limitations of GNN-based recommendations constrained by layer depth, hindering access to sufficient information and long-distance collaborative signals. We proposed GGCRF to augment recommendation systems with super-nodes, summarizing high-level characteristics and bolstering sparse nodes. Using a graph coarsening algorithm, we generated super-nodes and establish connections based on user-item interactions, enriching message passing paths beyond the original graph. Through experiments on five datasets, we demonstrated GGCRF's effectiveness in enhancing GNN-based recommendations, connecting users to a wider array of potentially interesting items.

Acknowledgments. This work was partially supported by Guangzhou-HKUST(GZ) Joint Funding Program(Grant No.2023A03J0008), Education Bureau of Guangzhou Municipality, Guangdong Science and Technology Department, Foshan HKUST Projects (FSUST21-FYTRI01A) and China Postdoctoral Science Foundation (Grant No.2023M730785).

References

1. A survey on causal inference for recommendation. The Innovation (2024)
2. Berg, R.v.d., Kipf, T.N., Welling, M.: Graph convolutional matrix completion. arXiv preprint arXiv:1706.02263 (2017)
3. Chen, L., Wu, L., Hong, R., Zhang, K., Wang, M.: Revisiting graph based collaborative filtering: A linear residual graph convolutional network approach. In: Proceedings of the AAAI Conference on Artificial Intelligence (2020)
4. Chen, M., Wang, C., Qin, C., Xu, T., Ma, J., Chen, E., Xiong, H.: A trend-aware investment target recommendation system with heterogeneous graph. In: 2021 International Joint Conference on Neural Networks (IJCNN). IEEE (2021)
5. Cho, E., Myers, S.A., Leskovec, J.: Friendship and mobility: user movement in location-based social networks. In: Proceedings of the 17th ACM SIGKDD International Conference on Knowledge Discovery and Data Mining (2011)
6. Fan, W., Liu, X., Jin, W., Zhao, X., Tang, J., Li, Q.: Graph trend filtering networks for recommendation. In: Proceedings of the 45th International ACM SIGIR Conference on Research and Development in Information Retrieval (2022)
7. Glorot, X., Bengio, Y.: Understanding the difficulty of training deep feedforward neural networks. In: Proceedings of the Thirteenth International Conference on Artificial Intelligence and Statistics. JMLR Workshop and Conference Proceedings (2010)
8. Hamilton, W., Ying, Z., Leskovec, J.: Inductive representation learning on large graphs. Advances in neural information processing systems (2017)
9. Harper, F.M., Konstan, J.A.: The movielens datasets: History and context. Acm transactions on interactive intelligent systems (tiis) (2015)

10. He, X., Deng, K., Wang, X., Li, Y., Zhang, Y., Wang, M.: Lightgcn: Simplifying and powering graph convolution network for recommendation. In: Proceedings of the 43rd International ACM SIGIR Conference on Research and Development in Information Retrieval (2020)
11. Kingma, D.P., Ba, J.: Adam: A method for stochastic optimization. arXiv preprint arXiv:1412.6980 (2014)
12. Liang, D., Krishnan, R.G., Hoffman, M.D., Jebara, T.: Variational autoencoders for collaborative filtering. In: Proceedings of the 2018 World Wide Web Conference (2018)
13. Loukas, A.: Graph reduction with spectral and cut guarantees. J. Mach. Learn. Res. (2019)
14. Mao, K., Zhu, J., Xiao, X., Lu, B., Wang, Z., He, X.: Ultragcn: ultra simplification of graph convolutional networks for recommendation. In: Proceedings of the 30th ACM International Conference on Information & Knowledge Management (2021)
15. McAuley, J., Targett, C., Shi, Q., Van Den Hengel, A.: Image-based recommendations on styles and substitutes. In: Proceedings of the 38th International ACM SIGIR Conference on Research and Development in Information Retrieval (2015)
16. Milojević, S.: Power law distributions in information science: Making the case for logarithmic binning. J. Am. Soc. Inf. Sci. Technol. (2010)
17. Mu, R.: A survey of recommender systems based on deep learning. IEEE Access (2018)
18. Parés, F., et al.: Fluid communities: a competitive, scalable and diverse community detection algorithm. In: Cherifi, C., Cherifi, H., Karsai, M., Musolesi, M. (eds.) COMPLEX NETWORKS 2017 2017. SCI, vol. 689, pp. 229–240. Springer, Cham (2018). https://doi.org/10.1007/978-3-319-72150-7_19
19. Qin, C., et al.: A comprehensive survey of artificial intelligence techniques for talent analytics. arXiv preprint arXiv:2307.03195 (2023)
20. Qin, C., Zhu, H., Shen, D., Sun, Y., Yao, K., Wang, P., Xiong, H.: Automatic skill-oriented question generation and recommendation for intelligent job interviews. ACM Trans. Inf. Syst. (2023)
21. Reddit: World news subreddit. https://www.reddit.com/r/worldnews/
22. Song, X., Feng, F., Han, X., Yang, X., Liu, W., Nie, L.: Neural compatibility modeling with attentive knowledge distillation. In: The 41st International ACM SIGIR Conference on Research & Development in Information Retrieval (2018)
23. Wang, C., Zhu, H., Hao, Q., Xiao, K., Xiong, H.: Variable interval time sequence modeling for career trajectory prediction: deep collaborative perspective. In: Proceedings of the Web Conference 2021 (2021)
24. Wang, C., Zhu, H., Zhu, C., Qin, C., Chen, E., Xiong, H.: Setrank: a setwise Bayesian approach for collaborative ranking in recommender system. ACM Trans. Inf. Syst. (2023)
25. Wang, X., He, X., Wang, M., Feng, F., Chua, T.S.: Neural graph collaborative filtering. In: Proceedings of the 42nd International ACM SIGIR Conference on Research and Development in Information Retrieval (2019)
26. Wang, X., Jin, H., Zhang, A., He, X., Xu, T., Chua, T.S.: Disentangled graph collaborative filtering. In: Proceedings of the 43rd International ACM SIGIR Conference on Research and Development in Information Retrieval (2020)
27. Wei, Y., et al.: Contrastive learning for cold-start recommendation. In: Proceedings of the 29th ACM International Conference on Multimedia (2021)
28. Wu, J., et al.: Self-supervised graph learning for recommendation. In: Proceedings of the 44th international ACM SIGIR Conference on Research and Development in Information Retrieval (2021)

29. Wu, L., et al.: A survey on large language models for recommendation. arXiv preprint arXiv:2305.19860 (2023)
30. Yelp: Yelp open dataset. https://www.yelp.com/dataset
31. Yu, X., et al.: Rdgt: enhancing group cognitive diagnosis with relation-guided dual-side graph transformer. IEEE Trans. Knowl. Data Eng. (2024)
32. Zha, R., et al.: Career mobility analysis with uncertainty-aware graph autoencoders: a job title transition perspective. IEEE Trans. Comput. Soc. Syst. (2023)

History Driven Sampling for Scalable Graph Neural Networks

Yang Li(✉), Bingbing Xu, Fei Sun, Qi Cao, Yige Yuan, Huawei Shen, and Xueqi Cheng

Institute of Computing Technology, Chinese Academy of Sciences, Beijing, China
{liyang21s,xubingbing,caoqi,yuanyige20z,shenhuawei,cxq}@ict.ac.cn

Abstract. Graph Neural Networks (GNNs) which achieve great success in numerous scenarios, suffer from unsustainable computational and storage costs due to the exponential increase in the neighbour size as the number of layers escalates. In this view, graph sampling is one of the prevalent methods to improve GNNs' scalability. Despite its benefits, the variance introduced by sampling severely decelerates the convergence speed and degrades the performance of GNNs. To address this, we leverage Lagrange multipliers to optimize the constrained non-linear variance reduction problem, yielding the theoretical minimum variance sampler. However, putting this sampler into practice faces a key challenge—circular dependency, i.e., sampling probability of the minimum variance sampler is determined by node embeddings while node embeddings can not be calculated until sampling is finished. Therefore, we propose a general framework named **H**istory **D**riven **S**ampling for scalable **GNN** (HDSGNN). HDSGNN estimates the minimum variance sampler with the historical nodes' embeddings to break the circular dependency and then employs this estimated sampler for computation graph sampling and representation learning. We implement HDSGNN in both layer-wise and subgraph sampling respectively. Comprehensive experiments on seven representative benchmarks verify HDSGNN's effectiveness and efficiency over the SOTA baselines.

Keywords: Graph Neural Network · Scalability · Circular Dependency · History Driven Sampling

1 Introduction

Graph Neural Networks (GNNs) [8,11], which show remarkable success in numerous scenarios, suffer from computational and storage costs due to the exponential increase in the neighbour size as the number of layers increases. Several strategies are advanced to enhance the scalability of GNNs. Among these, graph sampling, including node-wise sampling [1,4,8], layer-wise sampling [2,10,17] and subgraph sampling [3,16], becomes a prevalent method due to its considerable improvement in training efficiency.

M. Onizuka et al. (Eds.): DASFAA 2024, LNCS 14855, pp. 364–374, 2024.
https://doi.org/10.1007/978-981-97-5572-1_25

Although sampling methods alleviate scalability issues in GNNs, they inadvertently introduce substantial variance during training, which severely decelerates the convergence speed and degrades the performance of GNNs [1,2]. Unfortunately, minimizing variance by optimizing the sampling probability [5,7,12] manifests as a constrained non-linear optimization problem. Previous methods [2,10,16] often compromise by overlooking the probability constraint, i.e., the sum of the sampling probabilities for each node equals 1, to derive a sampler firstly which leads to a deviation from variance minimization. To address this, we introduce Lagrange multipliers [13] to integrate this probability constraint into optimization, yielding a theoretical minimum variance sampler.

However, implementing this theoretical optimal solution into practice confronts a key challenge, i.e., circular dependency. Specifically, the minimum variance sampling probability for a specific node is determined by its own node embeddings while the node embeddings are aggregated from the sampled computation graphs. Hence, the node embeddings are not available until the sampling process is fully completed. The key to breaking the circular dependency lies in how to obtain the nodes' embeddings learned by GNNs prior to the sampling process at each iteration. Considering that as the model converges [1], the historical nodes' embeddings will gradually approximate the current unknown nodes' embeddings. In this view, we propose to leverage historical nodes' embeddings to estimate the sampling probability, thereby breaking the circular dependency without significantly increasing computational and storage costs. Such a method can estimate the theoretical minimum variance sampler effectively and efficiently.

Therefore, we propose a general framework **H**istory **D**riven **S**ampling for scalable **GNN**s (HDSGNN). We implement HDSGNN under layer-wise sampling and subgraph sampling paradigm respectively given that node-wise sampling still faces exponentially increasing complexity [10]. **For layer-wise sampling**, HDSGNN establishes sequential relationships among the history at each iteration and leverages a combination of all historical nodes' embeddings to estimate the sampler. To encourage precise approximation, we employ optimistic initialization to boost early exploration. **For subgraph sampling**, following the previous studies [3,16] where sampling occurs only once before training, we regard the input node features as the initial history for estimating the sampler. Then HDSGNN samples the computation graphs and learns nodes' representations on these sampled graphs. Comprehensive experiments on seven prevalent benchmarks verify HDSGNN's effectiveness and efficiency over SOTA baselines.

2 Related Work

Overall, graph sampling consists of three sampling paradigms, i.e., node-wise sampling, layer-wise sampling and subgraph sampling.

Node-Wise Sampling. Node-wise sampling [1,8] controls the number of neighbors involved in embedding aggregation to a small constant for each node. However, this paradigm still faces the problem of exponentially increase in neighbour size, thus losing its scalability when the layers are stacked deeper.

Layer-Wise Sampling. Layer-wise sampling operates with fixed-size node sets sampled at each layer. To reduce the variance, FastGCN [2] analyzes the variance and detaches the sampling probability from dependency on hidden embeddings in the front layer for simplicity. LADIES [17] alleviates the sparseness problem caused by FastGCN. Additionally, ASGCN [10] approximates the minimum variance sampler with modified graph attention. But the computation of attention imposes significant time costs.

Subgraph Sampling. Subgraph sampling constructs several subgraphs from the original graph and learns the embeddings of nodes in each subgraph respectively. In the subgraph sampling paradigm, the sampling is executed only once before training. For instance, ClusterGCN [3] applies graph clustering algorithms for partition. Differently, GraphSAINT [16] develops several variance reduction samplers which significantly accelerate the partition.

3 Preliminaries

The traditional aggregation of vanilla GCN can be expressed as follows:

$$h_i^{(l)} = \sigma(F_i^{(l)}) = \sigma(\sum_{j \in N(v_i)} \hat{A}_{ij} h_j^{(l-1)} W^{(l)}), \quad l = 1, \cdots, L \tag{1}$$

where $\hat{A} \in \mathbf{R}^{n \times n}$ denotes the degree normalized adjacency matrix, n is the number of vertices, $N(v_i)$ denotes the set of v_i's neighbours, $W^{(l)}$ denotes the parameter matrix in the (l)-th layer, $h_i^{(l)}$ denotes the hidden embeddings of node v_i in the (l)-th layer. Specially, $h_i^{(0)}$ denotes the input features x_i. **For simplicity, we denote the transformed embeddings $h_j^{(l-1)} W^{(l)}$ as $\mathcal{H}_j^{(l-1)}$.**

To maintain the power of vanilla GCN, the aggregation formula of $\hat{F}_i^{(l)}$ in sampling based methods is envisaged as an unbiased estimator of $F_i^{(l)}$:

$$\hat{F}_i^{(l)} = \frac{1}{\left|V_i^{(l-1)}\right|} \sum_{j \in V_i^{(l-1)}} \frac{\hat{A}_{ij}}{q_{ij}^{(l-1)}} \mathcal{H}_j^{(l-1)} \tag{2}$$

where $q_{ij}^{(l-1)}$ denotes the sampling probability of node v_j in the $(l-1)$-th layer given node v_i in the (l)-th layer. $V_i^{(l-1)}$ denotes the sampled nodes set in the $(l-1)$-th layer for aggregating the embeddings of node v_i. Specifically, node-wise sampling limits the number of sampled neighbours for each node in each layer which implies that $q_{ij}^{(l-1)}$ is related to the v_i, v_j and l. Layer-wise sampling conducts sampling only once in each layer conditionally based on all the nodes in the next layer, where $q_{ij}^{(l-1)}$ is simplified to $q_j^{(l-1)}$. Subgraph sampling only partitions the original graph into several small subgraphs before training, where $q_{ij}^{(l-1)}$ can be simplified to q_j. Furthermore, $V_i^{(l-1)}$ can be simplified to $V^{(l-1)}$ for layer-wise sampling and V_s for subgraph sampling. V_s denotes the nodes set in each sampled subgraph s. $h_i^{(l)}$ can be approximated with $\sigma(\hat{F}_i^{(l)})$.

4 Variance Reduction Analysis

4.1 Minimum Variance Sampler

To obtain the minimum variance sampler, we first conduct the variance reduction analysis theoretically. Without loss of generality, we take layer-wise sampling as an example for derivation where $q_{ij}^{(l-1)}$ is simplified to $q_j^{(l-1)}$ and $V_i^{(l-1)}$ is simplified to $V^{(l-1)}$. In the (l)-th layer, we aim to minimize the sum of the variance of $\hat{F}_i^{(l)}$ for each node v_i. Meanwhile, the sampling probability must strictly satisfy the constraint $\sum_{j=1}^{n} q_j^{(l-1)} = 1$. Therefore, we modify the optimization objective into a constrained non-linear optimization problem and introduce Lagrange multipliers for derivation. The Lagrange function can be defined as follows:

$$L(\lambda_{l-1}, q_j^{(l-1)}) = \sum_{i \in V^{(l)}} Var_{q_j^{(l-1)}}(\hat{F}_i^{(l)}) + \lambda_{l-1}(1 - \sum_{j=1}^{n} q_j^{(l-1)}) \tag{3}$$

where λ_{l-1} denotes the Lagrange multiplier. Then we can determine the extreme value of $L(\lambda_{l-1}, q_j^{(l-1)})$ by letting $\frac{\partial L(\lambda_{l-1}, q_j^{(l-1)})}{\partial q_j^{(l-1)}} = 0$ and $\frac{\partial L(\lambda_{l-1}, q_j^{(l-1)})}{\partial \lambda_{l-1}} = 0$:

Then, we illustrate the minimum variance sampler under layer-wise sampling, node-wise sampling and subgraph sampling paradigm as follows:

$$\begin{cases} q_{ij}^{(l-1)*} = \dfrac{\hat{A}_{ij} \|\mathcal{H}_j^{(l-1)}\|_2}{\sum_{k=1}^{n} \hat{A}_{ik} \|\mathcal{H}_k^{(l-1)}\|_2} & \text{, node-wise sampling} \\ q_j^{(l-1)*} = \dfrac{(\sum_{i \in V^{(l)}} \hat{A}_{ij}^2)^{\frac{1}{2}} \|\mathcal{H}_j^{(l-1)}\|_2}{\sum_{k=1}^{n} (\sum_{i \in V^{(l)}} \hat{A}_{ik}^2)^{\frac{1}{2}} \|\mathcal{H}_k^{(l-1)}\|_2} & \text{, layer-wise sampling} \\ q_j{}^* = \dfrac{\sqrt{\sum_{l=1}^{L} \sum_{i=1}^{n} \hat{A}_{ij}^2 \|\mathcal{H}_j^{(l-1)}\|_2^2}}{\sum_{k=1}^{n} \sqrt{\sum_{l=1}^{L} \sum_{i=1}^{n} \hat{A}_{ik}^2 \|\mathcal{H}_k^{(l-1)}\|_2^2}} & \text{, subgraph sampling} \end{cases} \tag{4}$$

4.2 Circular Dependency

Although we have derived the theoretical minimum variance samplers through variance reduction analysis, it remains unattainable owing to the key challenge known as **circular dependency** regardless of the sampling paradigm employed.

Specifically, under layer-wise sampling paradigm, the minimum variance sampling probability $q_j^{(l-1)^*}$ of node v_j in the $(l-1)$-th layer depends on both $\mathcal{H}_j^{(l-1)}$ and node set $V^{(l)}$ in the (l)-th layer. Apparently, $\mathcal{H}_j^{(l-1)}$ of node v_j are aggregated from the node set sampled in the $(l-2)$-th layer $V^{(l-2)}$. However, only after sampling the node set $V^{(l-1)}$, the $(l-2)$-th layer nodes can be sampled, resulting in $\mathcal{H}_j^{(l-1)}$ cannot be calculated before sampling in the $(l-1)$-th layer. Similarly, node-wise sampler $q_{ij}^{(l-1)^*}$ also encounters the circular dependency similar to layer-wise sampling.

Under subgraph sampling paradigm, the acquisition of the minimum variance sampler $q_j{}^*$ is also hindered by circular dependency. Specifically, the minimum variance sampling probability $q_j{}^*$ is related to $\|\mathcal{H}_j^{(l-1)}\|_2$ for each node in each layer. However, the embeddings are unavailable until the completion of the sampling process.

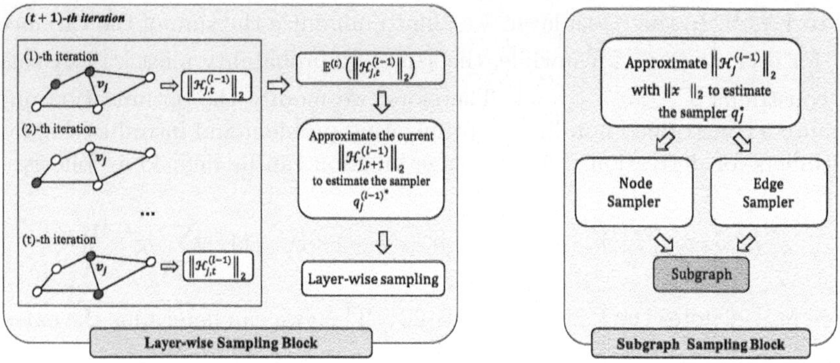

(a) Layer-wise sampling. (b) Subgraph sampling.

Fig. 1. Overall framework of HDSGNN.

5 Methodology

In this section, we introduce our **H**istory **D**riven **S**ampling for scalable **GNN**s (HDSGNN). Considering that node-wise sampling still struggle with exponentially increasing complexity, we implement HDSGNN under layer-wise sampling and subgraph sampling paradigm respectively.

5.1 HDSGNN in Layer-Wise Sampling

The implementation of HDSGNN in layer-wise sampling is illustrated in Fig. 1a.

For convenience, we denote $\|\mathcal{H}_j^{(l-1)}\|_2$ at (t)-th iteration $(t = 1, 2, \ldots, T)$ as $\|\mathcal{H}_{j,t}^{(l-1)}\|_2$. Note that since node v_j in the $(l-1)$-th layer might not be sampled in the last iteration, it would be unreasonable to only use the information from the last iteration. Meanwhile, considering potential fluctuations in the model's historical nodes' embeddings due to oscillations during iterations, we leverage a combination of all historical nodes' embeddings as an approximation for the current unknown $\|\mathcal{H}_{j,t}^{(l-1)}\|_2$ to facilitate faster convergence. Therefore, $q_j^{(l-1)^*}$ can be estimated as:

$$\hat{q}_j^{(l-1)} = \frac{(\sum_{i\in V^{(l)}} \hat{A}_{ij}^2)^{\frac{1}{2}}\mathbb{E}(\|\mathcal{H}_{j,t}^{(l-1)}\|_2)}{\sum_{k=1}^{n}(\sum_{i\in V^{(l)}} \hat{A}_{ik}^2)^{\frac{1}{2}}\mathbb{E}(\|\mathcal{H}_{k,t}^{(l-1)}\|_2)} \tag{5}$$

where t denotes the current iteration, $\mathbb{E}(\|\mathcal{H}_{j,t}^{(l-1)}\|_2)$ denotes the approximation of current $\|\mathcal{H}_{j,t}^{(l-1)}\|_2$ at the (t)-th iteration. To avoid significant additional storage costs, we propose to calculate $\mathbb{E}(\|\mathcal{H}_{j,t+1}^{(l-1)}\|_2)$ through an iterative process:

$$\mathbb{E}(\|\mathcal{H}_{j,t+1}^{(l-1)}\|_2)=\begin{cases}\mathbb{E}(\|\mathcal{H}_{j,t}^{(l-1)}\|_2) & \text{, if } \mathbb{I}(v_j, l-1, t) = 0 \\ \frac{N_{j,t}^{(l-1)}\mathbb{E}(\|\mathcal{H}_{j,t}^{(l-1)}\|_2)+\|\mathcal{H}_{j,t}^{(l-1)}\|_2}{N_{j,t+1}^{(l-1)}} & \text{, else}\end{cases} \tag{6}$$

where $\mathbb{I}(v_j, l-1, t)$ is an indicator function that equals 1 if v_j is sampled in $(l-1)$-th layer at (t)-th iteration, otherwise it is 0. $N_{j,t}^{(l-1)}$ denotes the times that v_j is sampled in $(l-1)$-th layer prior to the (t)-th iteration and we set $N_j^{(1)} = 0$. Additionally, $\mathbb{E}(\|\mathcal{H}_{j,1}^{(l-1)}\|_2)$ is the initial term set to a constant. On account that every node is expected to be sampled to obtain a precise approximation of $\|\mathcal{H}_j^{(l-1)}\|_2$ with Eq. (6), we apply optimistic initialization [14] for $\mathbb{E}(\|\mathcal{H}_{j,1}^{(l-1)}\|_2)$ of each node v_j with a large constant to encourage the early exploration.

5.2 HDSGNN in Subgraph Sampling

The implementation of HDSGNN in subgraph sampling is illustrated in Fig. 1b.

Following the same framework, we also attempt to leverage the historical nodes' embeddings to estimate the theoretical minimum variance sampler ${q_j}^*$ in subgraph sampling. Nevertheless, different from layer-wise sampling, subgraph sampling is conducted only once prior to the whole training of GNNs model. This implies that only input node features x_j can be used as historical nodes' embeddings to approximate the $\|\mathcal{H}_j^{(l-1)}\|_2$. As a result, ${q_j}^*$ is estimated:

$$\hat{q}_j = \frac{(\sum_{i=1}^{n} \hat{A}_{ij}^2)^{\frac{1}{2}} \|x_j\|_2}{\sum_{k=1}^{n}(\sum_{i=1}^{n} \hat{A}_{ik}^2)^{\frac{1}{2}} \|x_k\|_2} \tag{7}$$

Meanwhile, sampling is expected to preserve as much of the original graph's structural information as possible. Therefore, implementing an edge sampler and adding the nodes connected by edges into the computation graph proves beneficial in retaining the structural information [16]. Inspired by this, we also develop our own edge sampler to sample subgraphs. Intuitively, the edge sampling probability should be proportional to that of its two connected nodes respectively. Therefore, to satisfy the constraint that the sum of sampling probabilities equals 1, we induce the edge sampler from node sampling probability $\hat{q}_j$ as:

$$p_{ij} = \frac{\hat{q}_i}{D_i} + \frac{\hat{q}_j}{D_j} \tag{8}$$

where p_{ij} denotes the sampling probability of the edge connecting node v_i and v_j and D_i denotes the degree of node v_i.

Overall, we propose a subgraph node sampler with $\hat{q}_j$ and a subgraph edge sampler with p_{ij} respectively for subgraph sampling based HDSGNN.

6 Experiments

6.1 Benchmarks and Baselines

Benchmarks. These experiments are analyzed on seven node classification datasets including 1): Cora, CiteSeer and PubMed following the "fixed-partition" splits in [2], 2): Flickr, Reddit and Yelp following the "fixed-partition" splits in [16], 3): Ogbn-products following the "fixed-partition" splits in [9]. Detailed information of these seven datasets is listed in Table 1.

Baselines. We select various baselines such as **five mini batched GNNs** including GCN, GraphSAGE, GAT along with two scalable GNNs, GAS [6] and ShadowGCN [15], **three layer-wise sampling based GNNs** including FastGCN [2], LADIES [17] and ASGCN [10] and **three subgraph sampling based GNNs** including ClusterGCN [3], GraphSAINT (node sampler) [16] and GraphSAINT (edge sampler) [16].

Table 1. Dataset Statistics

Dataset	Nodes	Edges	Features	Classes
Cora	2,708	5,429	1,433	7
CiteSeer	3,327	4,732	3,707	6
PubMed	19,717	44,338	500	3
Flickr	89,250	899,756	500	7
Reddit	232,965	11,606,919	602	41
Yelp	716,847	6,977,410	300	100
Ogbn-products	2,449,029	61,859,140	100	47

6.2 Experimental Settings

Setup and Metrics. Same as previous studies [2,10,16] with an inductive setting, we apply sampling solely for training while all the topological information are leveraged for inference. We leverage F1-micro score to evaluate the classification performance and running time measured in seconds for efficiency.

Configurations. We implement our HDSGNN and all baselineson a server equipped with a NVIDIA V100 32 GB GPU. All the models are optimized with Adam optimizer using grid search for layer numbers, dropout rate, batch size and hidden dimension. We select the models yielding the highest average F1-micro score over 10 trials on the validation set respectively for all models.

6.3 Classification Performance Analysis

Table 2 presents a comparison between our proposed HDSGNN and other baselines. In Table 2, "L" denotes layer-wise sampling, "SN" denotes subgraph node sampling, "SE" denotes subgraph edge sampling and "—" denotes out of memory error. Horizontally, each of the three implementations of HDSGNN outperforms other baselines within their respective sampling paradigms. Generally, HDSGNN(L) outperforms SOTA models on sparse graphs and matches the performance of baselines on dense graph where other layer-wise sampling based baselines show poor performance. Additionally, subgraph sampling based HDSGNN further improves F1-micro score of SOTA models when it comes to dense graphs. Note that our HDSGNN(L) greatly reduces the training time cost compared with ASGCN while maintaining the same time complexity as the model with the fastest training speed. Detailed analysis of running time will be presented in Sect. 6.4.

Table 2. F1-micro Score Comparison between HDSGNN and Baselines

Paradigm	Method	Cora	CiteSeer	PubMed	Flickr	Reddit	Yelp	Ogbn-products
Mini Batched	GCN	0.851± 0.001	0.770± 0.003	0.867± 0.001	0.492 ± 0.003	0.933 ± 0.000	0.378± 0.001	0.758± 0.002
	GraphSAGE	0.822± 0.010	0.714± 0.008	0.873± 0.002	0.501± 0.010	0.953± 0.001	0.634± 0.010	0.782± 0.010
	GAT	0.863± 0.001	0.769± 0.003	0.870± 0.001	0.507± 0.002	—	—	—
	GAS	0.858± 0.001	0.774± 0.009	0.888± 0.001	0.513± 0.014	0.952± 0.001	0.620± 0.018	0.767± 0.001
	ShadowGCN	0.862± 0.001	0.780± 0.008	0.885± 0.001	0.510± 0.001	0.957± 0.001	0.609± 0.001	0.772± 0.001
Layer-wise Sampling	FastGCN	0.850± 0.005	0.776± 0.008	0.880± 0.010	0.504± 0.001	0.922± 0.001	0.261± 0.050	0.316± 0.167
	LADIES	0.854± 0.002	0.783± 0.002	0.883 ± 0.001	0.504 ± 0.002	0.941 ± 0.001	0.602 ± 0.020	0.652 ± 0.034
	ASGCN	0.861± 0.006	**0.792± 0.008**	0.896 ± 0.002	0.504 ± 0.002	0.945 ± 0.005	—	—
	HDSGNN(L)	**0.872 ± 0.001**	0.789± 0.004	**0.900± 0.003**	0.511± 0.002	0.947± 0.001	**0.652± 0.017**	0.764± 0.016
Subgraph Sampling	ClusterGCN	0.847± 0.005	0.717± 0.012	0.884± 0.005	0.481± 0.004	0.954± 0.001	0.609± 0.005	0.790± 0.003
	GraphSAINT(SN)	0.851± 0.002	0.766± 0.007	0.898 ± 0.001	0.507± 0.003	0.966± 0.001	0.641± 0.000	0.788± 0.002
	GraphSAINT(SE)	0.856± 0.003	0.753± 0.011	0.897± 0.002	0.510± 0.002	0.966± 0.001	0.648± 0.000	0.799± 0.001
	HDSGNN(SN)	0.860± 0.003	0.777± 0.007	0.898± 0.002	0.511± 0.003	0.970± 0.001	0.644± 0.000	0.794± 0.001
	HDSGNN(SE)	0.863± 0.001	0.784± 0.003	0.896± 0.001	**0.517± 0.003**	**0.971± 0.000**	0.650± 0.000	**0.806± 0.001**

[a] The best results on each benchmark are highlighted in bold while the best results on each benchmark under each sampling paradigm are underlined

6.4 Running Time Analysis

Layer-Wise Sampling. Table 3 shows the experimental and theoretical comparison results. In Table 3, d is the abbreviation for the initial node feature dimension $d^{(0)}$ and s denotes the size of sampled nodes set in each layer. The theoretical time complexity comparison results demonstrate that the time complexity of HDSGNN(L) is consistent with that of FastGCN and LADIES, and significantly surpasses ASGCN. This observation is corroborated by the experimental running time per epoch comparisons across each benchmark.

Subgraph Sampling. On account of the fact that subgraph sampling decouples the sampling from the training process by performing the sampling only prior to training, the most considerable difference among various subgraph sampling based methods lies in the time cost during the sampling process. As a result, we present a comparison of time cost for the sampling process in Table 4 where we can figure out that the costs of HDSGNN remain on par with the sampling cost of GraphSAINT. Meanwhile, the clustering algorithm utilized in ClusterGCN incurs a high sampling cost when dealing with large-scale graphs.

Table 3. Running Time Per Epoch and Time Complexity Comparison of Layer-wise Sampling based Methods

Method	Cora	CiteSeer	PubMed	Flickr	Reddit	Yelp	Ogbn-products	Time Complexity
FastGCN	0.026	0.033	0.291	0.904	5.752	27.561	7.012	$O(Lds^2 + Ld^2s)$
LADIES	0.029	0.040	0.382	1.219	8.131	32.244	10.537	$O(Lds^2 + Ld^2s)$
ASGCN	0.032	0.046	0.614	1.874	44.358	—	—	$O(Lds^2 + Ld^2s + Lds^3)$
HDSGNN(L)	0.028	0.037	0.408	1.124	7.085	33.183	8.652	$O(Lds^2 + Ld^2s)$

Table 4. Sampling Time Comparison of Subgraph Sampling based Methods

Method	Cora	CiteSeer	PubMed	Flickr	Reddit	Yelp	Ogbn-products
ClusterGCN	0.072	0.159	0.977	8.321	28.371	70.262	872.627
GraphSAINT(SN)	0.024	0.075	0.294	0.452	6.946	19.780	4.348
GraphSAINT(SE)	0.027	0.074	0.306	0.580	13.361	29.837	23.132
HDSGNN(SN)	0.025	0.072	0.306	0.507	7.275	18.791	4.424
HDSGNN(SE)	0.031	0.078	0.327	0.703	14.263	25.035	27.061

6.5 Ablation Studies

Initialization Analysis for HDSGNN(L). As mentioned in Sect. 5.1, we evaluate the influence of different initialization strategies in Fig. 2 where pessimistic initialization sets the initial term to 1 and optimistic initialization sets the initial term to 100. From Fig. 2 we can observe that the adoption of optimistic initialization prompts early exploration, increasing the likelihood of each node in every layer to be sampled, thereby capture more accurate representations.

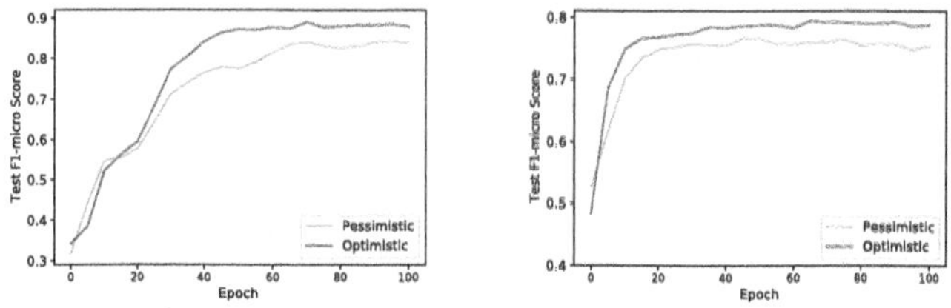

(a) Cora. (b) CiteSeer.

Fig. 2. Test F1-micro curves of HDSGNN(L) with different initializations.

Difference Between HDSGNN(SN) and HDSGNN(SE). Additionally, the performance difference between HDSGNN(SN) and HDSGNN(SE) is also worth pondering, as shown in Table 2. Both these two models perform well on Reddit while our HDSGNN(SE) outperforms our HDSGNN(SN) on Yelp. This difference can potentially be attributed to edge sampler's ensuring that each sampled node has at least one neighbour for message passing, a situation not necessarily guaranteed by the node sampler.

7 Conclusion

In this paper, to reduce the variance introduced by sampling based scalable GNNs, we derive the theoretical minimum variance sampler by leveraging Lagrange multipliers to take the inherent sampling constraint into account. Then we analyze the key challenge, namely circular dependency, for putting the theoretical minimum variance sampler into practice. To handle this challenge, we propose HDSGNN which leverages the historical nodes' embeddings to break the circular dependency for estimating the sampler effectively and efficiently. Comprehensive experiments confirm the superiority of HDSGNN compared to baselines.

Acknowledgement. This work was supported by the National Natural Science Foundation of China (Grant No. U21B2046, No. 62202448), the Strategic Priority Research Program of the CAS under Grants No. XDB0680302, and the National Key R&D Program of China (No. 2022YFB3103700, No. 2022YFB3103704).

References

1. Chen, J., Zhu, J., Song, L.: Stochastic training of graph convolutional networks with variance reduction. arXiv preprint arXiv:1710.10568 (2017)
2. Chen, J., Ma, T., Xiao, C.: Fastgcn: fast learning with graph convolutional networks via importance sampling. arXiv preprint arXiv:1801.10247 (2018)
3. Chiang, W.L., Liu, X., Si, S., Li, Y., Bengio, S., Hsieh, C.J.: Cluster-gcn: an efficient algorithm for training deep and large graph convolutional networks. In: Proceedings of the 25th ACM SIGKDD International Conference on Knowledge Discovery & Data Mining, pp. 257–266 (2019)
4. Cong, W., Forsati, R., Kandemir, M., Mahdavi, M.: Minimal variance sampling with provable guarantees for fast training of graph neural networks. In: Proceedings of the 26th ACM SIGKDD International Conference on Knowledge Discovery & Data Mining, pp. 1393–1403 (2020)
5. Etikan, I., Bala, K.: Sampling and sampling methods. Biometrics Biostatistics Int. J. **5**(6), 00149 (2017)
6. Fey, M., Lenssen, J.E., Weichert, F., Leskovec, J.: Gnnautoscale: scalable and expressive graph neural networks via historical embeddings. In: International Conference on Machine Learning, pp. 3294–3304. PMLR (2021)
7. Glynn, P.W., Iglehart, D.L.: Importance sampling for stochastic simulations. Manage. Sci. **35**(11), 1367–1392 (1989)
8. Hamilton, W., Ying, Z., Leskovec, J.: Inductive representation learning on large graphs. Advances in neural information processing systems 30 (2017)
9. Hu, W., et al.: Open graph benchmark: Ddatasets for machine learning on graphs. Adv. Neural. Inf. Process. Syst. **33**, 22118–22133 (2020)
10. Huang, W., Zhang, T., Rong, Y., Huang, J.: Adaptive sampling towards fast graph representation learning. Advances in neural information processing systems 31 (2018)
11. Kipf, T.N., Welling, M.: Semi-supervised classification with graph convolutional networks. arXiv preprint arXiv:1609.02907 (2016)
12. Owen, A.B.: Monte Carlo theory, methods and examples (2013)

13. Rockafellar, R.T.: Lagrange multipliers and optimality. SIAM Rev. **35**(2), 183–238 (1993)
14. Tijsma, A.D., Drugan, M.M., Wiering, M.A.: Comparing exploration strategies for q-learning in random stochastic mazes. In: 2016 IEEE Symposium Series on Computational Intelligence (SSCI), pp. 1–8. IEEE (2016)
15. Zeng, H., et al.: Decoupling the depth and scope of graph neural networks. Adv. Neural. Inf. Process. Syst. **34**, 19665–19679 (2021)
16. Zeng, H., Zhou, H., Srivastava, A., Kannan, R., Prasanna, V.: Graphsaint: graph sampling based inductive learning method. arXiv preprint arXiv:1907.04931 (2019)
17. Zou, D., Hu, Z., Wang, Y., Jiang, S., Sun, Y., Gu, Q.: Layer-dependent importance sampling for training deep and large graph convolutional networks. Advances in neural information processing systems **32** (2019)

Subgraph Patterns Enhanced Graph Neural Network for Fraud Detection

Yao Zou[1], Sheng Xiang[2], Qijun Miao[1], Dawei Cheng[1,3(✉)], and Changjun Jiang[1,3]

[1] Department of Computer Science and Technology, Tongji University, Shanghai, China
{sky_zy,dcheng,cjjiang}@tongji.edu.cn
[2] AAII, University of Technology Sydney, Sydney, Australia
sheng.xiang@student.uts.edu.au
[3] Shanghai Artifcial Intelligence Laboratory, Shanghai, China

Abstract. Credit card fraud is a serious problem that causes significant losses for cardholders and issuing banks. Current detection methods often focus on spatial-temporal anomalies in transactions but tend to overlook the disguising techniques of fraudsters. We notice that covert credit card fraud activities frequently manifest localized clustering characteristics, which are particularly evident in different subgraph patterns. To address this, we propose Subgraph Patterns enhanced Graph Neural Network (SPGNN), a novel approach that effectively harnesses specific traits to significantly improve fraud detection capabilities. This method employs subgraph pattern features to more sharply distinguish between fraudulent and legitimate nodes, aiding in the identification of disguised fraudsters. In particular, we devise a probabilistic neighbor selector to assist nodes in selecting more similar minority class nodes, effectively balancing data distribution and filtering out disguised nodes. Furthermore, we introduce a reinforcement learning module for supervised similarity measurement, further filtering out disguised fraudsters. Extensive experiments on several benchmark datasets demonstrate that SPGNN surpasses state-of-the-art models in detecting fraudulent activities, achieving the most advanced performance.

Keywords: Graph Neural Networks · Fraud Detection · Reinforcement Learning · subgraph patterns

1 Introduction

Financial fraud not only damages the financial well-being of individuals and businesses but also has a significant impact on the broader economy, erodes trust within the financial system, and disrupts the legal environment of society. Credit card fraud detection, a pivotal area of research, involves unauthorized fund usage, often via credit or debit cards [1]. Global card fraud losses are projected to reach \$397.40 billion over the next decade [15], highlighting

M. Onizuka et al. (Eds.): DASFAA 2024, LNCS 14855, pp. 375–384, 2024.
https://doi.org/10.1007/978-981-97-5572-1_26

its significant impact on financial markets. Various models, from rules-based to machine learning approaches, have been developed to tackle fraudulent transactions. Deep learning models [4,21], such as graph neural networks (GNNs), have emerged to capture abnormal patterns as they can identify sophisticated and covert transactions by analyzing relational graphs, leading to more accurate fraud detection [23].

However, existing fraudsters engage in various behaviors to disguise themselves and evade detection [3,6,7]. These disguising actions include feature disguise and relationship disguise. 1) Feature disguise involves fraudsters mimicking the transaction characteristics of legitimate entities, such as transaction frequency and preferences. 2) Relationship disguise involves engaging with many legitimate entities to dilute node information and avoiding direct transactions with other fraudsters to evade detection as part of a fraud ring, which is insidious as it prevents the aggregation and transmission of valuable information.

Some recent work has noticed similar challenges. PCGNN [11] reduces redundant edges to tackle disguise with benign entities. Liu et al. [12] identify feature-disguising fraudsters with a score and merge context embeddings. Yu et al. [19] decompose multi-layered tree subgraphs to alleviate information dilution. However, while the aforementioned methods address some fraudulent behavior aspects, they're inadequate when fraudsters intentionally avoid direct transactions. When fraudulent nodes are amidst benign entities, messages aren't effectively transmitted, rendering methods ineffective. This underscores a critical gap in current methodologies, where indirect transactional behaviors are insufficiently considered, potentially overlooking sophisticated fraudulent activities.

We have noticed that covert credit card fraud activities frequently manifest localized clustering characteristics, which are particularly evident in different subgraph patterns. This distinction remains evident even in scenarios where fraudsters avoid direct transactions. Therefore, we propose Subgraph Patterns enhanced Graph Neural Network (SPGNN) to effectively tackle the challenge posed by disguise behaviors. Our method focuses on subgraph patterns, particularly targeting the disguise of avoiding direct transactions. Additionally, we introduce a probabilistic neighbor selector that integrates label distribution and node similarity to address other disguise issues. We also incorporate a reinforcement learning (RL) module into the training phase. This module employs a supervised similarity metric to further filter disguised fraudsters and guide the adaptive exploration of the optimal neighbor threshold. The contributions of this paper are summarized as follows:

1) To the best of our knowledge, this is the first work that leverages the power of subgraph patterns, which have the ability to effectively characterize and discriminate fraudsters to address the challenge of disguised fraudsters in credit card fraud detection.
2) We designed a probabilistic neighbor selector as the first layer of filtering for feature disguise and relationship disguise connected to multiple benign entities. Additionally, we employ an RL module for supervised similarity measurement, further filtering out disguised fraudsters.

3) We conduct extensive experiments to compare our method with state-of-the-art baselines on both public and real-world datasets. The results demonstrate that our model significantly improves GNN performance on graphs containing disguised fraudsters.

2 Preliminaries

2.1 Subgraph Patterns Analysis

Figure 1 visualizes scaled subgraph pattern features, with the left section depicting fraudulent transactions and the right section showing legitimate ones. Detailed subgraph pattern extraction steps will be described later. Subgraph patterns of 2–3 nodes are shown on the right, with the central node represented by X. Neighboring nodes are labeled as 1 for fraudulent and 0 for benign, resulting in features like X-0. Statistical analysis found that most triangle subgraph patterns resulted in zeros for both fraudulent and benign nodes. There are significant differences in the color distribution of the feature heatmaps between fraudulent and benign nodes. In the case of fraudulent transactions, certain subgraph pattern features (such as X-0-1) exhibit more pronounced color blocks in fraud features, indicating that these subgraph pattern feature values are highly significant in fraudulent activities. We opt for subgraph pattern features of 2–3 nodes (excluding triangle subgraph patterns), denoted as $\mathcal{G}_{sub} = \{\mathcal{G}_1, \mathcal{G}_2 \cdots \mathcal{G}_n\}$.

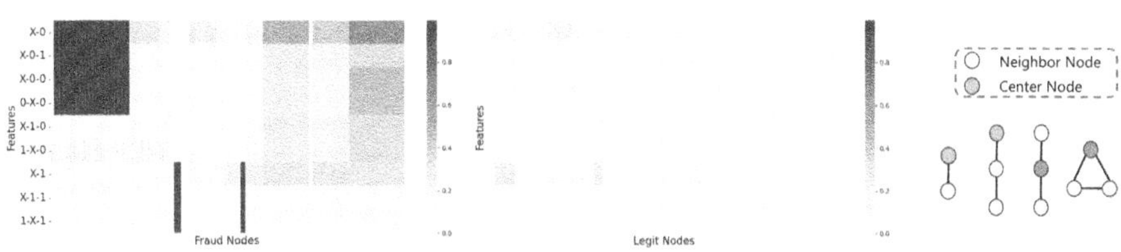

Fig. 1. Heat maps of subgraph pattern features from both fraudulent and legitimate transactions. The y-coordinate represents the corresponding subgraph pattern features. The central node is indicated by X, while neighboring nodes are denoted as 1 for fraudulent neighbors and 0 for benign neighbors.

2.2 Problem Definition

Definition 1. Transactions. A transaction record r can be defined as an attribute tuple $r = \{s, t, \boldsymbol{x}\}$ in the transaction payment process, where s denotes the transaction initiator, t represents the merchant or receiver, and $\boldsymbol{x}$ means the characteristic vector of the transaction.

Definition 2. Graph & Subgraph Pattern. A graph is an ordered pair $\mathcal{G} = (\mathcal{V}, \mathcal{E})$. $\mathcal{V}$ is a set, whose items are called vertices or nodes, and $\mathcal{E}$ is a set of unordered pairs of vertices. A graph $\mathcal{G}^{'} = (\mathcal{V}^{'}, \mathcal{E}^{'})$ is a subgraph pattern of the graph $\mathcal{G} = (\mathcal{V}, \mathcal{E})$ where $\mathcal{V}^{'} \subseteq \mathcal{V}$, $\mathcal{E}^{'} \subseteq \mathcal{E}$, and $(v_1, v_2) \in \mathcal{E}^{'} \rightarrow v_1, v_2 \in \mathcal{V}^{'}$.

Problem Definition. Given the original transaction records $\mathcal{R} = \{\mathcal{S}, \mathcal{T}, \boldsymbol{\mathcal{X}}\}$, we construct a transaction graph $\mathcal{G} = (\mathcal{V}, \boldsymbol{\mathcal{X}}, \mathcal{E}, \mathcal{Y})$. Firstly, we treat each transaction as a discrete node in the graph, $\mathcal{V} = \{v_1, \cdots, v_n\}$, $\mathcal{E} = \{\emptyset\}$. Each node v_i has a d-dimensional feature vector $\boldsymbol{x}_i \in \mathbb{R}^d$ and $\boldsymbol{\mathcal{X}} = \{\boldsymbol{x}_1, \cdots, \boldsymbol{x}_n\}$ represents a set of all node features. When the merchants or transaction initiators of two transactions v_i, v_j are the same, we add an edge between their corresponding nodes, $\mathcal{E} = \mathcal{E} \cup \{e_{i,j}\}$. $\mathcal{Y} = \{y_1, \cdots, y_n\}$, $y_i \in \{0, 1\}$ denotes the label of node v_i. If a transaction is reported by a cardholder or identified by financial experts as fraudulent, we label it as 1; otherwise, it is labeled as 0. For the set of subgraph patterns $\mathcal{G}_{sub} = \{\mathcal{G}_1, \mathcal{G}_2 \cdots \mathcal{G}_n\}$, subgraph pattern feature extraction is performed to obtain the subgraph pattern features which will be combined with the $\boldsymbol{\mathcal{X}}$ to form a new feature $\boldsymbol{\mathcal{X}}'$. We hope to infer the possibility of fraudulent transactions based $\mathcal{G}' = (\mathcal{V}, \boldsymbol{\mathcal{X}}', \mathcal{E}, \mathcal{Y})$.

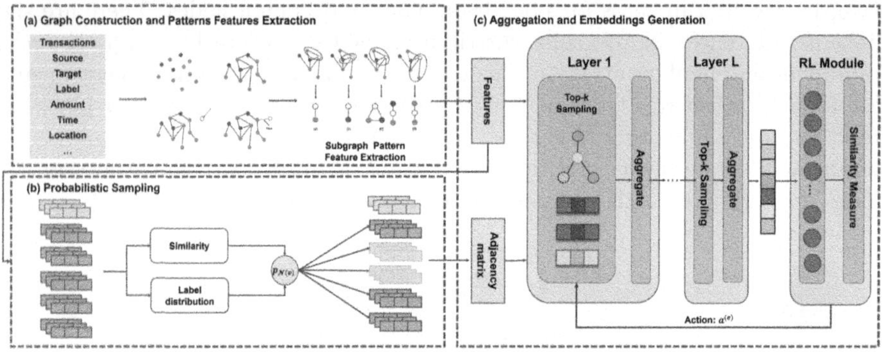

Fig. 2. The illustration of the proposed SPGNN.

3 The Proposed Approaches

3.1 Model Architecture

Figure 2 shows the overall architecture of SPGNN. We first generate complex transaction graphs based on original transaction records. Subsequently, subgraph pattern feature extraction is performed to obtain the subgraph pattern features, which are then combined with the original node attribute features. We implement a probabilistic neighbor selector that takes into account label distribution and node similarity to improve neighbor selection. This approach focuses on minority class nodes with high similarity, thereby achieving a balanced data distribution and effective identification of concealed nodes. Subsequently, these refined node features are fed into the GNN. Through the integration of multiple neighbor selectors and aggregation layers, the GNN effectively shares and compiles information across nodes. Simultaneously, the training process of the GNN serves as the environment for the RL module, in which the RL agent, incorporating supervised similarity measurement, further filters out disguised fraudsters.

3.2 Probabilistic Neighbor Selector and RL Module

We devise a probabilistic neighbor selector to pick more informative nodes to filter disguised fraudsters. The key idea lies in incorporating label distribution information and node similarity into the sampling process. For the neighbor selector, those minority class nodes with high similarity have a higher sampling probability. The sampling probability of neighbor node u can be defined:

$$p_u = \frac{S(v,u)}{\sum_{v' \in \mathcal{N}(v)} S(v,v') \cdot DF(v, C(u))}, \tag{1}$$

where v is the central node and $\mathcal{N}(v)$ represents the neighbor set of node v. $DF(v,(C(u)))$ denotes the label frequency of class $C(u)$ in $\mathcal{N}(v)$. $S(v,u)$ is the similarity between neighbor node u and the center node v (We set it as cosine similarity). To refine fraud detection, we integrate an RL module in training. It learns an adaptive parameter p, acting as a threshold, ensuring nodes aggregate only with similar neighboring nodes. Inspired by LAGCN [2], we incorporate a Single-layer Perceptron (SLP) as a new layer before the aggregation layer of GNN. This SLP predicts node labels based on their embeddings. For a center node v at the l-th layer and edge $(v,v') \in \mathcal{E}$, the distance between v and v' is defined as:

$$\mathcal{Dist}^{(l)}(v,v') = \left| \sigma\left(\mathrm{SLP}^{(l)}(\mathbf{h}_v^{(l-1)})\right) - \sigma\left(\mathrm{SLP}^{(l)}(\mathbf{h}_{v'}^{(l-1)})\right) \right|, \tag{2}$$

where $\mathbf{h}_v^{(l)}$ is the hidden embedding at the l-th layer of node v, $\mathbf{h}_v^{(0)} = \boldsymbol{x}_v$ is the input feature, and σ is a nonlinear activation function (we use sigmoid). The similarity measure is defined as:

$$\mathcal{Simi}^{(l)}(v,v') = 1 - \mathcal{Dist}^{(l)}(v,v'). \tag{3}$$

To optimize computational efficiency, we simplify the input by using only the node embedding, rather than combined embeddings as in LAGCN [2]. This reduces the time complexity of the proposed similarity measure from $O(|\mathcal{V}|\bar{D}d)$ to $O(|\mathcal{V}|d)$. We define the cross-entropy loss of the SLP at the l-th layer as:

$$\mathcal{L}_{\mathrm{Simi}}^{(l)} = \sum_{v \in \mathcal{V}} -\log\left(y_v \cdot \sigma\left(\mathrm{SLP}^{(l)}\left(\mathbf{h}_v^{(l)}\right)\right)\right), \tag{4}$$

where σ is the sigmoid activation function. We define the RL process as a Bernoulli Multi-armed Bandit (BMAB) $\mathcal{B}(A,R,T)$ between the neighbor selector and the GNN with the similarity measure. Here, A is the action space, R is the reward function, and T is the terminal condition.

Reward function. Optimal p minimizes distance (or maximizes similarity) between the central node and its neighbors in the first layer. Binary stochastic reward is based on average distance differences between consecutive epochs. Average neighbor distances for epoch e are calculated as:

$$O(\mathcal{Dist})^{(e)} = \frac{\sum_{v \in \mathcal{V}_{\mathrm{train}}} \mathcal{Dist}^{(1)}(v,v')^{(e)}}{|\mathcal{V}_{\mathrm{train}}|}. \tag{5}$$

denoted as $O^{(e)}$ for simplicity. The reward function for epoch e is defined as:

$$a^{(e)} = \begin{cases} +1, O^{(e-1)} - O^{(e)} \geq 0, \\ -1, O^{(e-1)} - O^{(e)} < 0. \end{cases} \tag{6}$$

Action: p adjusts incrementally based on the reward (Eq. (7)), aiming to reduce neighboring nodes when distance increases.

$$p = p + a^{(e)} \cdot \tau \tag{7}$$

Terminal: Eq. (8) defines a terminal condition, signifying RL module convergence in recent epochs, identifying an optimal threshold p.

$$\left| \sum_{e-10}^{e} a^{(e)} \right| \leq 1, \text{ where } e \geq 10. \tag{8}$$

3.3 Aggregation and Embedding Generation

We choose the mean operator as the aggregator of GNN, represented as:

$$\mathbf{h}_v^{(l)} = \sigma \left(W^{(l)} \cdot \text{MEAN} \left(\left\{ \mathbf{h}_v^{(l-1)} \right\} \cup \left\{ \mathbf{h}_u^{(l-1)}, \forall u \in \mathcal{N}(v) \right\} \right) \right), \tag{9}$$

where the $W^{(l)}$ is the l-th parameter matrix. Following the aggregation step, an MLP classifier is trained together with GNNs to minimize the cross-entropy loss.

$$p_v = \text{MLP}(\mathbf{h}_v^{(L)}), \tag{10}$$

where L is the number of layers. We define the graph-based loss function:

$$\mathcal{L}_{\text{gnn}} = - \sum_{v \in \mathcal{V}} \left[y_v \log p_v + (1 - y_v) \log (1 - p_v) \right] \tag{11}$$

Combined with the similarity measure loss, the loss of SPGNN is defined as:

$$\mathcal{L}_{\text{SPGNN}} = \lambda \cdot \mathcal{L}_{\text{Simi}}^{(1)} + \mathcal{L}_{\text{gnn}}, \tag{12}$$

where λ is a weighting parameter.

4 Experiments

4.1 Experiments Settings

Datasets. We collected real-world credit card transaction data from a major commercial bank spanning January 1 to December 31, 2021. The dataset, referred to as **CCDS** (**C**redit **C**ard Fraud **D**etection **D**ata**S**et), consists of 140,576 transactions involving 20,313 unique users. Transaction features include location, amount, and type. Additionally, we utilized two public datasets for experiments: the YelpChi graph dataset containing hotel and restaurant reviews and the Amazon graph dataset comprising product reviews of musical instruments. Basic statistics for the datasets are summarized in Table 2.

Compared Methods and Settings. We compare SPGNN with the baselines: GCN [8], GEM [10], GAT [16], FdGars [18], Play2Vec [20], GeniePath [13], Graphconsis [12], SEMIGNN [17], CAREGNN [5], PCGNN [11]. Sub-models of SPGNN are SPGNN-*nosample/nosubgragh/noRL*, in which probabilistic neighbor selector, subgraph pattern features, or RL module are not used. SPGNN is the full proposed model. Baseline methods utilize their originally proposed optimal parameters. SPGNN employs a two-layer aggregation scheme with an embedding size of 64. We use a learning rate of 0.01 and an RL step size of 0.01. Evaluation includes AUC and F1 metrics on three datasets.

Table 1. Fraud detection performance on three datasets.

Model	YelpChi		Amazon		CCDS	
	AUC	F1	AUC	F1	AUC	F1
GCN	0.5310	0.4614	0.5305	0.4413	0.5266	0.4175
GEM	0.5201	0.5017	0.5292	0.4983	0.5388	0.6599
GAT	0.5322	0.4634	0.5357	0.4822	0.5333	0.6624
FdGarS	0.5133	0.4304	0.6601	0.3757	0.5475	0.3689
Play2Vec	0.5231	0.4606	0.5205	0.4587	0.5380	0.5022
GeniePath	0.6761	0.5915	0.7832	0.7952	0.6150	0.6234
Graphconsis	0.7060	0.6041	0.8225	0.7766	0.5893	0.6402
SEMIGNN	0.5201	0.1045	0.8782	0.7819	0.5473	0.4485
CAREGNN	0.7934	0.6493	0.9115	0.8531	0.6534	0.5771
PCGNN	0.7987	0.6300	0.9405	0.8865	0.6795	0.6077
SPGNN-Nosample	0.7793	0.5796	0.9362	0.8929	0.7209	0.6222
SPGNN-Nosubgraph	0.7579	0.5897	0.9217	0.8971	0.6922	0.6467
SPGNN-NoRL	0.7825	0.5165	0.9513	0.8923	0.7189	0.6284
SPGNN-all	**0.8013**	**0.6517**	**0.9519**	**0.9159**	**0.7352**	**0.6654**

4.2 Fraud Detection Performance

We repeated the experiments ten times for each method and have shown the average performance in Table 1. The first five rows of Table 1 report the results of some classic graph-based methods, including GCN, GEM, GAT, FdGars, and Play2Vec. It is clear that the results of GCN and GEM are not satisfactory, demonstrating the limitations of shallow models in addressing complex fraud patterns. GAT introduces the attention mechanism to improve the indiscriminate aggregation of neighbor information compared with the first two methods, but the improvement is not significant. FdGars and Play2Vec improve performance, partially due to their enlarged model capacities. GeniePath and Graphconsis perform closely to each other and better than the first five methods. The results demonstrate the effectiveness of aggregating neighborhoods based on node similarity in detecting fraud transactions. SEMIGNN performs well in the Amazon dataset, while it performs poorly on the Yelp and CCDS datasets as it is an unsupervised method and data distribution can seriously affect its model performance. CAREGNN introduces RL module, resulting in improved performance. PCGNN achieved more competitive results by sampling nodes and edges.

Lines 9-12 show the results of SPGNN and its sub-models. It should be noted that SPGNN-*nosubgraph* performs considerably lower than the other two sub-models, which proves the superior performance of the subgraph pattern features. In general, SPGNN performed the best across all metrics.

Table 2. Node similarity statistics.

Dataset	YelpChi	Amazon	CCDS
Node	45,954	11,948	140,576
Edge (M)	3.85	4.40	19.96
Fraud (%)	14.5	9.5	7.8
Old	0.77	0.65	0.69
New	0.55	0.46	0.57

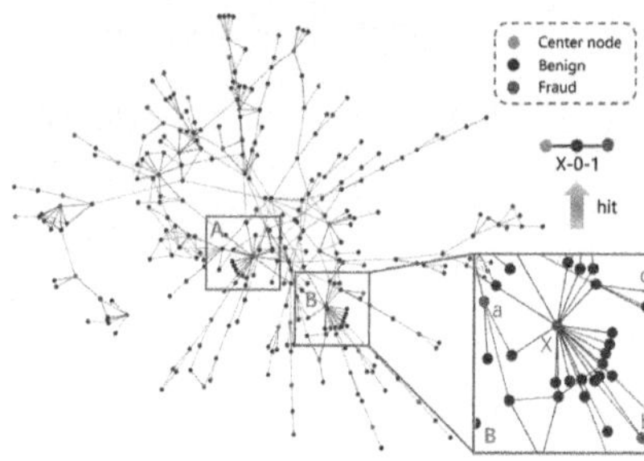

Fig. 3. The layout of a typical graph in Amazon.

4.3 Interpretative Analysis

We calculate the feature similarity of neighboring nodes based on the Euclidean distance of their feature vectors, ranging from 0 to 1. The average similarity is normalized with respect to the total number of edges, as shown in Table 2. Initially, the feature similarity was unusually high. However, after incorporating subgraph pattern features and re-measuring, we observed a decrease in similarity of at least 13.2%, indicating significant differences between fraudulent and legitimate nodes' subgraph pattern features, thus enhancing fraud detection accuracy. Figure 3 illustrates a typical case in the Amazon dataset, involving 296 users. Fraudulent nodes are marked in red, while normal nodes are in blue. Box A showcases the first type of relationship disguise: engaging with many legitimate entities. Box B illustrates the second type: avoiding direct transactions. Zooming in on box B, it shows a central node labeled X, surrounded by benign entities and not directly connected to other fraudulent nodes. However, this central node exists in a special subgraph pattern X-0-1 with nodes a, b, and c. This further demonstrates how subgraph pattern features aid in identifying disguised fraudsters even without direct connections.

5 Related Work

Credit card fraud detection has been a significant application area for machine learning techniques. Traditional methods often rely on time perception [22] to capture fraudulent transaction patterns. However, these approaches may overlook relevant characteristics or focus solely on temporal features. Recently, GNNs have shown promise in fraud detection [14], yet they struggle to effectively identify disguised fraudsters. To address this challenge, existing sampling methods are introduced [9]. Despite their advancements, these methods often suffer from high computational complexity and may struggle with complex disguise scenarios. In contrast, SPGNN combines a simpler similarity measure with lower computational complexity and leverages subgraph pattern features to effectively identify disguised fraudsters, enhancing the capabilities of GNNs.

6 Conclusion

This paper addresses credit card fraud detection, a critical real-world challenge. Recognizing the significant impact of fraudsters' disguise tactics on GNN-based detectors, we propose a probabilistic neighbor selector for initial disguise filtering. Furthermore, we integrate reinforcement learning to adjust neighbor selection based on node similarity, enhancing detection accuracy. Leveraging subgraph pattern features, we unveil intricate transaction dependencies, capturing complex interactions in the graph and offering insights into fraudulent activities. SPGNN's capability to identify and utilize subgraph patterns is pivotal for precise fraud detection, minimizing false positives. Extensive experiments across various fraud datasets validate SPGNN's superiority over baseline methods, underscoring its efficacy and practical applicability in real-world scenarios.

Acknowledgments. This work was supported by the National Key R&D Program of China (Grant no. 2022YFB4501704), the National Natural Science Foundation of China (Grant no. 62102287), and the Shanghai Science and Technology Innovation Action Plan Project (Grant no. 22YS1400600 and 22511100700).

References

1. Bhattacharyya, S., Jha, S., Tharakunnel, K., Westland, J.C.: Data mining for credit card fraud: a comparative study. Decis. Support Syst. **50**(3), 602–613 (2011)
2. Chen, H., Wang, L., Wang, S., Luo, D., Huang, W., Li, Z.: Label aware graph convolutional network - not all edges deserve your attention. CoRR abs/1907.04707 (2019)
3. Cheng, D., Chen, C., Wang, X., Xiang, S.: Efficient top-k vulnerable nodes detection in uncertain graphs. IEEE Trans. Knowl. Data Eng. **35**(2), 1460–1472 (2021)
4. Cheng, D., Wang, X., Zhang, Y., Zhang, L.: Risk guarantee prediction in networked-loans. In: IJCAI (2020)
5. Dou, Y., Liu, Z., Sun, L., Deng, Y., Peng, H., Yu, P.S.: Enhancing graph neural network-based fraud detectors against camouflaged fraudsters. In: CIKM, pp. 315–324 (2020)

6. Dou, Y., Ma, G., Yu, P.S., Xie, S.: Robust spammer detection by nash reinforcement learning. In: SIGKDD, pp. 924–933. Association for Computing Machinery, New York (2020)
7. Kaghazgaran, P., Caverlee, J., Squicciarini, A.: Combating crowdsourced review manipulators: a neighborhood-based approach. In: WSDM, pp. 306–314 (2018)
8. Kipf, T.N., Welling, M.: Semi-supervised classification with graph convolutional networks. In: ICLR (2017)
9. Li, F., Zhang, T., Cui, S., Liu, H., Ren, Z., Di, D., Wang, X., Zhang, P., Yu, G.: A sampling method based on forecasting and combinatorial optimization for high performance a/b testing. Front. Comp. Sci. **17**(6), 176616 (2023)
10. Lin, W., Lan, H., Li, B.: Generative causal explanations for graph neural networks. In: ICML, pp. 6666–6679 (2021)
11. Liu, Y., Ao, X., Qin, Z., Chi, J., Feng, J., Yang, H., He, Q.: Pick and choose: a gnn-based imbalanced learning approach for fraud detection. In: WWW, pp. 3168–3177 (2021)
12. Liu, Z., Dou, Y., Yu, P.S., Deng, Y., Peng, H.: Alleviating the inconsistency problem of applying graph neural network to fraud detection. In: SIGIR, pp. 1569–1572 (2020)
13. Liu, Z., Chen, C., Li, L., Zhou, J., Li, X., Song, L., Qi, Y.: Geniepath: graph neural networks with adaptive receptive paths. In: AAAI, vol. 33, pp. 4424–4431 (2019)
14. Ma, J., et al.: Fighting against organized fraudsters using risk diffusion-based parallel graph neural network. In: IJCAI, pp. 6138–6146 (2023)
15. Nilson Report: Card fraud losses (2022). https://nilsonreport.com/research/research-14th-edition/. Accessed 20 Dec 2023
16. Velickovic, P., Cucurull, G., Casanova, A., Romero, A., Lio, P., Bengio, Y.: Graph attention networks. In: ICLR (2018)
17. Wang, D., e al.: A semi-supervised graph attentive network for financial fraud detection. In: ICDM, pp. 598–607. IEEE (2019)
18. Wang, J., Wen, R., Wu, C., Huang, Y., Xiong, J.: Fdgars: fraudster detection via graph convolutional networks in online app review system. In: WWW, pp. 310–316 (2019)
19. Wang, Y., Derr, T.: Tree decomposed graph neural network. In: CIKM, pp. 2040–2049 (2021)
20. Wang, Z., Long, C., Cong, G., Ju, C.: Effective and efficient sports play retrieval with deep representation learning. In: SIGKDD, pp. 499–509 (2019)
21. Xiang, S., et al.: Semi-supervised credit card fraud detection via attribute-driven graph representation. In: AAAI, vol. 37, pp. 14557–14565 (2023)
22. Xie, Y., Liu, G., Yan, C., Jiang, C., Zhou, M., Li, M.: Learning transactional behavioral representations for credit card fraud detection. TNNLS (2022)
23. Zhang, R., et al.: Pre-trained online contrastive learning for insurance fraud detection. In: AAAI, pp. 22511–22519 (2024)

Advancing Latent Representation Ranking for Masked Graph Autoencoder

Yulan Hu[1,3], Ge Chen[2], Sheng Ouyang[1], Zhirui Yang[1], Junchen Wan[3], Fuzheng Zhang[3], Zhongyuan Wang[3], Zhao Cao[4], Shangquan Wu[4], and Yong Liu[1(✉)]

[1] Gaoling School of Artificial Intelligence, Renmin University of China, Beijing, China
{huyulan,ouyangsheng,yangzhirui,liuyonggsai}@ruc.edu.cn

[2] University of Chinese Academy of Sciences, Beijing, China
chenge221@mails.ucas.ac.cn

[3] Kuaishou Technology, Beijing, China
{wanjunchen,zhangfuzheng,wangzhongyuan}@kuaishou.com

[4] Huawei Technology, Shenzhen, China
{caozhao1,wushangquan}@huawei.com

Abstract. Generative graph self-supervised learning (SSL), represented by graph autoencoders (GAEs), has increasingly garnered interest and demonstrated its potential in graph representation. However, existing generative SSL methods primarily concentrate on refining model designs, such as masking strategies and reconstruction objectives, while the latent representation generated by the encoder is scarcely explored beyond decoding by the decoder. Motivated by this observation, we identify and analyze the issues faced by previous works in utilizing latent representation and introduce GMAERank, a masked autoencoder with ranking learning. GMAERank presents a novel ranking strategy that leverages the latent representation. Specifically, GMAERank posits that the similarity between two augmented latent views should consistently outrank that between any augmented view and an unrelated random view-a condition typically achievable. Subsequently, GMAERank focus on the optimization of more difficult samples with the ranking strategy. In contrast to earlier research that involves auxiliary models for optimization, GMAERank is simple and lightweight. We conduct comprehensive experiments across node, edge, and graph-level tasks, and the results reveal that GMAERank achieves competitive performance and retains versatility when compared to state-of-the-art (SOTA) baselines.

Keywords: graph neural networks · ranking learning · autoencoders

1 Introduction

Graph-structured data, such as biological, social, or academic networks, are prevalent in the real world. Recently, self-supervised learning (SSL) has emerged

M. Onizuka et al. (Eds.): DASFAA 2024, LNCS 14855, pp. 385–394, 2024.
https://doi.org/10.1007/978-981-97-5572-1_27

as an effective paradigm for graph representation learning, attracting significant interest [6,28]. Graph contrastive learning (GCL) is a prominent SSL approach that has attracted extensive research by constructing contrasting views to mine valuable information embedded. Concurrently, another category of graph SSL, generative SSL, has been progressively capturing researchers' attention.

Generative SSL aims to reconstruct the input graph. The graph autoencoder (GAE) is a representative structure of generative SSL. A GAE typically comprises an encoder to map the input graph into a latent representation and a decoder to reconstruct the graph from this latent representation [13]. While GAEs have achieved initial success, there has been limited work on enhancing graph representations through latent representation strategies. Recent studies have employed a variational GAE variant to generate samples for various tasks, thereby strengthening model capabilities [8,9]. Nonetheless, devising strategies based on latent representation remains challenging. ARGA [17] attempts to distinguish between latent representation and a Gaussian prior, while the inclusion of an extra discriminator results in additional complexity. Moreover, while HGC-VAE [9] enhances latent representation with contrastive learning, this strategy requires a large quantity of high-quality negative samples.

Motivation. Current GAEs chiefly focus on preserving graph attributes, such as node features and graph structures, while under-utilizing or neglecting latent codes. As observed in [17], unregulated latent codes could result in weak representations when handling noisy graph data. While some studies have recognized this problem and proposed certain solutions, as discussed above, these may still introduce drawbacks such as enhanced model complexity and the requirement for high-quality negative samples.

Contributions. This study aims at optimizing the use of latent representation in GAEs to improve overall training quality. To achieve this, we introduce GMAERank, a masked GAE that incorporates a novel ranking strategy for advanced graph representation learning. Extensive experiments validate the efficiency of GMAERank at different levels. In summary, our contributions include:

- We analyze existing GAEs and find that the latent representation is under-utilized, suggesting that prior methods such as ARGA are complex as they require an additional discriminator.
- We propose GMAERank, an innovative, simple and lightweight strategy that utilizes the latent representation of GAE for ranking.
- We conduct 21 experiments at different task levels, which show that GMAERank consistently outperforms SOTA baselines, highlighting its effectiveness and adaptability.

2 Methodology

In this section, we formally introduce GMAERank, which integrates a masked GAE with a ranking module. Figure 1 provides a overall overview of GMAERank. We will introduce the preliminary work for this study and discuss the encoding processes, the decoding processes and the ranking strategy. Finally, we will present the optimization objectives.

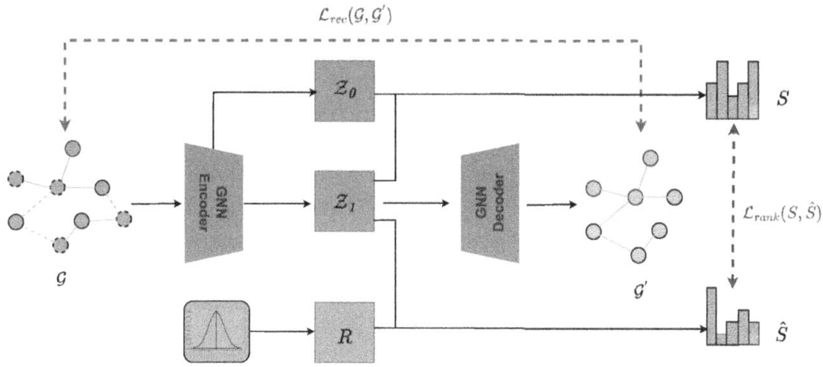

Fig. 1. The overview of GMAERank.

2.1 Preliminaries

Notations. Given a graph $\mathcal{G} = (\mathcal{V}, \mathcal{E}, \mathcal{X}, \mathcal{A})$, where $\mathcal{V}$ is the set of N nodes and $\mathcal{E} \subseteq \mathcal{V} \times \mathcal{V}$ is the set of edges. $\mathcal{X} \in \mathbb{R}^{N \times F_s}$ is the input node feature, F_s denotes the feature dimension. Each node $v \in \mathcal{V}$ is associated with a feature vector $x_v \in \mathcal{X}$, and each edge $e_{u,v} \in \mathcal{E}$ denotes a connection between node u and node v. The graph structure can also be represented by an adjacency matrix $\mathcal{A} \in \{0,1\}^{N \times N}$ with $A_{u,v} = 1$ if $e_{u,v} \in \mathcal{E}$, and $A_{u,v} = 0$ if $e_{u,v} \notin \mathcal{E}$. For masked GAE, it usually consists of an encoder f_E and a decoder f_D.

2.2 Masked Graph Autoencoder

The masked GAE aims to reconstruct the original graph from the corrupted input. While the reconstruction goal can be the node features, the adjacency matrix, or both, our focus is on feature reconstruction. A standard masked GAE involves masking, encoding and decoding, which will be discussed in the following sections.

Masking. Consider a graph $\mathcal{G} = (\mathcal{V}, \mathcal{E})$ with vertices $v_i \in \mathcal{V}$ and edges $\mathcal{E} = \{e_{i,j} \mid i,j = 1, \ldots, N\}$. In accordance with the MAE [4], we corrupt the input graph by masking only the node features. We use λ to represent the mask ratio, $\lambda \times N$ nodes are masked. Then we sample a random mask $m \in \{0,1\}^N$ according to the Bernoulli distribution, a masked node is replaced by a mask token $\boldsymbol{e}_{[\mathrm{M}]}$. We denote the masked nodes as $\tilde{\mathcal{V}}$ and generate the corrupted node feature as $\tilde{\boldsymbol{x}} \triangleq \left\{\tilde{\boldsymbol{x}}_i \mid (1 - m_i)\, \boldsymbol{x}_i + m_i \boldsymbol{e}_{[\mathrm{M}]}\right\}_{i=1}^{N}$. The encoder accepts $\mathcal{A}$ and $\tilde{\mathcal{X}}$ as input.

Encoder. The encoder is utilized to convert sparse input features, commonly characterized by onehot, into low-dimensional representation, and can be any kind of GNNs with multiple layers, e.g., GCN [12], GAT [22]. In this paper, we adopt a variational GAE f_E to encode the corrupted graph. In particular, for the ranking learning process described in Sect. 2.3, we perform the encoding process twice and obtain $\mathcal{H}_1$ and $\mathcal{H}_2$ as

$$\mathcal{H}_1 = f_E(\tilde{\mathcal{X}}, \mathcal{A}), \mathcal{H}_2 = f_E(\tilde{\mathcal{X}}, \mathcal{A}). \tag{1}$$

Due to the various stochastic modules in the encoder, e.g., the dropout operations, $\mathcal{H}_1$ and $\mathcal{H}_2$ are similar but not identical.

We employ variational inference in $\mathcal{H}_1$, $\mathcal{H}_2$ to generate the latent representations referred to as $\mathcal{Z}_1$ and $\mathcal{Z}_2$. Generating $\mathcal{Z}_1$ and $\mathcal{Z}_2$ requires learning the posterior distribution $p_\theta(\mathcal{Z}|\mathcal{H},\mathcal{A})$, given the encoded representation and the adjacency matrix. Two additional encoders are utilized to learn the mean (μ) and variance (σ) of $\mathcal{H}$. VGAE [11] pointed out that p_θ is intractable, we use a new approximation function q_ϕ, parameterized by ϕ, to estimate the latent representation:

$$q_\phi(\mathcal{Z} \mid \mathcal{H},\mathcal{A}) = \prod_{i=1}^{N} q\left(\mathbf{z}_i \mid \mathcal{H},\mathcal{A}\right) = \prod_{i=1}^{N} \mathcal{N}\left(\mathbf{z}_i \mid \boldsymbol{\mu}_i,\boldsymbol{\sigma}_i\right). \tag{2}$$

Once μ and σ are obtained, we can generate $\mathcal{Z}_1$ and $\mathcal{Z}_2$ using the reparametrization technique [11] as follows:

$$\mathcal{Z}_1 = \boldsymbol{\mu} + \boldsymbol{\sigma} \odot \rho_1, \mathcal{Z}_2 = \boldsymbol{\mu} + \boldsymbol{\sigma} \odot \rho_2; \rho_1 \sim \mathcal{N}(0,1), \rho_2 \sim \mathcal{N}(0,1), \tag{3}$$

where $\odot$ denotes element-wise multiplication, while ρ_1 and ρ_2 are random samples from the standard normal distribution. The decoder reconstructs the input graph using $\mathcal{Z}_1$ (or $\mathcal{Z}_2$), while the ranking learning acts on $\mathcal{Z}_1$ and $\mathcal{Z}_2$ to improve the overall performance of the model.

Decoder. The decoder is utilized to reconstruct the features of the input graph, aligning with the encoder's emphasis on node features rather than the graph structure. Specifically, we denote f_D as the decoder, which may consist of a single or multiple GNN layers. f_D accepts the latent representation $\mathcal{Z}$ and adjacency matrix $\mathcal{A}$ as input, resulting in the reconstructed node feature $\mathcal{X}_{rec}$:

$$\mathcal{X}_{rec} = f_D(\mathcal{Z},\mathcal{A}). \tag{4}$$

After obtaining the reconstructed node feature $\mathcal{X}_{rec}$, we adopt the scaled cosine error (SCE) [6] as a criterion to measure the discrepancy between $\mathcal{X}$ and $\mathcal{X}_{rec}$, result in the reconstruction loss $\mathcal{L}_{rec}$.

2.3 Rank With Latent Representation

GMAERank essentially utilizes latent representations, $\mathcal{Z}_1$ and $\mathcal{Z}_2$, to rank against random noise through marginal rank loss. In contrast to the extra optimization of adversarial module, which was introduced by ARGA [17], GMAERank simplifies the optimization process, eliminating the need for an auxiliary model.

Let v represent the i-th individual node, with $\mathcal{Z}_1^v$ and $\mathcal{Z}_2^v$ corresponding to the representation of v in $\mathcal{Z}_1$ and $\mathcal{Z}_2$ respectively. The reparameterization technique distinctly samples the noise twice from the Gaussian prior, consequently yielding similarities (although not identical) between $\mathcal{Z}_1^v$ and $\mathcal{Z}_2^v$. We calculate similarity S_v between $\mathcal{Z}_1^v$ and $\mathcal{Z}_2^v$ using an arbitrary apparent similarity function:

$$S_v = similarity(\mathcal{Z}_1^v,\mathcal{Z}_2^v). \tag{5}$$

Next, we randomly sample noise R with the same shape as the latent representation from a normal distribution. We then calculate the similarity of the corresponding data point in R with $\mathcal{Z}_1^v$ (or $\mathcal{Z}_2^v$) as:

$$\hat{S}_v = similarity(R_i, \mathcal{Z}_1^v). \tag{6}$$

Intuitively, $\mathcal{Z}_1^v$ and $\mathcal{Z}_2^v$ represent the embedding of a node in two views, thus the similarity score S_v should consistently rank higher than $\hat{S}_v$. For simplicity, we denote the pairwise similarity between $\mathcal{Z}_1$ and $\mathcal{Z}_2$ as S, and the pairwise similarity between $\mathcal{Z}_2$ and R as $\hat{S}$. For each node in $\mathcal{V}$, we enumerate similarity scores in $\hat{S}$ and supervise them with S as follows:

$$\mathcal{L}_{rank} = \sum_{v \in \mathcal{V}} max(0, -(S_v - \hat{S}_v) + \delta), \tag{7}$$

where δ denotes the margin hyper-parameter, which is typically a small value.

2.4 Optimization Objective

For GMAERank, the overall optimization objective encompasses both the reconstruction loss and margin ranking loss 7. Due to the variational property of GMAERank, we incorporate an additional variational lower bound to reconcile the posterior distribution with the estimated distribution:

$$\mathcal{L}_{kl} = \mathbb{E}_{z \sim q_\phi(\mathcal{Z}|\mathcal{H},\mathcal{A})} \log p(\mathcal{H} \mid \mathcal{Z}) - \mathbf{KL}(q_\phi(\mathcal{Z} \mid \mathcal{H}, \mathcal{A}) \| p_\theta(\mathcal{Z})), \tag{8}$$

where $\mathbf{KL}[q(\bullet)||p(\bullet)]$ signifies the Kullback-Leibler divergence between $q(\bullet)$ and $p(\bullet)$. In this instance, we use a Gaussian prior $p_\theta(\mathcal{Z}) = \prod_i^N \mathcal{N}(z_i|0, \mathbf{I})$. The overall optimization objective for GMAERank can be formulated as:

$$\mathcal{L} = \mathcal{L}_{rec} + \alpha \cdot \mathcal{L}_{rank} + \beta \cdot \mathcal{L}_{kl}, \tag{9}$$

where both α and β are mediation parameters.

3 Experiment

We validate the effectiveness and generality of GMAERank through experiments at the node, edge, and graph levels. We mainly focus on exploring the answers to the following four questions: **(RQ1)** How does GMAERank perform on node-level tasks? **(RQ2)** How does GMAERank perform on edge-level tasks? **(RQ3)** Is GMAERank capable of generalizing effectively to graph-level tasks? **(RQ4)** What impact do the core components of GMAERank, such as the ranking strategy and feature masking, have on model performance?

Our experimental approach strictly adheres to established procedures, including data splitting and evaluation protocols according to the standard settings referenced in works such as [6,14]. To ensure the reliability of the results, all experiments are repeated five times and the results are reported together with their standard deviations. The best result is marked in bold and the second best result is underlined.

Table 1. Summary of the results (±std) on node classification.

Dataset	Cora	Citeseer	Pubmed	A-Photo	A-Computer	C-CS	C-Physics
GCN [12]	80.84±0.44	70.46±0.85	79.02±0.26	90.79±2.47	85.34±1.64	92.10±0.16	95.41±0.11
GAT [22]	83.00±0.70	72.50±0.70	79.00±0.30	93.49±0.16	89.79±0.76	91.48±0.17	95.25±0.14
GRACE [28]	77.60±0.28	67.24±0.93	78.40±1.33	92.45±0.34	88.11±0.45	67.57±12.91	85.33±6.76
DGI [23]	82.30±0.60	71.80±0.70	76.80±0.60	91.61±0.22	83.95±0.47	92.15±0.63	94.51±0.52
CCA-SSG [27]	84.00±0.40	73.10±0.30	81.00±0.40	93.14±0.14	88.74±0.28	93.31±0.22	95.38±0.06
MVGRL [3]	83.50±0.40	73.30±0.50	80.10±0.70	91.74±0.07	87.52±0.11	92.11±0.12	95.33±0.03
BGRL [21]	82.70±0.60	71.10±0.80	79.60±0.50	93.17±0.30	90.34±0.19	93.31±0.13	95.73±0.05
InfoGCL [24]	83.50±0.30	73.50±0.40	79.10±0.20	-	-	-	-
GAE [13]	71.50±0.40	65.80±0.40	72.10±0.50	-	-	-	-
GPT-GNN [10]	80.10±1.0	68.40±1.60	76.30±0.80	-	-	-	-
GraphMAE [6]	84.20±0.40	73.40±0.40	81.10±0.40	92.98±0.35	88.34±0.27	93.08±0.17	95.30±0.12
SeeGera [14]	**84.30±0.40**	73.00±0.80	80.40±0.40	92.81±0.45	88.39±0.26	93.84±0.11	95.39±0.08
GMAERank	83.55±0.25	**74.10±0.20**	**81.80±0.10**	**94.37±0.13**	**90.45±0.02**	**94.50±0.27**	**96.27±0.00**

The results not reported are due to unavailable code or out-of-memory.

3.1 Node Classification (RQ1)

Setup. The node classification experiments are conducted on Cora, CiteSeer, PubMed, Coauthor CS, Coauthor Physics, Amazon Computer, and Amazon Photo. We use accuracy as the evaluation metric.
Results. GMAERank is compared against ten SOTA SSL methods. The detailed results are shown in Table 1. GMAERank consistently outperforms all twelve methods or achieves comparable results, demonstrating its exceptional efficacy in node-level tasks as a generative model.

3.2 Link Prediction (RQ2)

Setup. For link prediction, we evaluate on datasets identical to those for node classification. Following the protocol from [14], we report AUC and AP to assess model performance.
Results. We compare GMAERank with nine SOTA SSL methods, and summarize the results in Table 3. GMAERank performs well across all methods except SeeGera [14], a method that excels in reconstructing adjacency matrices and is thus highly relevant to link prediction. However, it is worth noting that SeeGera's performance in node classification is comparatively less impressive. In contrast, GMAERank shows strong performance on both node and edge tasks.

3.3 Graph Classification (RQ3)

Setup. For graph classification, we use seven benchmark datasets (MUTAG, IMDB-B, IMDB-M, PROTEINS, COLLAB, REDDIT-B, and NCI1). We use accuracy as the evaluation metric.
Results. GMAERank is compared with ten SOTA methods. The results show that GMAERank is very competitive in all these categories and especially outperforms the recent generative SSL methods(GraphMAE [6] and S2GAE [20]).

Table 2. Summary of the results (±std) on link prediction.

Metrics	Method	Cora	Citeseer	Pubmed	A-Photo	A-Computer	C-CS	C-Physics
AUC	DGI	93.88±1.00	95.98±0.72	96.30±0.20	80.95±0.39	81.27±0.51	93.81±0.20	93.51±0.22
	MVGRL	93.33±0.68	88.66±5.27	95.89±0.22	69.58±2.04	92.37±0.78	91.45±0.67	OOM
	GRACE	82.67±0.27	87.74±0.96	94.09±0.92	81.72±0.31	82.94±0.20	85.26±2.07	83.48±0.96
	GCA	81.46±4.86	84.81±1.25	94.20±0.59	70.02±9.66	89.92±0.91	84.35±1.13	85.24±5.41
	CCA-SSG	93.88±0.95	94.69±0.95	96.63±0.15	73.98±1.31	75.91±1.50	96.80±0.16	96.74±0.05
	CAN	93.67±0.62	94.56±0.68	-	97.00±0.28	96.03±0.37	-	-
	SIG-VAE	94.10±0.68	92.88±0.74	85.89±0.54	94.98±0.86	91.14±1.10	95.26±0.36	98.76±0.23
	GraphMAE	90.70±0.01	70.55±0.05	69.12±0.01	77.42±0.02	75.14±0.02	91.47±0.01	87.61±0.02
	SeeGERA	95.50±0.71	97.04±0.47	97.87±0.20	**98.64±0.05**	97.70±0.19	98.42±0.13	**99.03±0.05**
	GMAERank	**98.11±0.07**	**99.27±0.08**	**97.94±0.04**	96.99±0.07	**97.77±0.10**	**99.05±0.01**	95.46±0.04
AP	DGI	93.60±1.14	96.18±0.68	95.65±0.26	81.01±0.47	82.05±0.50	92.79±0.31	92.10±0.29
	MVGRL	92.95±0.82	89.37±4.55	95.53±0.30	63.43±2.02	91.73±0.40	89.14±0.93	OOM
	GRACE	82.36 0.24	86.92±1.11	93.26±1.20	81.18 ± 0.37	83.12±0.23	83.90±2.20	82.20±1.06
	GCA	80.87±4.11	81.93±1.76	93.31±0.75	65.17±10.11	89.50±0.64	83.24±1.16	82.80±4.46
	CCA-SSG	93.74±1.15	95.06 ± 0.91	95.97±0.23	67.99±1.60	69.47±1.94	96.40±0.30	96.26±0.10
	CAN	94.49±0.60	95.49±0.61	-	96.68±0.30	95.96±0.38	-	-
	SIG-VAE	94.79±0.71	94.21±0.53	85.02±0.49	94.53±0.93	91.23±1.04	94.93±0.37	98.85±0.12
	GraphMAE	89.52±0.01	74.50±0.04	87.92±0.01	77.18 0.02	75.80±0.01	83.58±0.01	86.44±0.03
	SeeGERA	95.92±0.68	97.33±0.46	**97.87±0.20**	**98.48±0.06**	97.50±0.15	98.53±0.18	**99.18±0.04**
	GMAERank	**97.92±0.08**	**99.28±0.09**	97.79±0.02	96.44±0.03	**97.50±0.14**	**98.91±0.02**	96.06±0.06

The results not reported are due to unavailable code or out-of-memory.

This underscores GMAERank's effective generalization capabilities to graph-level tasks, although it was not explicitly designed for this purpose.

Table 3. Summary of the results (±std) on graph classification.

Dataset	IMDB-B	IMDB-M	PROTEINS	COLLAB	MUTAG	REDDIT-B	NCI
graph2vec [15]	71.10±0.54	50.44±0.87	73.30±2.05	-	83.15±9.25	75.78±1.03	73.22±1.81
Infograph [19]	73.03±0.87	49.69±0.53	74.44±0.31	70.65±1.13	89.01±1.13	82.50±1.42	76.20±1.06
GraphCL [26]	71.14±0.44	48.58±0.67	74.39±0.45	71.36±1.15	86.80±1.34	**89.53±0.84**	77.87±0.41
JOAO [25]	70.21±3.08	49.20±0.77	74.55±0.41	69.50±0.36	87.35±1.02	85.29±1.35	78.07±0.47
GCC [18]	72.0	49.4	-	78.9	-	89.8	-
MVGRL [3]	74.20±0.70	51.20±0.50	-	-	89.70±1.10	84.50±0.60	-
InfoGCL [24]	75.10±0.90	51.40±0.80	-	80.00±1.30	**91.20±1.30**	-	80.20±0.60
GraphMAE [6]	75.52±0.66	51.63±0.52	75.30±0.39	80.32±0.46	88.19±1.26	88.01±0.19	80.40±0.30
S2GAE [20]	75.76±0.62	51.79±0.36	**76.37±0.43**	81.02±0.53	88.26±0.76	87.83±0.27	**80.80±0.24**
GMAERank	**76.05±0.15**	**52.83±0.03**	75.38±0.27	**82.46±0.23**	89.36±0.00	88.40±0.65	77.15±0.40

The results not reported are due to unavailable code or out-of-memory.

3.4 Ablation Studies (RQ4)

In this section, we conduct ablation studies across three graph tasks, each on two datasets, to assess the impact of the ranking strategy and the masking mechanism. To explore the influence of the ranking strategy further, we substitute it with a contrastive learning approach. This involves treating the augmented views as anchor, positive and negative views, with InfoNCE [16] as the loss function.

Table 4 reveals three key findings. First, every component significantly influences the performance across all tasks. Second, the different strategies contribute differently across task levels, with the input graph masking having the most impact and the ranking strategy notably affecting graph-level tasks. Lastly, replacing the ranking strategy with contrastive learning minimally affects node and edge levels, but significantly alters the graph-level tasks. Considering contrastive learning's limitations, ranking strategies prove advantageous in utilizing within GAE.

Table 4. Ablation studies

Tasks	Node-level		Edge-level		Graph-level	
Datasets	PubMed	Computer	Citeseer	CS	IMDB-B	PROTEINS
GMAERank	**81.80**	**90.45**	**99.27**	**99.05**	**76.05**	**75.38**
w.o. Mask	78.95	89.43	96.68	97.63	74.90	74.88
w.o. Ranking	80.85	90.31	98.57	98.81	74.75	74.84
Rank. → Contra.	81.25	90.11	98.64	98.72	74.35	74.66

4 Related Work

In this section, we present related work on generative SSL and rank learning. Generative SSL methods have been developed to reconstruct input data using the input data itself as supervision signals. Following autoencoders, generative graph SSL approaches typically use an encoder-decoder framework, in particular graph autoencoders (GAEs). These GAEs first use an encoder to map the input graph to a latent representation, which is then used to restore the input graph by the decoder. The earliest instances of GAEs can be traced back to GAE/VGAE [13], which adopted a Graph Convolutional Network (GCN) encoder and a simple inner dot product decoder for link prediction. Following GAE/VGAE, SIG-VAE [2] incorporated hierarchical variational inference to improve the quality of generative representations for graph data. Subsequently, ARGA/ARVGA [17] proposed a method for robust GAE learning by regularizing latent representations. Later, GraphMAE [6] introduced the idea of MAE to graph learning with auxiliary strategies and achieved promising results on multiple tasks. GraphMAE2 [5] introduces the prediction of latent representation and randomness during the decoding phase, which proves to be effective in several large-scale datasets. Ranking learning is used to evaluate and rank multiple model outputs. LLP [1] and GraphRank [7] utilize ranking-based techniques for knowledge distillation and contrastive learning, respectively, achieving competitive performance. Inspired by LLP and GraphRank, we propose a ranking learning strategy tailored for generative SSL.

5 Conclusion And Discussion

In this paper, we propose GMAERank, a simple but effective model. GMAERank introduces a novel ranking strategy, which ranks the similarity scores based on the latent and random views and updates it in conjunction with the reconstruction loss. The experiments conducted at node, edge, and graph levels demonstrate its effectiveness and versatility.

References

1. Guo, Z., et al.: Linkless link prediction via relational distillation (2023)
2. Hasanzadeh, A., Hajiramezanali, E., Narayanan, K., Duffield, N., Zhou, M., Qian, X.: Semi-implicit graph variational auto-encoders. In: Advances in Neural Information Processing Systems, vol. 32 (2019)
3. Hassani, K., Khasahmadi, A.H.: Contrastive multi-view representation learning on graphs. In: International Conference on Machine Learning, pp. 4116–4126. PMLR (2020)
4. He, K., Chen, X., Xie, S., Li, Y., Dollár, P., Girshick, R.: Masked autoencoders are scalable vision learners. In: Proceedings of the IEEE/CVF Conference on Computer Vision and Pattern Recognition, pp. 16000–16009 (2022)
5. Hou, Z., et al.: GraphMAE2: a decoding-enhanced masked self-supervised graph learner (2023)
6. Hou, Z., et al.: GraphMAE: self-supervised masked graph autoencoders. In: Proceedings of the 28th ACM SIGKDD Conference on Knowledge Discovery and Data Mining (2022)
7. Hu, Y., et al.: Do we really need contrastive learning for graph representation? (2023)
8. Hu, Y., Ouyang, S., Yang, Z., Liu, Y.: VIGraph: self-supervised learning for class-imbalanced node classification. arXiv preprint arXiv:2311.01191 (2023)
9. Hu, Y., Yang, Z., Ouyang, S., Liu, Y.: HGCVAE: integrating generative and contrastive learning for heterogeneous graph learning. arXiv preprint arXiv:2310.11102 (2023)
10. Hu, Z., Dong, Y., Wang, K., Chang, K.W., Sun, Y.: GPT-GNN: generative pre-training of graph neural networks. In: Proceedings of the 26th ACM SIGKDD International Conference on Knowledge Discovery & Data Mining, pp. 1857–1867 (2020)
11. Kingma, D.P., Welling, M.: Auto-encoding variational Bayes. arXiv preprint arXiv:1312.6114 (2013)
12. Kipf, T.N., Welling, M.: Semi-supervised classification with graph convolutional networks. arXiv preprint arXiv:1609.02907 (2016)
13. Kipf, T.N., Welling, M.: Variational graph auto-encoders. arXiv preprint arXiv:1611.07308 (2016)
14. Li, X., Ye, T., Shan, C., Li, D., Gao, M.: SeeGera: self-supervised semi-implicit graph variational auto-encoders with masking. In: Proceedings of the ACM Web Conference 2023, pp. 143–153 (2023)
15. Narayanan, A., Chandramohan, M., Venkatesan, R., Chen, L., Liu, Y., Jaiswal, S.: graph2vec: Learning distributed representations of graphs. arXiv preprint arXiv:1707.05005 (2017)

16. Oord, A.v.d., Li, Y., Vinyals, O.: Representation learning with contrastive predictive coding. arXiv preprint arXiv:1807.03748 (2018)
17. Pan, S., Hu, R., Long, G., Jiang, J., Yao, L., Zhang, C.: Adversarially regularized graph autoencoder for graph embedding (2019)
18. Qiu, J., et al.: GCC: graph contrastive coding for graph neural network pre-training. In: Proceedings of the 26th ACM SIGKDD International Conference on Knowledge Discovery & Data Mining, pp. 1150–1160 (2020)
19. Sun, F.Y., Hoffmann, J., Verma, V., Tang, J.: InfoGraph: unsupervised and semi-supervised graph-level representation learning via mutual information maximization. arXiv preprint arXiv:1908.01000 (2019)
20. Tan, Q., et al.: S2GAE: self-supervised graph autoencoders are generalizable learners with graph masking. In: Proceedings of the Sixteenth ACM International Conference on Web Search and Data Mining, pp. 787–795 (2023)
21. Thakoor, S., et al.: Large-scale representation learning on graphs via bootstrapping. arXiv preprint arXiv:2102.06514 (2021)
22. Velickovic, P., Cucurull, G., Casanova, A., Romero, A., Lio, P., Bengio, Y., et al.: Graph attention networks. Statistics **1050**(20), 10–48550 (2017)
23. Veličković, P., Fedus, W., Hamilton, W.L., Liò, P., Bengio, Y., Hjelm, R.D.: Deep graph infomax. arXiv preprint arXiv:1809.10341 (2018)
24. Xu, D., Cheng, W., Luo, D., Chen, H., Zhang, X.: InfoGCL: information-aware graph contrastive learning. Adv. Neural. Inf. Process. Syst. **34**, 30414–30425 (2021)
25. You, Y., Chen, T., Shen, Y., Wang, Z.: Graph contrastive learning automated. In: International Conference on Machine Learning, pp. 12121–12132. PMLR (2021)
26. You, Y., Chen, T., Sui, Y., Chen, T., Wang, Z., Shen, Y.: Graph contrastive learning with augmentations. Adv. Neural. Inf. Process. Syst. **33**, 5812–5823 (2020)
27. Zhang, H., Wu, Q., Yan, J., Wipf, D., Yu, P.S.: From canonical correlation analysis to self-supervised graph neural networks. Adv. Neural. Inf. Process. Syst. **34**, 76–89 (2021)
28. Zhu, Y., Xu, Y., Yu, F., Liu, Q., Wu, S., Wang, L.: Deep graph contrastive representation learning. arXiv preprint arXiv:2006.04131 (2020)

Path-Aware Co-contrastive Learning for Signed Directed Network Embedding

Yuechen Tang[1], Huifang Ma[1,2(✉)], Ke Shu[1], Zhixin Li[2], and Liang Chang[3]

[1] College of Computer Science and Engineering, Northwest Normal University, Lanzhou 730070, Gansu, China
mahuifang@yeah.net

[2] Key Lab of Education Blockchain and Intelligent Technology, Ministry of Education, Guangxi Normal University, Guilin 541004, Guangxi, China

[3] School of Computer Science and Information Security, Guilin University of Electronic Technology, Guilin 541004, Guangxi, China

Abstract. Signed directed network embedding (a.k.a. representation) aims to generate low-dimensional node representations by considering both edge sign and direction. Previous studies predominantly focused on applying a singular social theory to derive node embeddings. It overlooks network structure and lacks comprehensive methods that integrate different theories, resulting in less thorough node representations. Recently, contrastive learning has gained popularity as a self-supervised method, showing great potential in learning from diverse perspectives. Inspired by this, we introduce **P**ath-**A**ware Co-Contrastive Learning for **S**igned **D**irected Network Embedding (PASD). PASD is specifically crafted to capture the intricate interplay between edge sign and direction in both local and global structures. It intentionally incorporates both long and short paths modeling approaches to fully leverage the benefits of balance and status theories. To refine node embeddings, PASD integrates a dual-path co-contrastive learning mechanism, enabling the exploration of directional influences within the network, encompassing positive and negative relationships. Extensive experimental results demonstrate that PASD outperforms state-of-the-art methods on four widely-used datasets, validating the effectiveness of considering balance theory and status theory based on network structure in the task of signed directed network embedding.

Keywords: Sign · Direction · Dual-Path · Node representations · Co-contrastive learning

1 Introduction

A signed network serves as a crucial structure, portraying positive or negative connections between entities. For instance, in social networks, links represent friendships (positive) or enmities (negative). Signed directed networks are more intricate, preserving both relations and adding directional details. Directional

M. Onizuka et al. (Eds.): DASFAA 2024, LNCS 14855, pp. 395–406, 2024.
https://doi.org/10.1007/978-981-97-5572-1_28

edges in trust networks convey trust (positive) or distrust (negative), as shown in Fig. 1 (left). Research focuses on embedding signed directed networks, capturing directional features alongside positive/negative relations. These embeddings aid in pattern recognition, information extraction, and various tasks like link prediction and community detection [12]. Thus, they enhance understanding and decision-making in diverse contexts.

Recent studies on signed graphs employ Graph Neural Networks (GNNs) to understand signed semantics and structural complexity deeply. Guided by social theories, they utilize balance theory and status theory as foundations. Balance theory (e.g., SGCN [2] and SNEA [9]) uncovers 'friend-enemy' relationships via long-distance positive/negative edges (i.e., long paths in Fig. 1(mid)), suggesting that triads with an even number of negative edges are prevalent in the real world [5]. Status theory (e.g., SiGAT [6] and S-NN [10]) integrates signed and directional information to assess node status comprehensively [8], guiding information aggregation from direct neighbors and shaping node representations (i.e., short paths in Fig. 1(right)). Existing studies demonstrate balance and status theories' efficacy in modeling various paths. However, relying solely on one theory is insufficient [13]. For instance, BESIDE [1] uses balance theory for triangles and status theory for 'bridge' edges, overlooking the complementary nature of signed and directional information. A holistic approach is needed, combining both theories for effective modeling of signed directed networks.

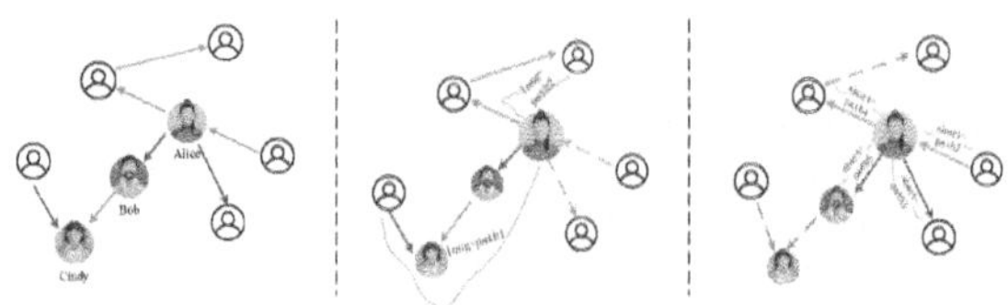

Fig. 1. Toy example of signed directed network. Figure 1(left) depicts trust network with red edges as positive interactions. Figure 1(middle) and Fig. 1(right) show prevalent paths (Color figure online)

Based on the above analysis, an important question arises: **(Q1) How to represent long-path and short-path in signed directed networks?** Long and short paths are common in these networks. Balance theory captures multi-hop neighbor relationships, aiding long-path modeling. Meanwhile, status theory clarifies one-hop neighbor connections, supporting short-path modeling. Combining both theories for path representation raises another question: **(Q2) How to fully exploit the complementarity of signs and directions in balance and status theories?** Balance theory focuses on sign-based node relationships, while status theory emphasizes directional semantics, complementing each other. Thus, an effective optimization scheme is necessary to continuously learn sign and direction information, enhancing node representations.

To answer both of the aforementioned questions, we introduce PASD (Path-Aware Co-Contrastive Learning for Signed Directed Network Embedding). **For**

Q1, we use balance theory for long paths and status theory for short paths, capturing global and local network structures, respectively. **For Q2**, we introduce co-contrastive learning, training separate classifiers for each path to enrich information for the opposing path. This improves node representation quality by iteratively integrating sign-direction information.

2 Preliminaries

Structural Balance Theory [5]: The theory sorts cycles in signed networks as balanced or unbalanced based on the number of negative edges. As Fig. 2 (left) shows, a balanced triad with zero '–' means 'a friend of a friend is my friend', while two '–' suggest 'an enemy of an enemy becomes my friend'. Conversely, an odd number of '–' indicates an unbalanced triad.

Status Theory [8]: The theory defines the core principles guiding the relationships represented by signed links in a directed network. In status theory, a positive link $e_{u \longrightarrow v}$ implies v is perceived with higher status from u's viewpoint, while a negative link $e_{u \longrightarrow v}$ indicates v is viewed with lower status, as shown in Fig. 2 (right).

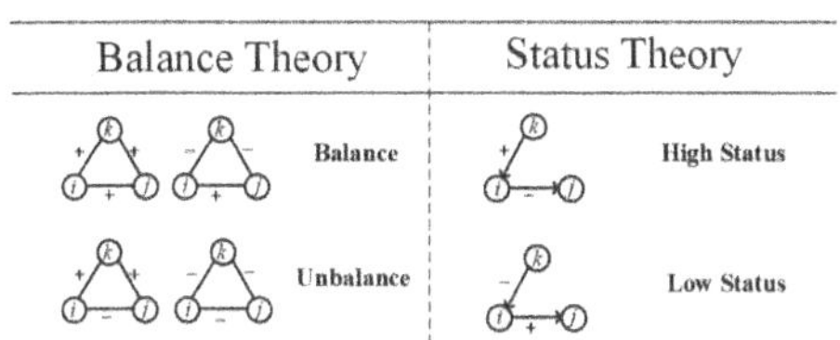

Fig. 2. Schematic diagram of balance theory (left) and status theory (right)

Problem Formalization: In this paper, we consider a signed directed network/graph, denoted as $G = (V, E, s)$, where V is the set of nodes in a graph G, and E is the edge list with signs s and directions. E consist of E^+ and E^- while $E^+ \bigcap E^- = \emptyset$, E^+ and E^- denote the sets of positive and negative links (i.e. while a link $e_{i \to j} \in \{E^+ \bigcup E^-\}$ denotes a directed link from i to j with a positive or negative sign), respectively.

Task Description: Given $G = (V, E, s)$, the goal is to learn a function f to to low-dimensional vectors $z_i \in \mathbb{R}^d$, where d is the dimension of node embeddings, so that the embeddings are useful in downstream tasks such as link sign prediction.

3 Method

We propose the model PASD as shown in Fig. 3 and describe the modules in detail in this section.

3.1 Long-Path View

Using balance theory, we classify node paths in signed networks as friend or enemy paths based on positive or negative link counts. Friend paths use a friend aggregator, and enemy paths use an enemy aggregator.

Friend Aggregator. Therefore, for the Friend aggregator, we employ $\mathbf{h}_i^0 \in \mathbb{R}^{d^{in}}$ to represent the initial d^{in}-dimensional node features for node i. Thus, we define the friend aggregator as follows:

$$\mathbf{h}_i^{F(1)} = \sigma(\mathbf{W}^{F(1)}[(\sum_{j\in N_i^+} \frac{\mathbf{h}_j^0}{|N_i^+|}); \mathbf{h}_i^0]), \tag{1}$$

$$\mathbf{h}_i^{F(l)} = \sigma(\mathbf{W}^{F(l)}[(\sum_{j\in N_i^+} \frac{\mathbf{h}_j^{F(l-1)}}{|N_i^+|}); (\sum_{k\in N_i^-} \frac{\mathbf{h}_k^{E(l-1)}}{|N_i^-|}); \mathbf{h}_i^{F(l-1)}]). \tag{2}$$

where $\mathbf{h}_i^{F(l)}$ is the friend representation of node i under l-th layer, $\mathbf{h}_k^{E(l-1)}$ is the enemy representation under $(l-1)$-th layer, σ is a non-linear activation function, $\mathbf{W}^{F(1)} \in \mathbb{R}^{d^{out}\times 2d^{in}}$, $\mathbf{W}^{F(l)} \in \mathbb{R}^{d^{out}\times 3d^{in}}$ are learnable weight matrices, N_i^+, N_i^- are the set of positive and negative neighbors of node i, respectively.

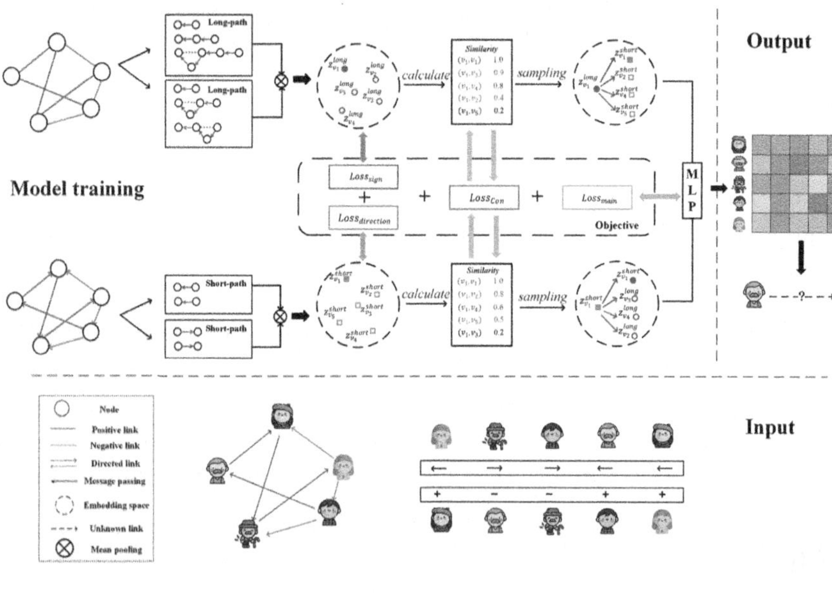

Fig. 3. The framework of PASD

Enemy Aggregator. The aggregation process of the Enemy aggregator is similar to that of the Friend aggregator. Therefore, we employ a similar formalization process and utilize Eqs. (3) and (4) to illustrate the concepts of 'enemies of friends are enemies' and 'friends of enemies are enemies' as follows:

$$\mathbf{h}_i^{E(1)} = \sigma(\mathbf{W}^{E(1)}[(\sum_{j \in N_i^-} \frac{\mathbf{h}_j^0}{|N_i^-|}); \mathbf{h}_i^0]), \tag{3}$$

$$\mathbf{h}_i^{E(l)} = \sigma(\mathbf{W}^{E(l)}[(\sum_{j \in N_i^+} \frac{\mathbf{h}_j^{E(l-1)}}{|N_i^+|}); (\sum_{k \in N_i^-} \frac{\mathbf{h}_k^{F(l-1)}}{|N_i^-|}); \mathbf{h}_i^{E(l-1)}]). \tag{4}$$

Long-Path Representation. We use the standard Full Contact layer (FC layer) to combine $\mathbf{h}_i^{F(l)}$ and $\mathbf{h}_i^{E(l)}$, the final representation of node i in the long-path view is generated as follows:

$$\mathbf{z}_i^{long} = \mathbf{W}^{long} \cdot [\mathbf{h}_i^{F(l)}; \mathbf{h}_i^{E(l)}] + \mathbf{b}^{long}, \tag{5}$$

where $\mathbf{W}^{long} \in \mathbb{R}^{d^{out} \times 2d^{in}}$ and $\mathbf{b}^{long}$ are the weight and bias in long-path view.

3.2 Short-Path View

In signed directed networks, nodes receive positive/negative edges. Using status theory, credibility is assessed by accumulating high/low status. We develop high/low status aggregators integrating diverse signed and directional information, starting with one-hot encoding for signs.

$$\mathbf{sign}_{e_{i \to j}, e_{i \leftarrow j}} = \begin{cases} [0,1]^T, + \\ [1,0]^T, - \end{cases}, \tag{6}$$

High Status Aggregator. Node i's high status is determined by aggregating info from positive and negative incoming neighbors. We convert sign coding into dense vectors and concatenate previous layer representations, yielding two types of incoming representations.

$$\mathbf{h}_{I(j)}^{(l)+} = \left[\mathbf{h}^{(l-1)}[j]; \left(\mathbf{W}_{i \leftarrow j}^{(l)} \cdot \mathbf{sign}_{e_{i \leftarrow j(+)}}\right)\right], \tag{7}$$

$$\mathbf{h}_{I(j)}^{(l)-} = \left[\mathbf{h}^{(l-1)}[j]; \left(\mathbf{W}_{i \leftarrow j}^{(l)} \cdot \mathbf{sign}_{e_{i \leftarrow j(-)}}\right)\right], \tag{8}$$

where [;] represents the concatenate operation, $\mathbf{h}^0[j]$ is the initial feature vector of node j, and $\mathbf{W}_{i \leftarrow j} \in \mathbb{R}^{d^{in} \times 2}$ is the learnable weight matrix, $\mathbf{sign}_{e_{i \leftarrow j}(+)}, \mathbf{sign}_{e_{i \leftarrow j}(-)}$ are one-hot representations for the observed signed link.

Then, we integrate the two representation of node i to obtain the high status representation of the node as $\mathbf{h}_i^{H(l)}$:

$$\mathbf{h}_i^{H(l)} = Mean\left[\left(\frac{1}{\left|N_{I(i)}^{+}\right|}\sum_{j\in I_{I(i)}^{+}} \mathbf{h}_{I(j)}^{(l)+}\right), \left(\frac{1}{\left|N_{I(i)}^{-}\right|}\sum_{j\in N_{I(i)}^{-}} \mathbf{h}_{I(j)}^{(l)-}\right)\right], \tag{9}$$

where $Mean[,]$ represents the mean function, $N_{I(i)}^{+}$ and $N_{I(i)}^{-}$ are positive and negative incoming neighbors of node i.

Low Status Aggregator. Similar to the high status aggregator, we utilize a similar formalization process to combine sign and outgoing information for aggregating neighbor data. Following Eqs. (10)–(11), we separately aggregate negative and positive outgoing neighbors. Then, we integrate these two types of outgoing representations using Eq. (12).

$$\mathbf{h}_{O(j)}^{(l)-} = \left[\mathbf{h}^{(l-1)}[j]; \left(\mathbf{W}_{i\to j}^{(l)} \cdot \mathbf{sign}_{e_{i\to j(-)}}\right)\right], \tag{10}$$

$$\mathbf{h}_{O(j)}^{(l)+} = \left[\mathbf{h}^{(l-1)}[j]; \left(\mathbf{W}_{i\to j}^{(l)} \cdot \mathbf{sign}_{e_{i\to j(+)}}\right)\right], \tag{11}$$

$$\mathbf{h}_i^{L(l)} = Mean\left[\left(\frac{1}{\left|N_{O(i)}^{-}\right|}\sum_{j\in N_{O(i)}^{-}} \mathbf{h}_{O(j)}^{(l)-}\right), \left(\frac{1}{\left|N_{O(i)}^{+}\right|}\sum_{j\in N_{O(i)}^{+}} \mathbf{h}_{O(j)}^{(l)+}\right)\right]. \tag{12}$$

Short-Path Representation. We obtain the short-path representation of node i in the same way as obtaining the long-path representation:

$$\mathbf{z}_i^{short} = \mathbf{W}^{short} \cdot [\mathbf{h}_i^{H(l)}; \mathbf{h}_i^{L(l)}] + \mathbf{b}^{short}. \tag{13}$$

where $\mathbf{W}^{short} \in \mathbb{R}^{d^{in}\times 2}$ and $\mathbf{b}^{short}$ are the model trainable parameters.

3.3 Co-contrastive Learning for Dual Views

Pseudo-labels Generation. In the long-path view, node i uses its representation from the short-path view to select positive and negative samples. Based on short-path representations, similarity between node i and others is ranked to choose the top-K nodes as positive samples. This process is denoted as c_{long}^{pos}.

$$score_i^{short} = \mathbf{Z}^{short} \cdot (\mathbf{z}_i^{short})^T, \tag{14}$$

$$c_{long}^{pos} = top - K[softmax(score_i^{short})], \tag{15}$$

where z_i^{short} is the representation of node i learned in the short-path view and $\mathbf{Z}^{short} \in \mathbb{R}^{|V| \times d^{out}}$.

From the top 10% similar nodes, excluding them as positive samples $Y^{short'}$, we randomly pick K nodes as negative samples (negative pseudo-labels) for the long-path view. Designated as c_{long}^{neg}, they serve as hard negatives, enriching information while reducing false negatives risk.

$$c_{long}^{neg} = random - K(Y^{short'}). \tag{16}$$

Similarly, the same approach can be employed to identify positive and negative samples for the short-path view as c_{short}^{pos} and c_{short}^{neg}, respectively.

Contrastive Learning of Dual Paths. We adopt InfoNCE [11] to enhance agreement among positives and minimizing it among negatives.

$$L_{\mathrm{Con}} = \begin{array}{l} \sum_{i \in V} - \log \frac{\sum_{j \in c_{long}^{pos}} \exp\left(\mathrm{sim}\left(\mathbf{z}_i^{long}, \mathbf{z}_j^{long}\right)/\tau\right)}{\sum_{j \in c_{long}^{pos}} \exp\left(\mathrm{sim}\left(\mathbf{z}_i^{long}, \mathbf{z}_j^{long}\right)/\tau\right) + \sum_{k \in c_{long}^{neg}} \exp\left(\mathrm{sim}\left(\mathbf{z}_i^{long}, \mathbf{z}_k^{long}\right)/\tau\right)} \\ + \sum_{i \in V} - \log \frac{\sum_{j \in c_{short}^{pos}} \exp\left(\mathrm{sim}\left(\mathbf{z}_i^{short}, \mathbf{z}_j^{short}\right)/\tau\right)}{\sum_{j \in c_{short}^{pos}} \exp\left(\mathrm{sim}\left(\mathbf{z}_i^{short}, \mathbf{z}_j^{short}\right)/\tau\right) + \sum_{k \in c_{short}^{neg}} \exp\left(\mathrm{sim}\left(\mathbf{z}_i^{short}, \mathbf{z}_k^{short}\right)/\tau\right)} \end{array}, \tag{17}$$

where L_{Con} is the contrastive learning loss, $sim(\cdot)$ is the similarity between two vectors, which is set to the cosine similarity function in this paper, and τ is the temperature hyperparameter.

In light of the demonstrated simplicity and expressiveness of the concatenation method in previous GNNs studies [3], we concatenate the representations of node i's long-path and short-path to obtain its final representation:

$$\mathbf{z}_i = \left[\mathbf{z}_i^{long}; \mathbf{z}_i^{short}\right]. \tag{18}$$

3.4 Model Optimization

To refine our model parameters, we've developed four optimization objectives and strategies.

Definition of Link Prediction. Formalizing link prediction loss based on 'friends' representations should be similar, enemies' representations should be distant' concept:

$$L_{main} = \sum_{i \in V} \left[- \sum_{j \in N(i)^+} \log\left(\sigma\left(\mathbf{z}_i^T \mathbf{z}_j\right)\right) - Q \sum_{k \in N(i)^-} \log\left(\sigma\left(-\mathbf{z}_i^T \mathbf{z}_k\right)\right) \right], \tag{19}$$

where σ is the sigmoid function, Q is the hyperparameter used to balance the imbalance between the number of positive and negative neighbors.

Optimization with Respect to Sign. Modeling link's sign relationships using cross-entropy loss:

$$L_{sign} = \sum_{e_{ij} \in E} -sign_{e_{ij}} \log \left(\sigma \left(\mathbf{z}_i^T \mathbf{z}_j\right)\right) - \left(1 - sign_{e_{ij}}\right) \log \left(1 - \sigma \left(\mathbf{z}_i^T \mathbf{z}_j\right)\right), \quad (20)$$

where $sign_{e_{ij}}$ is the sign ground truth, and E is the edge list with signs.

Optimization in the Context of Direction. Modeling link's direction relationships using cross-entropy loss:

$$L_{direction} = \sum_{e_{i \to j} \in E^+} (sign_{e_{i \to j}}) \cdot \widehat{sign}_{e_{i \to j}} + \sum_{e_{i \to j} \in E^-} (sign_{e_{i \to j}}) \cdot \widehat{sign}_{e_{i \to j}} + \lambda \left\|\theta\right\|_2^2, \quad (21)$$

where $e_{i \to j} \in E^+$ (or E^-) denotes a positive (or negative) edge that node i target to node j. $sign_{e_{i \to j}}$ is the sign ground truth, and use an MLP to encode vectors into scores, $\widehat{sign}_{e_{i \to j}} = sigmoid(\text{MLP}(\mathbf{z}_i) - \text{MLP}(\mathbf{z}_j))$, and $\lambda \cdot \left\|\theta\right\|_2^2$ is a regularization term, λ controls the L_2 regularization strength to prevent overfitting.

Ultimately, the sign prediction for link $e_{i \to j}$ can be predicted as:

$$\widehat{sign}_{e_{i \to j}} = \begin{cases} -1, & \sigma \left(\text{MLP}\left(\mathbf{z}_i\right) - \text{MLP}\left(\mathbf{z}_j\right)\right) > 0.5. \\ +1, & \text{otherwise.} \end{cases} \quad (22)$$

This leads to the formulation of a comprehensive loss:

$$Loss = L_{main} + \alpha \cdot L_{sign} + \beta \cdot L_{direction} + \gamma \cdot L_{Con}. \quad (23)$$

where α, β and γ are hyperparameters.

4 Experiments

In this section, we evaluate the performance of link sign prediction of our proposed method PASD on four real-world datasets through answering the following questions:

- RQ1: How does PASD perform as compared with state-of-the-art GNN methods?
- RQ2: How well PASD performs under random perturbations?
- RQ3: How effective are the different modules?

4.1 Datasets

We adopt four real-world signed directed datasets for experiments. **Bitcoin-alpha and Bitcoin-otc** are extracted from Bitcoin trading platforms and **WikiRfa and WikiElec** are defined by voting for candidates for Wikipedia administrator. The detailed statistics of the datasets are shown in Table 1.

Table 1. Description of the dataset

Dataset	# Of Nodes	# Of Edges	+ Edges(%)	– Edges(%)
Bitcoin-alpha	3783	24186	93.65	6.25
Bitcoin-otc	5881	35592	89.99	10.01
WikiRfa	7634	173332	78.32	21.68
WikiElec	7126	104167	78.78	21.22

Baselines. We compare PASD with seven representive baselines. The first two baselines (i.e. **SGCN** [2], **SNEA** [9]) are designed for signed undirected networks with balance theory. The following three baselines (i.e. **SiGAT** [6], **SDGNN** [7], **S-GNN** [10]) are designed for signed directed networks, employing status theory. And the rest baseline(i.e. **BESIDE** [1]) apply two theories for signed directed networks. The specific methods of the first four methods can be found in [4].

Parameter Settings. We split each dataset into two parts: 80% edges for training and 20% for testing. For each run, we run with different train-test splits. Our model is implemented by PyTorch with the Adam optimizer (Learning Rate $= 10^{-2}$, Weight Decay $= 10^{-5}$). In addition, the epoch is set as 100. We use the 2-layer-GCN aggregators for each path to build our model and set $\alpha = 1$, $\beta = 0.4$, $\gamma = 0.9$. All experiments run on a computer with Intel(R) Xeon(R) Platinum 8255C CPU @ 2.50 GHz.

4.2 Experiment Results

Performance Comparison (RQ1). Table 2 reports the performance comparison results, where the bold scores are the best and the ones underlined are the second best. We have the following observations.

Table 2. Overall performance comparison on ACC, AUC, F1 and Macro-F1

	Bitcoin-alpha				Bitcoin-otc				WikiRfa				WikiElec			
	ACC	AUC	F1	Macro-F1	ACC	AUC	F1	Macro-F1	ACC	AUC	F1	Macro-F1	ACC	AUC	F1	Macro-F1
SGCN	64.7	87.5	70.2	68.9	65.6	88.8	82.3	74.1	71.3	84.7	83.9	71.1	72.3	85.3	90.7	74.3
SNEA	64.1	81.6	92.7	73.3	67.7	81.8	92.4	78.6	79.5	87.8	86.4	73.2	76.3	77.8	86.9	74.1
SiGAT	62.9	86.5	97.1	68.6	62.8	88.9	96.0	73.8	85.9	87.6	90.4	74.7	77.2	63.9	88.2	79.5
SDGNN	91.5	88.3	**97.3**	73.7	65.3	90.8	**96.3**	**79.6**	92.1	89.5	91.3	78.1	92.5	90.0	91.2	80.3
S-GNN	85.2	85.0	91.9	**75.4**	87.6	**93.7**	89.3	77.2	95.3	89.8	91.4	78.5	94.5	90.1	91.9	79.0
BESIDE	91.2	**89.8**	**97.3**	73.0	70.1	91.5	96.1	78.4	91.5	89.7	91.0	77.4	94.3	90.0	91.5	77.4
PASD	**91.7**	84.6	91.2	74.6	**89.7**	86.8	88.8	79.2	**99.7**	**99.9**	**99.9**	**99.8**	**94.9**	**90.4**	**92.1**	**81.8**

We observe that PASD outperforms SGCN, SNEA in signed undirected networks. Compared to methods for signed directed networks, PASD shows competitive performance, excelling on large datasets (WikiRfa, WikiElec). Relying solely on balance theory, as with methods for signed undirected graphs, leads to inferior performance. Thus, using balance theory alone for modeling signed directed networks is inadequate.

Compared to models like SiGAT, SDGNN, S-GNN, and BESIDE for signed directed graphs, PASD achieves superior accuracy on the Bitcoin dataset, crucial for financial transactions. It outperforms existing methods across all metrics on WikipRfa and WikiElec, notably enhancing AUC by 11.24% and Macro-F1 by 27.13% on WikiRfa.

Despite both utilizing balance and status theories, BESIDE neglects sign and direction semantics, using simple concatenation for node representations. In contrast, PASD continuously learns complementary sign and direction information during training, significantly outperforming BESIDE across most metrics.

Robustness of PASD to Random Perturbations (RQ2). We evaluate PASD's robustness by perturbing it with link or node additions/removals (10%, 30%, and 50%) on Bitcoin-alpha and WikiRfa, similar to [14]. Dataset statistics are shown in Fig. 4(a) and Fig. 4(b).

Figures 5(a) and 5(b) display performance on perturbed datasets. Bitcoin-alpha shows a notable performance drop with node deletion, likely due to limited negative edge information hindering accurate predictions. Conversely, WikiRfa's network structure is notably affected by edge addition, leading to performance degradation with higher perturbation ratios. This emphasizes the benefit of negative edges in improving prediction accuracy.

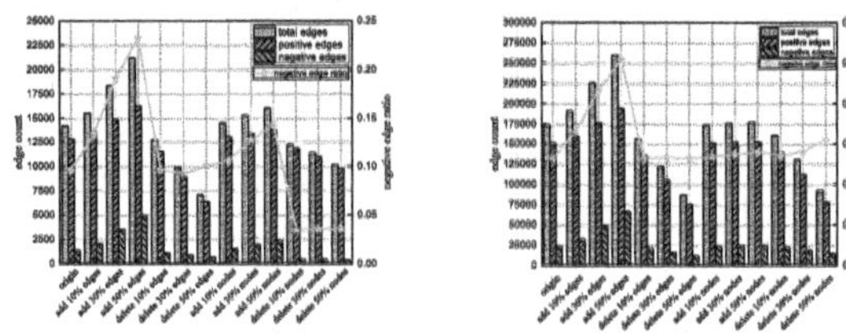

(a) Bitcoin-alpha (b) WikiRfa

Fig. 4. Bitcoin-alpha and WikiRfa after perturbations

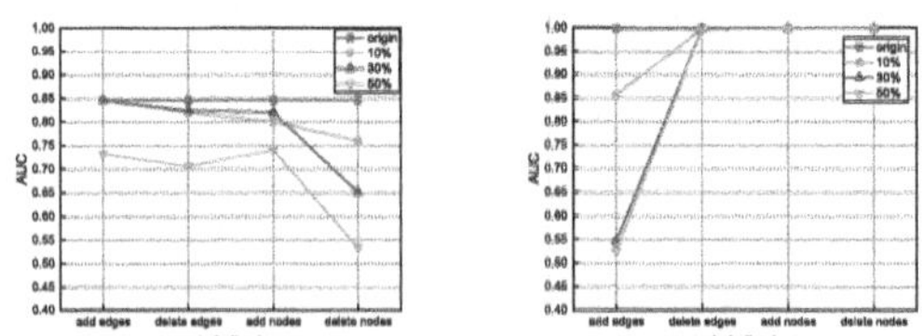

(a) Bitcoin-alpha (b) WikiRfa

Fig. 5. The performance for Bitcoin-alpha and WikiRfa of four perturbations

Ablation Study on Different Modules (RQ3). In this subsection, we conduct ablation studies to examine the design of three modules. Similar to the previous subsection, we utilize the Bitcoin-alpha and WikiRfa datasets to analyze the effects of different module designs.

Table 3. Ablation study on different modules

	Bitcoin-alpha		WikiRfa	
	ACC	Macro-F1	ACC	Macro-F1
PASD (w/o long-path)	82.1	70.3	97.2	97.0
PASD (w/o short-path)	72.5	68.2	93.6	86.7
PASD (w/o co-contrastive learning)	<u>86.5</u>	<u>70.4</u>	<u>99.2</u>	<u>98.4</u>
PASD	**91.7**	**74.6**	**99.8**	**99.9**

Table 3 shows all ablations perform worse than the full model. Poor performance of PASD(w/o long-path) confirms balance theory's significance. Worst performance of PASD(w/o short-path) emphasizes status theory's importance in directional modeling. Dual-path integration (PASD(w/o co-contrastive learning)) is better than single-path, but co-contrastive learning enriches information continuously, yielding optimal performance in the full model.

5 Conclusion

This paper explores dual-path modeling in signed directed networks, addressing challenges in representing various path types. We introduce two social theories to guide modeling and propose PASD, utilizing co-contrastive learning for enhanced performance. Extensive experiments validate PASD's superiority over baselines.

Acknowledgements. This study is supported by the National Natural Science Foundation of China (U22A2099,61762078 and 62276073), Industrial Support Project of Gansu Colleges, China (No. 2022CYZC11), and Guangxi Key Laboratory of Trusted Software (kx202302).

References

1. Chen, Y., Qian, T., Liu, H., Sun, K.: "Bridge" enhanced signed directed network embedding. In: Proceedings of the 27th ACM International Conference on Information and Knowledge Management, pp. 773–782 (2018)
2. Derr, T., Ma, Y., Tang, J.: Signed graph convolutional networks. In: 2018 IEEE International Conference on Data Mining (ICDM), pp. 929–934. IEEE (2018)
3. Hamilton, W., Ying, Z., Leskovec, J.: Inductive representation learning on large graphs. Advances in neural information processing systems **30** (2017)

4. He, Y., Zhang, X., Huang, J., Rozemberczki, B., Cucuringu, M., Reinert, G.: Pytorch geometric signed directed: A software package on graph neural networks for signed and directed graphs. arXiv preprint arXiv:2202.10793 (2022)
5. Heider, F.: Attitudes and cognitive organization. J. Psychol. **21**(1), 107–112 (1946)
6. Huang, J., Shen, H., Hou, L., Cheng, X.: Signed graph attention networks. In: Artificial Neural Networks and Machine Learning–ICANN 2019: Workshop and Special Sessions: 28th International Conference on Artificial Neural Networks, Munich, Germany, September 17–19, 2019, Proceedings 28, pp. 566–577. Springer (2019)
7. Huang, J., Shen, H., Hou, L., Cheng, X.: Sdgnn: learning node representation for signed directed networks. In: Proceedings of the AAAI Conference on Artificial Intelligence, vol. 35, pp. 196–203 (2021)
8. Leskovec, J., Huttenlocher, D., Kleinberg, J.: Signed networks in social media. In: Proceedings of the SIGCHI Conference on Human Factors in Computing Systems, pp. 1361–1370 (2010)
9. Li, Y., Tian, Y., Zhang, J., Chang, Y.: Learning signed network embedding via graph attention. In: Proceedings of the AAAI Conference on Artificial Intelligence, vol. 34, pp. 4772–4779 (2020)
10. Lin, W., Li, B.: Status-aware signed heterogeneous network embedding with graph neural networks. IEEE Trans. Neural Networks Learn. Syst. (2022)
11. Oord, A.v.d., Li, Y., Vinyals, O.: Representation learning with contrastive predictive coding. arXiv preprint arXiv:1807.03748 (2018)
12. Yang, F., Ma, H., Wang, W., Li, Z., Chang, L.: Adaptive fusion of structure and attribute guided polarized communities search. Front. Comp. Sci. **18**(1), 181337 (2024)
13. Yang, F., Ma, H., Yan, C., Li, Z., Chang, L.: Polarized communities search via co-guided random walk in attributed signed networks. ACM Trans. Internet Technol. (2023)
14. Zhang, Z., Liu, J., Zhao, K., Yang, S., Zheng, X., Wang, Y.: Contrastive learning for signed bipartite graphs. In: Proceedings of the 46th International ACM SIGIR Conference on Research and Development in Information Retrieval, pp. 1629–1638 (2023)

NGDRL: A Dynamic News Graph-Based Deep Reinforcement Learning Framework for Portfolio Optimization

Yuxuan Bian[1], Haoyu Sun[1], Yang Lei[1], Peng Zhu[1], and Dawei Cheng[1,2(✉)]

[1] Tongji University, Shanghai, China
{treast_bean,the_shy,leiy1103,pengzhu,dcheng}@tongji.edu.cn
[2] Shanghai Artificial Intelligence Laboratory, Shanghai, China

Abstract. News is widely employed in studying the intrinsic financial market dynamics due to its excellent timeliness and coverage. However, existing studies treat news either as discrete trading signals or as continuous text features, leading to challenges such as human prior bias, difficulties in aligning different modalities, and neglecting the dynamic multi-dimensional logic relations among equities within news. Furthermore, there is a growing emphasis on multi-granularity information and deep reinforcement learning for their contributions to precise market estimation and alignment with investor objectives. Therefore, in this paper, we introduce NGDRL, a dynamic **N**ews **G**raph-based **D**eep **R**einforcement **L**earning framework for portfolio optimization. Specifically, we identify three key dimensions of the latent logical structure and construct corresponding dynamic contextual relation graphs, serving as representatives to capture the core of the inter-equity relations. And we introduce a novel multi-granularity information integration structure to harness multi-granularity information. Extensive experiments on real-world datasets from the United States and China show the superiority of our proposed framework over current state-of-the-art baselines, highlighting its effectiveness for practical applications in portfolio optimization.

Keywords: Graph Neural Network · Dynamic News Graph · Deep Reinforcement Learning · Portfolio Optimization

1 Introduction

The complex nature and nonlinear dynamics inherent in the movements of stock prices present a formidable challenge for portfolio management [1]. Recent studies, however, have shed new light on this by revealing a discernible connection between stock market dynamics and the content of financial news, albeit with potential lag effects [14]. The correlation is grounded in the recognition that financial news contains valuable insights into financial markets, macroeconomic trends, and policy shifts, thereby playing a pivotal guiding role in market analysis and the formulation of portfolio strategies [17,26]. Specifically, the crucial

M. Onizuka et al. (Eds.): DASFAA 2024, LNCS 14855, pp. 407–417, 2024.
https://doi.org/10.1007/978-981-97-5572-1_29

insights regarding the future market movement trends often manifest in dynamic, multi-level, and contextual relations among different stocks [3,15,20].

Based on whether news is utilized, current portfolio optimization methods can be primarily categorized into two types: (1)**News-based Methods**: They treat news merely as continuous text features [2,5,20] or discrete quantifiable signals [13,27], resulting in challenges such as complex alignment of different modalities, slow optimization, and low signal-to-noise ratio in the features. And the hidden logical structural relations among assets embedded in the news are neglected. (2)**Non-news-based Methods**: After modeling prices as time series [18,23], these methods recognized the potential performance improvement in portfolio optimization by considering inter-equity relations and began describing inter-equity relations using historical prices for dynamic relations [25], industry classifications [24,26], and similar approaches for static relations [30]. However, these methods are constrained by challenges like information redundancy in historic features for dynamic relation modeling [22,25], poor timeliness for static relation modeling [2,24], leading to sub-optimal model performance.

In addition, multi-granularity market features bring identification of systemic risks, local fluctuations, and potential opportunities [10,26]. But most researches only focus on a single granularity, including the stock temporal properties [18,23,29], or the high-level market state [24], lacking comprehensive utilization of multi-granularity information of the market, leading to incomplete estimations of market states. Simultaneously, the traditional two-phase approach, forecasting equity trends and subsequently weighting them through combinatorial optimization, impairs the portfolio due to the misalignment between investor goals and model optimization objectives. And deep reinforcement learning (DRL) methods [8,22–24] have shown advances in reward chasing with optimal policies.

To overcome previous limitations, we introduce a novel Dynamic News Graph-based Deep Reinforcement Learning (NGDRL) framework. This framework is tailored to uncover multi-dimensional logical inter-equity relations within the news, leveraging dynamic relations and combined market dynamics, local environment, and individual stock behaviors to enhance investment performance. Initially, we extract sentiment, sector, and concept entities from raw news. Then, diverse inter-stock relation graphs are constructed. Furthermore, we implement three modules: Dynamic Contextual Relation Perception, Global Hypergraph Relation Mining, and Individual Stock Representation Mining, to extract information from dynamic graphs and temporal features. A Multi-granularity Portfolio Generator is employed to synthesize impacts across different granularities for accurate market estimates. Finally, we train the model interactively through reinforcement learning to align investor goals with model optimization objectives, conduct comprehensive experiments to validate module contributions and explore how dynamic relations enhance the portfolio.

2 Related Work

The relation between assets is pivotal in portfolio optimization [22–26]. Current methods mainly fall into two categories: static construction methods [2,24,26], which remain unchanged throughout the entire period, and dynamic construction methods [22,25]. For instance, DeepTrader [24] establishes a static graph based on industry classification, while THGNN [25] crafts a dynamic relation graph by assessing historical price sequence similarities. However, static graphs suffer from low timeliness, and historical data-based dynamic graph methods may introduce information redundancy. In contrast, the proliferation of online information alongside advances in natural language processing and text mining empowers investors to extract market trends and volatility from web content [5,20,31]. Particularly noteworthy is news, renowned for its timeliness and broad coverage. However, existing research often treats news as discrete events [2,4,13,27] or utilizes continuous text features [9,28,30], neglecting to fully explore the rich dynamic multidimensional relationships inherent in news logical structures.

3 Method

3.1 Stock Relation Graph Preparation

We choose three pivotal relation entities to represent news's dynamic contextual relations, including sentiment, sectors, and concepts. Sentiment functions as an indicator of the market's focus on stocks, sectors define the particular industry to which stocks belong, and concepts are the current trending topics linked to these stocks. The construction of these graphs follows the same pipeline. For sentiment graphs, on investment day t, we identify the n_g nearest neighboring stocks $u_i^t \in \mathcal{N}_g^t(v^t)$ based on the closest predicted scores for each financial asset v^t. Here, $\mathcal{N}_g^t(v^t)$ signifies the neighbor set of v^t in graph g^t. Consequently, directed edges e_{v,u_i}^t are established from node v^t to all nodes u_i^t within $\mathcal{N}_g^t(v^t)$. This construction method highlights the similarity in market sentiment attention among these stocks, effectively capturing the interplay between long and short forces in the market and their substantial impact on financial assets. As a result, we derive the sets $\mathcal{V}_{pos/neg} = \{v^{t_1}, v^{t_2}, \ldots, v^T\}$ and $\mathcal{E}_{pos/neg} = \{e^{t_1}, e^{t_2}, \ldots, e^T\}$. The construction methodology for concept and sector graphs mirrors this approach. On investment day t, for each financial asset v^t, within a look-back window T_w, we identify the n_g nearest neighboring financial stocks $\mathcal{N}_g^t(v^t)$ that belong to the same sector or are associated with the same hot concept. Subsequently, undirected edges $e_{v^t,u_i^t}^t$ are established from node v^t to all nodes u_i^t within $\mathcal{N}_g^t(v^t)$. This construction method inherently captures sector rotation, hot concepts, and the unique contextual relations of stocks. Thus, we obtain the sets $\mathcal{V}_{concept/sector} = \{v^{t_1}, v^{t_2}, \ldots, v^T\}$ and $\mathcal{E}_{concept/sector} = \{e^{t_1}, e^{t_2}, \ldots, e^T\}$.

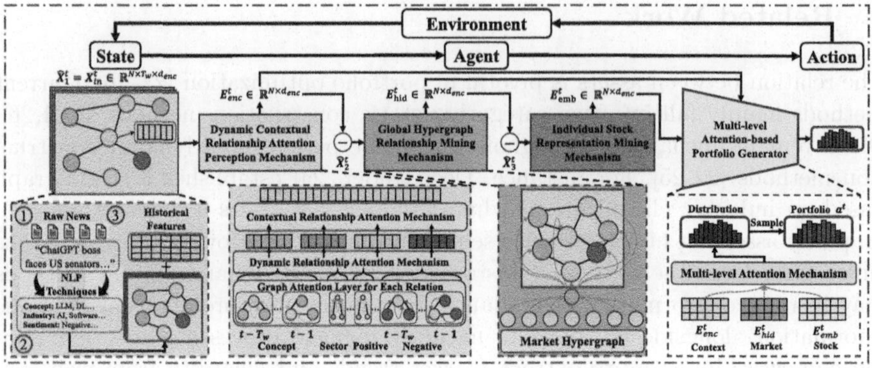

Fig. 1. The proposed NGDRL framework.

3.2 Method Overview

In this section, we introduce the NGDRL framework. As depicted in Fig. 1, **NGDRL comprises four main components and a residual Information Transmission Structure (ITS).** To effectively integrate features of different granularity, we adopted a simple and effective residual information transmission structure. Given input $\mathbf{X}_i$ and $\mathbf{Y}_i$ at a certain granularity i, the input features for the next granularity are as Eq. 1. Through this orthogonal information transmission, the overlapping of information across different granularity is minimized.

$$\mathbf{X}_{i+1} = \mathcal{F}(\mathbf{X}_i, \mathbf{Y}_i) = \mathbf{X}_i - \frac{\mathbf{X}_i \cdot \mathbf{Y}_i}{|\mathbf{X}_i|} \tag{1}$$

The first module, **Dynamic Contextual Relation Perception Module (DCRPM)**, takes $\mathbf{X}_{in} \in \mathbb{R}^{N \times d_{in} \times T}$ and $\hat{\mathcal{G}}$ for N stocks over T days as input and extracts $\mathbf{E}^t_{enc} \in \mathbb{R}^{N \times d_{enc}}$ using $\hat{\mathbf{X}}^t_1 = \mathbf{X}^t_{in} = \{\mathbf{s}_{t-T_w}, \ldots, \mathbf{s}_{t-1}\} \in \mathbb{R}^{N \times T_w \times d_{in}}$ and $\hat{\mathcal{G}}_t$. To enhance our comprehension of the market's impact, we incorporate the **Global Hypergraph Relation Mining Module (GHRMM)** as our second module. This module utilizes $\hat{\mathbf{X}}^t_2 = \mathcal{F}(\mathbf{X}^t_1, \mathbf{E}^t_{enc})$ as input to extract market-level insights while mitigating the influence of local relations, ultimately yielding the global market representation $\mathbf{E}^t_{hid} \in \mathbb{R}^{N \times d_{enc}}$. The third module, **Individual Stock Representation Mining Model (ISRMM)**, utilizes $\hat{\mathbf{X}}^t_3 = \mathcal{F}(\mathbf{X}^t_2, \mathbf{E}^t_{hid})$ as input to extract individual stock-level information $\mathbf{E}^t_{emb} \in \mathbb{R}^{N \times d_{enc}}$ while eliminating local and global market relations. The fourth module, **Multi-granularity Portfolio Generator (MPG)**, integrates information from prior modules to derive comprehensive final representations $\mathbf{E}^t_{out}$, and yield the portfolio action distribution $\boldsymbol{a}_t$ and the estimated reward score v_t.

3.3 Dynamic Contextual Relation Perception Module

The DCRPM is designed to effectively capture the changing dynamics of stocks by learning from dynamic multi-dimensional inter-asset relations $\hat{g}^t \in \hat{\mathcal{G}}^t$. Dynamic relations $\mathbf{E}^{\hat{t}}_{\hat{g}^t}$ for each $\hat{g}^t$ are modeled within a look-back window T_w:

$$\mathbf{E}_{\hat{g}^t}^{\hat{t}} = \text{Concat}\,(\text{TgaHead}_{\hat{g}^t}^1, \ldots, \text{TgaHead}_{\hat{g}^t}^{n_{tga}})\mathbf{W}_{\hat{g},o}, \tag{2}$$

where $\mathbf{E}_{\hat{g}^t}^{\hat{t}} \in \mathbb{R}^{N \times d_{enc}}$ denotes the output of the graph attention layer, and $\mathbf{W}_{\hat{g},o} \in \mathbb{R}^{n_{tga} \times d_{enc} \times d_{enc}}$ is the convolution matrix for graph $\hat{g}$, n_{tga} denotes the number of attention heads, and each head $\text{TgaHead}_{\hat{g}^t}^i \in \mathbb{R}^{N \times d_{enc}}$ is defined as:

$$\begin{aligned} \text{TgaHead}_{\hat{g}^t}^i &= \sum_{v^{\hat{t}} \in \hat{V}} \text{LeakyRelu}\Big(\sum_{u^{\hat{t}} \in \mathcal{N}_g^t(v^{\hat{t}})} \alpha_{u^{\hat{t}},v^{\hat{t}}}^{i,\hat{g}} \boldsymbol{e}_{u^{\hat{t}}} \Big), \\ \alpha_{u^{\hat{t}},v^{\hat{t}}}^{i,\hat{g}} &= \frac{\exp\,(\text{LeakyRelu}(\boldsymbol{a}_{\hat{g},i}^{\mathrm{T}}[\boldsymbol{e}_{u^{\hat{t}}} \parallel \boldsymbol{e}_{v^{\hat{t}}}]))}{\sum_{k^{\hat{t}} \in \mathcal{N}_g(v^{\hat{t}})} \exp\,(\text{LeakyRelu}(\boldsymbol{a}_{\hat{g},i}^{\mathrm{T}}[\boldsymbol{e}_{k^{\hat{t}}} \parallel \boldsymbol{e}_{v^{\hat{t}}}]))}, \end{aligned} \tag{3}$$

where $\boldsymbol{e}_{u^{\hat{t}}} \in \mathbb{R}^{d_{enc}}$ denotes the $u^{\hat{t}}$-th row of historical feature embedding, $\alpha_{u^{\hat{t}},v^{\hat{t}}}^{i,\hat{g}}$ is the importance of dynamic graph edge $(u^{\hat{t}}, v^{\hat{t}})$ in i-th head for $\hat{g}$, and $\boldsymbol{a}_{\hat{g},i} \in \mathbb{R}^{2d_{enc}}$ denotes the trainable dynamic graph attention weight vector of graph $\hat{g}$ and i-th head. Then we utilize the dynamic relation attention mechanism to capture the final dynamic representation $\mathbf{E}_{\hat{g}^t}^t$ for each relation $\hat{g}^t$. First, we employ a position encoding layer to obtain $\mathbf{H}_{\hat{g}^t}^t \in \mathbb{R}^{N \times T_w \times d_{enc}}$, then we leverage the temporal attention mechanism to encode $\mathbf{H}_{\hat{g}^t}^t$ in each $\hat{t}$ and obtain the final dynamic representation $\mathbf{E}_{\hat{g}^t}^t \in \mathbb{R}^{N \times d_{enc}}$, where $\mathbf{E}_{\hat{g}^t}^t = \text{Attention}(\mathbf{Q}_{\hat{g}^t}^t, \mathbf{K}_{\hat{g}^t}^t, \mathbf{V}_{\hat{g}^t}^t)$:

$$\begin{aligned} \mathbf{H}_{\hat{g}^t}^t &= \mathbf{W}_{tem}\,\text{Stack}_{\hat{t}=t-T_w}^{t-1}(\mathbf{E}_{\hat{g}^t}^{\hat{t}}) + \text{PE}, \\ \mathbf{Q}_{\hat{g}^t}^t &= \mathbf{H}_{\hat{g}^t}^t \mathbf{W}_{Q,\hat{g}}, \mathbf{K}_{\hat{g}^t}^t = \mathbf{H}_{\hat{g}^t}^t \mathbf{W}_{K,\hat{g}}, \mathbf{V}_{\hat{g}^t}^t = \mathbf{H}_{\hat{g}^t}^t \mathbf{W}_{V,\hat{g}}, \end{aligned} \tag{4}$$

where PE is the position embedding, $\mathbf{W}_{tem} \in \mathbb{R}^{d_{enc} \times d_{enc}}, \mathbf{W}_{Q,\hat{g}}, \mathbf{W}_{K,\hat{g}}, \mathbf{W}_{V,\hat{g}} \in \mathbb{R}^{d_{enc} \times d_{enc}}$ denote the projection weights, and d_{enc} denotes the dimension of hidden layer. We have acquired four dynamic relation representations $\mathbf{E}_{\hat{g}^t}^t$ from positive, negative, concept and industry neighbors, respectively, where $\hat{g}^t \in \{\hat{g}_{Pos}^t, \hat{g}_{Neg}^t, \hat{g}_{Concept}^t, \hat{g}_{Sector}^t\}$. Then we apply the contextual relation attention module to adaptively synthesize the contextual representation $\mathbf{E}_{enc}^t \in \mathbb{R}^{N \times d_{enc}}$:

$$\begin{aligned} \mathbf{E}_{enc}^t &= \text{LeakyRelu}(\mathbf{W}_{graph} \mathbf{H}_{g^t}^t + \boldsymbol{b}_{graph}), \\ \mathbf{H}_{g^t}^t &= \text{LeakyRelu}(\text{Attention}(\mathbf{Q}_{g^t}^t, \mathbf{K}_{g^t}^t, \mathbf{V}_{g^t}^t)), \\ \mathbf{Q}_{g^t}^t &= \mathbf{E}_{g^t}^t \mathbf{W}_{Q,g}, \mathbf{K}_{g^t}^t = \mathbf{E}_{g^t}^t \mathbf{W}_{K,g}, \mathbf{V}_{g^t}^t = \mathbf{E}_{g^t}^t \mathbf{W}_{V,g}, \\ \mathbf{E}_{g^t}^t &= \text{Stack}(\mathbf{E}_{\hat{g}_{pos}^t}^t, \mathbf{E}_{\hat{g}_{neg}^t}^t, \mathbf{E}_{\hat{g}_{concept}^t}^t, \mathbf{E}_{\hat{g}_{industry}^t}^t) \end{aligned} \tag{5}$$

where $\mathbf{W}_{Q,g}, \mathbf{W}_{K,g}, \mathbf{W}_{V,g} \in \mathbb{R}^{d_{enc} \times d_{enc}}$ denote the projection matrices, $\mathbf{W}_{graph} \in \mathbb{R}^{4 * d_{enc} \times d_{enc}}$ denotes the synthesizing projection matrix.

3.4 Global Hypergraph Relation Mining Module

Effectively modeling the market is crucial for avoiding extreme situations and enhancing strategy robustness. GHRMM abstracts the hierarchical market relations as a hypergraph [6], considering all stocks connecting each other.

The module takes $\hat{\mathbf{X}}_2^t = \mathcal{F}(\mathbf{X}_1^t, \mathbf{E}_{enc}^t)$ and derive the market representation $\mathbf{E}_{hid}^t \in \mathbb{R}^{N\times d_{enc}}$ following an inductive representation learning method on large graphs [7]:

$$\mathbf{E}_{hid}^t = \text{LeakyRelu}(\mathbf{W}_{1,m}\mathbf{X}_2^t + \mathbf{W}_{2,m} \cdot \underset{v\in\mathcal{V}}{\text{Mean}}\, \hat{\mathbf{X}}_{2,v}^t), \tag{6}$$

where $\mathbf{W}_{1,m}, \mathbf{W}_{2,m}, \in \mathbb{R}^{d_{enc}\times d_{enc}}$ denote the transformation parameters.

3.5 Individual Stock Representation Mining Module

Apart from shaped by market trends and local dynamics, stocks exhibit distinctive traits, which may impede stock differentiation and potentially diminish overall investment returns when neglected. So we propose ISRMM, utilizing $\hat{\mathbf{X}}_3^t = \mathcal{F}(\mathbf{X}_2^t, \mathbf{E}_{hid}^t)$ to generate the individual stock representation $\mathbf{E}_{emb}^t \in \mathbb{R}^{N\times d_{enc}}$:

$$\mathbf{E}_{emb}^t = \text{LeakyRelu}(\mathbf{W}_{emb}\hat{\mathbf{X}}_3^t + \boldsymbol{b}_{emb}), \tag{7}$$

where $\mathbf{W}_{emb} \in \mathbb{R}^{d_{enc}\times d_{enc}}$, $\boldsymbol{b}_{emb} \in \mathbb{R}^{d_{enc}}$ denote the parameters.

3.6 Multi-granularity Portfolio Generator

We have extracted dynamic contextual relation representations $\mathbf{E}_{enc}^t$, overall market status $\mathbf{E}_{hid}^t$, and individual stock properties information $\mathbf{E}_{emb}^t$. Employing these embeddings, we utilize an attention mechanism to produce adaptive dynamic weights (α_{enc}^t, α_{hid}^t, α_{emb}^t) corresponding to the three granularity:

$$\begin{aligned} \alpha_l^t &= \frac{\exp(\mathbf{H}_l^t)}{\sum_{l\in\{enc,hid,emb\}} \exp(\mathbf{H}_l^t)}, \\ \mathbf{H}_l^t &= \underset{v^t\in\mathcal{V}}{\text{Mean}}(\boldsymbol{q}^{\mathrm{T}} \cdot \text{LeakyRelu}(\mathbf{W}_{att}\mathbf{E}_{l,v}^t + \boldsymbol{b}_{att})), \end{aligned} \tag{8}$$

where we employ an attention vector $\boldsymbol{q} \in \mathbb{R}^{d_{enc}}$ to adaptively learn the dynamic weights for various granularity. Then we assess the contribution of each granularity and compute the final features $\mathbf{E}_{out}^t \in \mathbb{R}^{N\times denc} = \sum_{l\in\{enc,hid,emb\}} \alpha_l \cdot \mathbf{E}_l^t$. Then we obtain the final portfolio probability $\boldsymbol{a}_t \in \mathbb{R}^N$:

$$\boldsymbol{a}_t = \mathbf{W}_p\mathbf{E}_{out}^t + \boldsymbol{b}_p, \tag{9}$$

where $\mathbf{W}_p \in \mathbb{R}^{d_{enc}\times 1}$, $\boldsymbol{b}_p \in \mathbb{R}$ are both trainable parameters.

Finally, we incorporate NGNRL into the Actor ($\mathcal{A}$)-Critic ($\mathcal{C}$) Reinforcement Learning (RL) framework [12] to achieve end-to-end optimization.

4 Experiments

Datasets. For data consistency and relevance, we utilized historical market features and news from stocks in the CSI300 index of China and stocks in the NASDAQ index of the United States. The training set encompasses data from 2019 to 2021, while the data from 2022 constitutes the test set.

Baseline Comparison. We compare various baselines. Traditional approaches include reversal strategy BLSW [11] and momentum strategy CSM [19]. DL models include TRA [16], THGNN [25], and CGM [30], representing sequence modeling, dynamic relationship modeling, and news modeling, respectively. RL models under consideration are the classic PPO [21], attention-based AlphaStock [23], static-graph-based DeepTrader [24] and dynamic-graph-based DeepPocket [22].

Evaluation Metrics. Evaluation metrics include Annualized Rate of Return (ARR), Annualized Volatility (AVol), Maximum Draw Down (MDD), Annualized Sharpe Ratio (ASR), Calmar Ratio (CR), and Information Ratio (IR).

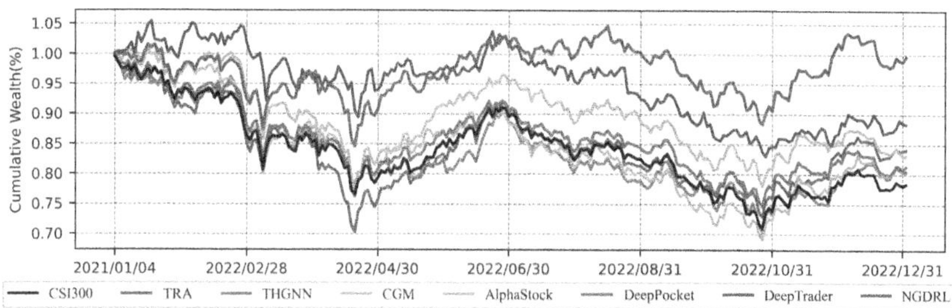

Fig. 2. The accumulated excess returns gained each day in the test set (2022) of our model, CSI300 index, and selected baselines.

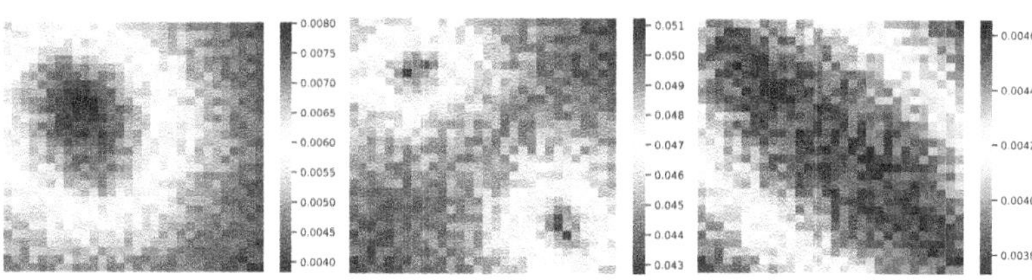

Fig. 3. Visualization of relation attention weights, X-axis and Y-axis both denote the node degree in each news relation graph. The x-axis from left to right, and the y-axis from bottom to top, represent the gradual increase of degrees.

4.1 Results and Analysis

NGDRL consistently outperforms all baselines across two datasets, with quantitative results in Table 1 and the visualization in Fig. 2. The model achieves significant portfolio improvement under the reward function that maximizes cumulative returns. **As shown in Table 1, while risk-related metrics only achieve second place, the most important metrics that measure profit per unit of risk, far exceed other baselines.** This demonstrates its ability to navigate market trend risks and consistent performance across diverse market conditions. The cumulative return curve vividly illustrates NGDRL's robust profitability and sensitivity to market opportunities. Even in an adverse market characterized by an overall decline tendency, it demonstrated resilience with limited drawdown and the ability to withstand downward pressure, showcasing NGDRL's nuanced understanding of stock dynamic contextual relations. In summary, NGDRL consistently delivers excellent profitability and risk resilience in the Chinese Stock Market and the U.S. Stock Market.

Table 1. Performance of all models on CSI300 and NASDAQ 100, with bold highlighting for the best and underlining for the second-best results.

Datasets	CSI300						NASDAQ					
Models	ARR ↑	AVol ↓	MDD ↑	ASR ↑	CR ↑	IR ↑	ARR ↑	AVol ↓	MDD ↑	ASR ↑	CR ↑	IR ↑
BLSW	−0.108	0.136	−0.131	−0.793	−0.824	−0.712	0.280	0.212	−0.117	1.325	2.389	1.207
CSM	0.012	0.146	−0.150	0.083	0.081	0.301	−0.121	0.451	−0.392	−0.269	−0.310	−0.102
TRA	0.080	0.094	−0.095	0.851	0.842	0.833	0.259	**0.086**	−0.049	3.011	5.280	2.682
THGNN	0.040	0.103	−0.083	0.388	0.483	0.388	0.182	0.132	−0.104	1.374	1.754	1.334
CGM	0.033	0.089	−0.103	0.373	0.321	0.377	0.149	0.101	−0.058	1.474	2.568	1.325
PPO	−0.031	0.282	−0.109	−0.110	−0.285	0.011	0.171	0.458	−0.230	0.373	0.742	0.574
AlphaStock	0.062	0.083	−0.079	0.746	0.785	0.706	0.288	0.477	−0.171	0.604	1.681	0.770
DeepPocket	0.027	**0.040**	**−0.030**	0.675	0.902	0.817	0.206	0.090	−0.047	2.889	4.382	2.315
DeepTrader	0.083	0.309	−0.135	0.268	0.615	0.399	0.244	0.469	−0.105	0.520	2.323	0.702
NGDRL	**0.281**	0.082	−0.089	**3.427**	**3.157**	**3.034**	**0.320**	0.088	**−0.045**	**3.635**	**7.111**	**3.211**

Table 2. Abalation performance for multi-granularity modules, and the RL training framework, with bold highlighting for the best.

Models	ARR↑	AVol↓	MDD↑	ASR↑	CR↑	IR↑
NGDRL (w/o ISRMM)	0.159	0.087	**−0.054**	1.828	2.944	1.697
NGDRL (w/o DCRPM)	0.068	0.084	−0.076	0.810	0.894	0.769
NGDRL (w/o GHRMM)	0.223	0.128	−0.079	1.742	2.822	1.602
NGDRL (w/o ITS)	0.231	0.085	−0.092	2.718	2.511	3.017
NGDRL (DL)	0.185	**0.089**	−0.065	2.313	2.846	2.063
NGDRL	**0.281**	**0.082**	−0.089	**3.427**	**3.157**	**3.034**

4.2 Ablation Study

We conducted experiments to validate the significance of multi-granularity modules and the rationale for utilizing RL in portfolio optimization. From Table 2, it is evident that removing any granularity-specific feature learning module or our designed Residual Information Transmission Structure (ITS) results in the inability to learn information for the specified granularity or the failure to orthogonally fuse multi-granularity features across different levels. Changing to DL necessitates adjusting the objective function to predict future stock prices, deviating from investor goals, and ultimately compromising model performance.

4.3 Interpretability of the Portfolio

To delve deeper into understanding multi-dimensional dynamic relationships, we conducted a thorough analysis of attention weights across three types of relation graphs. From Fig. 3(a), it is evident that Concept Relation predominantly impacts nodes with lower degrees, facilitating information exchange among smaller-scale startups during shifts in hot-topic areas. Meanwhile, Sector Relation demonstrates a noticeable size effect, revealing two dynamic centers in the graph. These centers are situated within node clusters with varying degrees of connectivity, while other nodes are comparatively less influenced by sector relations, as depicted in Fig. 3(b). Additionally, Fig. 3(c) highlights a noteworthy size-matching effect within Sentiment Relation; nodes with similar degrees contribute more information to each other sentiment-wise. This observation aligns with the strategies of human experts to closely examine benchmark companies or competitors closely associated with the investment target.

5 Conclusion

In this paper, we propose a dynamic news graph-based deep reinforcement learning framework for enhancing portfolio strategies. Our model extracts multi-dimensional logical information from news texts and constructs dynamic stock contextual relation graphs accordingly. Subsequently, we design effective modules to extract information from the dynamic graphs and multi-granularity temporal features and directly optimize portfolio returns. We conduct extensive experiments and extensive ablation studies on two representative real-world stock datasets, demonstrating the state-of-the-art performance.

Acknowledgments. This work was supported by the National Key R&D Program of China (Grant no. 2022YFB4501704), the National Natural Science Foundation of China (Grant no. 62102287), and the Shanghai Science and Technology Innovation Action Plan Project (Grant no. 22YS1400600 and 22511100700).

References

1. Adam, K., Marcet, A., Nicolini, J.P.: Stock market volatility and learning. The Journal of finance (2016)
2. Cheng, D., Yang, F., Wang, X., Zhang, Y., Zhang, L.: Knowledge graph-based event embedding framework for financial quantitative investments. In: SIGIR (2020)
3. Ding, X., Zhang, Y., Liu, T., Duan, J.: Using structured events to predict stock price movement: an empirical investigation. In: EMNLP (2014)
4. Ding, X., Zhang, Y., Liu, T., Duan, J.: Deep learning for event-driven stock prediction. In: IJCAI (2015)
5. Du, X., Tanaka-Ishii, K.: Stock embeddings acquired from news articles and price history, and an application to portfolio optimization. In: ACL (2020)
6. Gao, Y., Feng, Y., Ji, S., Ji, R.: Hgnn+: general hypergraph neural networks. TPAMI (2022)

7. Hamilton, W., Ying, Z., Leskovec, J.: Inductive representation learning on large graphs. In: NeurIPS (2017)
8. Han, L., Ding, N., Wang, G., Cheng, D., Liang, Y.: Efficient continuous space policy optimization for high-frequency trading. In: KDD (2023)
9. Hu, Z., Liu, W., Bian, J., Liu, X., Liu, T.Y.: Listening to chaotic whispers: a deep learning framework for news-oriented stock trend prediction. In: WSDM (2018)
10. Huang, H., Gao, J., Xu, C., Ying, X.: Pattern adaptive specialist network for learning trading patterns in stock market. In: CIKM (2022)
11. Jegadeesh, N., Titman, S.: Returns to buying winners and selling losers: implications for stock market efficiency. J. Finance (1993)
12. Konda, V., Tsitsiklis, J.: Actor-critic algorithms. In: NeurIPS (1999)
13. Liang, Q., Zhu, M., Zheng, X., Wang, Y.: An adaptive news-driven method for cvar-sensitive online portfolio selection in non-stationary financial markets. In: IJCAI (2021)
14. Liang, X., Cheng, D., Yang, F., Luo, Y., Qian, W., Zhou, A.: F-hmtc: detecting financial events for investment decisions based on neural hierarchical multi-label text classification. In: IJCAI (2020)
15. Liang, Y., Song, Q., Zhao, Z., Zhou, H., Gong, M.: BA-GNN: behavior-aware graph neural network for session-based recommendation. Frontiers Comput. Sci. (2023)
16. Lin, H., Zhou, D., Liu, W., Bian, J.: Learning multiple stock trading patterns with temporal routing adaptor and optimal transport. In: KDD (2021)
17. Liu, Q., Cheng, X., Su, S., Zhu, S.: Hierarchical complementary attention network for predicting stock price movements with news. In: CIKM (2018)
18. Niu, H., Li, S., Li, J.: Metatrader: an reinforcement learning approach integrating diverse policies for portfolio optimization. In: CIKM (2022)
19. Poterba, J.M., Summers, L.H.: Mean reversion in stock prices: evidence and implications. J. Financ. Econ. (1988)
20. Sawhney, R., Wadhwa, A., Agarwal, S., Shah, R.R.: FAST: financial news and tweet based time aware network for stock trading. In: ACL (2021)
21. Schulman, J., Wolski, F., Dhariwal, P., Radford, A., Klimov, O.: Proximal policy optimization algorithms. arXiv preprint arXiv:1707.06347 (2017)
22. Soleymani, F., Paquet, E.: Deep graph convolutional reinforcement learning for financial portfolio management. Expert Syst. Appl. (2021)
23. Wang, J., Zhang, Y., Tang, K., Wu, J., Xiong, Z.: Alphastock: a buying-winners-and-selling-losers investment strategy using interpretable deep reinforcement attention networks. In: KDD (2019)
24. Wang, Z., Huang, B., Tu, S., Zhang, K., Xu, L.: Deeptrader: a deep reinforcement learning approach for risk-return balanced portfolio management with market conditions embedding. AAAI (2021)
25. Xiang, S., Cheng, D., Shang, C., Zhang, Y., Liang, Y.: Temporal and heterogeneous graph neural network for financial time series prediction. In: CIKM (2022)
26. Xu, W., et al.: Hist: a graph-based framework for stock trend forecasting via mining concept-oriented shared information. arXiv preprint arXiv:2110.13716 (2021)
27. Xu, W., Liu, W., Xu, C., Bian, J., Yin, J., Liu, T.: REST: relational event-driven stock trend forecasting. In: WWW (2021)
28. Xu, Y., Cohen, S.B.: Stock movement prediction from tweets and historical prices. In: ACL (2018)
29. Yang, M., Zheng, X., Liang, Q., Han, B., Zhu, M.: A smart trader for portfolio management based on normalizing flows. In: IJCAI (2022)

30. Zhao, L., Li, W., Bao, R., Harimoto, K., Wu, Y., Sun, X.: Long-term, short-term and sudden event: trading volume movement prediction with graph-based multi-view modeling. In: IJCAI (2021)
31. Zou, J., Cao, H., Liu, L., Lin, Y., Abbasnejad, E., Shi, J.Q.: Astock: a new dataset and automated stock trading based on stock-specific news analyzing model. In: ACL (2022)

CGCL: Collaborative Graph Contrastive Learning Without Handcrafted Graph Data Augmentations

Tianyu Zhang[1], Yuxiang Ren[2(✉)], Wenzheng Feng[2], Weitao Du[2,3], and Xuecang Zhang[2]

[1] Department of Automation, Tsinghua University, Beijing, China
[2] Huawei Technologies, Shenzhen, China
{renyuxiang1,fengwenzheng,zhangxuecang}@huawei.com,
[3] Chinese Academy of Sciences, Beijing, China
duweitao@amss.ac.cn

Abstract. Existing graph contrastive learning (GCL) aims to learn the invariance across multiple augmentation views, which renders it heavily reliant on the handcrafted graph augmentations. However, inappropriate graph data augmentations can potentially jeopardize such invariance. In this paper, we show the potential hazards of inappropriate augmentations and then propose a novel Collaborative Graph Contrastive Learning framework (CGCL). This framework harnesses multiple graph encoders to observe the graph. Features observed from different encoders serve as the contrastive views in contrastive learning, which avoids inducing unstable perturbation and guarantees the invariance. To ensure the collaboration among diverse graph encoders, we propose the concepts of asymmetric architecture and complementary encoders as the design principle. To further prove the rationality, we utilize two quantitative metrics to measure the assembly of CGCL respectively. Extensive experiments demonstrate the advantages of CGCL in unsupervised graph-level representation learning and the potential of collaborative framework. The source code is available at https://github.com/zhangtia16/CGCL.

Keywords: Graph Representation Learning · Contrastive Learning · Collaborative Framework

1 Introduction

Graph neural networks (GNNs) [4,14,16] have demonstrated excellent performance for graph representation learning in supervised or semi-supervised scenarios. However, task-specific labels could be scarce and unsupervised graph representation learning has emerged as the critical technology to achieve breakthroughs. Recently, contrastive learning has shown potential in unsupervised

T. Zhang and Y. Ren—Contribute equally to this paper.

M. Onizuka et al. (Eds.): DASFAA 2024, LNCS 14855, pp. 418–428, 2024.
https://doi.org/10.1007/978-981-97-5572-1_30

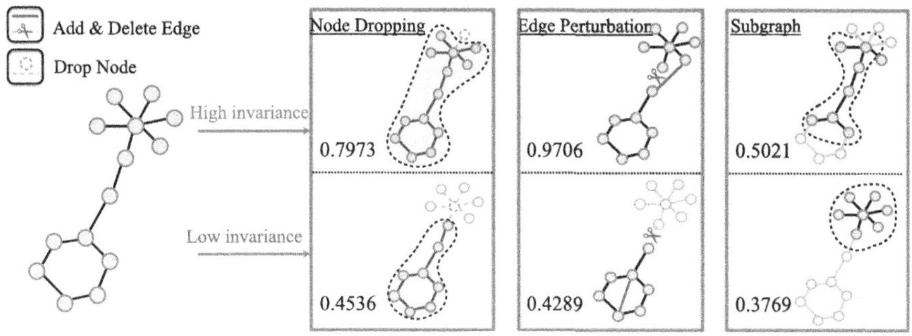

Fig. 1. An illustration of the unstable invariance of three graph augmentation strategies. The value of each augmented graph is its cosine similarity to the original graph. The upper part of each augmentation strategy shows augmented graphs preserving high invariance, while the lower part's augmentations bring low invariance.

graph representation learning [6,11,15,17,21], tending to maximize feature consistency under differently augmented views to learn the desired invariance. As a result, data augmentation strategies have a strong influence on contrastive learning performance. Data augmentation has achieved great success in image data. However, due to the complex structural information, the changes induced by the data augmentation to graphs are not easy to measure. Specifically, removing a critical edge is enough to change a graph from a connected graph to a disconnected one, making the augmented graph and the original graph have little learnable invariance. In Fig. 1, we provide an illustration to show that the impact of handcrafted augmentations on invariance is uncontrollable, even under the same augmentation strategy.

To remedy the issue of unstable invariance from inappropriate data augmentations, we propose a novel graph-level contrastive learning framework named CGCL, where no handcrafted graph augmentation is needed. CGCL uses multiple GNN-based graph encoders to enforce contrastive learning collaboratively. To cope with model collapse, we devise the asymmetric structure for CGCL. The asymmetry lies in the differences of graph encoders. Besides, graph encoders are supposed to be complementary for a stronger fitting ability. Specifically, high complementarity indicate that encoders together carry less redundant parameters. Furthermore, we propose two metrics to measure the asymmetry and complementarity of the collaborative framework quantitatively. Compared with the existing methods, CGCL demonstrates better generalization on various datasets and achieves better results. In addition, we implement experiments with the two quantitative metrics and find that the assembly with high asymmetry and complementarity has a better performance. The contributions of our work are summarized as follows: (1) We propose a novel **C**ollaborative **G**raph **C**ontrastive **L**earning (CGCL) to reinforce unsupervised graph-level representation learning. (2) We propose the concepts of asymmetric structure and complementary encoders as foundational principles for CGCL. Besides, we design two quantitative metrics to assess both the asymmetry and complementarity inherent in the collaborative framework. (3) Extensive experiments show that CGCL has advantages in graph classification task. Empirical evidence also validates that the architecture of CGCL, characterized by its inherent asymmetry and complementarity, indeed yields enhanced performance outcomes.

2 Related Work

Graph-Level Representation Learning. Early researches could track back to kernel-based methods, including GK [13], WL [12], and DGK [20]. Later, the idea of neural networks enables researchers to develop some graph embedding-based methods, such as Node2vec [3], Graph2vec [8] and Sub2vec [1]. Recently, GNNs have shown awe-inspiring capabilities. Classic GNNs include GCN [4], GAT [16], GIN [19] and etc. Recently, due to the insufficient labeled data, unsupervised setting attracts researchers' attention. Some representative methods include InfoGraph [14], GraphCL [21], AD-GCL [15] and RGCL [6].

Graph Contrastive Learning. DGI [17] and InfoGraph [14] use the idea of mutual information maximization in graph representation learning. More recently, GCL methods with data augmentations have achieved impressive performance, such GraphCL [21], AD-GCL [15], LCGNN [10] and etc. Those methods heavily rely on the choices of handcrafted augmentations, which may bring unstable performance or clumsy pipelines. Some latest researches have been dedicating to explore the non-necessity of graph data augmentation, including AF-GRL [5], SimGCL [22] and etc. By contrast, our proposed CGCL learns graph-level representations through a novel collaborative contrastive framework.

3 Methodology

3.1 GNN-Based Graph Encoders

In Fig. 2, we illustrate the overview of CGCL. Multiple graph encoders process input graphs, yielding embeddings for each graph. Every graph encoder updates its parameters by contrasting its learned embeddings to the outputs from other encoders. We mainly employ GNNs as graph encoders. The graph encoder candidates include GIN [19], GCN [4], GAT [16], DGCNN [23], and Set2Set [18] in this paper. Graph encoders can be combined into what we term the CGCL assembly, and can be trained in the collaborative contrastive learning framework.

3.2 Collaborative Contrastive Learning

In this section we first explain the essence of collaborative framework. Subsequently, we delve into the two characteristics inherent to CGCL: an asymmetric structure and the complementary encoders. To augment our theoretical exposition, we introduce two quantitative metrics tailored to assess both the asymmetry and complementarity of the collaborative framework.

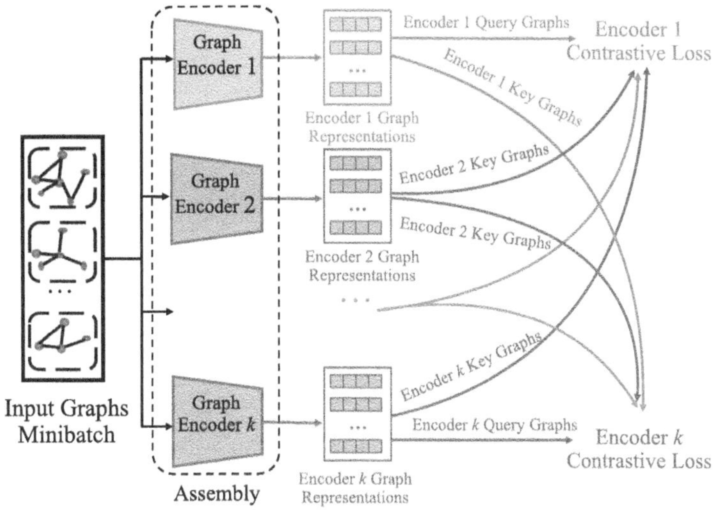

Fig. 2. Framework overview of CGCL.

Encoder Perspective for Generating Contrastive Views. It is pivotal to recognize that the quintessence of contrastive learning is to learn invariance between different contrastive views inconsistent to the original graph. This underscores that CGCL innovatively generates different contrastive views from the encoder perspective, eschewing traditional data augmentations. Given that variations in either encoder configurations or data invariably manifest in the resultant embeddings, collaborative contrastive learning parallels the efficacy of augmentation-based methods but is free from the often cumbersome or resource-intensive augmentation phase. With the multi contrastive views, encoders collaboratively interact. Specifically, each encoder's output, serving as both query and key graphs, influences its inherent loss and those of its counterparts. This process fosters a mutual distillation of knowledge on graph latent representations.

Asymmetric Architecture. In contrastive learning, model collapse presents a significant challenge. Existing methods employ the Exponential Moving Average or Stop Gradient to avoid model collapse. Those techniques can be concluded as introducing asymmetry into model architecture. Reflecting on CGCL, diverse encoders sustain the asymmetry of the framework. Naturally, the extent of asymmetry can be reflected on the differences between encoders. Thus, we introduce the Representational Dissimilarity Matrix (RDM), which is a prevalent tool in cognitive neuroscience [9]. Model discrepancies can be measured by calculating the correlation between two RDMs. For clarification, we illustrate the process of calculating the correlation between RDMs in Fig. 3. Based on the concept of RDM, we define the Asymmetry Coefficient of a certain CGCL's assembly as follows:

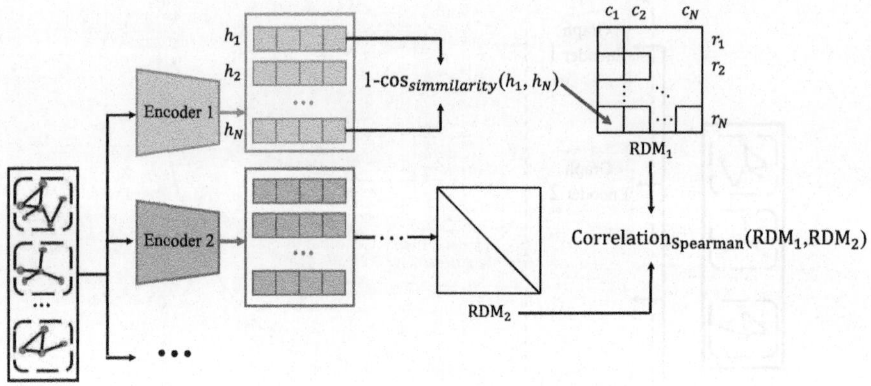

Fig. 3. Calculation of correlation between RDMs. With k encoders, Asymmetry Coefficient is calculated by averaging the correlation between any pair of encoders.

Definition 1. *Let k denote the number of trained encoders, and RDM_i^D be the Representational Dissimilarity Matrix defined by encoder i and data D. Then the Asymmetry Coefficient (AC) is formulated as:*

$$AC = 1 - \frac{1}{C_k^2} \sum_{(i,j) \in Combination(k)} Correlation(RDM_i^D, RDM_j^D) \tag{1}$$

A high AC indicates that CGCL's assembly holds high asymmetry and vice versa.

Complementary Encoders. Within CGCL, multiple graph encoders observe input graphs to yield contrastive views. Ideally, these encoders should exhibit complementarity to enhance fitting capability. This notion of complementarity in CGCL mirrors the diversity imperative of base learners in ensemble learning, where distinct learners better capture varied information. Inspired by [2], we introduce a loss-centric metric to measure the complementarity of CGCL's encoders. For clarification, we refer the training loss upon completion as the stopping loss. Given a consistent dataset, a smaller stopping loss signifies the assembly's better fitting ability. With the above intuition, we define the Complementarity Coefficient of a certain CGCL's assembly as follows:

Definition 2. *Let k denote the number of encoders, and L_i^{Stop} be the loss of encoder i when collaborative training is stopped. The Complementarity Coefficient (CC) is given by:*

$$CC = -\frac{1}{k} \sum_{i \in 1,2,\ldots,k} L_i^{Stop} \tag{2}$$

With this definition, CGCL's encoders are more complementary with a higher CC. In Eq. (2), we introduce normalization to get rid of the effect of k.

3.3 Batch-Wise Contrastive Loss

Given a minibatch of N graphs $\{\mathcal{G}_1, \mathcal{G}_2, \ldots, \mathcal{G}_N\}$ as the input of k graph encoders $\{M_1, M_2, \ldots, M_k\}$, we utilize the training batch to conduct contrastive learning. We take M_1 as an example. The representations $\{\boldsymbol{h}_{\mathcal{G}_1}^{M_1}, \boldsymbol{h}_{\mathcal{G}_2}^{M_1}, \ldots, \boldsymbol{h}_{\mathcal{G}_N}^{M_1}\}$ learned by M_1 act as query graphs, while the ones learned by other encoders are key graphs. The representations learned by different graph encoders of the same graph serve as positive pairs. Thus, the contrastive loss $\mathcal{L}_{M_p}(M_q)$ between two encoders M_p and M_q can be computed as:

$$\mathcal{L}_{M_p}(M_q) = -\sum_{j=1}^{N} \log \frac{\exp\left(\boldsymbol{h}_{\mathcal{G}_j}^{M_p} \cdot \boldsymbol{h}_{\mathcal{G}_j}^{M_q} / \tau\right)}{\sum_{n=1}^{N} \exp\left(\boldsymbol{h}_{\mathcal{G}_j}^{M_p} \cdot \boldsymbol{h}_{\mathcal{G}_n}^{M_q} / \tau\right)} \tag{3}$$

Here, τ is the temperature hyper-parameter. In this contrastive loss, M_p acts as the query encoder, while M_q provides key graphs for contrast. When considering all collaborative graph encoders, the contrastive loss of M_p can be denoted as:

$$\mathcal{L}_{M_p} = \sum_{M_q \in \{M_i\}_{i=1}^{k} \setminus M_p} \mathcal{L}_{M_p}(M_q) \tag{4}$$

M_p can be updated based on the contrastive loss $\mathcal{L}_{M_p}$ through back-propagation. Other graph encoders follow the same process to compute their own contrastive loss and get updated by minibatch training iteratively.

4 Experiments

4.1 Datasets and Baselines

To evaluate CGCL for graph-level representation learning, we conduct experiments on nine widely used benchmark datasets: NCI1, PROTEINS, D&D, MUTAG, COLLAB, IMDB-BINARY, IMDB-MULTI, REDDIT-BINARY, REDDIT-MULTI-5K. All of them come from the benchmark TUDataset [7]. Following [21], we select three categories of baseline methods: kernel-based methods including **WL** [12], **GK** [13], **DGK** [20], graph embedding-based methods including **Sub2vec** [1], **Node2vec** [3], **Graph2vec** [8] and GCL methods including **InfoGraph** [14], **GarphCL** [21], **AD-GCL** [15], **RGCL** [6].

4.2 Unsupervised Graph Representation Learning

We first evaluate CGCL in the setting of unsupervised learning. Following InfoGraph [14], the representations are learned by models without any labels and then fed into a SVM to evaluate the graph classification performance. CGCL$_{GIN}$ denotes GIN trained under the proposed CGCL. The graph encoders of CGCL include GIN, GAT, GCN, and DGCNN. The representations learned by CGCL$_{GIN}$ are extracted for the downstream graph classification. We

Table 1. Performance on graph classification accuracy. The best result is **bolded**.

Methods	NCI1	PROTEINS	D&D	MUTAG	COLLAB
WL	76.65 ± 1.99	72.92 ± 0.56	76.44 ± 2.35	80.72 ± 3.00	-
GK	62.48 ± 2.11	72.23 ± 4.49	72.54 ± 3.83	81.66 ± 2.11	72.84 ± 0.28
DGK	**80.31 ± 0.46**	73.30 ± 0.82	71.12 ± 0.21	87.44 ± 2.72	73.09 ± 0.25
Node2vec	54.89 ± 1.61	57.49 ± 3.57	67.12 ± 4.32	72.63 ± 10.20	-
Sub2vec	52.84 ± 1.47	53.03 ± 5.55	59.34 ± 8.01	61.05 ± 15.80	-
Graph2vec	73.22 ± 1.81	73.30 ± 2.05	71.98 ± 3.54	83.15 ± 9.25	-
InfoGraph	76.20 ± 1.06	74.44 ± 0.31	72.85 ± 1.78	89.01 ± 1.13	70.65 ± 1.13
GraphCL	77.87 ± 0.41	74.39 ± 0.45	78.62 ± 0.40	86.80 ± 1.34	71.36 ± 1.15
AD-GCL	73.91 ± 0.77	73.28 ± 0.46	75.79 ± 0.87	88.74 ± 1.85	72.02 ± 0.56
RGCL	78.14 ± 1.08	75.03 ± 0.43	78.86 ± 0.48	87.66 ± 1.01	70.92 ± 0.65
$CGCL_{GIN}$	77.89 ± 0.54	**76.28 ± 0.31**	**79.37 ± 0.47**	**89.05 ± 1.42**	**73.28 ± 2.12**
Methods	IMDB-B	IMDB-M	RDT-B	RDT-M	
WL	72.30 ± 3.44	46.95 ± 0.46	68.82 ± 0.41	46.06 ± 0.21	
GK	65.87 ± 0.98	-	77.34 ± 0.18	41.01 ± 0.17	
DGK	66.96 ± 0.56	44.55 ± 0.52	78.04 ± 0.39	41.27 ± 0.18	
Node2vec	61.03 ± 7.13	-	-	-	
Sub2vec	55.26 ± 1.54	36.67 ± 0.83	71.48 ± 0.41	36.68 ± 0.42	
Graph2vec	71.10 ± 0.54	50.44 ± 0.87	75.78 ± 1.03	47.86 ± 0.26	
InfoGraph	73.03 ± 0.87	49.69 ± 0.53	82.50 ± 1.42	53.46 ± 1.03	
GraphCL	71.14 ± 0.44	-	89.53 ± 0.84	55.99 ± 0.28	
AD-GCL	70.21 ± 0.68	46.60 ± 0.01	90.07 ± 0.85	54.33 ± 0.32	
RGCL	71.85 ± 0.84	49.27 ± 0.00	90.34 ± 0.58	**56.38 ± 0.40**	
$CGCL_{GIN}$	**73.11 ± 0.74**	**51.73 ± 1.37**	**91.31 ± 1.22**	54.47 ± 1.08	

test the task 10 times and report the average accuracy and the standard deviation. From the results in Table 1, CGCL achieves the best result on seven out of nine datasets. Noting that GraphCL and RGCL have better result on RDT-M, where average node degree is extremely large compared to other datasets. On the contrary, those two methods with data augmentations perform worse then the proposed CGCL on other datasets with a low average node degree. This observation is in line with our illustration in Fig. 1, since when the graph is small and not dense enough, the perturbation from the data augmentation is more likely to damage the invariance.

4.3 Asymmetry and Complementarity of CGCL's Assembly

For a further exploration of CGCL's working mechanism, we test various assembly with respect to the two quantitative metrics AC and CC. The candidate encoders include GIN, GCN, GAT and Set2Set. Specifically, we combine two

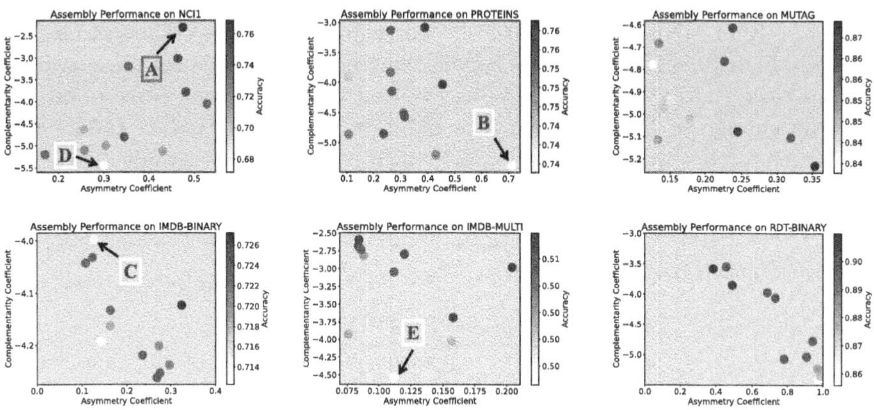

Fig. 4. The performance of CGCL's assembly w.r.t. asymmetry and complementarity over multiple datasets. Point A indicates an example of the assembly with a high AC and CC. While point B has a high AC and a low CC, point C is with a low AC and a high CC high. Point D an E refer to the examples whose AC and CC are both low.

or three of graph encoders on six datasets and use the best result of multiple encoders as this assembly's result. Each point represents a certain assembly and its color indicates the performance. This experiment is implemented in the setting of unsupervised graph representation learning. As illustrated in Fig. 4, the best assembly generally appears in the top right-hand corner, for example point A. This indicates that the assembly with both high AC and CC generally performs better, which further justifies the rationality of the design of asymmetric architecture and complementary encoders for CGCL. Besides, the assembly holding only one property may fail the graph representation learning and result in the bad performance of downstream classification. For instance, the assembly in the bottom right-hand corner (point B) on PROTEINS has high asymmetry, however performs bad because of its very low complementarity. Another bad example is the assembly in the top left-hand corner of IMDB-BINARY (point C). Though it has a high CC, the AC of this assembly is fairly low. In addition, there also exist assembly, whose AC and CC are both at low levels, such as point D on NCI1 and point E on IMDB-MULTI. With the above analysis, we further confirm the effectiveness of our proposed framework.

4.4 Convergence Analysis

To check the reliability of collaborative mechanism, we empirically analyze the convergence in the optimization process of each encoder on PROTEINS and IMDB-BINARY. The assembly includes GIN, GCN and GAT. Figure 5 shows that each graph encoder converges synchronously on the two datasets. Besides, we list the RDMs correlation between pairs of GIN, GCN and GAT in Table 2 for reference. According to the figure and the table, the trend of three encoders' contrastive losses is in accord with their RDMs' correlation of each other. In terms of PROTEINS, for example, the correlation between GIN and GCN is relatively high (0.8620), while the correlations between GAT and those two are low and close (0.5732/0.6308). We observe that those relations correspond well with the change of losses, which also exhibits the distinction of GAT. Such accordance further proves the rationality of introducing the concept of RDM.

Table 2. The RDMs Correlation between graph encoders in $\text{CGCL}_{GIN/GCN/GAT}$ on PROTEINS and IMDB-BINARY.

RDMs Correlation	PROTEINS	IMDB-B
GIN-GCN	0.8620	0.7164
GIN-GAT	0.5732	0.6891
GCN-GAT	0.6308	0.7038

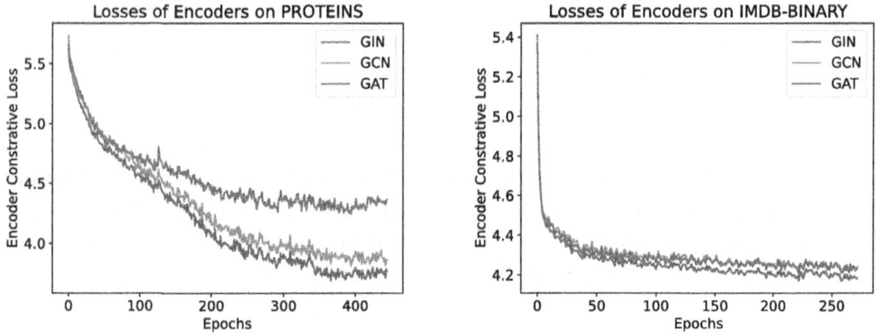

Fig. 5. Empirical convergence study of different graph encoders in CGCL.

5 Conclusion

This paper introduces CGCL, a novel collaborative graph contrastive learning framework, designed to address the invariance challenge encountered in current GCL methods. CGCL employs multiple GNN-based encoders to generate multiple contrastive views. We then propose the concepts of asymmetric structure and complementary encoders as the design principle for the collaborative framework. Furthermore, we propose two quantitative metrics to measure the asymmetry and complementarity of the collaborative framework. Extensive experiments substantiate the advantages of CGCL and underscores the potential of collaborative framework in the field of GCL.

References

1. Adhikari, B., Zhang, Y., Ramakrishnan, N., Prakash, B.A.: Sub2Vec: feature learning for subgraphs. In: Phung, D., Tseng, V.S., Webb, G.I., Ho, B., Ganji, M., Rashidi, L. (eds.) PAKDD 2018. LNCS (LNAI), vol. 10938, pp. 170–182. Springer, Cham (2018). https://doi.org/10.1007/978-3-319-93037-4_14
2. Gontijo-Lopes, R., Smullin, S.J., Cubuk, E.D., Dyer, E.: Affinity and diversity: Quantifying mechanisms of data augmentation. arXiv preprint arXiv:2002.08973 (2020)
3. Grover, A., Leskovec, J.: node2vec: scalable feature learning for networks. In: Proceedings of the 22nd ACM SIGKDD International Conference on Knowledge Discovery and Data Mining, pp. 855–864 (2016)
4. Kipf, T.N., Welling, M.: Semi-supervised classification with graph convolutional networks. arXiv preprint arXiv:1609.02907 (2016)
5. Lee, N., Lee, J., Park, C.: Augmentation-free self-supervised learning on graphs. In: Proceedings of the AAAI Conference on Artificial Intelligence, vol. 36, pp. 7372–7380 (2022)
6. Li, S., Wang, X., Zhang, A., Wu, Y., He, X., Chua, T.S.: Let invariant rationale discovery inspire graph contrastive learning. In: International Conference on Machine Learning, pp. 13052–13065. PMLR (2022)
7. Morris, C., Kriege, N.M., Bause, F., Kersting, K., Mutzel, P., Neumann, M.: Tudataset: a collection of benchmark datasets for learning with graphs. arXiv preprint arXiv:2007.08663 (2020)a
8. Narayanan, A., Chandramohan, M., Venkatesan, R., Chen, L., Liu, Y., Jaiswal, S.: graph2vec: learning distributed representations of graphs. arXiv preprint arXiv:1707.05005 (2017)
9. Popal, H., Wang, Y., Olson, I.R.: A guide to representational similarity analysis for social neuroscience. Soc. Cognitive Affective Neurosci. **14**(11), 1243–1253 (2019)
10. Ren, Y., Bai, J., Zhang, J.: Label contrastive coding based graph neural network for graph classification. In: Database Systems for Advanced Applications (2021)
11. Ren, Y., Liu, B., Huang, C., Dai, P., Bo, L., Zhang, J.: Heterogeneous deep graph infomax. arXiv preprint arXiv:1911.08538 (2019)
12. Shervashidze, N., Schweitzer, P., Van Leeuwen, E.J., Mehlhorn, K., Borgwardt, K.M.: Weisfeiler-lehman graph kernels. J. Mach. Learn. Res. **12**(9) (2011)
13. Shervashidze, N., Vishwanathan, S., Petri, T., Mehlhorn, K., Borgwardt, K.: Efficient graphlet kernels for large graph comparison. In: Artificial Intelligence and Statistics, pp. 488–495. PMLR (2009)
14. Sun, F.Y., Hoffmann, J., Verma, V., Tang, J.: Infograph: unsupervised and semi-supervised graph-level representation learning via mutual information maximization. arXiv preprint arXiv:1908.01000 (2019)
15. Suresh, S., Li, P., Hao, C., Neville, J.: Adversarial graph augmentation to improve graph contrastive learning. Adv. Neural. Inf. Process. Syst. **34**, 15920–15933 (2021)
16. Veličković, P., Cucurull, G., Casanova, A., Romero, A., Lio, P., Bengio, Y.: Graph attention networks. arXiv preprint arXiv:1710.10903 (2017)
17. Veličković, P., Fedus, W., Hamilton, W.L., Liò, P., Bengio, Y., Hjelm, R.D.: Deep graph infomax. arXiv preprint arXiv:1809.10341 (2018)
18. Vinyals, O., Bengio, S., Kudlur, M.: Order matters: Sequence to sequence for sets. arXiv preprint arXiv:1511.06391 (2015)
19. Xu, K., Hu, W., Leskovec, J., Jegelka, S.: How powerful are graph neural networks? arXiv preprint arXiv:1810.00826 (2018)

20. Yanardag, P., Vishwanathan, S.: Deep graph kernels. In: Proceedings of the 21th ACM SIGKDD International Conference on Knowledge Discovery and Data Mining, pp. 1365–1374 (2015)
21. You, Y., Chen, T., Sui, Y., Chen, T., Wang, Z., Shen, Y.: Graph contrastive learning with augmentations. In: Advances in Neural Information Processing Systems **33** (2020)
22. Yu, J., Yin, H., Xia, X., Chen, T., Cui, L., Nguyen, Q.V.H.: Are graph augmentations necessary? simple graph contrastive learning for recommendation. In: ACM SIGIR, pp. 1294–1303 (2022)
23. Zhang, M., Cui, Z., Neumann, M., Chen, Y.: An end-to-end deep learning architecture for graph classification. In: Proceedings of the AAAI Conference on Artificial Intelligence, vol. 32 (2018)

SymphoNEI: Symphony of Node and Edge Inductive Representations on Large Heterophilic Graphs

Kyusik Kim and Bongwon Suh(✉)

Department of Intelligence and Information, Graduate School of Convergence Science and Technology, Seoul National University, Seoul, South Korea
{kyu823,bongwon}@snu.ac.kr

Abstract. Homophily-based GNNs do not adequately interpret heterophilic graphs, i.e., classes or attributes between adjacent nodes are often different. Therefore, GNNs for heterophilic graphs have recently been proposed, but several limitations remain. Applying GNNs to large-scale graphs is performed in inductive settings with several sampling methods. However, most heterophilic GNNs do not perform well when combined with sampling methods and do not generalize to unseen nodes because they only consider the transductive setting. In addition, heterophilic GNNs only extract information from nodes and graph structures. They do not represent edges, which could be essential elements in the graph. To challenge the issues, we propose a new model SymphoNEI. SymphoNEI successfully embeds large heterophilic graphs in the inductive setting by learning both node and edge representations. To the best of our knowledge, SymphoNEI is the first of the heterophilic GNNs to represent the edges and evaluate inductive node classification on large heterophilic graphs with and without edge features. We used eight large heterophilic graphs for model evaluation: five without edge features and three with edge features. Extensive experimental results indicate that SymphoNEI achieves state-of-the-art performance for inductive node classification on large heterophilic graphs.

Keywords: Graph Neural Networks · Inductive Learning · Large-scale Graphs · Heterophilic Graphs

1 Introduction

Many GNNs [6,10,17] were developed on the premise that the graph is homophilic. Homophily is where nodes connected by edges have similar classes or attributes. However, many real-world graphs are not homophilic. For instance,

---

Supplementary Information The online version contains supplementary material available at https://doi.org/10.1007/978-981-97-5572-1_31.

M. Onizuka et al. (Eds.): DASFAA 2024, LNCS 14855, pp. 429–439, 2024.
https://doi.org/10.1007/978-981-97-5572-1_31

individuals on dating websites connect more with the opposite gender, and fraudsters on purchasing platforms primarily contact non-fraudulent users. Graphs where nodes frequently connect with those of differing properties, are heterophilic graphs. Most homophilic GNNs have poor node classification performance in the heterophilic graphs. Thus, models [5,11–13,21] with improved node classification performance for heterophilic graphs have been developed. However, heterophilic GNNs have some limitations in large-scale graphs. Most heterophilic GNNs underperform with sampling methods for large graphs and face challenges in generalizing to unseen nodes, a critical requirement for applications where complete graph is often unfeasible. Furthermore, heterophilic GNNs mainly focus on node and graph structure, neglecting edge representations. This oversight is critical in heterophilic graphs, where relying solely on node information may be insufficient due to the likelihood of neighboring nodes having different properties. Addressing these gaps necessitates an approach that incorporates edge information for more comprehensive graph representation.

In this paper, we propose a new model SymphoNEI (Symphony of Node and Edge Inductive representantions) that improves inductive node classification performance in large-scale heterophilic graphs to overcome these limitations. To the best of our knowledge, SymphoNEI is the first among heterophilic GNNs to separately obtain edge representations to utilize not only node and graph structure information but also edge information. SymphoNEI was evaluated on eight large-scale heterophilic graphs belonging to various domains. Five are graphs without edge features, and the remaining three have edge features. Moreover, to our knowledge, SymphoNEI was evaluated on large-scale heterophilic graphs with edge features for the first time. Through extensive experiments, we showed that SymphoNEI achieved state-of-the-art performance in node classification for various large heterophilic graphs.

In summary, the crucial contributions of this paper are as follows:

- We first studied inductive node classification on large heterophilic graphs.
- To our knowledge, we used edge representations and evaluated node classification on large-scale heterophilic graphs with edge features for the first time among heterophilic GNNs.
- We conducted extensive experiments on eight large-scale heterophilic graphs. As a result, SymphoNEI reached state-of-the-art performance in inductive node classification on large-scale heterophilic graphs.

2 Related Works

2.1 Heterophilic Graph Representation Learning

Several metrics [12,14,21] exist to measure the homophily of a graph. These metrics exhibit shortcomings, as they can be influenced by the number of classes and nodes belonging to each class. Platonov et al. [15] introduced adjusted homophily ratio (h_{adj}) and Label Informativeness (LI). h_{adj} is less influenced by the number of classes or class size balance compared to other metrics. LI evaluates the informational value a neighboring node's label offers regarding the focal node's label. We employed h_{adj} and LI as metric for homophily.

Many GNNs [6,10,17] have been developed on the premise that graphs are homophilic. Due to their poor node classification performance on heterophilic graphs [21], models [5,11–13,21] that use various methods have recently been developed to embed heterophilic graphs well. H2GCN [21] uses multi-hop neighbors to obtain the representation of the ego node. ACM [13], and GloGNN [11] change the aggregation method. H2GCN and ACM obtain the ego-node and aggregated representations of neighbor nodes separately. H2GCN and GPR-GNN [5] get the final node representation by merging the output of each layer. LINKX [12] embeds adjacency matrix and node feature matrix separately through different multi-linear perceptron (Fig. 1).

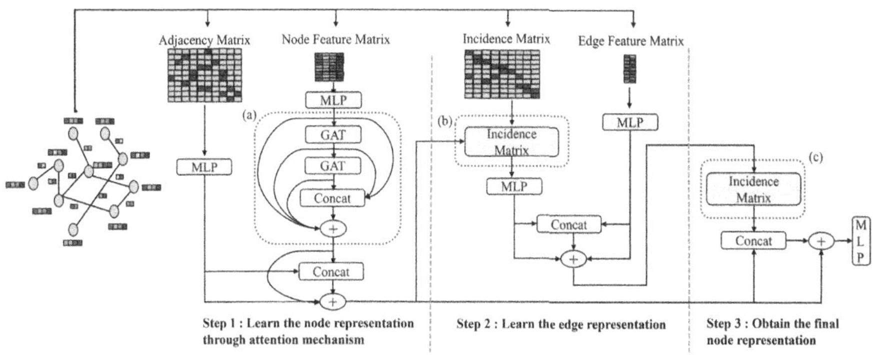

Fig. 1. The overall framework of SymphoNEI. (a) is the attention mechanism used in the model. In (b), the incidence matrix is used to concatenate the representations of two nodes connected by the edge. In (c), the edge representations obtained in Step 2 are multiplied with transposed incidence matrix. All residual connections in SymphoNEI are ReZero connections.

2.2 Inductive and Edge Representations in GNNs

Several GNNs [4,7,20] have been developed to embed large graphs well in inductive settings. GraphSAGE [7] samples nodes for each layer of GNNs. Cluster-gcn [4] and GraphSAINT [20] learn each batch by sampling the subgraph of the input graph. However, they are developed under the assumption that the graph is homophilic. Moreover, many GNNs [6,7,10] focus on learning information from nodes considering the connectivity of nodes. However, the edge is also a crucial factor in the graph. There are methods [3,9,17] to calculate the attention weight of an edge when using edge information, but not many tasks represent the edge. A few models [2,8,19] have obtained edge representations, but all are limited to homophilic graphs with small sizes.

3 Proposed Framework - SymphoNEI

3.1 Simple Minibatching

To learn large-scale graphs inductively, subgraph sampling methods and methods for sampling neighbor nodes for each GNN layer have been introduced. In the graph data, as nodes connected by edges influence each other, existing sampling methods consider graph topology. Thus as the number of layers increases, the number of neighbor nodes may explode, and the hyperparameters used in the sampling method should be optimized. To respond to these challenges, Lim et al. [12] introduced a simple minibatching technique which is also applied to SymphoNEI. This method selects i.i.d. samples of nodes from the adjacency matrix. It is much simpler and takes less time compared to other sampling techniques.

Moreover, simple minibatching method is effective in heterophilic graphs. Traditional sampling methods assume homophilic graphs, employing techniques where a subset of neighboring nodes surrounding an ego node is sampled. However, in heterophilic graphs, distant nodes from the ego node could potentially impart vital information. Leveraging conventional sampling methods may result in the less frequent sampling of such remote nodes due to inherent biases. Simple minibatching method provides a solution to this challenge by choosing nodes from the graph with uniform probability. This uniform selection ensures that nodes located further from the ego node receive equal representation during the sampling process.

3.2 Details of SymphoNEI

Learn the Node Representation Through Attention Mechanism. $A \in R^{n \times n}$ is an adjacency matrix, and $X \in R^{D \times n}$ is a feature matrix, where n is the number of nodes and D is the number of node features. Before calculating node representations, we embed an adjacency matrix A and a feature matrix X separately through different MLP, as expressed in Eq. 1.

$$h_A = MLP_A(A) \in R^{d \times n}; \qquad h_X = MLP_X(X) \in R^{d \times n} \tag{1}$$

Here, d is the size of the hidden dimension, h_A is the hidden representation of an adjacency matrix, h_X is the hidden representation of a node feature matrix, and h_X is composed of vectors such as $h_X = \{\boldsymbol{h_1}, \boldsymbol{h_2}, ..., \boldsymbol{h_n}\}$, $\boldsymbol{h_i} \in R^d$. The two types of information can be maintained without mixing by embedding an adjacency matrix and a node feature matrix separately. In traditonal GNNs, information from the adjacency matrix and the node feature matrix is mixed in message passing and aggregation. As the two matrices are components of the graph, they are inevitably influenced by each other. However, preserving them is vital because each matrix can contain meaningful information. Furthermore, the incorporation of information from distant nodes through the MLP allows for a more comprehensive understanding of heterophilic graphs.

$$h'_X = \alpha_1 W_{GAT} \left[GAT_1(h_X) \,\|\, GAT_2(h_X) \,\|\, h_X\right] + \alpha_2 GAT_1(h_X) + \alpha_3 GAT_2(h_X) + h_X \tag{2}$$

Subsequently, we use the attention mechanism in the embedded feature matrix to reflect the graph topological information. The corresponding attention mechanism consists of two layers of GAT [17] and ReZero connections [1]. GAT_k denotes the kth layer of GAT, $\|$ means concatenation, and h'_X is the result of the attention mechanism. $W_{GAT} \in R^{d\times 3d}$ is a weight matrix, and α_1, α_2, α_3 are learnable scalar parameters. The method in Eq. 2 brings two advantages. First, the method effectively leverages global information. SymphoNEI is able to extract vital information from the combined local and global information in the neighbor node representations through GAT. Moreover, the method preserves the ego-node information. In heterophilic graphs, an ego node may have different properties from one-hop neighbor nodes. Hence, if they are simply merged, information loss may occur. It is essential to conserve ego-node information to prevent information loss [21]. Concatenating the ego-node with aggregated neighbor node representations inhibits their intermixing. As detailed in Sect. 3.2.4, the ReZero connection is incorporated to retain ego-node information.

GAT used in Eq. 2 is divided into two types according to the method of calculating attention scores $\alpha_{i,j}$ between node i and j. For graphs with edge features, GAT follows Eq. 3, and for graphs without edge features, it follows Eq. 3 without concatenating $\mathbf{W_e} e_{i,j}$. $\mathcal{N}(i)$ is neighbor nodes of node i. $\boldsymbol{e}_{i,j}$ is features of the edge that connects node i and j. Weight vector $\boldsymbol{a}$ is $\boldsymbol{a} \in R^{2d}$ for graphs without edge features and $\boldsymbol{a} \in R^{3d}$ for graphs with edge features. $W \in R^{d\times d}$ is the shared weight matrix for every node, and $W_e \in R^{d\times d'}$ is the weight matrix for edge features, where d' is the number of edge features.

$$\alpha_{i,j} = \frac{\exp\left(\text{LeakyReLU}\left(\boldsymbol{a}^\top [\mathbf{W}\boldsymbol{h}_i \,\|\, \mathbf{W}\boldsymbol{h}_j \,\|\, \mathbf{W}_e \boldsymbol{e}_{i,j}]\right)\right)}{\sum_{k\in\mathcal{N}(i)} \exp\left(\text{LeakyReLU}\left(\boldsymbol{a}^\top [\mathbf{W}\boldsymbol{x}_i \,\|\, \mathbf{W}\boldsymbol{x}_k \,\|\, \mathbf{W}_e e_{i,k}]\right)\right)} \tag{3}$$

After computing the attention score, we average multi-head attention for neighbor node representations and apply the ELU as in Eq. 4. K is the number of heads in multi-head attention, and W^k is the weight matrix.

$$\boldsymbol{h}_i = ELU\left(\frac{1}{K}\sum_{k=1}^{K}\left(\sum_{j\in\mathcal{N}(i)} \alpha_{i,j}^k W^k \boldsymbol{h}_j\right)\right) \tag{4}$$

We follow Eq. 5 to get a node representation $h_{nr} \in \mathbb{R}^{d\times n}$ by combining the representation of adjacency matrix h_A and that of node features h'_X. They are concatenated and linearly transformed through a weight matrix $W_{nr} \in \mathbb{R}^{d\times 2d}$. By adding h'_X and h_A through the residual connection, node features and adjacency matrix information are further preserved. As the residual connection is the ReZero connection, β_1 and β_2 are learnable scalar parameters.

$$h_{nr} = ReLU(W_{nr}\,[h_A \,\|\, h'_X] + \beta_1 h_A + \beta_2 h'_X) \tag{5}$$

Learn the Edge Representation. Unlike existing heterophilic GNNs, SymphoNEI represents edges. The node and graph structure information can be sufficient to represent homophilic graph as properties between connected nodes are similar. However, node and graph structure information alone may not be enough to embed heterophilic graphs, as there is a high probability that the attributes between the connected nodes are different. To compensate, we represent edges that are important elements in graphs. As shown in Eq. 6, the representations of the two nodes connected by the edge are concatenated.

$$h_e = \{\boldsymbol{h}_{e_{ij}} = [\boldsymbol{h}_{n_i r} \parallel \boldsymbol{h}_{n_j r}] \mid j \in \mathcal{N}(i), i \in [1, 2, ..., n]\} \tag{6}$$

$h_{nr} = \{\boldsymbol{h}_{n_1 r}, \boldsymbol{h}_{n_2 r}, ..., \boldsymbol{h}_{n_n r}\}$ is node representations from Eq. 5 and $\boldsymbol{h}_{n_i r}$ is the representation of ith node. $h_e \in \mathbb{R}^{2d \times n'}$ is the initial representation of edges, where n' is the number of edges. In graphs with edge features, the edge feature matrix is defined as $EX \in \mathbb{R}^{d' \times n'}$. If a graph does not have edge features, we concatenate the feature representations of the two nodes connected by the edge, as in Eq. 7. Through this method, information from the two nodes of an edge is reflected within the edge representation, thereby ensuring optimal preservation of node representations.

$$EX = \left\{\overrightarrow{EX}_{ij} = [\boldsymbol{h'}_i \parallel \boldsymbol{h'}_j] \mid j \in \mathcal{N}(i), i \in [1, 2, ..., n]\right\} \in \mathbb{R}^{2d \times n'} \tag{7}$$

$h'_X = \{\boldsymbol{h}'_1, \boldsymbol{h}'_2, ..., \boldsymbol{h}'_n\}$ is representations of feature matrix and $\boldsymbol{h}'_i$ is representation of ith node features. The edge representation obtained by concatenating the node representation and that of the edge features are embedded through MLP, as in Eq. 8. By embedding the edge information obtained from the node representations and edge features separately, the two types of information can be maintained without mixing. As the two matrices are elements of the graph, they are affected by each other. However, preserving them is also important because each matrix can contain meaningful information.

$$h_{e'} = MLP_e(h_e) \in \mathbb{R}^{d \times n'}; \qquad h_{EX} = MLP_{EX}(EX) \in \mathbb{R}^{d \times n'} \tag{8}$$

To obtain the final edge representation $h_{er} \in \mathbb{R}^{d \times n'}$, we concatenate edge embedding $h_{e'}$ from node representation and edge feature representation h_{EX} and then linearly transform with a weight matrix $W_{er} \in \mathbb{R}^{d \times 2d}$. $h_{e'}$ and h_{EX} are then added through residual connection according to Eq. 9. γ_1 and γ_2 are learnable scalar parameters. We represent edges by combining $h_{e'}$ and h_{EX} and preserving each information appropriately.

$$h_{er} = ReLU(W_{er}[h_{e'} \parallel h_{EX}] + \gamma_1 h_{e'} + \gamma_2 h_{EX}) \tag{9}$$

Obtain the Final Node Representation We use both node and edge representations to obtain the final node representation. First, node representation h'_{nr} is calculated by multiplying edge representation h_{er} from Sect. 3.2.2 and transposed incidence matrix $I \in \mathbb{R}^{n \times n'}$, as in Eq. 10. Incidence matrix I is a method of expressing a graph like an adjacency matrix. Each node is set in a row, and an edge connected to the corresponding node is set in a column. By multiplying the incidence matrix and the edge representation, the node is embedded by aggregating representations of all edges connected to the node.

$$h'_{nr} = h_{er} I^T \in \mathbb{R}^{d \times n}; \tag{10}$$

Subsequently, we follow Eq. 11 to obtain the final node representation $h''_{nr} \in \mathbb{R}^{d \times n}$. We concatenate two different node representations h_{nr}, which is the result of the attention mechanism, and h'_{nr} obtained from edge representation h_{er}. Concatenation of node representations is linearly transformed through a weight matrix $W'_{nr} \in \mathbb{R}^{d \times 2d}$. In addition, h_{nr} is added through the ReZero connection, and δ is a learnable scalar parameter.

$$h''_{nr} = ReLU(h_{nr} + \delta W'_{nr} [h_{nr} \parallel h'_{nr}]) \tag{11}$$

ReZero Connection. The residual connection used in SymphoNEI is the ReZero connection [1], which is a residual connection multiplied by a residual parameter. The residual parameter is a learnable scalar parameter, and the initial value is set 0. The ReZero connection has the following advantages [1]. The ReZero connection can replace learning rate warmup strategies and normalization-common methods to optimize learning rates and mitigate overfitting. Further, the ReZero connection accelerates model convergence compared to the residual connection. As SymphoNEI uses the ReZero connection, SymphoNEI inherits these advantages. Moreover, SymphoNEI with the ReZero connection reflects the influence of various types of information used to represent nodes and edges better than the residual connection due to a residual parameter.

4 Experiments

4.1 Dataset and Baseline Methods

SymphoNEI was evaluated on eight large heterophilic graphs. Five large heterophilic graphs without edge features are Penn94, pokec, arxiv-year, snap-patents, and twitch-gamer [12]. Three large heterophilic graphs with edge features are Bitcoin-2014, Bitcoin-2015, and Bitcoin-2016 from the bitcoin transaction dataset [18]. Bitcoin-2014, Bitcoin-2015, and Bitcoin-2016 represent bitcoin transaction snapshots spanning six months from their respective years. Within these graphs, bitcoin addresses are nodes, while transactions form edges. The task of the three datasets is to find the mixing address. Detailed statistics for the datasets are shown in Table 1.

We compared MLP, homophilic GNNs, and heterophilic GNNs combined with several sampling methods to evaluate the superiority of SymphoNEI in inductive node classification. Unlike GNNs, MLP does not consider graph topology and uses node features only. Homophilic GNNs used are GCN [10] and GAT [17]. GPR-GNN [5], H2GCN [21], ACM-GCN [13], GloGNN++ [11] and LINKX

[12] are used as heterophilic GNNs. The sampling methods used are a simple i.i.d. node minibatching [12] and subgraph sampling methods, GraphSAINT [20] by node-based and random walk-based.

4.2 Results

By combining various sampling methods and GNNs, 27 methods were tested. The experimental results are in Table 2. According to the results, some GNNs performed worse than MLPs in heterophilic graphs. Moreover, heterophilic GNNs performed comparably to homophilic GNNs when integrated with various sampling methods. The results also indicated that SymphoNEI achieved state-of-the-art performance in eight datasets. Among the datasets without edge features, the difference in accuracy between SymphoNEI and the runner-up result in snap-patents was the largest as 8%. In the datasets with edge features, which are highly imbalanced, SymphoNEI and LINKX had remarkable performance differences from others. SymphoNEI and LINKX only exceeded 70% G-Mean when combined with sampling methods. Meanwhile, SymphoNEI exceeded G-Mean 96% in both Bitcoin-2014 and Bitcoin-2015, which have high heterophily degrees. Moreover, SymphoNEI exhibited lower performance on Bitcoin-2016 compared to Bitcoin-2014 and Bitcoin-2015. Despite the datasets being structurally similar, the reduced LI in Bitcoin-2016 likely hindered effective information retrieval from neighboring nodes, leading to the decreased performance.

The performance of baseline methods without LINKX showed minimal variation with different sampling techniques. While simple minibatching accesses more distant nodes, most baseline methods generally failed to effectively utilize the global information. In contrast, SymphoNEI efficiently leverages global information from distant nodes through an MLP applied to the full node feature matrix. By incorporating attention mechanisms, ReZero connections, and edge representations, SymphoNEI effectively integrates local and global information, outperforming baseline methods across diverse graph-related tasks.

Table 1. Statistics of datasets. $|\nu|$, $|\varepsilon|$, Nf, Ef, c, h_{adj} and LI mean the number of nodes, edges, number of node features, number of edge features, classes, adjusted homophily ratio and label informativeness.

Dataset	$\|\nu\|$	$\|\varepsilon\|$	Nf	Ef	c	Class types	h_{adj}	LI
Penn94	41,554	1,362,229	5	X	2	gender	−0.06	0.00
pokec	1,632,803	30,622,564	65	X	2	gender	−0.15	0.02
arxiv-year	169,343	1,166,243	128	X	5	pub year	0.01	0.04
snap-patents	2,923,922	13,975,788	269	X	5	time granted	0.00	0.02
twitch-gamer	168,114	6,797,557	7	X	2	mature content	0.09	0.01
Bitcoin-2014	1,989,888	15,639,323	6	5	2	fraudulence	0.38	0.56
Bitcoin-2015	1,967,475	13,047,111	6	5	2	fraudulence	0.33	0.48
Bitcoin-2016	1,859,821	9,241,065	6	5	2	fraudulence	0.31	0.05

Table 2. Results of several models evaluated on large heterophilic graphs. Accuracy was employed for datasets without edge features, while Geometric Mean (G-Mean) [16,18] was utilized for those with edge features. The best results are highlighted in bold. Mb, SN, SRW mean minibatching, GraphSaint-Node, GraphSaint-RandomWalk. The size of minibatch for all methods is 10% of the graph.

Model	Penn94	pokec	arxiv-year	snap-patents	twitch-gamer	Bitcoin-2014	Bitcoin-2015	Bitcoin-2016
MLP_Mb	74.31 ± 0.68	62.17 ± 0.02	37.01 ± 0.06	24.46 ± 1.52	60.56 ± 0.07	60.72 ± 0.65	62.39 ± 1.29	67.49 ± 0.32
MLP_SN	73.09 ± 0.49	61.58 ± 0.05	33.63 ± 0.38	25.39 ± 0.67	59.27 ± 0.09	52.39 ± 3.29	52.64 ± 4.29	58.32 ± 7.74
MLP_SRW	73.97 ± 0.39	61.66 ± 0.10	30.32 ± 2.30	29.06 ± 0.17	59.50 ± 0.16	58.88 ± 4.53	57.96 ± 3.41	67.91 ± 0.65
GCN_Mb	68.96 ± 0.41	61.34 ± 0.29	37.51 ± 0.30	21.46 ± 0.85	60.10 ± 0.72	52.17 ± 0.74	54.42 ± 0.80	55.63 ± 4.89
GCN_SN	69.04 ± 0.41	60.16 ± 0.18	30.19 ± 0.53	23.97 ± 0.38	58.48 ± 0.15	50.35 ± 0.19	50.41 ± 0.44	49.99 ± 0.00
GCN_SRW	70.29 ± 0.21	60.38 ± 0.18	35.69 ± 0.63	31.18 ± 0.10	59.33 ± 0.28	51.24 ± 0.08	51.36 ± 1.00	51.57 ± 0.44
GAT_Mb	67.87 ± 0.51	61.93 ± 0.67	38.86 ± 0.23	21.59 ± 0.29	59.27 ± 0.46	52.49 ± 1.59	53.31 ± 0.59	52.27 ± 1.47
GAT_SN	73.07 ± 0.33	62.17 ± 0.06	33.53 ± 0.00	24.52 ± 0.43	58.28 ± 0.72	51.65 ± 0.05	51.84 ± 0.19	50.34 ± 0.48
GAT_SRW	73.68 ± 0.11	62.03 ± 0.18	37.26 ± 0.29	32.78 ± 0.05	58.94 ± 0.10	52.12 ± 0.69	51.77 ± 1.15	51.75 ± 0.14
GPR-GNN_Mb	73.38 ± 0.54	63.66 ± 0.05	35.42 ± 0.30	21.76 ± 0.26	59.44 ± 0.21	57.81 ± 3.33	58.79 ± 4.53	61.71 ± 6.80
GPR-GNN_SN	73.66 ± 0.95	63.68 ± 0.08	36.11 ± 0.02	29.20 ± 0.78	58.94 ± 0.28	50.87 ± 1.06	55.96 ± 3.73	55.76 ± 8.15
GPR-GNN_SRW	74.44 ± 0.85	63.74 ± 0.15	35.31 ± 0.65	31.58 ± 0.02	59.21 ± 0.07	60.73 ± 2.53	57.69 ± 3.00	61.33 ± 1.93
H2GCN_Mb	74.09 ± 0.55	61.54 ± 0.05	35.5910.47	25.70 ± 0.36	60.75 ± 0.24	62.05 ± 0.05	62.37 ± 1.28	67.56 ± 1.03
H2GCN_SN	73.96 ± 0.64	61.39 ± 0.05	33.31 ± 0.29	24.05 ± 0.03	59.42 ± 0.00	53.71 ± 0.35	52.36 ± 0.05	67.35 ± 0.30
H2GCN_SRW	74.46 ± 0.37	61.57 ± 0.02	32.66 ± 0.62	28.46 ± 0.03	60.55 ± 0.03	63.19 ± 0.10	60.58 ± 0.37	67.82 ± 0.01
ACM-GCN_Mb	74.01 ± 0.42	60.43 ± 1.54	37.55 ± 0.16	29.91 ± 1.66	60.06 ± 0.39	63.32 ± 0.41	61.49 ± 3.74	55.41 ± 6.13
ACM-GCN_SN	73.60 ± 0.91	59.61 ± 3.07	31.99 ± 0.85	29.25 ± 0.09	59.37 ± 0.05	70.95 ± 3.47	53.66 ± 0.44	54.97 ± 8.23
ACM-GCN_SRW	74.32 ± 0.03	59.72 ± 2.22	32.59 ± 0.48	32.80 ± 0.03	60.06 ± 0.40	64.25 ± 3.41	56.23 ± 3.22	52.41 ± 0.99
GloGNN++_Mb	74.48 ± 0.62	65.12 ± 0.15	36.72 ± 0.26	21.25 ± 0.93	60.76 ± 0.26	56.69 ± 3.35	58.07 ± 5.30	59.87 ± 7.49
GloGNN++_SN	62.56 ± 16.1	61.40 ± 0.07	31.72 ± 0.15	23.23 ± 0.07	59.39 ± 0.07	58.37 ± 5.03	53.62 ± 3.37	53.68 ± 0.60
GloGNN++_SRW	74.24 ± 0.17	61.01 ± 0.03	36.95 ± 0.07	30.55 ± 0.08	59.26 ± 0.40	55.00 ± 1.24	55.54 ± 4.81	56.20 ± 3.72
LINKX_Mb	84.39 ± 0.70	81.08 ± 0.30	53.89 ± 0.19	60.02 ± 0.23	65.89 ± 0.37	98.20 ± 0.55	93.61 ± 1.02	75.11 ± 1.10
LINKX_SN	73.17 ± 0.72	60.97 ± 0.09	34.56 ± 0.18	21.38 ± 0.05	59.83 ± 0.40	55.97 ± 3.84	51.39 ± 0.68	63.98 ± 1.49
LINKX_SRW	74.20 ± 0.05	60.78 ± 0.03	27.51 ± 1.00	29.47 ± 0.14	59.72 ± 0.29	54.69 ± 1.05	59.20 ± 1.64	65.58 ± 0.78
SymphoNEI_Mb	$\mathbf{84.65 \pm 0.41}$	$\mathbf{82.06 \pm 0.10}$	$\mathbf{55.45 \pm 0.24}$	$\mathbf{68.01 \pm 1.75}$	$\mathbf{66.11 \pm 0.25}$	$\mathbf{98.81 \pm 0.68}$	$\mathbf{96.26 \pm 0.89}$	$\mathbf{76.67 \pm 1.51}$
SymphoNEI_SN	72.94 ± 0.58	60.55 ± 0.38	33.84 ± 0.42	31.04 ± 0.41	60.40 ± 0.10	66.01 ± 4.18	56.61 ± 3.21	60.96 ± 2.62
SymphoNEI_SRW	73.29 ± 0.41	60.38 ± 0.08	31.71 ± 0.20	34.45 ± 0.36	60.66 ± 0.25	75.94 ± 1.28	62.53 ± 1.41	67.19 ± 0.48

5 Conclusion

In this paper, we propose a new model SymphoNEI. SymphoNEI, combined with a simple i.i.d. node minibatching, successfully embeds large heterophilic graphs inductively by learning both node and edge representations. The major contributions of this paper are the followings. We first study inductive node classification on large heterophilic graphs. Moreover, we obtain the edge representations and use large heterophilic graphs with edge features for the first time among heterophilic GNNs. Extensive experiments on various datasets show that SymphoNEI achieves state-of-the-art performance on inductive node classification in large heterophilic graphs.

Acknowledgement. This work was partly supported by Institute of Information & communications Technology Planning & Evaluation (IITP) grant funded by the Korea government(MSIT) [NO.2021-0-01343-004, Artificial Intelligence Graduate School Program (Seoul National University)] and Basic Science Research Program through the National Research Foundation of Korea(NRF) funded by the Ministry of Education(2022R1A6A1A03063039).

References

1. Bachlechner, T., Majumder, B.P., Mao, H., Cottrell, G., McAuley, J.: Rezero is all you need: fast convergence at large depth. In: Uncertainty in Artificial Intelligence, pp. 1352–1361. PMLR (2021)
2. Bandyopadhyay, S., Biswas, A., Murty, M.N., Narayanam, R.: Beyond node embedding: a direct unsupervised edge representation framework for homogeneous networks. arXiv preprint arXiv:1912.05140 (2019)
3. Brody, S., Alon, U., Yahav, E.: How attentive are graph attention networks? In: International Conference on Learning Representations (2021)
4. Chiang, W.L., Liu, X., Si, S., Li, Y., Bengio, S., Hsieh, C.J.: Cluster-GCN: an efficient algorithm for training deep and large graph convolutional networks. In: Proceedings of the 25th ACM SIGKDD, pp. 257–266 (2019)
5. Chien, E., Peng, J., Li, P., Milenkovic, O.: Adaptive universal generalized pagerank graph neural network. In: International Conference on Learning Representations (2020)
6. Defferrard, M., Bresson, X., Vandergheynst, P.: Convolutional neural networks on graphs with fast localized spectral filtering. Adv. Neural Inf. Process. Syst. **29** (2016)
7. Hamilton, W., Ying, Z., Leskovec, J.: Inductive representation learning on large graphs. Adv. Neural Inf. Process. Syst. **30** (2017)
8. Jiang, X., Ji, P., Li, S.: Censnet: convolution with edge-node switching in graph neural networks. In: IJCAI, pp. 2656–2662 (2019)
9. Kim, D., Oh, A.: How to find your friendly neighborhood: graph attention design with self-supervision. In: 9th International Conference on Learning Representations, ICLR 2021, Virtual Event, Austria, 3–7 May 2021 (2021)
10. Kipf, T.N., Welling, M.: Semi-supervised classification with graph convolutional networks. In: 5th International Conference on Learning Representations, ICLR 2017, Toulon, France, 24–26 April 2017, Conference Track Proceedings (2017)
11. Li, X., et al.: Finding global homophily in graph neural networks when meeting heterophily. In: International Conference on Machine Learning, pp. 13242–13256. PMLR (2022)
12. Lim, D., et al.: Large scale learning on non-homophilous graphs: new benchmarks and strong simple methods. Adv. Neural Inf. Process. Syst. **34** (2021)
13. Luan, S., et al.: Is heterophily a real nightmare for graph neural networks to do node classification? arXiv preprint arXiv:2109.05641 (2021)
14. Pei, H., Wei, B., Chang, K.C., Lei, Y., Yang, B.: Geom-GCN: geometric graph convolutional networks. In: 8th International Conference on Learning Representations, ICLR 2020, Addis Ababa, Ethiopia, 26–30 April 2020 (2020)
15. Platonov, O., Kuznedelev, D., Babenko, A., Prokhorenkova, L.: Characterizing graph datasets for node classification: homophily-heterophily dichotomy and beyond. In: The Second Learning on Graphs Conference (2023)
16. Tang, B., He, H., Baggenstoss, P.M., Kay, S.: A bayesian classification approach using class-specific features for text categorization. IEEE Trans. Knowl. Data Eng. **28**(6), 1602–1606 (2016)
17. Veličković, P., Cucurull, G., Casanova, A., Romero, A., Liò, P., Bengio, Y.: Graph attention networks. In: International Conference on Learning Representations (2018)
18. Wu, J., Liu, J., Chen, W., Huang, H., Zheng, Z., Zhang, Y.: Detecting mixing services via mining bitcoin transaction network with hybrid motifs. IEEE Trans. Syst. Man Cybern. Syst. **52**, 2237–2249 (2021)

19. Yang, Y., Li, D.: Nenn: incorporate node and edge features in graph neural networks. In: Asian Conference on Machine Learning, pp. 593–608. PMLR (2020)
20. Zeng, H., Zhou, H., Srivastava, A., Kannan, R., Prasanna, V.K.: Graphsaint: graph sampling based inductive learning method. In: 8th International Conference on Learning Representations, Addis Ababa, Ethiopia, 26–30 April 2020 (2020)
21. Zhu, J., Yan, Y., Zhao, L., Heimann, M., Akoglu, L., Koutra, D.: Beyond homophily in graph neural networks: current limitations and effective designs. Adv. Neural. Inf. Process. Syst. **33**, 7793–7804 (2020)

HopLearn: Leveraging Multi-Hop Neighbors and Learnable Parameters for GNNs with Missing Node Features

Kyusik Kim and Bongwon Suh(✉)

Department of Intelligence and Information, Graduate School of Convergence Science and Technology, Seoul National University, Seoul, South Korea
{kyu823,bongwon}@snu.ac.kr

Abstract. Graph neural networks (GNNs) are becoming increasingly central to various applications. A common presumption underpinning their utility is the completeness of node features in graphs. However, this often proves impractical in real-world scenarios where acquiring comprehensive feature information for every node remains a challenge. To address this problem in node classification task, we present a new method called *Leveraging Multi-Hop Neighbors and Learnable Parameters* (HopLearn) which is simple and easily implementable. In contrast to conventional methods that substitute missing node features with features extracted from multi-hop neighbors, HopLearn not only performs this substitution but also incorporates certain replaced node features as learnable parameters. We evaluate HopLearn using six benchmark datasets across both homophilic and heterophilic graphs, i.e., classes or attributes are often different between adjacent nodes. The results demonstrate that HopLearn exhibits robust efficacy across diverse graph structures. Moreover, HopLearn is compatible with various GNNs and offers rapid processing speeds. Given these attributes, HopLearn is beneficial for applications harnessing GNNs with incomplete graph datasets.

Keywords: Graph Neural Networks · Missing Node features · Heterophilic Graphs

1 Introduction

Graph Neural Networks (GNNs) have gained extensive research attention in recent years and have been utilized in many areas [6,11,18]. Generally, GNNs [3,5,8,22] embed nodes through message-passing and aggregation. In message-passing, nodes send their feature representations to their neighbor nodes. Nodes' feature representations are initialized to their original features, and are updated

Supplementary Information The online version contains supplementary material available at https://doi.org/10.1007/978-981-97-5572-1_32.

M. Onizuka et al. (Eds.): DASFAA 2024, LNCS 14855, pp. 440–450, 2024.
https://doi.org/10.1007/978-981-97-5572-1_32

by aggregating messages passed by their neighbor nodes. The features of all nodes must be present in GNNs in order to obtain proper node representations.

However, it is often difficult to obtain complete feature information on all nodes in a graph in real-world scenarios. The co-purchase network may not provide a complete description of all products. A growing awareness of digital privacy makes it increasingly difficult to obtain personal information from social networks. As GNNs are typically designed to process data containing a full set of features, this presents a significant challenge. Several methods [2,4,7,17,19,21,23] have been proposed for handling the absence of node features in GNNs. However, the current methods suffer from some limitations. Several methods are not as computationally efficient when dealing with large datasets. SAT [2], GCNMF [19], and PaGNN [7] exhibit longer run-time than traditional GNNs because they modify the structure of GNN to address missing node features. In addition, certain methods may not be easily combined with other GNNs, limiting their utility in various contexts. For example, GCNMF and PaGNN, which are modified versions of GCN, might not be easily integrated with other GNNs. Furthermore, a prevalent challenge is the accurate embedding of heterophilic graphs. Heterophilic graphs, characterized by a higher likelihood of connecting dissimilar nodes, frequently manifest in real-world scenarios. For example, molecular protein networks or dating websites exemplify settings where such heterophilic connections are routine. However, FP [17] is not optimal in heterophilic graphs, as it functions as a low-pass filter.

To tackle these challenges, we propose a novel method *Leveraging Multi-Hop Neighbors and Learnable Parameters* (HopLearn), to handle missing node features in node classifiation task with GNNs. For the first time among methods addressing missing node features, we consider them as learnable parameters. In order to evaluate the effectiveness of HopLearn, we conducted experiments on six benchmark datasets for node classification task, including two homophilic graphs and four heterophilic graphs with varying ratios of missing node features. HopLearn was tested with combined two GNNs, GCN [8] and GAT-sep [16]. Overall, HopLearn outperforms state-of-the-art performance in various missing ratios. The results indicate the effectiveness of HopLearn in handling missing node features in GNNs for both homophilic and heterophilic graphs. Additionally, HopLearn is fast and easy to implement. HopLearn shows robust performance in combination with a variety of GNNs regardless of their structure. Due to the above advantages, HopLearn is useful in various applications that use GNNs with incomplete data.

In summary, the crucial contributions of this paper are as follows:

- We propose a novel strategy HopLearn to handle the missing features of nodes in the graph for GNNs.
- Extensive experiments with various missing ratios of node features justify the effectiveness of HopLearn on both homophilic and heterophilic graphs.
- HopLearn is fast and capable of integration with diverse GNNs.

2 Related Works

Heterophily: Several metrics [10,14,24] exist to quantify graph homophily. However, these often falter due to susceptibility to the number of classes and node distribution across classes. Adjusted homophily ratio (h_{adj}) and Label Informativeness (LI) were introduced to address the challenges [15]. h_{adj} is resilient to class variations and imbalances, making it superior to existing homophily metrics. Moreover, LI delves into the informational value of neighbor labels in predicting the focal node's label. In light of these advancements, we employ both h_{adj} and LI to assess the dataset's characteristics. Many GNNs [3,8,22] have been developed on the premise that graphs are homophilic. Due to their poor node classification performance on heterophilic graphs [24], several models [9,12,24] have recently been developed to embed heterophilic graphs well. However, they all assume that the node features of the graph are complete.

GNNs with Missing Node Features: Several traditional methods are applied to GNNs when node features are missing. Traditional methods include replacing missing node features with zeros, random values, global averages, and neighbors' averages. Various methods [2,4,7,17,19,23] have recently been proposed to handle missing node features. PaGNN [7] propagates only features that are not missing in a GCN-like structure. FP [17] reconstructs missing features through the iterative multiplication of the feature matrix with the normalized adjacency matrix. PCFI [21] is a confidence-based feature imputation scheme for graphs with missing node features that leverages the shortest path distance between nodes to compute pseudo-confidence. However, these methods still have limitations. SAT [2], GCNMF [19], and PaGNN are not suitable for large datasets because they take a long time to run. GCNMF and PaGNN are specific GCN-type models, making them difficult to apply to other GNNs. In heterophilic graphs with missing node features, FP and PCFI is unable to embed nodes successfully. FP works as a low-pass filter to make adjacent nodes that have different characteristics from the ego-node become similar.

3 Methods

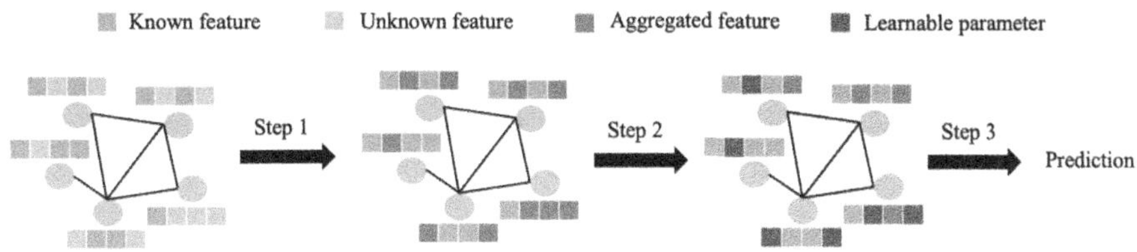

Fig. 1. Overview of HopLearn. In Step 1, features of multi-hop neighbors are aggregated to replace the unknown features. In Step 2, some of replaced features are allocated as learnable parameters. In Step 3, the learnable parameters are trained with downstream GNN.

3.1 Aggregating Features of Multi-Hop Neighbors

In order to address the absence of node features, we leverage the features of multi-hop neighbors. Initially, absent node features are initialized to zero. Subsequently, the features of i-hop neighbors are aggregated to replace the missing features of the focal node through a process known as message passing.

A notable aspect of our approach is the simultaneous use of both low-pass (LP) and high-pass (HP) filters during message passing. Message passing of traditional GNNs [5,8] predominantly utilizes adjacency matrix as a low-pass filter. This strategy primarily serves to attenuate noise emanating from neighboring nodes, and it has demonstrated efficacy, particularly in homophilic graphs. On the other hand, several prior studies [1,12] have underscored the significance of high-frequency graph signals. These signals, which are harnessed using a high-pass filter, play a pivotal role in confronting challenges posed by heterophily. The high-pass filter emphasizes differences between nodes, thereby counteracting detrimental heterophily.

Majority of GNNs [8,22], predominantly adopt a uni-channel filtering. This often entails exclusively relying on either a low-pass or high-pass channel. In contrast, we employed both low-pass and high-pass filters to extract balanced information from multi-hop neighbor nodes, aiming to effectively impute missing node features in both homophilic and heterophilic graphs. As $LP + HP = I$ ensures no information loss from the input signal [12], we use the normalized adjacency matrix $\tilde{A}$ as low-pass filter and $I - \tilde{A}$ as high-pass filter. Here, $\tilde{A}$ denotes $D^{-1/2}AD^{-1/2}$ where A and D symbolize the adjacency matrix and degree matrix respectively. I means identity matrix. During message passing, $i - 1$ hop neighbors' features undergoes multiplication with both filters. Subsequently, known features are reverted to their initial values.

In the subsequent phase, it is imperative to compute a normalized weighted summation derived from the outputs of both high-pass and low-pass filters. The inherent advantage of employing weighted summation lies in its ability to adaptively emphasize results based on the characteristics of neighboring nodes relative to the focal node. Specifically, when numerous neighboring nodes possess attributes distinct from the focal node, the output from the high-pass filter is accentuated. Conversely, when neighboring nodes bear a resemblance to the focal node, it results in the emphasis of the output from the low-pass filter. Such adaptive emphasis endows the mechanism with the capability to effectively respond to a diverse range of graph topologies. The incorporation of normalization in the weighting is not merely a design choice but a necessity. When applying a high-pass filter through multi-hop without utilizing normalized weights, there is a potential risk of accumulated information disparities between nodes leading to divergence. Hence, it is essential to employ normalized weights to mitigate the risk.

3.2 Learnable Parameters for Missing Node Features

Language models, such as Llama2 [20] and GPT-4 [13], have consistently showcased superior performance across a diverse range of linguistic tasks. Notably, these models excel even when processing words or phrases that do not possess inherent unique features. The foundational principle of these language models lies in their capacity to convert words, which intrinsically lack specific features, into high-dimensional vectors known as word embeddings. These embeddings act as primary inputs for the models and are continuously refined during the training phase to effectively encapsulate the intricate semantic and syntactic relationships exhibited between words.

Drawing inspiration from the success of word embeddings in natural language processing models, we propose a novel method where missing node features are integrated as learnable parameters within the GNN framework. Throughout the training process, these parameters undergo continuous refinement, allowing GNNs to efficiently capture and represent complex graph topologies. The efficacy of our proposed method is particularly underscored when addressing various graph structures, characterized by unique homophily ratios and LI values. Such various graph structures inherently pose challenges in terms of feature interactions and propagation patterns. Our approach leverages learnable parameters to stand in for these missing features, endowing GNNs with the capacity to responsively adjust to diverse graph configurations. Throughout the training phase, these parameters iteratively refine, grasping the latent characteristics inherent in the disparate graphs, thereby effectively filling the information gaps. Progressive fine-tuning facilitates an adaptive and robust representation, which is essential for comprehending diverse graph structures.

However, treating all missing node features as learnable parameters or training them throughout every epoch can lead to overfitting. It's worth noting that the primary design purpose of GNNs is geared towards graph embeddings rather than continuous feature adjustments. Overemphasizing the latter might divert the focus of GNNs, leading to issues like overfitting or model instability. To address these concerns, as demonstrated in Step 2 of Fig. 1, only a random subset of the replaced missing node features is considered as learnable parameters. They are trained for a specific number of epochs and are frozen for the remainder of the training epochs. By updating some learnable parameters over certain epochs, GNNs can adeptly embed complex graph topology. This approach assures that, once feature values are optimized, GNNs can remain focused on its primary task: graph representation learning.

4 Experiments

4.1 Datasets

For our experiments, we used six node classification benchmark datasets to test the efficacy of the methods intended to resolve missing node features in GNNs on both homophilic and heterophilic graphs. For this purpose, two of the six datasets, Cora and PubMed are homophilic graphs and the remaining four are heterophilic graphs. Existing solutions [7,17,21] for addressing missing node features have not been validated on heterophily benchmark datasets. Furthermore, GNNs [9,12,24] have employed datasets like Texas, Wisconsin, and Cornell [14] as heterophilic graph benchmarks. However, these datasets are notably small in size, and their classes exhibit severe imbalances. This imbalance leads to results that are unstable and statistically inconclusive, making them unsuitable as representative benchmark datasets [16]. Therefore, we opted to utilize the heterophilic graph benchmark dataset, amazon, minesweeper, tolokers, and questions [16]. Further information regarding the statistical details of the datasets can be found in the Table 1.

Table 1. Statistics of datasets. amazon., mine., $|\nu|$, $|\varepsilon|$, c, d and h_{adj} mean amazon-ratings, minesweeper, the number of nodes, edges, classes and features and adjusted homophily ratio.

	$\lvert\nu\rvert$	$\lvert\varepsilon\rvert$	c	d	h_{adj}	LI
Cora	2,708	5,278	7	1,433	0.77	0.59
PubMed	19,717	44,324	3	500	0.69	0.41
amazon.	24,492	93,050	5	300	0.14	0.04
mine.	10,000	39,402	2	7	0.01	0.00
tolokers	11,758	519,000	2	10	0.09	0.01
questions	48,921	153,540	2	301	0.02	0.00

4.2 Baseline Methods and Experiment Settings

We used one traditional method for imputing missing node features: setting them to zero. Furthermore, we utilized PaGNN, FP and PCFI which are methods for resolving missing node features in GNNs. Other methods [2,4,19,23] were not used since they had previously demonstrated inferior performance to PaGNN, FP and PCFI [17,21].

To evaluate the robustness of the proposed techniques in handling missing node features across diverse GNNs, we employed GCN [8] and GAT-sep [16]. GCN is the most representative GNNs and has been used as baseline methods in numerous experiments. GAT-sep strategically distinguishes between ego and neighbor node embeddings within GAT [22], demonstrating state-of-the-art performance, particularly in heterophilic graph node classification [16]. We conducted experiments on six datasets in conjunction with GCN. Excluding PaGNN, a specific GCN-type model, the remaining four methods were tested with GAT-sep on three heterophilic datasets.

4.3 Results

We examined the effectiveness of the proposed method, HopLearn, in handling missing node features in GNNs. We conducted a comprehensive set of experiments with two different downstream GNNs, GCN and GAT-sep. For GCN, we

Table 2. Result of six benchmark datasets with varying missing ratios of node features. Accuracy is reported for Cora, Pubmed, amazon-ratings and ROC AUC is reported for minesweeper, questions and tolokers. Z, Pa, HL mean Zero, PaGNN, HopLearn. Zero, FP, PCFI and HopLearn are combined with GCN. Best methods are in bold and the runner-ups are underlined.

Dataset	Cora					PubMed					amazon-ratings				
Method	Z	Pa	FP	PCFI	HL	Z	Pa	FP	PCFI	HL	Z	Pa	FP	PCFI	HL
10%	76.83	**78.59**	76.95	77.40	78.51	78.66	77.75	79.02	78.94	**80.69**	47.16	36.85	46.95	47.42	**47.50**
20%	74.91	**78.11**	76.35	76.63	77.26	78.42	77.79	78.57	78.72	**80.25**	46.25	36.58	46.20	46.42	**47.33**
30%	74.71	**76.63**	75.36	75.99	76.22	78.42	77.77	78.40	78.74	**79.76**	45.55	36.45	45.72	45.75	**47.30**
40%	73.80	**76.97**	75.48	76.17	75.76	77.22	77.32	77.12	78.31	**79.12**	44.90	36.41	44.95	45.03	**47.46**
50%	72.56	**74.96**	73.64	74.65	74.40	77.39	77.19	77.22	77.86	**78.50**	44.20	36.46	44.37	44.25	**47.19**
60%	72.48	**75.76**	73.75	74.76	72.99	76.77	76.19	77.32	**77.41**	77.24	43.29	36.57	43.44	43.12	**47.07**
70%	68.34	73.79	72.86	**74.08**	70.43	75.99	75.49	76.96	**77.05**	75.67	42.55	36.50	42.63	42.13	**46.99**
80%	66.72	70.33	71.44	**73.26**	67.96	72.92	73.59	**76.49**	76.46	73.39	42.51	36.57	41.29	41.74	**46.82**
90%	61.58	68.06	72.76	**73.79**	67.99	68.64	71.25	**75.84**	**75.84**	68.96	36.60	36.61	36.61	36.67	**46.85**
Dataset	minesweeper					questions					tolokers				
Method	Z	Pa	FP	PCFI	HL	Z	Pa	FP	PCFI	HL	Z	Pa	FP	PCFI	HL
10%	85.80	60.70	84.75	84.73	**85.95**	72.37	45.78	72.89	72.47	**74.45**	79.71	52.91	79.10	**80.00**	79.24
20%	83.69	60.24	82.89	82.93	**84.30**	71.71	50.17	73.00	71.81	**73.69**	79.41	52.87	79.05	**79.48**	79.03
30%	82.21	59.69	80.46	80.50	**82.72**	71.31	50.26	72.55	70.71	**73.32**	78.25	52.79	78.28	**78.88**	78.26
40%	80.09	58.42	77.34	77.34	**80.78**	69.67	50.15	72.23	69.62	**72.61**	77.60	52.53	77.45	**78.23**	78.19
50%	77.29	57.92	73.77	73.91	**78.42**	69.54	50.10	**72.14**	68.76	72.05	77.10	54.58	77.43	77.46	**77.88**
60%	75.04	57.36	70.01	69.98	**76.34**	68.24	49.79	**71.85**	67.31	71.10	75.95	52.00	75.96	74.05	**76.93**
70%	71.24	56.54	65.99	65.93	**71.69**	68.23	49.90	**72.66**	66.42	69.83	**76.61**	53.17	75.91	72.24	76.43
80%	66.68	55.99	60.63	60.77	**67.45**	65.97	45.46	**71.77**	64.12	68.43	75.05	52.62	74.91	70.29	**76.39**
90%	58.82	54.84	55.89	56.02	**61.13**	64.18	45.93	**71.36**	63.00	67.68	71.26	52.11	72.57	68.70	**75.96**

Table 3. Result of three benchmark datasets with varying missing ratios of node features. Accuracy is reported for amazon-ratings and ROC-AUC is reported for minesweeper and questions. Z and HL mean Zero and HopLearn. Zero, FP, PCFI and HopLearn are combined with GAT-sep. Best methods are in bold and the runner-ups are underlined.

Dataset	amazon-ratings				minesweeper				questions			
Method	Z	FP	PCFI	HL	Z	FP	PCFI	HL	Z	FP	PCFI	HL
10%	46.27	46.29	46.55	**47.48**	87.47	86.96	87.02	**88.08**	75.64	75.61	75.58	**76.60**
20%	45.69	45.40	45.92	**47.48**	85.60	85.34	85.41	**86.58**	75.51	75.69	75.54	**76.38**
30%	44.52	43.96	45.37	**46.70**	84.24	82.90	83.03	**84.81**	74.84	75.17	74.50	**76.15**
40%	43.92	44.57	45.08	**45.94**	82.01	79.80	79.72	**82.60**	72.73	73.66	72.84	**75.05**
50%	43.92	44.20	44.84	**45.63**	79.68	77.34	77.31	**80.05**	71.92	73.93	72.25	**74.65**
60%	43.04	43.53	44.00	**45.39**	76.68	74.15	74.21	**77.02**	69.82	73.11	69.37	**73.49**
70%	42.28	42.52	43.38	**45.27**	71.96	70.31	70.31	**72.59**	70.37	**73.40**	67.37	72.92
80%	41.61	42.32	42.47	**44.78**	66.41	64.59	64.97	**67.57**	66.68	70.67	64.06	**71.08**
90%	36.61	36.60	36.60	**43.13**	58.61	58.63	58.89	**61.27**	60.45	**70.17**	61.93	65.61

evaluated 54 scenarios, encompassing two homophilic graphs, four heterophilic graphs, and nine distinct missing node feature ratios. For GAT-sep, 27 scenarios were examined, consisting of three heterophilic graphs and nine varied ratios of missing node features.

The results of these experiments demonstrate the robustness and superiority of HopLearn when it comes to handling missing node features in GNNs. Table 2 displays experimental results using GCN as the downstream GNN. Based on the results, HopLearn has an average ranking of 1.81, followed by PCFI at 2.64, FP at 2.81, Zero at 3.35, and PaGNN at 4.29. Among the five methods, HopLearn achieved the highest average rank across 54 settings. Notably, in amazon-ratings and minesweeper, HopLearn outperformed all other methods at every missing ratio, with a performance margin exceeding 10% at a 90% missing ratio in amazon-ratings. Table 3 presents results when using GAT-sep as the downstream GNN. We compared four methods excluding PaGNN, which are GCN-specific methods. The average ranking of each method in Table 3 is as follows: HopLearn with 1.07, FP with 2.88, PCFI with 2.92, Zero with 3.03. HopLearn achieved the highest performance in 25 out of the 27 settings. Considering Tables 2 and 3, HopLearn consistently delivered optimal outcomes when paired with both GCN and GAT-sep. The results highlight HopLearn's consistent superiority and adaptability across various graph settings when integrated with different GNNs.

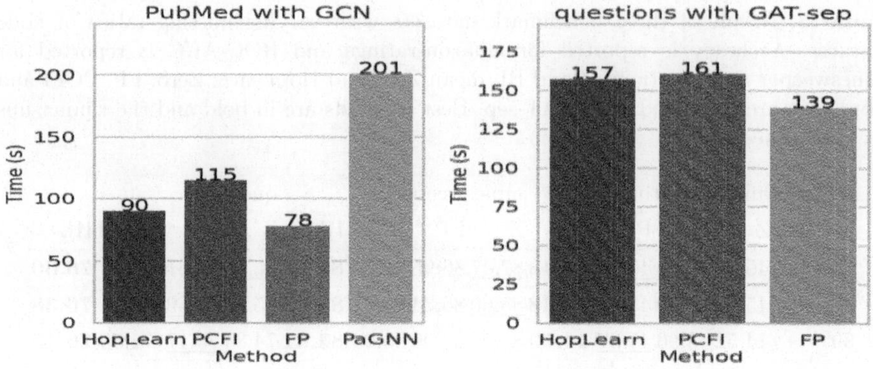

Fig. 2. Run time of HopLearn, PCFI, FP and PaGNN.

4.4 Run-Time Analysis

The run-time of HopLearn, PCFI, FP, and PaGNN was evaluated. We experimented with combining HopLearn, PCFI, and FP with GCN on PubMed to compare with the GCN-specific model, PaGNN. In addition, to measure the run-time when integrated with GAT-sep, HopLearn, PCFI, and FP were also combined with GAT-sep and tested on questions. Both experiments were conducted with a missing ratio of 90%. For a fair comparison, the computation time of HopLearn, PCFI, and FP encompassed the entire period until the downstream GNN models were trained. As in Fig. 2, HopLearn is the second fastest after FP. FP employs a very simple approach of repetitively using only a low-pass filter. When compared to this method, the run-time difference of HopLearn is not significant, and among the methods excluding FP, HopLearn is the fastest. From the run-time measurements, HopLearn was found to be, on average, 14% slower than FP, while PCFI was 31% slower. Moreover, when combined with GCN, PaGNN had a run-time more than double that of FP. Based on these results, it can be concluded that HopLearn is a considerably fast method.

5 Conclusion

In this paper, we present a novel method, HopLearn, for handling missing node features in GNNs. HopLearn leverages features of multi-hop neighbor nodes and learnable parameters to replace missing node features. Our empirical evaluations on six benchmark datasets, spanning both homophilic and heterophilic graphs, affirm the method's consistent performance and adaptability across various graph topologies. Morevoer, HopLearn is compatible with a range of GNN architectures, making it versatile for diverse applications. The rapid processing capability of HopLearn further enhances its practicality. Thus, HopLearn is beneficial for various applications that make use of GNNs with incomplete data.

Acknowledgement. This work was partly supported by Institute of Information & communications Technology Planning & Evaluation (IITP) grant funded by the Korea government(MSIT) [NO.2021-0-01343-004, Artificial Intelligence Graduate School Program (Seoul National University)] and Basic Science Research Program through the National Research Foundation of Korea(NRF) funded by the Ministry of Education (2022R1A6A1A03063039).

References

1. Bo, D., Wang, X., Shi, C., Shen, H.: Beyond low-frequency information in graph convolutional networks. In: Proceedings of the AAAI Conference on Artificial Intelligence, vol. 35, pp. 3950–3957 (2021)
2. Chen, X., Chen, S., Yao, J., Zheng, H., Zhang, Y., Tsang, I.W.: Learning on attribute-missing graphs. IEEE Trans. Pattern Anal. Mach. Intell. **44**(2), 740–757 (2020)
3. Defferrard, M., Bresson, X., Vandergheynst, P.: Convolutional neural networks on graphs with fast localized spectral filtering. Adv. Neural Inf. Process. Syst. **29** (2016)
4. Diederik, P.K., Welling, M., et al.: Auto-encoding variational bayes. In: Proceedings of the International Conference on Learning Representations (ICLR), vol. 1 (2014)
5. Hamilton, W., Ying, Z., Leskovec, J.: Inductive representation learning on large graphs. Adv. Neural Inf. Process. Syst. **30** (2017)
6. Huang, M., et al.: Auc-oriented graph neural network for fraud detection. In: Proceedings of the ACM Web Conference 2022, pp. 1311–1321 (2022)
7. Jiang, B., Zhang, Z.: Incomplete graph representation and learning via partial graph neural networks. arXiv preprint arXiv:2003.10130 (2020)
8. Kipf, T.N., Welling, M.: Semi-supervised classification with graph convolutional networks. In: 5th International Conference on Learning Representations, ICLR 2017, Toulon, France, 24–26 April 2017, Conference Track Proceedings (2017)
9. Li, X., et al.: Finding global homophily in graph neural networks when meeting heterophily. In: International Conference on Machine Learning, pp. 13242–13256. PMLR (2022)
10. Lim, D., et al.: Large scale learning on non-homophilous graphs: new benchmarks and strong simple methods. Adv. Neural Inf. Process. Syst. **34** (2021)
11. Lin, X., Quan, Z., Wang, Z.J., Ma, T., Zeng, X.: KGNN: knowledge graph neural network for drug-drug interaction prediction. In: IJCAI, vol. 380, pp. 2739–2745 (2020)
12. Luan, S., et al.: Revisiting heterophily for graph neural networks. Adv. Neural. Inf. Process. Syst. **35**, 1362–1375 (2022)
13. OpenAI: Gpt-4 technical report (2023)
14. Pei, H., Wei, B., Chang, K.C., Lei, Y., Yang, B.: GEOM-GCN: geometric graph convolutional networks. In: 8th International Conference on Learning Representations, ICLR 2020, Addis Ababa, Ethiopia, 26–30 April 2020 (2020)
15. Platonov, O., Kuznedelev, D., Babenko, A., Prokhorenkova, L.: Characterizing graph datasets for node classification: homophily-heterophily dichotomy and beyond. In: The Second Learning on Graphs Conference (2023)
16. Platonov, O., Kuznedelev, D., Diskin, M., Babenko, A., Prokhorenkova, L.: A critical look at the evaluation of GNNs under heterophily: are we really making progress? In: The Eleventh International Conference on Learning Representations, ICLR 2023, Kigali, Rwanda, 1–5 May 2023 (2023)

17. Rossi, E., Kenlay, H., Gorinova, M.I., Chamberlain, B.P., Dong, X., Bronstein, M.M.: On the unreasonable effectiveness of feature propagation in learning on graphs with missing node features. In: Learning on Graphs Conference, pp. 11–1. PMLR (2022)
18. Schlichtkrull, M.S., Cao, N.D., Titov, I.: Interpreting graph neural networks for NLP with differentiable edge masking. In: 9th International Conference on Learning Representations, ICLR 2021, Virtual Event, Austria, 3–7 May 2021 (2021)
19. Taguchi, H., Liu, X., Murata, T.: Graph convolutional networks for graphs containing missing features. Futur. Gener. Comput. Syst. **117**, 155–168 (2021)
20. Touvron, H., et al.: Llama 2: open foundation and fine-tuned chat models. arXiv preprint arXiv:2307.09288 (2023)
21. Um, D., Park, J., Park, S., Choi, J.Y.: Confidence-based feature imputation for graphs with partially known features. In: The Eleventh International Conference on Learning Representations, ICLR 2023, Kigali, Rwanda, 1–5 May 2023 (2023)
22. Veličković, P., Cucurull, G., Casanova, A., Romero, A., Liò, P., Bengio, Y.: Graph attention networks. In: International Conference on Learning Representations (2018)
23. Yoon, J., Jordon, J., Schaar, M.: Gain: missing data imputation using generative adversarial nets. In: International Conference on Machine Learning, pp. 5689–5698. PMLR (2018)
24. Zhu, J., Yan, Y., Zhao, L., Heimann, M., Akoglu, L., Koutra, D.: Beyond homophily in graph neural networks: current limitations and effective designs. Adv. Neural. Inf. Process. Syst. **33**, 7793–7804 (2020)

Hardware Acceleration

Redesigning Data and Metadata Updates in PM File Systems with Persistent CPU Caches

Congyong Chen[1], Shengan Zheng[2(✉)], Yuhang Zhang[1], and Linpeng Huang[1(✉)]

[1] Shanghai Jiao Tong University, Shanghai, China
{ndsffx304ccy,carlislefelix,lphuang}@sjtu.edu.cn
[2] MoE Key Lab of Artificial Intelligence, AI Institute, Shanghai Jiao Tong University, Shanghai, China
shengan@sjtu.edu.cn

Abstract. The recent advent of persistent memory has revolutionized file system design by enabling efficient, fast, and durable data updates. However, to cope with mismatched access granularities and volatile CPU caches, existing file systems resort to costly data and metadata update approaches, leading to high write-back latency and excessive PM bandwidth consumption. Recent cache persistence techniques, such as Intel's eADR, address this issue by automatically flushing data from CPU caches to PM during power failures, making CPU caches effectively persistent. Persistent CPU caches provide both cacheline access granularity and data durability, acting as a new storage medium that can greatly reduce the latency and throughput overhead of data persistence. We present FusionFS, a file system that leverages persistent CPU caches to redesign data and metadata update approaches. FusionFS employs an *adaptive data update* approach that chooses the most effective mechanism based on file access patterns, minimizing PM bandwidth consumption and update latency. FusionFS also adopts an *aggregated metadata update* approach that consolidates small entries in persistent log buffers before appending them to PM logs, minimizing small random writes to PM. Experimental results show that FusionFS outperforms existing PM file systems in terms of latency and throughput in various scenarios.

Keywords: Persistent memory · File system · CPU cache

1 Introduction

Emerging byte-addressable persistent memory (PM) offers DRAM-like performance and disk-like durability, enabling in-memory system designs with high throughput and low persistence overhead. However, commercially available PM products, such as Intel Optane DC Persistent Memory, do not meet the expectations of system designers in terms of data flushing and access granularity [7,19,28]. Applications must issue flush instructions and memory barriers on

M. Onizuka et al. (Eds.): DASFAA 2024, LNCS 14855, pp. 453–462, 2024.
https://doi.org/10.1007/978-981-97-5572-1_33

platforms with volatile CPU caches (e.g., ADR-based platforms) to guarantee data persistence. This results in prolonged critical path latency and high consumption of PM write bandwidth. Moreover, PM's internal XPLine access granularity (256B) mismatches with CPU caches' cacheline access granularity (64B), causing small random PM writes to trigger additional read-modify-write traffic. Therefore, accessing PM with blocks aligned to 256B can maximize the utilization of PM bandwidth.

These challenges force existing systems to adopt expensive data and metadata update approaches. Existing data update approaches are mainly based on three types of mechanisms: i) *active-flush update* eagerly persists data synchronously with flush instructions and memory barriers [10,20,26], ii) *non-temporal update* directly writes data to PM bypassing CPU caches [8,18,27], iii) *asynchronous update* buffers data in DRAM and persists updates asynchronously [9,14,20,25,30]. The first two lead to high synchronous data persistence overhead and write amplification, while the third cannot guarantee immediate data consistency. Performance degradation occurs in access modes less conducive to these approaches. Similarly, metadata update approaches fall into two main categories: i) *in-place approach* involves direct modifications to persistent indexes, making it susceptible to write amplification [9,10,16,22,25,29], ii) *log-structured approach* places append-only logs in PM and maintains volatile indexes in DRAM. However, the small update granularity in both methods introduces write amplification.

Fortunately, recent cache persistence techniques (e.g., Intel's eADR [17], battery-backed cache [1], CXL's Global Persistent Flush [6]) enable automatic data flushing from CPU caches to PM during power failures. This feature provides persistence to byte-addressable CPU caches, enabling them to operate as a novel storage medium that can absorb writes to the hot PM region. By leveraging persistent CPU caches, data update mechanisms that are usually meant for volatile data structures can be applied to update persistent data. These mechanisms include: i) *pinned-cache update* pins data to a designated cache space within persistent CPU caches [31] using cache allocation technologies [2,15], ii) *in-place update* does not explicitly issue flush instructions after writes [21,29]. In addition, hardware transactional memory (HTM) can now be used to transactionally update multiple cachelines in PM. Previously, the use of HTM was limited by the need to flush data from volatile CPU caches for persistence, causing HTM transactions to abort.

However, existing PM systems, including those based on eADR, fail to fully harness the potential of persistent CPU caches. For data updates, ADR-based systems use *active-flush/non-temporal update* to ensure consistency. On eADR-enabled platforms, they completely switch to *in-place update* to avoid persistence overhead [21,29]. However, such a fixed approach can lead to write amplification as writes larger than a cacheline may be split into multiple cacheline-sized random writes due to the cacheline eviction policy (e.g., LRU algorithm) [13]. For metadata updates, *in-place approach* often writes with atomic instructions, whose granularity mismatches with PM's internal access granularity, resulting in write amplification. Even when using *log-structured approach* to write sequentially and taking advantage of PM's internal buffer [3,27], massive metadata (e.g., NOVA's per-inode logs, KucoFS's per-partition-tree logs) can overwhelm the small buffers (about 16KB per Optane DIMM) [28].

In this paper, we present FusionFS, a file system that harnesses persistent CPU caches to optimize data and metadata operations. FusionFS uses an *adaptive data update* approach to select the most suitable data update mechanism based on file access patterns (e.g., data hotness, update size, and consistency requirements). FusionFS also includes a hotspot detector that considers access counts and last access timestamps, supporting both system calls and memory mapping accesses. Moreover, FusionFS proposes an *aggregated metadata update* approach to consolidate small metadata log entries within log buffers located in persistent CPU caches. In summary, the contributions of this paper include:

- We perform an in-depth analysis of the impact of persistent CPU caches on PM, highlighting their potential in optimizing the data and metadata update approaches of existing PM systems.
- We propose FusionFS, a PM file system that integrates both *adaptive data update* and *aggregated metadata update* to leverage the capabilities of persistent CPU caches, thereby reducing latency and minimizing PM bandwidth consumption.
- We implement FusionFS as a POSIX-compliant file system for Linux. Performance results show that FusionFS achieves significantly lower latency and higher throughput than state-of-the-art PM file systems.

2 Design and Implementation

In this section, we introduce FusionFS, a novel PM file system that exploits persistent CPU caches to redesign data and metadata update approaches. We first propose an *adaptive data update* approach to select the optimal data update mechanism from the five update mechanisms based on access patterns (Sect. 2.1). We then design an *aggregated metadata update* approach to minimize small metadata writes to PM (Sect. 2.2).

2.1 Adaptive Data Update

Update Policy. FusionFS employs an adaptive approach that dynamically selects the most suitable data update mechanism during system calls and memory mapping accesses, as shown in Fig. 1. FusionFS uses radix trees as file indexes for efficient file management. In addition, FusionFS uses HTM transactions to replace inode locks, ensuring consistency of in-place writes and allowing simultaneous writes to different sections of the same file.

For large ($\geq 256B$) hot data, FusionFS employs *in-place update* to buffer writes in persistent CPU caches. This is especially effective for hot data that

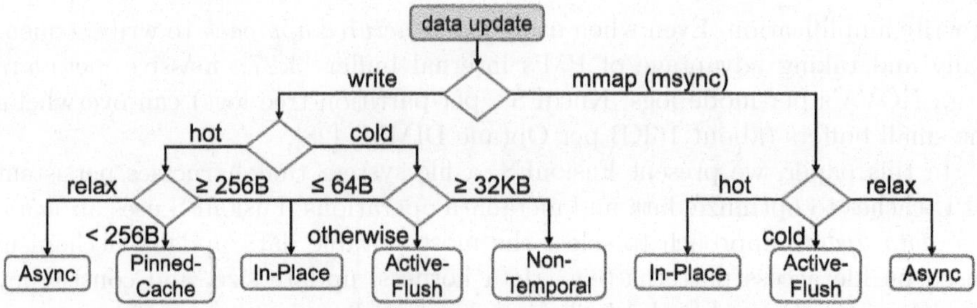

Fig. 1. Adaptive Data Update

is likely to be accessed again before eviction, reducing writes to PM and leveraging the benefits of cache hits. For hot data that is frequently updated in small granularity ($< 256B$), FusionFS uses *pinned-cache update*. It protects these small data segments from cacheline eviction, ensuring their persistence in CPU caches while minimizing CPU cache usage. For hot data with relaxed consistency, including recoverable data and data flagged as relaxed consistent by the user, FusionFS chooses *asynchronous update* to buffer writes in DRAM and persist them asynchronously. This is because it is free of write amplification on cacheline eviction and allows the high performance of DRAM to be exploited.

For small ($\leq$ 64B) cold data, FusionFS uses *in-place update* to minimize the cost of explicit flush instructions, since actively flushing a single cacheline will not mitigate write amplification caused by cacheline eviction. Conversely, for large ($\geq$ 32KB) cold data, FusionFS employs *non-temporal update* due to its lower latency and higher bandwidth. FusionFS ensures compatibility by writing shadow pages outside of HTM transactions and achieves consistency by modifying indexes within HTM transactions. Recognizing that the overhead of index modification and copying unaligned parts is non-trivial for smaller updates, FusionFS switches to *active-flush update* for medium-sized cold data. FusionFS ensures compatibility by delaying flushes until HTM transactions are complete and achieves consistency by performing in-place writes to data pages within HTM transactions.

FusionFS extends support for the *adaptive data update* approach to memory-mapped files. For hot data, FusionFS uses *in-place update* by omitting flush instructions in synchronization calls (e.g., *fsync*, *fdatasync*, and *msync*) to reduce writes to PM. For cold data, FusionFS uses *active-flush update* to aggregate the sequential writes in XPBuffer. For hot data with relaxed consistency, FusionFS uses *asynchronous update* by buffering data pages in DRAM and persisting them to PM during synchronization calls to take advantage of the high performance of DRAM.

By dynamically selecting the most effective mechanism based on access patterns, FusionFS optimizes data updates during system calls and memory mapping accesses. The adaptive approach of FusionFS ensures better PM bandwidth utilization and lower latency across various scenarios, addressing the limitations observed in conventional approaches.

Hotspot Detector. We design a lightweight yet effective hotspot detector to measure the hotness of data pages during system calls and memory mapping accesses. For system call accesses, the detector relies on the number of accesses to each data page within a designated period. After profiling the access counts for all data pages during this period, FusionFS classifies pages that exceed a certain threshold as hot data. The size of the identified hot data set matches the specified CPU cache size.

For memory mapping accesses, obtaining access count information directly is challenging. To address this, FusionFS proposes an approach inspired by the kernel's automatic NUMA balancing. FusionFS periodically adjusts the permissions of data pages in the background. This allows the next access to be captured and recorded by the page fault handler function. By recording the first access time of all pages during a scan period, FusionFS categorizes pages whose difference between access time and scan time is less than a certain threshold as hot data.

By analyzing both the number of accesses to data pages and the timestamp of the last access, FusionFS effectively identifies hot data in both system call and memory mapping scenarios. The lightweight design of this detector avoids the use of locks, minimizing update overhead.

2.2 Aggregated Metadata Update

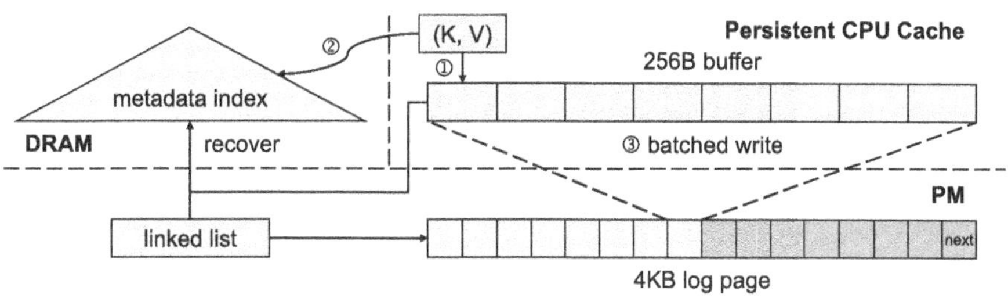

Fig. 2. Aggregated Metadata Update

FusionFS introduces the *aggregated metadata update* approach to increase the size of metadata writes to PM, as shown in Fig. 2. FusionFS maintains individual logs for each directory. During directory system calls (e.g., *create*, *mkdir*, *unlink*, and *rmdir*) that modify directory hash tables, FusionFS inserts corresponding log entries into the directory inode's log. In contrast to *log-structured approach*, FusionFS caches small metadata log entries ($< 256B$) in buffers pinned to persistent CPU caches (Step①). FusionFS also maintains metadata indexes in DRAM to speed up lookup operations (Step②). These entries are batched and copied to PM logs with *active-flush update* when the accumulated size exceeds 256B (Step③). While FusionFS writes to CPU caches, DRAM, and PM during metadata updates, the benefit of writing in XPLine granularity to PM outweighs the small latency caused by redundant writes to the high-speed CPU cache and DRAM. PM logs in FusionFS are designed to be extendable and organized as linked lists of 4KB pages. To ensure crash consistency, FusionFS records the used space of each buffer and log page, and the pointer to the next log page. In the event of system failures, these indexes can be reconstructed by replaying log entries from both PM logs and buffers. This aggregated approach significantly reduces the frequency of small random writes to PM during metadata updates, while ensuring immediate consistency of log entries.

3 Evaluation

In this section, we evaluate the performance of FusionFS and answer the following questions:

- Can FusionFS leverage the *adaptive data update* approach to optimize file updates?
- Can FusionFS utilize the *aggregated metadata update* approach to improve metadata updates?
- Can FusionFS exhibit optimal performance in various application scenarios?

3.1 Experimental Platform

We evaluate FusionFS on a server equipped with two 28-core Intel Xeon Gold 6348 (42M cache), with four 128 GB Intel Optane and four 16 GB DDR4 DRAM on each node. All experiments are performed on one NUMA node to avoid NUMA effects. To evaluate the performance of FusionFS, we compare it to PMFS, NOVA, SoupFS, and EXT4-DAX. For micro-benchmarks, we use Fio [11] to generate various access patterns to test the effectiveness of FusionFS's design. For macro-benchmarks, we use several workloads of Filebench [24]. For application benchmarks, we use YCSB [4] on LevelDB [12] and TPC-C [5] on SQLite [23] to evaluate the performance of FusionFS.

3.2 Micro-benchmarks

We use Fio to demonstrate how FusionFS benefits from the *adaptive data update* approach in file writes. We vary the Zipf parameter θ to issue uniform or skewed ($\theta = 0.99$) random write requests. We use 20 threads for the experiments to saturate the PM bandwidth, each thread performs 4KB I/Os to a private file of 64MB, and all writes are synchronous.

Figure 3(a) shows that FusionFS achieves good performance in both scenarios by using the optimal data update mechanism for the access pattern during system calls. Figure 3(b) shows similar results using memory mapping. PMFS and NOVA exhibit near-zero throughput due to their suboptimal synchronization performance. In contrast, FusionFS maintains high performance by selectively excluding hot data flushes from `fdatasync()` calls to minimize PM writes, while retaining them for cold data to prevent write amplification caused by random cacheline evictions.

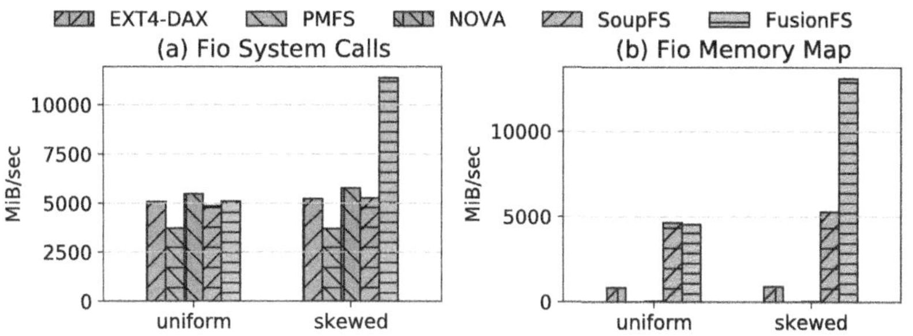

Fig. 3. Results of Fio

3.3 Macro-benchmarks

We evaluate the performance of FusionFS through Filebench, using three workloads to evaluate the performance of data and metadata updates. The results are shown in Fig. 4.

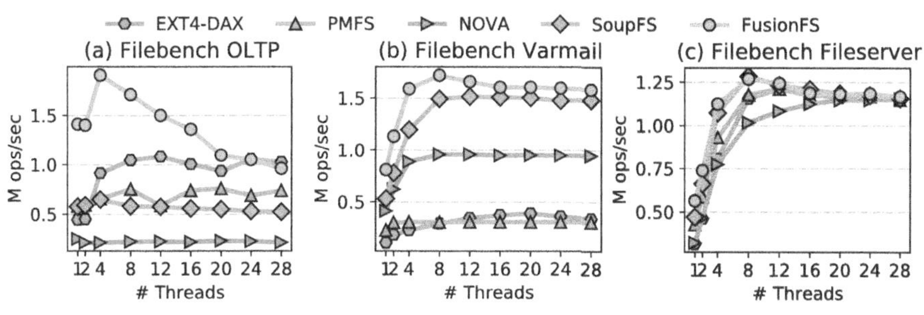

Fig. 4. Results of Filebench Workloads

OLTP[1] emulates database transactions at the file system level by issuing 2 KB asynchronous writes to a small data file and periodic 256 KB synchronous writes to a small log file. Given the high hotness of the files, FusionFS uses *in-place update*, allowing writes to hit CPU caches with high probability and outperforming other file systems by up to 8.9×.

Varmail emphasizes metadata updates by creating and deleting many small files in a single directory. In this workload, SoupFS, which is optimized for metadata updates, outperforms EXT4-DAX, PMFS, and NOVA. However, SoupFS executes data persistence in background threads, while Varmail frequently calls `fsync()`, incurring synchronized metadata persistence overhead that affects its throughput. FusionFS, in contrast, employs the *aggregated metadata update* approach, guaranteeing immediate consistency of metadata updates. This approach

[1] We increase the frequency of OLTP-initiated transactions to better match the high performance of PM.

leads to superior performance, outperforming SoupFS by up to 1.5× and other PM file systems by up to 7.5×.

Fileserver focuses on both file and metadata updates by keeping the number of files in each directory small and setting small file sizes. FusionFS and SoupFS show slightly better performance in this workload because they do not have to persist small writes immediately. While ADR-based SoupFS does not provide immediate consistency after writes, FusionFS outperforms other PM file systems by up to 1.2× and offers protection against data loss during power failures thanks to persistent CPU caches enabled by eADR.

3.4 Application Benchmarks

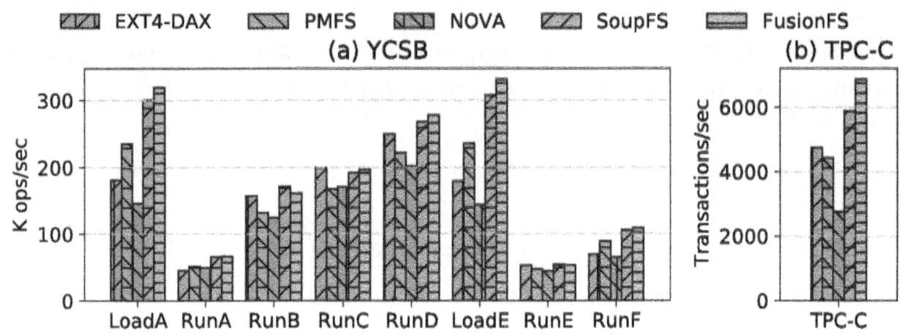

Fig. 5. Results of Application Benchmarks

Figure 5 summarizes real-world application performance across different file systems. We run YCSB workloads on LevelDB by issuing five million requests with a single thread, where each request contains a 16B key and a 64B value. We then measure the performance of TPC-C on SQLite in the Write-Ahead-Logging (WAL) mode. Overall, FusionFS achieves up to 2.5× the performance of other PM file systems in write-intensive workloads, including LoadA, LoadE, and TPC-C, while consistently maintaining optimal performance across all scenarios. This is realized by adaptively choosing the most appropriate data update mechanism for the access pattern.

4 Conclusions

Recent cache persistence techniques allow CPU caches to be considered persistent. In this paper, we propose FusionFS, a PM file system that leverages persistent CPU caches to redesign data and metadata update approaches. FusionFS employs an *adaptive data update* approach that chooses the most effective mechanism based on file access patterns. FusionFS also adopts an *aggregated metadata update* approach that consolidates small entries in persistent log buffers before appending them to PM logs. These novel approaches help reduce PM bandwidth consumption and update latency. Performance results show that FusionFS outperforms existing PM file systems in terms of latency and throughput in various scenarios.

Acknowledgements.. This work is supported by National Key Research and Development Program of China (Grant No. 2022YFB4500303), National Natural Science Foundation of China (NSFC) (Grant No. 62227809), the Fundamental Research Funds for the Central Universities, Shanghai Municipal Science and Technology Major Project (Grant No. 2021SHZDZX0102), Natural Science Foundation of Shanghai (Grant No. 22ZR1435400), and Huawei Innovation Research Plan.

References

1. Alshboul, M., Ramrakhyani, P., Wang, W., Tuck, J., Solihin, Y.: Bbb: Simplifying persistent programming using battery-backed buffers. In: 2021 IEEE International Symposium on High-Performance Computer Architecture (HPCA), pp. 111–124. IEEE (2021)
2. AMD: Amd64 technology platform quality of service extensions (2022). https://www.amd.com/content/dam/amd/en/documents/processor-tech-docs/other/56375/_1/_03/_PUB.pdf
3. Chen, Y., Lu, Y., Zhu, B., Arpaci-Dusseau, A.C., Arpaci-Dusseau, R.H., Shu, J.: Scalable persistent memory file system with {Kernel-Userspace} collaboration. In: 19th USENIX Conference on File and Storage Technologies (FAST 21), pp. 81–95 (2021)
4. Cooper, B.F., Silberstein, A., Tam, E., Ramakrishnan, R., Sears, R.: Benchmarking cloud serving systems with ycsb. In: Proceedings of the 1st ACM symposium on Cloud computing, pp. 143–154 (2010)
5. Council, T.P.P.: TPC benchmark c, standard specification version 5 (2001)
6. Compute express link^{TM}: The breakthrough cpu-to-device interconnect. https://www.computeexpresslink.org/
7. Daase, B., Bollmeier, L.J., Benson, L., Rabl, T.: Maximizing persistent memory bandwidth utilization for olap workloads. In: Proceedings of the 2021 International Conference on Management of Data, pp. 339–351 (2021)
8. Dong, M., Bu, H., Yi, J., Dong, B., Chen, H.: Performance and protection in the zofs user-space nvm file system. In: Proceedings of the 27th ACM Symposium on Operating Systems Principles, pp. 478–493 (2019)
9. Dong, M., Chen, H.: Soft updates made simple and fast on non-volatile memory. In: 2017 uSENIX Annual Technical Conference (uSENIX ATC 2017), pp. 719–731 (2017)
10. Dulloor, S.R., et al.: System software for persistent memory. In: Proceedings of the Ninth European Conference on Computer Systems, pp. 1–15 (2014)
11. Fio (flexible i/o tester). https://github.com/axboe/fio
12. Google: Leveldb. https://github.com/google/leveldb
13. Gugnani, S., Kashyap, A., Lu, X.: Understanding the idiosyncrasies of real persistent memory. Proc. VLDB Endow. **14**(4), 626–639 (2020)
14. He, K., An, Y., Luo, Y., Liu, X., Wang, G.: Flatlsm: write-optimized lsm-tree for pm-based kv stores. ACM Trans. Storage **19**(2), 1–26 (2023)
15. Intel: Improving real-time performance by utilizing cache allocation technology (2015). https://www.intel.com/content/dam/www/public/us/en/documents/white-papers/cache-allocation-technology-white-paper.pdf

16. Intel: Persistent memory development kit (2020). https://pmem.io/
17. Intel: eadr: New opportunities for persistent memory applications (2021). https://www.intel.com/content/www/us/en/developer/articles/technical/eadr-new-opportunities-for-persistent-memory-applications.html
18. Kadekodi, R., Lee, S.K., Kashyap, S., Kim, T., Kolli, A., Chidambaram, V.: Splitfs: reducing software overhead in file systems for persistent memory. In: Proceedings of the 27th ACM Symposium on Operating Systems Principles, pp. 494–508 (2019)
19. Kim, W.H., Krishnan, R.M., Fu, X., Kashyap, S., Min, C.: Pactree: a high performance persistent range index using pac guidelines. In: Proceedings of the ACM SIGOPS 28th Symposium on Operating Systems Principles, pp. 424–439 (2021)
20. Ou, J., Shu, J., Lu, Y.: A high performance file system for non-volatile main memory. In: Proceedings of the Eleventh European Conference on Computer Systems, pp. 1–16 (2016)
21. PMDK: libpmem. https://pmem.io/pmdk/libpmem/
22. Schwalb, D., Berning, T., Faust, M., Dreseler, M., Plattner, H.: nvm malloc: memory allocation for nvram. Adms@ Vldb **15**, 61–72 (2015)
23. SQLite: Sqlite transactional SQL database engine. https://www.sqlite.org/
24. Vasily Tarasov, E.Z., Shepler, S.: Filebench: a flexible framework for file system benchmarking. USENIX Login **41**(1), 6–12 (2016)
25. Woo, H., Han, D., Ha, S., Noh, S.H., Nam, B.: On stacking a persistent memory file system on legacy file systems. In: 21st USENIX Conference on File and Storage Technologies (FAST 2023), pp. 281–296 (2023)
26. Wu, X., Reddy, A.N.: SCMFS: a file system for storage class memory. In: Proceedings of 2011 International Conference for High Performance Computing, Networking, Storage and Analysis, pp. 1–11 (2011)
27. Xu, J., Swanson, S.: {NOVA}: a log-structured file system for hybrid {Volatile/Non-volatile} main memories. In: 14th USENIX Conference on File and Storage Technologies (FAST 2016), pp. 323–338 (2016)
28. Yang, J., Kim, J., Hoseinzadeh, M., Izraelevitz, J., Swanson, S.: An empirical guide to the behavior and use of scalable persistent memory. In: 18th USENIX Conference on File and Storage Technologies (FAST 2020), pp. 169–182 (2020)
29. Yi, J., Dong, M., Wu, F., Chen, H.: {HTMFS}: strong consistency comes for free with hardware transactional memory in persistent memory file systems. In: 20th USENIX Conference on File and Storage Technologies (FAST 2022), pp. 17–34 (2022)
30. Zheng, S., Hoseinzadeh, M., Swanson, S.: Ziggurat: a tiered file system for {Non-Volatile} main memories and disks. In: 17th USENIX Conference on File and Storage Technologies (FAST 2019), pp. 207–219 (2019)
31. Zhong, Y., Shen, Z., Yu, Z., Shu, J.: Redesigning high-performance lsm-based key-value stores with persistent cpu caches. In: 2023 IEEE 39th International Conference on Data Engineering (ICDE), pp. 1098–1111. IEEE (2023)

Reusing Your Prepared Data: An Informed Cache for Accelerating DNN Model Training

Ke Han, Wen Cheng, Yong Li(✉), Yunxiang Wu, Lingfang Zeng(✉), and Guang Chen

Zhejiang Lab, Hangzhou, China
{hanke,chengwen,yonglii,wuyunxiang,zenglf,chenguang}@zhejianglab.com

Abstract. In deep learning training, CPU-intensive data preprocessing often leads to CPU bottlenecks, and expensive GPUs cannot be fully utilized, thus degrading end-to-end training performance. In general, CPUs are used to doing preprocessing and GPUs are used to training the model. We propose a new caching algorithm for AI model training, named HCache, which uses the DLT-Informed caching approach to improve the reuse of cached data and the usage of memory. DLT-Informed Caching approach can make intelligent caching decisions by identifying data discrepancies in different preprocessing stages. Our evaluation shows that HCache can achieves about 2× speedup compared with the state-of-the-art CoorDL and Quiver in the training of computer vision models while maintaining comparable accuracy.

Keywords: data load · preprocess · deep neural networks · cache

1 Introduction

Data is the fuel powering machine learning. Large training datasets are empowering state-of-the-art accuracy for several machine learning tasks. Particularly, Deep Neural Networks (DNNs), have gained prominence, as they allow us to tackle problems that were previously intractable, such as image classification, translation, speech recognition, video captioning, and even predictive healthcare. Empowering DNNs to push state-of-the-art accuracy requires the model to be trained with a large volume of data. During training, the model predicts the output given training data; based on the output, the models weights are tuned. This happens iteratively, in many rounds called epochs. However, DNN training is data-hungry, resource-intensive, and time-consuming. It involves the holistic use of all the resources in a server from storage and CPU for fetching and pre-processing the dataset to the GPUs that perform computation on the transformed data. In the last few years, however, DL model and system developers have noted that the fast speed of the GPU does not always guarantee fast DL training, owing to CPU bottlenecks in the DL input pipeline. In the DL input

M. Onizuka et al. (Eds.): DASFAA 2024, LNCS 14855, pp. 463–473, 2024.
https://doi.org/10.1007/978-981-97-5572-1_34

pipeline, input data are decoded, preprocessed, and augmented on CPUs to create tensors for computations on GPUs. Although CPU and GPU operations can overlap, the time spent on CPUs for preprocessing can be longer than that spent on GPUs for tensor computations. As a result, expensive GPUs wait for CPU computations, leading to preprocessing (prep) stalls and GPU under-utilization. Recent researches [7–9,16,17] have attempted to alleviate CPU-bound problem by increasing the number of CPUs, but these approaches undoubtedly increase the cost.

To address this problem, we propose a novel sample reuse approach that reuses prepared data to speed up the preparation of DNN training data. We design HCache (hybrid cache), a novel caching and data loading system, to realize the usage of prepared data. Due to the use of a mixture of cached and non-cached samples for gradient computation in prepared data reusing, non-cached samples that need to be fetched from storage and undergo the complete preprocessing pipeline can become bottlenecks in the pipeline. This can hinder the overlap of computations between the CPU and DL accelerators. To further alleviate the I/O overhead of non-cached data, HCache is divided into two regions: T-Cache and O-Cache, storing preprocessed tensor data and raw unprocessed data respectively. Both of them are managed by key-value mechanism. T-Cache prioritizing samples with high I/O and preprocessing costs. When T-Cache is full, the data items within T-Cache remain unchanged during the entire training cycle, avoiding redundant I/O and preprocessing operations. O-Cache stores raw unprocessed data. When a client request arrives within a single epoch, if the data is not available in the HCache, the system randomly retrieves an unaccessed from either the O-Cache or the T-Cache and return.

The main contributions of this paper can be summarized as follows:

- Designing a caching system called HCache that carefully caches preprocessed data and reduce CPU overhead.
- Optimizing the memory inflation issue caused by caching preprocessed data to improve training throughput.
- By employing a hybrid caching design and a global replacement cache strategy, HCache ensures model performance while reducing training time. Compared to CoorDL [13] and Quiver [11], HCache improves training performance by a factor of 2 while achieving similar accuracy.

2 Background and Motivation

2.1 DNN Training

DNN training "trains" a DL model to perform a certain task by repeatedly adjusting the model weights with regard to the given set of training samples. Broadly speaking, training DNNs consists of two steps: data preparation and gradient computation. Until a certain termination condition is satisfied (e.g. target validation accuracy is met), the two steps are repeated for multiple epochs.

It is important to note that the data preparation and the gradient computation steps can be overlapped, as they are typically executed on different hardware. Thus, ideally the processing time of data preparation can be hidden by that of gradient computation, and thus the former has not been considered to have a significant impact on the performance of DNN training. However, recent development of specialized hardware accelerators [1,3] has dramatically reduced the processing time required for the gradient computation. Accordingly, the data preparation step is becoming the bottleneck of DNN training as pointed out in the recent works [5,12,14,15] .

2.2 Balance CPU Overhead and Memory Consumption

We observed that some preprocessing steps can lead to significant sample enlargement. For example, using images in JPEG format can increase memory consumption by 5 to 7 times after decoding. Additionally, the pixel centered operation converts each pixel from uint8 to float32, effectively quadrupling the memory consumption. On the other hand, there are some preprocessing operations that can effectively reduce memory consumption, such as resize and crop. Before loading batch data onto the GPU for computation, it is necessary to ensure that the samples have the same size, always using resize and crop to normalize the data size.

When the cache is sufficiently large, reusing previously prepared output data can reduce CPU computation and increase overall throughput. However, determining the optimal cache point becomes crucial when the cache cannot accommodate the entire dataset. In the example illustrated in Fig. 1, caching the cropped output results reduces CPU load. Nonetheless, the enlarged memory footprint results in fewer stored samples in the cache, thereby failing to offset the additional I/O overhead.

Our goal in caching is to maximize CPU throughput as much as possible. The strategy needs to satisfy the following:

- The cached preprocessed output results must be deterministic. Caching and reusing non-deterministic data would lead to a reduction in sample diversity and diminish the model's generalization capability.
- Balancing memory consumption and throughput.

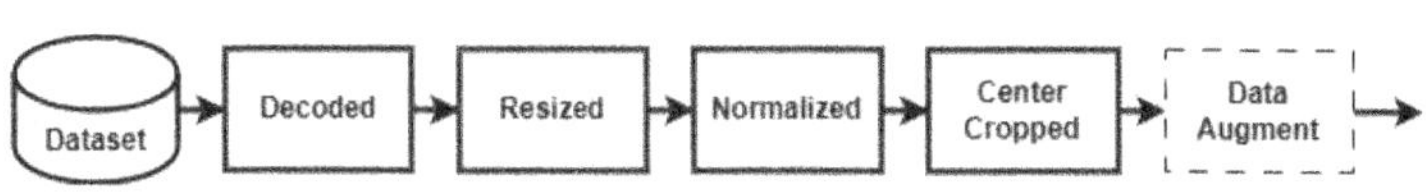

Fig. 1. CV preprocessing pipeline.

To analyze the memory consumption at various stages of data preprocessing, we utilized the ResNet-50 [10] model and the ImageNet [6] dataset and employed a sample data preprocessing pipeline shown in Fig. 1 to test training throughput.

We have analyzed the memory usage and throughput when caching the output results of different data preprocessing stages. As shown in Fig. 2 and Fig. 3, we have tested the throughput of different cache stages on A100 and V100, respectively. The test result include the overhead of various preprocessing steps. And, the I/O overhead is largely mitigated as the caching data. As illustrated in Fig. 4, caching decoded data increases memory consumption by a factor of 5× compared to caching raw data, The throughput only improves by around 10%. Memory consumption increases by 50% until resized, but the performance improvement of 50%, relative to no caching. The caching of data effectively masks I/O overhead, and throughput increases incrementally as preprocessing computational overhead decreases. Due to the relative scarcity of memory resources compared to storage resources, it's often preferred to cache resized or cropped output results.

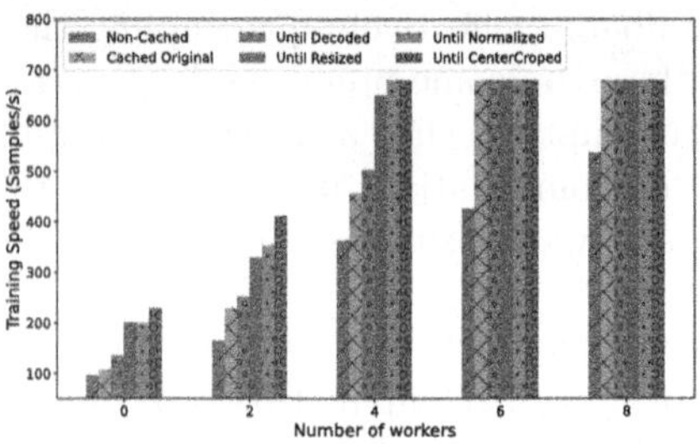

Fig. 2. Pre-processing cache on A100

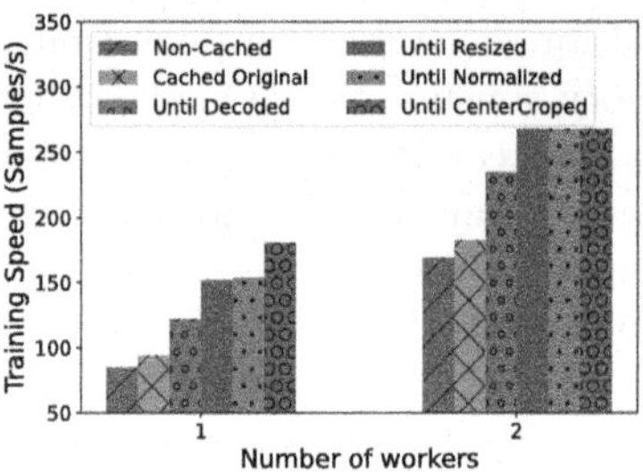

Fig. 3. Pre-processing cache on V100

2.3 Different Sample Sizes

We observed that the sample size in the dataset are non-uniform, so it is necessary to make adjustments during preprocessing to ensure that the batch tensor sizes are equal in the DL accelerator. In the ImageNet-22K dataset [2], the distribution of sample sizes is as follows: 0–10k samples account for 10% of the dataset, 10–100k samples account for 20%, 100k-1M samples account for 20%, etc.

While datasets with smaller samples hinder multi-threading performance during the training process, after preprocessing, the sample sizes become consistent. Therefore, for larger datasets with larger samples, more I/O and CPU overhead are required before caching. We conducted tests on images with the same storage size, adjusting the number of pixels, ranging from 150 × 200 pixels to 1600 × 2400 pixels. Figure 5 illustrates the impact of CPU processing on throughput for samples of the same data volume but different resolutions. The processing time for datasets composed of high-resolution samples (1600 × 2400) is over 7 times that of low-resolution samples (150 × 200). Additionally, we also examined the effect of sample storage size on throughput, with individual sample sizes ranging from 10 KB to 16 MB, the results show that there is almost no impact on throughput.

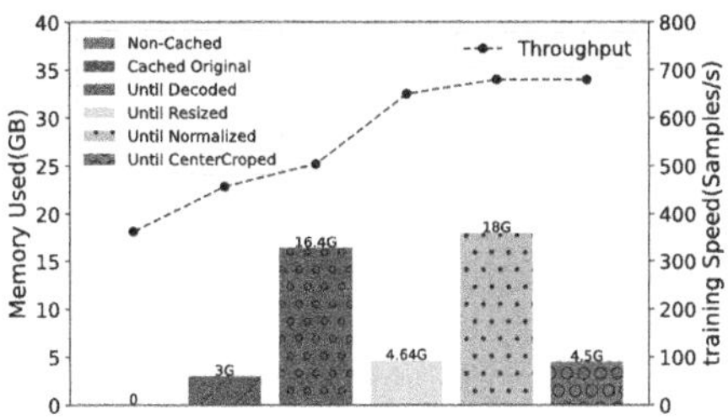

Fig. 4. Memory Consumption and Throughput for Different Caching Strategies.

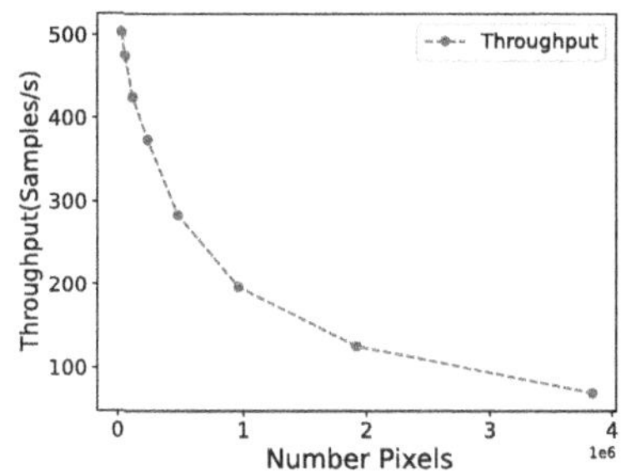

Fig. 5. Loading a dataset of 30,000 synthetic images.

3 HCache Design

3.1 Overview

HCache is an efficient data loading system that supports single-node multi-GPU training and multi-node distributed training. To illustrate the details, we take the example of the single-node architecture shown in Fig. 6. HCache consists of client modules, cache managers, cache servers, and storage systems.

HCache Client and Server Layers. The client module is integrated within deep learning frameworks (such as PyTorch and TensorFlow) and is primarily responsible for gradient computation and requesting data from the server layer. When the data loader of deep neural network application begins randomly selecting samples for reading, the client forwards the requests to the HCache server layer through API calls. Each client data corresponds to an independent deep neural network training task. The HCache server layer maintains two cache pools, namely T-Cache and O-Cache. T-Cache is used to cache preprocessed output tensor data, while O-Cache is used to cache raw data.

T-Cache. T-Cache operates on a simple and intuitive principle: once an item is cached, it remains in the cache until the end of training without replacement. The operation of T-Cache is as follows: during the initial epoch of training, HCache reads random samples from storage and obtains the size of each sample (referring to resolution in this context). Before the start of the second epoch,

T-Cache is populated in descending order of sample size until its capacity limit is reached. Once the cache capacity is reached, T-Cache does not evict any cached items. By employing a cache-fixed strategy throughout the entire training cycle, T-Cache effectively reduces redundant I/O and preprocessing overhead.

O-Cache. .To alleviate the I/O overhead of fetching non-cached data, we allocate a portion of memory to cache original samples. O-Cache is used to cache samples that haven't been loaded into T-Cache. Given that the capacity of O-Cache is smaller compared to T-Cache, and in scenarios where cache capacity is smaller than the dataset size, O-Cache only caches unprocessed original samples to ensure its data replacement speed. To ensure a good hit rate for HCache, the uncached data in T-Cache is organized into blocks, with each block being referred to as a partition. Data within each partition is further divided into serveral logical blocks. O-Cache uses a technique similar to double buffering, with data loading being done in units called "chunks," where each chunk comprises logical blocks selected from each partition.

Cache Manager. It maintains an index hash list that holds the IDs of uncached samples in T-Cache, providing help with the data loading and search scope for O-Cache.

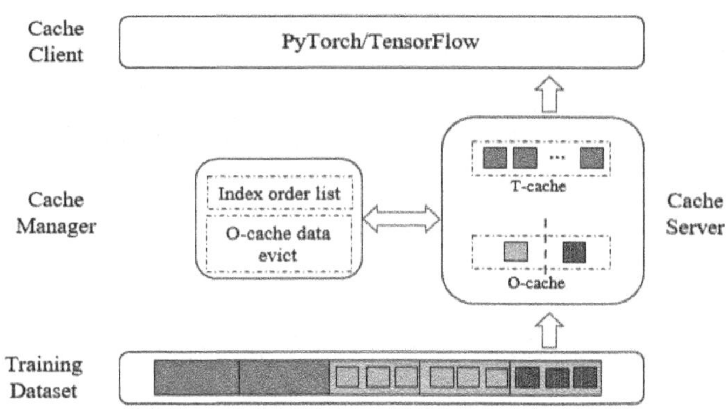

Fig. 6. The architecture of HCache and its end-to-end data preparation procedures.

3.2 Global Substitutable Cache Algorithm

We utilize key-value storage to manage the data entries in both T-Cache and O-Cache. For T-Cache, the key corresponds to the sample's index ID, and the value represents the preprocessed tensor data. For O-Cache, the key is the sample's index ID, and the value represents the raw, unprocessed sample data.

The specific algorithm is as described in Algorithm 1. HCache employs the concept of replaceable caching between the T-Cache and O-Cache cache pools to enhance cache hit rates. O-Cache implements optimization strategies aimed at minimizing I/O blockages, preventing T-Cache from being prioritized in each epoch to avoid weakening the effect of random access.

Algorithm 1: Global Substitutable Cache hit Algorithm

```
Input: id, Sample id;
for epoch in total_epochs do
    for ID in sample_dataset do
        if T-cache_hit then
            T-cache.get(id)
        if O-cache_hit then
            O-cache.get(id)
        else if HCache_miss and O-cache.exist(unaccessed_data) then
            O-cache.get(Random_unaccessed_id)
        else if HCache_miss and !O-cache.exist(unaccessed_data) and
        T-cache.exist(unaccessed_data) then
            T-cache.get(Random_unaccessed_id)
        else
            read_from_storage(id)
```

4 Evaluation

We conducted evaluations on a dedicated training server equipped with 2 Intel Xeon(R) Gold 6248R CPUs (24 cores, 2.10 GHz, 45 MB cache), 128 GB of memory, one NVIDIA A100 GPU, and an NFS file system. We adjusted the ratio between CPU cores and GPU count by setting different numbers of CPU cores using Docker containers [4] while keeping the GPU count fixed. By default, we set the CPU-GPU ratio to four, but we also evaluated Hache at different CPU-GPU ratios. Our evaluations were conducted using PyTorch 2.0, and we replaced PyTorch's dataloader with HCache for data preparation.

Baselines. We evaluate data reusing implemented in HCache against the following baselines.

- **Standard:** The standard setup represents the prescribed DNN training process with no caching reuse mechanism. The model accuracy achieved under this training serves as the target precision for other data loading methods.

- **Cache original:** We employ a fixed caching approach to cache the original data, where in our experiments the cache size allocated for Cache original is 1.5 times that of HCache to ensure a fair throughput comparison under the same cache size conditions.
- **TCache-Non-Sort:** As a ablation experiment, TCache-Non-Sort caches only preprocessed intermediate data, without considering the original sample resolutions, and does not incorporate the O-cache and global substitutability cache mechanism.
- **TCache-Sort:** To investigate the impact of caching higher-resolution samples on end-to-end testing, this configuration prioritizes caching higher-resolution samples within the T-Cache, while still not incorporating the O-cache and global substitutability cache strategy.

We use the identical model hyperparameters (e.g., the number of training epochs, learning rate, batch size, and optimizer) for each setting. All preprocessing steps follow standard procedures.

4.1 Comparison with Baselines

Table 1. The training throughput of ResNet50 on ImageNet In different caching strategies.

Config	Throughput (samples/s)
Non-Cache	290
Cache-Original	352
TCache-Non-Sort	525
TCache-Sort	648
HCache	720

In this evaluation, we demonstrate the superiority of HCache in data preprocessing. Table 1 shows that when training ResNet-50 with the same cache capacity, HCache provides over two times the throughput improvement compared to Cache-Original. In our strategy, T-Cache prioritizes caching samples with higher resolution, which has no impact on memory consumption. It can improve throughput by 20% compared to random caching. Finally, using the global alternative cache strategy, the throughput improvement is over 10% compared to setting only T-Cache.

4.2 Accuracy with Global Substitutability

In this evaluation, we demonstrate that the global substitutable cache used in HCache (i.e., replacing cache missed with data from T-cache or O-cache) has no impact on model accuracy. As shown in Fig. 7, HCache closely matches the top-1 accuracy curve under the standard setting. Table 2 displays the final top-1 and top-5 accuracies in both configurations.

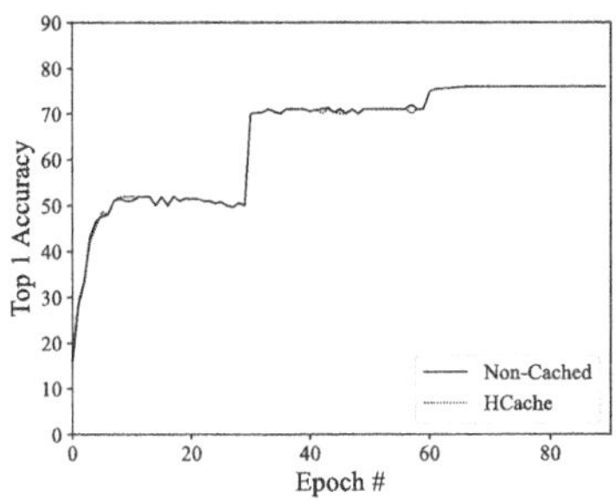

Fig. 7. Top-1 Accuracy of ResNet50 ImageNet model under globally random sampling, show that substitutable caching in Hcache has no impact on accuracy.

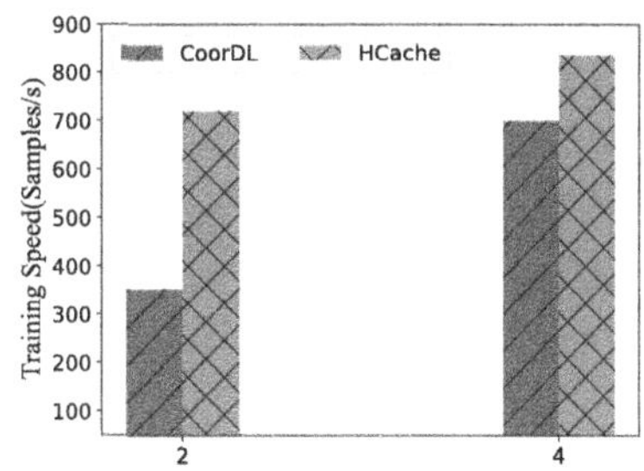

Fig. 8. The training throughput of ResNet50 on ImageNet for varying CPU-GPU ratios.

Table 2. ResNet50 on ImageNet: Final Accuracy after 90 epochs Average of two runs.

Config	Top-1 ACC.(%)	Top-5 ACC.(%)
Baseline sampling	75.67	92.88
HCache sampling	75.81	92.77

4.3 CPU-GPU Ratio

We evaluated the training throughput changes of training ResNet50 on the ImageNet dataset with CPU-GPU ratios varying from two to four. As summarized in Fig. 8, the performance improvement from Hcache is maximized in training environments with fewer CPUs, provided it is not bottlenecked by deep learning accelerators.

5 Conclusion

In deep learning training (DLT), CPU preprocessing presents is a major bottleneck. Simply caching samples can only mitigate bottlenecks on the I/O pipeline side. To address this issue, in this paper, we propose reusing "deterministic" preprocessed data to reduce redundant data processing. We achieve this through the design and implementation of HCache, a novel caching and data loading system aimed at addressing system-level challenges related to caching. HCache has demonstrated a 2× speedup in DNN training while maintaining accuracy levels comparable to the baseline.

Acknowledgement. This work is supported in part by the National Science and Technology Major Project (2021ZD0114300), the National Natural Science Foundation of China (U22A6001), the National Key R&D Program of China (2022YFB4500405, and 2023YFB4503005), and the Zhejiang provincial "Ten Thousand Talents Program" (2021R52007).

References

1. Cloud tensor processing units (tpu). https://cloud.google.com/tpu/docs/tpus
2. Imagenet22 dataset. https://opendatalab.com/OpenDataLab/ImageNet-21k
3. Nvidia a100. https://www.nvidia.com/en-us/data-center/a100/
4. Runtime options with memory, cpus, and gpus. https://docs.docker.com/config/containers/resource_constraints/
5. Choi, D., Passos, A., Shallue, C.J., Dahl, G.E.: Faster neural network training with data echoing. arXiv preprint arXiv:1907.05550 (2019)
6. Deng, J., Dong, W., Socher, R., Li, L.J., Li, K., Fei-Fei, L.: Imagenet: a large-scale hierarchical image database. In: 2009 IEEE Conference on Computer Vision and Pattern Recognition, pp. 248–255. IEEE (2009)
7. Graur, D., Aymon, D., Kluser, D., Albrici, T., Thekkath, C.A., Klimovic, A.: Cachew: machine learning input data processing as a service. In: 2022 USENIX Annual Technical Conference (USENIX ATC 2022), pp. 689–706 (2022)
8. Gu, R., et al.: Fluid: dataset abstraction and elastic acceleration for cloud-native deep learning training jobs. In: 2022 IEEE 38th International Conference on Data Engineering (ICDE), pp. 2182–2195. IEEE (2022)
9. Guirao, J.A., et al.: Fast AI data preprocessing with nvidia dali. In: GPU Technology Conference (2019)
10. He, K., Zhang, X., Ren, S., Sun, J.: Deep residual learning for image recognition. In: Proceedings of the IEEE Conference on Computer Vision and Pattern Recognition, pp. 770–778 (2016)
11. Kumar, A.V., Sivathanu, M.: Quiver: an informed storage cache for deep learning. In: 18th USENIX Conference on File and Storage Technologies (FAST 2020), pp. 283–296 (2020)
12. Lee, G., et al.: Refurbish your training data: reusing partially augmented samples for faster deep neural network training. In: 2021 USENIX Annual Technical Conference (USENIX ATC 2021), pp. 537–550 (2021)
13. Mohan, J., Phanishayee, A., Raniwala, A., Chidambaram, V.: Analyzing and mitigating data stalls in DNN training. arXiv preprint arXiv:2007.06775 (2020)

14. Naman Agarwal, R.A., Koren, T., Talwar, K., Zhang, C.: Stochastic optimization with laggard data pipelines. Adv. Neural Inf. Process. Syst. **33** (2020)
15. Park, P., Jeong, H., Kim, J.: Trainbox: an extreme-scale neural network training server architecture by systematically balancing operations. In: 2020 53rd Annual IEEE/ACM International Symposium on Microarchitecture (MICRO), pp. 825–838. IEEE (2020)
16. Um, T., Oh, B., Seo, B., Kweun, M., Kim, G., Lee, W.Y.: Fastflow: accelerating deep learning model training with smart offloading of input data pipeline. Proc. VLDB Endow. **16**(5), 1086–1099 (2023)
17. Zhao, M., et al.: Understanding data storage and ingestion for large-scale deep recommendation model training: industrial product. In: Proceedings of the 49th Annual International Symposium on Computer Architecture, pp. 1042–1057 (2022)

HyBuffer: A Distributed Hybrid Shared Buffer for Multi-master Databases

Rongrong Zhang[1], Zhiwei Ye[2], Peng Cai[1(✉)], Xuan Zhou[1], Aoying Zhou[1], Dunbo Cai[2], and Ling Qian[2]

[1] East China Normal University, Shanghai 200062, China
rrzhang@stu.ecnu.edu.cn, {pcai,xzhou,ayzhou}@dase.ecnu.edu.cn
[2] China Mobile Cloud Center, Suzhou 215000, China
{yezhiwei,caidunbo,qianling}@cmss.chinamobile.com

Abstract. Distributed shared buffer (DSB) is a well-known solution to support multi-master database systems. DSB involves controlling access to shared data among different nodes through lock-based cache coherence protocols. However, existing DSB implementations resolve conflicts at the page level, lacking the flexibility required by modern cloud database systems. In this paper, we present HyBuffer which integrates both rows and pages within the DSB framework. It enables multiple masters to independently modify different rows on the same page, thereby enhancing concurrency and performance without altering the disk-based storage structure. Optimized hybrid row and page buffers offer concurrent access to multiple versions in memory while facilitating the asynchronous consolidation of modified rows into their remote page owners. The optimized log subsystem in HyBuffer enables efficient detection of log flushing dependencies among masters, ensuring fast transaction committing. Experiments demonstrate that HyBuffer achieves better scalability and performance compared to existing approaches.

Keywords: Multi-master · Buffering · Shared-storage

1 Introduction

Multi-master transaction processing, which allows multiple computing nodes to execute read-write transactions over the whole database without the need for user-defined partitioning and two-phase commit (2PC), first supported by the cloud database Amazon Aurora [10]. However, Aurora's design exhibits limitations in write scalability. It detects transaction conflict based on the remote disaggregated storage, which currently only scales out to four master nodes [1]. This approach is optimistic, where transactions from different masters initially execute without inter-master conflict detection at the computation layer, and conflicts are detected on the storage layer.

Another multi-master scheme based on pessimistic locking utilizes a distributed shared buffer (DSB) to detect conflicting transactions between nodes.

M. Onizuka et al. (Eds.): DASFAA 2024, LNCS 14855, pp. 474–483, 2024.
https://doi.org/10.1007/978-981-97-5572-1_35

Oracle RAC [7], IBM DB2 pureScale [2], and the recently published Taurus MM [4] build a distributed shared **page** buffer for multiple masters and employ a pessimistic locking protocol to lock the cached pages, enabling runtime detection of multi-master transaction conflicts. The optimism-based approach, with deferred detection, performs well in conflict-free workloads. In scenarios with a high degree of conflict, pessimistic multi-master transaction processing exhibits higher performance, decreased latency, and reduced abort rates than optimistic.

In the design of a global buffer pool for disk-oriented traditional databases like Oracle RAC, DB2 pureScale, Taurus MM and cloud-native database PolarDB [13], the minimum granularity of cache items is constrained to pages to align with the disk storage structure, referred to as ***page-level*** *shared buffer*. The granularity of transaction concurrency control operates at the **row level**. When transactions from different masters access the same page, the concurrency granularity is elevated to the page level. Caching only pages in a DSB system exacerbates the mismatch between row-level transaction concurrency control and page-based buffer sharing. This issue turns transactions from different masters which have no real transaction conflicts, into sequential execution just because they access the same physical page.

In this work, we present HyBuffer, a distributed hybrid shared buffer. It enables transactions from different masters to write distinct rows on the same page while preserving the cache coherency, referred to as *hybrid-level shared buffer*. By inheriting the page-level shared buffer, we demonstrate that two masters can write different records on the same page by extending traditional DSB in HyBuffer. The main challenge is that the records modified by a master should be *synchronously* merged to another master which is the page owner. This introduces high latency for transactions requiring write access to records on a page owned by other masters. To address this issue, HyBuffer utilizes additional versioned row buffers in each master to cache the modified records. This enables the *asynchronous* merging of modified records to their respective remote cached pages. The contributions of this paper are summarized as follows:

- We present a distributed hybrid shared buffer, HyBuffer, to facilitate concurrent writes by multiple masters to the same page.
- HyBuffer employs additional versioned row buffers to ensure fine-grained sharing of cached data, reducing the latency of multi-master transactions by asynchronously merging the modified records to the remote page buffer.
- To ensure physical page consistency, HyBuffer adopts a scalable WAL log scheme to avoid expensive coordination.
- The experimental results demonstrate that HyBuffer exhibits better scalability and achieves excellent performance.

The remainder of the paper is organized as follows: Sect. 2 presents the detailed design of HyBuffer. Section 3 detailed explanation on how to handle lock requests in a hybrid shared buffer. Section 4 describes the optimization for log flushing dependency. The experimental results are reported in Sect. 5. Finally, Sect. 6 provides an overview of related works and Sect. 7 concludes this paper.

2 Design of HyBuffer

The multi-master database system is equipped with a local buffer for each node and combined with a locking protocol to control the concurrent access of cache items by different nodes. The current databases only control the concurrency at different nodes at page granularity, referred to as the *page owner model*. All the metadata for locating cached pages is stored in a data directories service (DDS).

2.1 Hybrid Data Owner Model

In the page owner model, a master exclusively owns a page, even in the case of Taurus MM [4] which supports a hybrid page-row locking protocol. As a result, it exhibits suboptimal performance. Meanwhile, many research-oriented databases that support multi-master often adopt key-value storage engines like [14], which do not align well with page-oriented structures. We refer to this as the *row owner model*. HyBuffer presents a *hybrid data owner model* specifically designed for DSB architecture with a page-oriented storage layer. In HyBuffer, a master can cache entire pages and individual rows within the local buffer through the *Owner Mapping Table* as shown in Fig. 1.

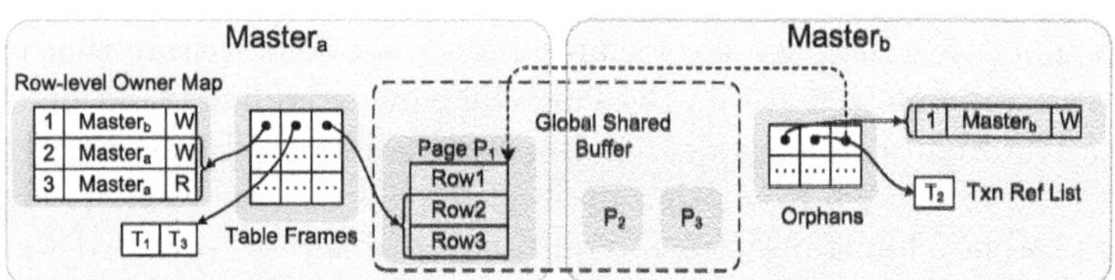

Fig. 1. Owner Mapping Table in different masters

Owner Mapping Table is mainly divided into two components, the *table frame* (or *orphans*) and the *row-level owner map*. The row-level owner map interacts directly with transactions, while the table frame associates row data and entity pages through pointers to achieve efficient hybrid granularity buffer management. The management of shared page buffer is similar to the traditional designs. At a given moment, a page can only be managed by one master, which is referred to as the owner of the page. For example, in Fig. 1, page P_1 is owned by $Master_a$, and P_2, P_3 are owned by $Master_b$. In contrast to traditional shared page buffers, we also support row-level data sharing. Although $master_b$ is not the owner of P_1, it can still cache $Row1$ on P_1 locally and has write access until $Row1$ is recycled (see Sect. 2.3). All scattered rows are managed by two types of frames, called table frames and orphans, depending on whether the corresponding page is currently owned by the master.

In the case of table frames, as long as the page is pinned in the local buffer of a master, the corresponding table frame stores a pointer exclusively to that page. Subsequent row lock request for the rows on that page no longer requires a remote operation. By decoupling row-level access management from the page buffer, this design allows transactions to focus on concurrency control at the record level. In the case of orphans, the pointer to the page is not physical but logical, represented by a dashed line in Fig. 1. After the active transactions are committed, the data rows need to be merged into the page's owner (Sect. 2.2).

Remote Locking Avoidance. An item of the row-level owner map at least three attributes: the identifier of the accessed row, the current accessing master, and the access mode. These attributes ensure that a data row is accessed by transactions globally in a compatible mode. When a transaction accesses data rows on a local page, it only needs to add an item in the row-level owner map associated with the table frames and then add a new transaction identifier to the list *txn_ref_list*. For remote transaction requests, it requires recording twice, in both the local and remote master's row-level owner map. The row-level owner map caches multiple row ownership, particularly when the current master is not the owner of certain pages, avoiding the network overhead caused by a significant number of remote requests on these rows.

2.2 Asynchronous Merging

While the owner mapping table allows multiple master nodes to concurrently modify different rows on the same page, the modified rows still need to be merged onto their associated pages synchronously when all active transactions on these rows are committed. A master can perform any compatible operation on data within the local row-level owner map, which establishes a connection between the local transaction space and the versioned row buffer (see Sect. 2.3). The dirty rows from the commit transaction are first temporarily stored in the row buffer and asynchronously merged to the corresponding page with the garbage collection thread. The row-level owner map guides the process of merging. After completing the merge, the items corresponding to the rows will be removed from the row-level owner map.

2.3 Versioned Row Buffer and Garbage Collection

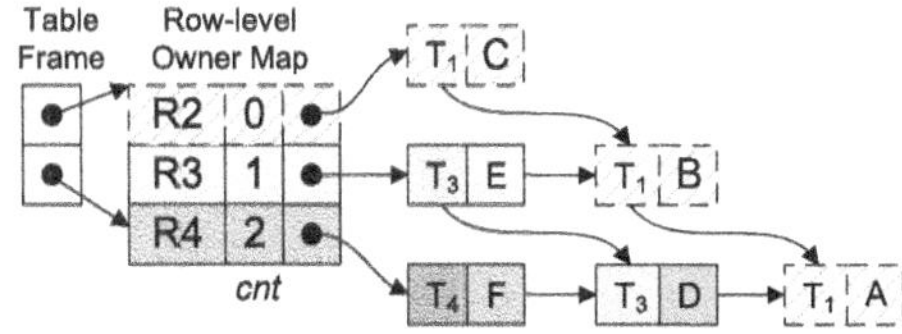

Fig. 2. The versioned row buffer

Innovatively, HyBuffer integrates lock-based multi-version concurrency control (MVCC) with row-level owner map to enhance the efficiency of multi-master transaction processing and hybrid rows/pages cache management. As shown in Fig. 2, we establish a three-level buffer, extending from page to row, and further to the multi-version space of transactions. The head of the version chain records the most recent change, and as traverses backward along the linked list, the recorded versions become older. The shaded area indicates the expired versions, and they are currently/will be garbage-collected. This structure is designed only for concurrent local transactions of a master and does not support creating version chains for the same data row across nodes.

The version chain is cleaned up when it is no longer needed for currently active transactions. We use pointers to link the historical versions modified by the same transaction, enabling faster lookups. In addition to cleaning up expired version chains, the latest changes need to be synchronously merged into their respective pages. This may trigger remote merging. For each row, we maintain a counter *cnt*, which tracks the number of writes and reads on the row. The background garbage collection thread is responsible for scanning entries with a *cnt* equal to 0, and triggering the merging operation.

3 Lock Processing

The mapping table can be seamlessly integrated into the row lock manager, ensuring the logical consistency required by transaction concurrency control. HyBuffer provides a local lock table for each master to ensure the ordered changes of transactions to the database, while the consistency of physical pages is guaranteed through logical clocks. This section delves into the adaptive handling of row locks and the transfer of ownership for pages through DDS.

Recalling the owner mapping table, table frames (or orphans) serve as a local row lock table through locking states (see the flags W and R in Fig. 1). When a local transaction initiates a row lock request, three following scenarios may occur: 1) If a master is the owner of the page, it means that any lock corresponding to a row on the page can be granted immediately, as long as compatibility is satisfied. 2) If a master is not the owner of the page, it will check for an item for the corresponding row in orphans. 3) If none of the above conditions are met, a remote row lock request needs to be sent to DDS.

Lock Processing in the DDS. DDS is responsible only for recording the location information of pages, indicating which master manages them. It does not assume the role of a global lock manager because a centralized lock manager can potentially become a bottleneck (e.g., network and computational [4]) when dealing with a large number of row and page lock requests. A remote lock request from a master will first query DDS to determine the master managing the page where the requested row is located. If there is a migration of ownership for a page during the lock processing, the corresponding location information in DDS will also be updated.

Row Locks Granting. Local row lock granting has been discussed earlier, and we only address how to handle row lock requests from other masters. When the requested master receives a remote row lock request, it does not simply grant or suspend based on compatibility. If compatibility checks pass and there are other unreleased row X-locks on the page, the row-level owner map inserts this lock request, including the requester's ID and the lock mode. Then the requested master directly returns the corresponding row lock.

Another scenario occurs when all row locks on the page have been released. This situation triggers a page migration operation. Page migration returns the entire page's ownership, along with the page itself, to the requester.

Page Lock Processing and Page Split. Certain operations require adjustments to the positions of tuples within the page, such as **INSERT** and **DELETE** statements. Especially in the case of page splitting, when row locks on a page are distributed among multiple masters, the splitting of the page can result in incorrect results when merging, resulting in the physical inconsistency of the page. Recalling the mapping table, a master manages its exclusively owned pages through the table frames and rejects incompatible requests for X-lock and S-lock from other masters when they apply for these pages. However, these requests cannot be indefinitely deferred. HyBuffer blocks subsequent row lock requests on a page when it receives a lock request for the page.

4 Ordering and Logging

4.1 Timestamp

The conventional ARIES protocol [9], due to its reliance on a centralized sequencer, would have poor scalability. We chose vector clocks [5] to record each change operation on a page, ensuring a global order. It not only guarantees the physical consistency of pages but also provides the global consistent snapshot required for certain isolation levels, such as Snapshot Isolation. We maintain a vector clock VC_i on each $Master_i$ to capture the state of all nodes in the system and synchronize these vectors during cross-node communication.

4.2 Log Flushing Dependency

HyBuffer provides separate logical log L_i for each $Master_i$ (all of which are flushed to the same shared storage). When flushing these log records, despite they belong to different logical logs, it is still necessary to check dependencies between logs from different masters to ensure: *multiple changes to the same page are logged in order.*

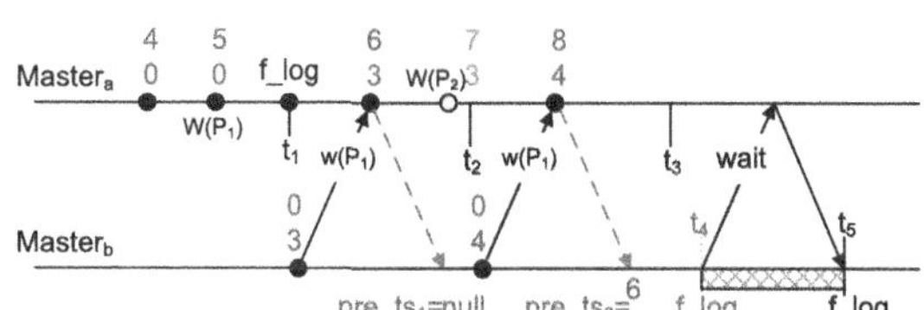

Fig. 3. An example of a log flushing dependency

Example 2.1. Let's illustrate why flushing dependencies exist: consider transaction Txn_1 from $Master_a$ updates a tuple on page P_1 and records this change with ts=(5, 0) and flushes log into L_1 at time t_1 before commit. Then transaction Txn_2 from $Master_b$ also inserts and updates two different tuples on page

P_1. According to vector clocks, these changes will be recorded with higher timestamps, for instance, (6, 3) and (8, 4). When Txn_2 commits, it must ensure that all preceding changes on this page have been persisted in the log system. It therefore needs to flush all other masters' logs up to (8, 4).

Figure 3 illustrates the process of this example. T_2 waits for a long time until all masters have flushed all log records before (8,4). Only then can T_2 flush its log. The flushing of T_2 at time t_5 is unnecessary since the modification of P_1 by T_1 was already logged at t_1. It is evident that there are the following constraints among multiple transactions with flushing dependent: (1) They modify some same pages and (2) They both have not flushed logs yet. During each remote clock synchronization, we return the timestamp of the most recently unflushed log for the modified page, denoted as *pre_ts*. This timestamp is used to determine whether a transaction is concurrently modifying a page with other transactions, and none of them have flushed logs. With this technique, T_2 in the example can detect earlier that pre_ts_1 is *null*, and all subsequent changes return *pre_ts* generated by T_2 (pre_ts_2) transaction itself. Then T_2 flushs directly at time t_4.

4.3 Versiond Read

If two database changes a and b occur globally, and a happens before b, we denote it as $a \rightarrow b$. It is easy to deduce $V_a \rightarrow V_b$ according to logical clocks, the reverse is not necessarily true. The vector clocks address this limitation, allowing deducibility in both directions based on $a \rightarrow b$ and $V_a \rightarrow V_b$. This capability enables the creation of a global consistent snapshot. HyBuffer does not provide multi-version snapshots on a global scale. However, this does not hinder its ability to provide snapshot reads to other masters, alignment of transaction timestamps with vector clocks is sufficient.

5 Evaluation

We implemented the prototype of HyBuffer and deployed it on 33 servers (32 masters and a DDS node) with Intel Xeon Silver 4100 processors and 160GB RAM. We compared HyBuffer with four baselines, with three using pessimistic locking protocols and one optimistically validating for inter-node transaction conflicts.

We implemented two coherence protocols, referred to as *Distributed* and *Centralized* used in Oracle RAC and DB2 pureScale respectively. The *Distributed* provides distributed management for both buffer pool and page directories, similar to Oracle RAC. The *Centralized* manages the shared buffer and page directories in a centralized node, similar to DB2 pureScale. HyBuffer architecture is similar to Taurus MM in that it uses a centralized lock manager. The difference is that Taurus MM suspends all lock requests to the page when starting a page transmission, and transmits all lock information on the page to the destination node along with the page, which greatly limits the concurrency of the system. Except for the comparative experiment with Starry, the benchmark driver is directly linked to the engine, so the results are unimpeded by network overhead.

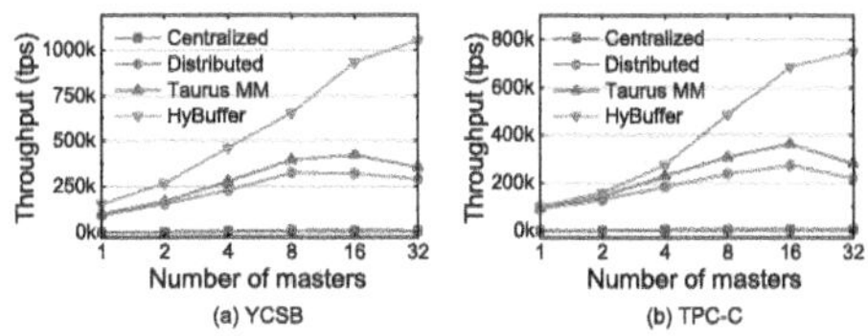

Fig. 4. Performance under different lock-based protocols

5.1 Performance Under Lock-Based Protocols

We compare the performance of Hybuffer and pessimistic locking protocols in a multi-master architecture using YCSB and TPC-C workloads. The coefficient θ of the YCSB is set to 0.5 and the read/write ratio is configured as 5:5. Under TPC-C workloads, each master maintains 20 warehouses. 85% of transactions access the local warehouse, while 15% of transactions access a remote warehouse.

As depicted in Fig. 4(a) and Fig. 4(b), HyBuffer exhibits a significant performance advantage over the other protocols. The throughput of *Centralized* is almost incomparable with others. Transactions from each master always retrieve page data and lock from the central node at the beginning and release them upon committing, thus consuming a considerable amount of network bandwidth. The performance of the *Distributed* is slightly inferior to that of Taurus MM, even though they shared data at the page level. The locking scheme utilized by the *Distributed* involves additional RPC calls, it must first locate the owner of the required pages by invoking the distributed page directory service.

5.2 Performance Compared with Starry

We used the synthetic workload Retwis for our comparative analysis with Starry. Due to the differences in architecture, we introduced additional client nodes to simulate the computing nodes in Starry. To make the results more intuitive, we refer to Starry's validation nodes and the transaction processing nodes in HyBuffer (i.e., nodes deployed with HyBuffer) as masters. Both coefficient Zipf was set to 0.75.

Figure 5(a) shows the throughput performance of the two systems. Starry's throughput slows down and reaches its peak at 9 nodes. HyBuffer exhibits a linear growth curve. At 15 computing nodes, HyBuffer achieves a throughput that is 1.77× higher than Starry. Figure 5(b) provides a breakdown of the average transaction time for the two systems. As the number of computing nodes increases, the average read/write time and prepare time in Starry increase, thereby impacting the overall throughput. HyBuffer's read/write latency does not increase with the number of computing nodes due to the absence of 2PC overhead and a high cache hit rate. We fixed the number of computing nodes at 12 and varied the Zipf coefficient to compare the abort rate of the two systems. The results are illustrated in Fig. 5(c). The abort rate of HyBuffer remains below 11% even when the Zipf coefficient is 1. In Starry, the abort rate increases sharply once the Zipf coefficient exceeds 0.7, reaching a peak of 29%.

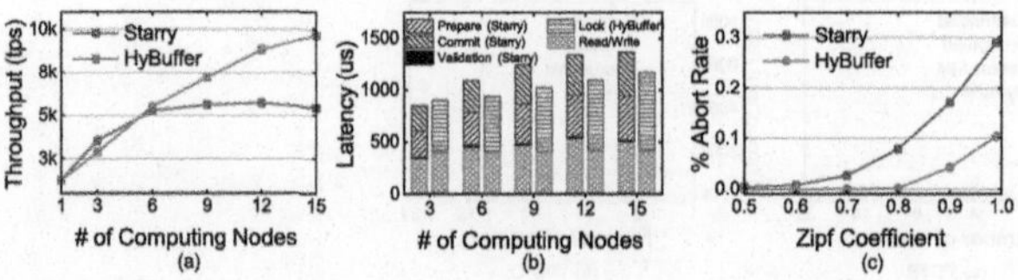

Fig. 5. Performance compared with Starry

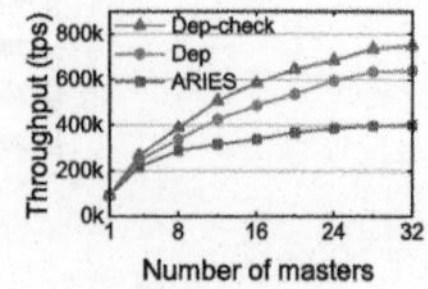

Fig. 6. Scalability of Logging

5.3 Scalability of Logging System

We conducted a comparative analysis of throughput and scalability, evaluating our proposed approach (referred to as Dep-Check) in contrast to ARIES and a distributed log flushing that lacks dependency verification (referred to as Dep). The experiment used a TPC-C workload, with 20 warehouses per master.

As illustrated in Fig. 6, our Dep-Check approach demonstrates notable scalability. It attains a transaction throughput of 93k transactions per second with a single master and scales impressively to 752k txn/s with 32 masters. In comparison, the ARIES-style logging, constrained by the global log, reaches its peak performance at 402k txn/s. The Dep benefit from the absence of synchronization during log entry creation, reaching a peak of 642k txn/s with 32 masters.

6 Related Work

Two approaches have been proposed to enhance the performance of standalone databases. Firstly, the shared-nothing partitioned database has emerged as the most popular solution [3,6,12] for database scaling. It supports both read and write transactions for each shard, but it suffers from the 2PC distributed transaction concurrency protocol and the low scalability of monolithic servers, where compute and storage are tightly coupled. Secondly, with the development of cloud computing, the architecture of compute and storage separation is widely adopted, such as Amazon Aurora [10], Taurus MM [4] and PolarDB [13]. However, they only support one single primary read/write (RW) master node and multiple read-only (RO) nodes. Aurura currently only supports a maximum of 4 write nodes.

Distributed shared buffer (DSB) is considered the next wave in database innovation, as mentioned in [11,15]. Notably, several research efforts such as PolarDB [13] and Microsoft Azure [8] are exploring the utilization of DSB to enhance memory utilization, reduce costs, and improve reliability. They are trying to integrate the shared buffer system with the transaction engine to explore the benefits of DSB within the database.

7 Conclusions

In this paper, we design and implement a multi-master database prototype, HyBuffer. It utilizes a hybrid row and page shared buffer, enabling masters to concurrently modify different records on the same page. Compared to previous coherence protocols that share only pages, HyBuffer exhibits excellent scalability and delivers superior performance, particularly under workloads with conflicts.

Acknowledgements. We thank the anonymous reviewers for their insightful comments and feedback. This work was supported by the National Natural Science Foundation of China under Grant No. U22B2020. Peng Cai is the corresponding author.

References

1. Amazon: Aurora multi-master clusters (2023). https://docs.aws.amazon.com/
2. Barshai, V., Chan, Y., Lu, H., Sohal, S.: Delivering continuity and extreme capacity with the IBM DB2 pureScale feature. In: IBM Redbooks (2012)
3. Corbett, J.C., et al.: Spanner: google's globally distributed database. TOCS **31**(3), 1–22 (2013)
4. Depoutovitch, A., et al.: Taurus mm: bringing multi-master to the cloud. VLDB **16**(12), 3488–3500 (2023)
5. Fidge, C.J.: Timestamps in message-passing systems that preserve the partial ordering (1987)
6. Huang, D., et al.: TIDB: a raft-based htap database. VLDB **13**(12), 3072–3084 (2020)
7. Klots, B.: Cache coherency in oracle parallel server. VLDB, 583 (1996)
8. Li, H., et al.: First-generation memory disaggregation for cloud platforms. CoRR (2022)
9. Mohan, C., Haderle, D., Lindsay, B., Pirahesh, H., Schwarz, P.: Aries: a transaction recovery method supporting fine-granularity locking and partial rollbacks using write-ahead logging. TODS **17**(1), 94–162 (1992)
10. Verbitski, A., et al.: Amazon aurora: design considerations for high throughput cloud-native relational databases. In: SIGMOD, pp. 1041–1052 (2017)
11. Wang, R., Wang, J., Idreos, S., Özsu, M.T., Aref, W.G.: The case for distributed shared-memory databases with RDMA-enabled memory disaggregation. VLDB **16**(1), 15–22 (2022)
12. Yang, Z., et al.: Oceanbase: a 707 million TPMC distributed relational database system. VLDB **15**(12), 3385–3397 (2022)
13. Zhang, Y., et al.: Towards cost-effective and elastic cloud database deployment via memory disaggregation. PVLDB **14**(10), 1900–1912 (2021)
14. Zhang, Z., Hu, H., Zhou, X., Wang, J.: Starry: multi-master transaction processing on semi-leader architecture. VLDB **16**(1), 77–89 (2022)
15. Ziegler, T., Bernstein, P., Leis, V., Binnig, C.: Is scalable OLTP in the cloud a solved problem. In: CIDR (2023)

Author Index

M. Onizuka et al. (Eds.): DASFAA 2024, LNCS 14855, pp. 485–487, 2024.
https://doi.org/10.1007/978-981-97-5572-1

The manufacturer's authorised representative in the EU is Springer Nature Customer Service Centre GmbH, Europaplatz 3, 69115 Heidelberg, Germany. If you have any concerns regarding our products, please contact ProductSafety@springernature.com

Printed and bound by CPI Group (UK) Ltd, Croydon, CR0 4YY
31/07/2026
02182956-0005